I0625250

Genocide Against Americans:
The Russian Government, the Russian Mafia, their 2023 Israel-Hamas Proxy War, & their Coup Against Americans -- Tips to Stay Alive

by Stacy A. Hackney

RussnCoupAttpt Publishing Co
March 2024
6th Edition

Dedicated to Paul, an entrapped American who refused to be a weapon for the Russian Mafia & who figured out a way to help me warn the world; and to the billion people worldwide under attack by the Russian Mafia, many of them children, who the Russian Mafia murders with impunity

Table of Contents

Table of Contents

Table of Contents

Table of Contents

Table of Contents

Table of Contents

Table of Contents

Table of Contents

Table of Contents

Table of Contents

Table of Contents

Table of Contents

Table of Contents

Table of Contents

Table of Contents

Table of Contents

Table of Contents

Table of Contents

Table of Contents

Table of Contents

Table of Contents

Table of Contents

Table of Contents

Table of Contents

Table of Contents

Table of Contents

Table of Contents

Table of Contents

Table of Contents

Table of Contents

Table of Contents

Table of Contents

Table of Contents

Table of Contents

Chapter 1 – The Russian Mafia is determined to seize world dominance in November 2024 & install one of their puppets into the our Presidency. To achieve this I believe they ordered Hamas to attack Israelis to begin the 2023 Israel-Hamas War. Russian General Gerasimov admitted in 2013 that the Russian government deploys protesters to attack their enemies. At least some (not all) of the pro-Palestinian protesters who harass our President at campaign rallies are deployed by the Russian Mafia in order to unseat him[1]

It's hard to believe but I'm 100% positive that it's true:

(1) that the Russian Mafiosi began the 2023 Israel-Hamas War as a proxy war to unseat a rival, a legally elected American President they don't like, in order to install one of their American puppets into that Presidency in 2024;

(2) that the Russian Mafiosi deploy some of their protesters worldwide to blame Americans and our government for a war the Russian Mafia itself ordered, using Hamas as it's proxy;

(3) that the Russian Mafia controls, owns, and/or blackmails so many of our ordinary people, media, newspapers, think tanks, etc. in America that I've read no one but me publicly accuse the Russian Mafia of ordering this proxy war attacking Americans;

---

1   General Valery Gerasimov, Chief of the General Staff of the Russian Federation,"The Value of Science in Prediction," *Military-Industrial Kurier* (February 27, 2013). General Valery Gerasimov states: *"... wars are no longer declared…" "...The very "rules of war" have changed…." "...The role of nonmilitary means of achieving political and strategic goals has grown, and, in many cases, they have exceeded the power of force of weapons in their effectiveness…." "...applied in coordination with the **protest potential of the population.…"*

(4) that although our President and government are working unceasingly towards a cease-fire behind the scenes, we need a cease-fire that doesn't destroy our multi-decades long alliance with the Israeli people (we have legal, contractual obligations to defend the Israeli people when they are attacked by a foreign enemy). Reneging on that deal would destroy our nation's defense strategy in that region, which the Russian Mafia well knows, and which is why they designed the war the way that they have, giving yet another, really loud, indication that the 2023 Israel-Hamas War is a proxy war; and

(5) that evil men (the Russian Mafiosi) can murder and/or injure over 20,000 innocent children (including newborns) in six months and no one can immediately stop them.

We cannot allow this evil to stand.

Chapter 1.1 – Anatomy of a Genocide: the Russian Mafia is working to annihilate Americans & our allies deploying a hard-wired-based psychological weapon called Active Measures. We must immediately launch an effective defense

The Problem - the Russian Mafia is working an undeclared, highly effective genocide attack against Americans called Active Measures, which is a psychological-based bomb keyed to human hard-wiring. Short term solution: dismantle the bomb before the Russian Mafia installs one of their many American puppets into the U.S. Presidency, where, if successful, they will annihilate all people living in America. Long term

<u>solution</u>: our species must decide how to effectively counter the Russian Mafia's worldwide, unrelenting attacks.[2]

The Russian Mafia (also known as the Russian government,[34] headed by Mr. Putin[5]), is running an undeclared, highly effective genocide[6] against Americans called Active Measures.[7] The part of their attack requiring immediate

---

2   Just a few of the Russian Mafia's ongoing major attacks against the innocent include: Active Measures, a secret genocide against everyone in America; the Russia-Ukraine War attacking Ukrainians; the Israel-Hamas proxy war attacking Gaza Palestinian civilians, using their babies as blood pawns to blame Americans for the war the Russian Mafia initiated in order to unseat President Biden and install one of their puppets into the White House in 2024; and their attack against Israeli civilians meant isolate Israel and drive it's citizenry to the right, to willingly align with the Russian Mafia.

3   According to Karen Dawisha in her book *Putin's Kleptocracy* (Simon & Schuster, 2015) "the FSB (Federal Security Services) has 'absorbed' organized crime …" She also states "….Russia under Putin had become a virtual 'mafia state' in which state structures operate hand in glove with criminal structures to their mutual benefit, ..."

4   According to Craig Unger in his book *House of of Trump House of Putin* (Penguin Random House UK 2018) Oleg Kalugin, former head of counterintelligence for the KGB told Mr. Unger…. "the Mafia is one of the branches of the Russian government today."

5   The Russian government and it's Mafia are one of the most successful illegal drug dealer systems in the world; targeting anyone, including children in West aligned nations, and is part of the Russian government's & it's allies' strategy to destroy us. According to Craig Unger in his book *House of of Trump House of Putin*, "...Putin and his cronies [...] were allowed to make billions through drug trafficking, extortion, elaborate financial schemes, the sex trade, arms deals, and the like…." Mr. Unger also states "...made Vladimir Putin the richest man in the world, with wealth, according to Hermitage Capital Management CEO Bill Browder, approaching $200 billion." Craig Unger, "*House of Trump House of Putin*," (Penguin Random House UK 2018).

6   Destruction of us, our families, our culture, our society and our government.

7   A stealth, subversion-based attack designed to secretly work against our government's, our nation's, our people's, including working against individual ordinary people's best self interests in America. An example of a recent big win for the Russian Mafia's Active Measures stealth war here was their successful striping of reproductive rights from American women. In the 1980s, the USSR working Active Measures ordered one

attention is a bomb keyed to human hard-wiring, our hierarchy (social organization), and our evolution, set to detonate in America this November 2024. I describe many components of Active Measures in play on the ground here, and other attacks I've endured and/or witnessed that were committed by the Russian Mafia's and it's predecessor government, the USSR, including attacks against civilians, children as young as newborns, and senior citizens into their 80s and/or 90s. My book is told from a noncriminal civilian victim's point of view, someone untrained in espionage: me. I am an ordinary woman and don't work for the government but I am fiercely loyal to my government, my people, my nation, the world's children,

---

of the teaching assistants they owned in a math class I was taking in engineering school in my mid-20s, and another operative fronting as a student in the class, to conspire to have me accused of cheating on a math test and kicked out of the test. Even at the time the compromised teaching assistant accused me of cheating, which I immediately denied, he said he saw the other student cheating off of my test not me cheating from anybody, but instead of throwing out the cheater he caught in the act, he insisted we both leave the test, despite my protests. That was one of a series of devastating USSR proxy attacks in my 20s and, not knowing the USSR had targeted me, I dropped out of school, having no idea how to counter such an accusation or who to appeal to. Even now, when I enrolled in city college programming classes, the Russian Mafia (who took up attacks against me after the USSR's demise), unrelentingly stealth attacks me in multiple ways, from having groups of operatives come stand and talk loudly outside my apartment just as I began my C programming final, to having an operative come stand outside my door this Spring 2023 as I was writing/studying to complete my finals program, and noise and startled me to such an extent, he precipitated my heart rate to 200bpm, necessitating trips to the ER, during final's week. After I left the ER and was taking a quiz the next day, the Russian Mafia had one of my neighbors suddenly blast a movie they were watching as soon as I began to take the quiz. I've endured a great many stealth attacks. The Russian Mafia does not want me to study difficult and/or complex subjects and they harass me when I attempt to do so. Active Measures is designed to secretly subvert our government and our people.

4

innocent people everywhere, and to earth's continued habitability, all of which are under threat by the anti-life Russian Mafia.

Chapter 1.2 – The Russian Mafia depresses President Biden's polling numbers by (1) forcing entrapped parents to groom & entrap their own children, (2) revealing the betrayal to the adult children, (3) preventing entrapped parents from explaining because of confidential plea deals, (4) thus destroying the family, (5) compromised grandparents are outed from their children's, grandchildren's & great grandchildren's lives, (6) compromised American law enforcement tell the extended family that our legitimate government is running these entrapments, leading (7) the entire family to loathe our legitimate government & vote in a pro-Kremlin candidate

The Russian Mafia intends to overthrow our government and destroy the American people. Ordinary people are the low hanging fruit since the KGB and the Russian Mafia were/are nuclear avoiding. By attacking our families and lying and saying our legitimate government is the culprit, while simultaneously preventing victim-families from complaining to our government, our government doesn't know this specific attack exists.

The Russian Mafia depresses President Biden's polling numbers by:

(1) forcing entrapped parents to groom & entrap their own children;

(2) revealing the betrayal to the adult children,

(3) preventing entrapped parents from explaining why they did what they did because of confidential plea deals they accepted to stay out of jail/prison when they were initially entrapped,

(4) betrayal destroys the family,

(5) compromised grandparents are outed from their children's, grandchildren's & great grandchildren's lives (that's what my mother is largely experiencing),

(6) compromised American law enforcement tell the extended family the lie that our legitimate government is running these entrapments, and

(7) the entire family loathes our legitimate government and votes in a pro-Kremlin candidate, never knowing they've been lied to all along by Russian Mafia controlled compromised law enforcement.

This is the Russian Mafia's endgame for us. We refuse Door Number 1, 2 or 3, and tell our people they've been victimized by a hate-filled enemy government.

A major component of Active Measures is entrapping noncriminal civilians, particularly parents, then forcing them to entrap their children. I've been targeted by the Russian Mafia who deployed both my parents in this way but I evaded entrapment because of my young mother's quality parenting during my childhood decades before the KGB entrapped her. Most if not all of my siblings are entrapped. The siblings my mother was forced to entrap loathe her. The Russian Mafia revealed her betrayal to them, and they blame her for entrapping them. She is not free to explain to them she was ordered under pain of imprisonment to entrap them, her plea deal enforces her silence. By deliberately revealing to her children that she entrapped them, the Russian Mafia destroyed our family. It interests me (and hurts me), that the Russian Mafia forced my mother to betray her sons, men she adores, but apparently didn't force her to betray her daughters. I would've been the first daughter she entrapped had she been able to do so. The Russian Mafia is ensuring that African American males hate African American females and they're doing so because they're gearing up to have African American men slaughter African American women if the Russian Mafia seizes control of America, which they expect to accomplish this November 2024. Another very important thing, crucially important to why President Biden's poll numbers are so low: the Russian Mafia lies to victims and tells them our legitimate government is running these entrapment attacks. Then the Russian Mafia deliberately destroys the family, causing adult children to loathe their parents, which causes compromised grandparents to lose access to beloved grandchildren, and causes all of the victims to loathe our government. The Russian Mafia is evil and intends to destroy us. It considers lying fair play, aiming our entrapment system against ordinary people as fair play, destroying our families as fair play, and committing genocide against us as fair play. The Russian Mafia is deliberately breaking American families and telling entrapped family members that our legitimate government is behind this whole scam. Nobody's talking because everyone is silenced by plea deals, blackmail, and mafia-level 'I'll rape your 90 year old grandmother and 92 year old grandfather' threats.[8]

I was first attacked by the USSR in Southern California, U.S.A. when I was approximately 11, around 1971. I'm typing these words in America in March 2024, age 64. I am harassed by Russian Mafia-controlled American operatives on a near daily basis.

My book is intended to help ordinary people worldwide and our governments worldwide survive the serious threat that is the Russian Mafia, by describing how the Russian Mafia stealth attacks ordinary people and is committing genocide against us and our government, how and why the Russian Mafia is destabilizing the quality of life of civilians, and how it is actively working to destroy our government.

I also describe how to disable the Russian Mafia's dominance-keyed hard-wiring precision bomb they've set to detonate this November 2024 in America.

Definitions

I define **subordinates** as ordinary people, including children, with the legal right to control themselves, and **dominants** as ordinary people, including children, who've received societal approval under specific circumstances to physically, psychologically control, direct, and/or order other individuals.

---

8   According to Janis Berzins in his paper "The New Generation of Russian Warfare" "...the Russian view of modern warfare is based on the idea that the main battle space is in the mind." Aspen Institute | Prague, *Aspen Review* (March 2014). New Generation Warfare: https://en.m.wikipedia.org/wiki/new-generation-warfare.

A child is a subordinate unless they are a babysitter or a worker who controls other workers, whereupon I call them a child-dominant-by-work. A parent is by law the dominant in their child's life until that child is 18 (or the age of adulthood in their country) but in many cultures parents remain a major dominant in their adult child's life until the parent's death. A parent's dominant status over their child and themselves may be the only legal dominance they have, or they may also work outside the home and control the work flow of others, or assist their society by aiding in the delivery of essential services,[9] whereupon they are an adult-dominant-by-work. Other important words are bolded when initially used.

Chapter 1.3 – Instead of hacking our voting machines, the Russian Mafia has hacked our people – they've set millions of dominant 'bomb-lets' to go off in November 2024 – we must warn our nation & the world

Dominants in our species are the glue that holds us together. They are our parents, grandparents, doctors, teachers, law enforcement, story tellers, utilities delivery managers, bus

---

[9] Essential services workers are involved in the manufacture, processing and/or delivery of essential services in their society. I designate them in my definitions as dominant-workers with elevated authority/responsibilities. In addition to the Russian Mafia attacking ordinary families as part of their genocide, and stealth attacking the extended family members of government officials to destabilize and/or control their families, the Russian Mafia is more likely to target and entrap dominant-workers because these people have an enhanced ability because of their work to weaken our society. A partial list of employees who's work category designate them as essential service providers and dominant-workers include: medical-related personnel; food delivery, stocking, and food preparation personnel; water, power, communications, sewage control, human safety, security, law enforcement, and fire prevention personnel.

drivers, supervisors, military, grocery stores managers, landlords, government, etc. They are homeless parents taking care of their children as best they can, to people earning a modest income to support their families, all the way to the wealthiest of us. They are men and women, gay, straight, trans and non-binary of all ethnicity's. They own or control most of the world's assets, tell the rest of us what to do, and form the substructure for our species' societies and people. They are the through-line in every person's life: we are all controlled by dominants, we just don't think about it because they form and maintain the reality we are born into. They are like air to our species: without them we would not be alive.

Dominants birth us and nurture us from the first moments of life. Babies are hard-wired to obey and accommodate dominant's requests and demands because at birth babies know they cannot survive without them. The Russian Mafia knows this too and is using their knowledge to seize control of our species in November 2024, by targeting dominants in America and aiming them at us and our allies, as if our dominants are a virus and we are vulnerable, unprotected cells.

The Russian Mafia has secretly enslaved many, probably many millions of our dominants (by illegally entrapping them) and is using them as weapons against us, exploiting the following information which they know, but which most of us don't:

(1) we are a hard-wired species but don't know that we are;

(2) the KGB, and later the Russian Mafia, has been in our country actively destroying ordinary people's lives, and their families, since at least 1971. I know because mine was one of the families they destabilized. They made my strong, 29 year old mother cry.[10] Active Measures, a bomb intending to annihilate Americans, is a psychological war, hard-wiring and soft-wiring-based weapon developed and deployed by the KGB last century. Mr. Putin and the Russian Mafia inherited it when they came to power in the 1990s, and intend to Close it this November 2024 when they expect to install one of their American puppets into our Presidency.

---

10 The KGB stealth attacked me and my family when I was 11, approximately in 1971 and had made me and my family homeless by the time I was approximately 12. They deployed proxies and illegally installed a KGB-controlled operative into my mother's bed, in violation of the Geneva Conventions of 1949, Additional Protocols. Specifically, Protocol 1 of the Geneva Conventions, Article 37. Prohibition against Perfidy. They set up my mother as an accomplice to a crime. She told me at that time that she was innocent but she couldn't really explain what had been done to her. I theorize that she'd been coerced into accepting a plea deal which she took to prevent her children being placed into foster care, which confidential plea deal prevented her discussing the 'crime' with anyone but her KGB-controlled handlers. She didn't know the American law enforcement officers handling her case were KGB-controlled because they represented themselves as American law enforcement. The KGB operated here illegally – it was not in their best interests to explain their attacks to their victims. My mother, and the compromised law enforcement who oversee entrapment attacks, are victims, lied to by a enemy foreign government working an evil, stealth genocide against us. It's possible some compromised law enforcement know it's not our legitimate government working these entrapments, but if that's the case only they and the Russian Mafia know the reasons.

We are in pre-Holocaust – countdown to Holocaust. We must launch an effective defense, and our people and the world must be warned;

(3) our legitimate government operates an entrapment system intended to capture criminals but the Russian Mafia[11] has commandeered part of it and uses it to illegally entrap noncriminal civilians to destabilize them and their families. Compromised law enforcement manage this system;[12]

(4) our species heeds the warnings of dominants sounding the alarm about non-overt threats and ignores subordinate's warnings about non-overt threats – a blind spot in our evolutionary hard-wiring defense system the KGB/FSB and the Russian Mafia are exploiting to attempt to annihilate us.[13] While our hard-wiring evolved, dominants (females and males) of that time were the ones most likely to spot a non-overt threat, sound the alarm, help their people effectively counter the threat, and survive to reproduce. A variation of

---

11 And before it the USSR.

12 My mother worked as a surgical technician and she is entrapped; my father worked as a police officer and he is entrapped. Many of my supervisors and co-workers when I worked at a studio, at a production company, and at multiple law firms, were entrapped. Many of my neighbors are entrapped.

13 From birth we are hard-wired to recognize and respond to overt threats (our pain receptors are part of that system, a baby cries when they feel pain). We've little problem recognizing overt threats (someone attacking you with knife), but non-overt threat recognition wasn't hard-wired in because hard-wiring is expensive and difficult for a life form to evolve, so it's a life-or-death recognition gift. A person needs to understand quickly that a tiger chasing them is an overt threat, while knowledge that a territory is prone to quicksand, a non-overt threat, can be tacked onto our hard-wired elevated valuation of dominants since we are social animals willing to teach our young about environmental other threats.

that demonstrated ability (recognize a non-overt threat, effective defense, survive to reproduce) happened so many times over millennia that all modern humans are descendants of ancestors with that demonstrated ability, or we're descendants of ancestors who had access to dominants with that ability. We are hard-wired (our genetic material predisposes us) to 'respond in an appropriately defensive manner' when dominants sound the alarm about a non-overt threat. Active Measures is genocide and a coup threat combination designed to disable our species' non-overt threat alarm system. Fewer non-compromised dominants means fewer people Americans have available we'll listen to warning us about Active Measures. Active Measures is a multi-generational, non-overt coup designed to not be noticeable (the Russian Mafia are experts in preventing victims from talking. I'm not entrapped yet they harass and/or threaten me on a near daily basis, so no telling how they control their entrapped slaves. They're brutally effective. Not even compromised law enforcement officers are sounding the alarm and we have some brave law enforcement, and millions of very brave ordinary people in America. My guess is people's loved ones are threatened in a way they believe).

So, our dominants have been off-lined after the KGB and the Russian Mafia worked decades destabilizing them, and since they're the people we're hard-wired to listen to regarding non-overt threats, nobody is sounding the alarm about this non-overt threat in the way we need to become activated. I've seen no editorials outing the Russian Mafia about this war, no guest essays by noted ordinary people accusing the Russian Mafia of murdering babies by proxy. When people do not effectively respond to a threat, they are destroyed. I refuse to let us and our allies, and Gaza Palestinians, be destroyed;

(5) the Russian Mafia will force their dominant slaves to tell everyone in the slave's circle of influence to vote for whoever the Russian Mafia wants voted in. Dominants impact every individual and/or family in this country and in the world. When compromised dominants tell their children, grandchildren, great grandchildren, parishioners, patients, employees, customers, students, etc. to vote for a pro-Kremlin candidate, unless we warn our people the Russian Mafia will use Active Measures to destroy us;

(6) the Russian Mafia deploys enslaved dominants to entrap, corrupt and/or influence subordinates and anyone in the dominant's circle of influence. Slaves are also used to gather intel a person would only tell someone they trust.[14] The

---

14 An entrapped loved one was deployed by the Russian Mafia to get a list of new medications I began taking the summer of 2023 after the Russian Mafia necessitated trips to the ER, and a medical care provider. When the Russian Mafia wants to know the medications you're taking it's not for a pro-you reason. They used this same loved one to order and/or bring me toilet paper, unasked. I believe the Russian Mafia has designed a weapon embedded into toilet paper. We're experiencing a spike in colon cancer and I wouldn't be at all surprised that the Russian Mafia assassinates people using toilet paper. After I dared to complain about a medical care provider who gave me shoddy care (I complained in a letter and it was a fair representation) but the Russian Mafia didn't like it, apparently the doctor belongs to them. After I dropped off the letter to the doctor's office, so he'd know what I said, a Russian Mafia controlled operative followed me into the check out line at a grocery store and activated a hand-held weapon he had in his coat pocket. He was behind me and I was the only one in line, on my mobility device.

Russian Mafia then uses that intel to destroy the target. One of my parents, a life-long democrat who is now entrapped, voted for former President Trump. She and one of my half-siblings, also entrapped, told me they voted for Mr. Trump in 2020. I hadn't asked. I rarely raise the subject of Mr. Trump but I am left-leaning, and the family members who I have access to much less so, the longer they're entrapped and the closer we get to November 2024. They try to convince me of the error of my ways from their point of view but what I believe is actually happening is the Russian Mafia is ordering slaves to tell those in their circle that they're voting for Mr. Trump to prepare our nation for the win the Russian Mafia is expecting this November. Unless we free those enslaved by the Russian Mafia, the Russian Mafia will install one of their puppets in the White House and they will annihilate us. My guess is 90% of enslaved dominants in America believe our legitimate government is their puppet-master, that most don't know the Russian Mafia has entrapped them as part of their genocide against us;

(7) Americans don't know the Russian Mafia is running a genocide attack against us. They don't know that (a) gerrymandering, (b) our Supreme Court lurching right, (c) our dysfunctional GOP, and (d) our police, jail and prison systems

---

Suddenly I felt as if I were about to pass out. He stared at me, standing a foot or so away from me, obviously wanting me to know he was attacking me even though he wasn't showing me how he was doing it. After a minute or so I was able to be functional and paid for my purchases. The Russian Mafia has plenty of secret weapons, I believe, based on the one or two they've deployed against me. Cancer causing toilet paper would not surprise me at all.

who can't seem to get out of the 1950s, (e) our overdoses, (f) our mass murders, (g) our children committing suicide due to bullying, (h) our homeless problem, (i) our unaffordable housing problem, (j) our species' inability to effectively counter the climate catastrophe, (k) our student's score slide, (l) our longevity step-back, (m) our people of color's inability to get ahead, all this, and more, are indicative of how successful a weapon Active Measures has been against us. It's taken Active Measures, from 1965 (to pick a year) about 60 years to get us to the point where the Russian Mafia fully expects to install one of their puppets this November 2024. We launch an effective defense else the Russian Mafia will succeed. This book is to help our people and our allies get on the same page so we all see the actual perpetrator, and effectively defend ourselves; and

(8) people don't know that we're hard-wired to not report dominants in our lives who try to corrupt us. The KGB/FSB and the Russian Mafia know it, and exploits our hard-wiring. As babies we interpret dominants as good because they help us live. Long after we no longer depend on them to give us a bottle, we protect them because in our baby-heart we feel they must exist or we can't survive. We know that's not true and we're not conscious of this feeling, but it's hard-wired in and controls our behavior. The Russian Mafia camouflages itself in them by enslaving them. Then it illegally tries to lead us to sin by stealing the skin of our loved one, and using our love for our loved one as weapon against us. They use this

attack against our children too. I estimate over 1,000 children have been deployed by the Russian Mafia to harass me, most of them with their parents. Non-compromised parents wouldn't use their child or conspire with their child to harass someone on a mobility device, using a cane, or walking unassisted, yet I've directly experienced, rough estimate 1,000 kid attackers, starting from when I was 11.

Because we don't know they've been enslaved by the Russian Mafia, and because most of us have no intention of following their law-breaking advice, suggestions, demands, orders, crazy ideas; and because we still 'see' them with our infant's-heart (which heart subconsciously insists they are essential to our survival) we'd never consider reporting what they do or say to law enforcement. Which the Russian Mafia knows. And exploits. Our government knows the Russian Mafia is here destabilizing but it is prevented by our law from monitoring private and legal relationships. When people don't report the dominant's trying to groom them to law enforcement, because we're hard-wired to protect our dominants, that gives the Russian Mafia a free lunch to enslave dominants and use them to destroy their families. And to install a pro-Kremlin candidate into the White House. If the Russian Mafia installs a pro-Kremlin candidate, it's game over for us. Forget all the other pro-Kremlin governments in the world which seem to be functional, sort of. The Russian Mafia will destroy us and we won't be coming back from that. Far better to just not go there.

The Russian Mafia knows that after they compromise a dominant they can drip poison the family using that slave/influencer until everyone in grandfather's and grandmother's, or mother's and father's, or mentor's, circle of influence is compromised. After victims are compromised the Russian Mafia prevents them from speaking out. The KGB deployed this weapon last century, I believe prioritizing the targeting of people of color (since my family was targeted), to get the kinks out. The difference I see between how enslaved parents engage with their children now versus how my mother did is now the Russian Mafia forces entrapped parents who accept a plea deal to immediately begin using their children as beards/co-comrades. When my mother was destabilized she wasn't immediately forced to harm us. That was 53+ years ago. The Russian Mafia has evolved their attack to begin grooming entrapped parents to as soon as possible view their children as weapons and tools. I've been harassed-engaged by entrapped parents with their newborn in tow. When a mafia can force post-natal women, still bleeding a little from the after-effects of labor[15] out of bed to harass a target, we've got a problem.

The Russian Mafia doesn't tell family members that grandmother is now an enemy operative trying to entrap her grandson, or that grandfather is being forced to entrap his granddaughter. The Russian Mafia is working to destroy us, not save us from them, so our children aren't safe in their own family. We're hard-wired to see our dominants as "friendlies," not as enemies. The Russian Mafia attacks and confuses our hard-wiring. One of their favorite ways to do this is by forcing the dominant offer 'help' when the dominant is actually offering entrapment. From our first moments of life dominants 'help' us. When the Russian Mafia has it's slave offer 'help,' when they're really offering entrapment, the labeling of the attack as 'help,' offered from the dominant who has 'helped us' all our lives, confuses us, and we default to trusting that person. That's disastrous for the victim because this dominant

---

15 Women continue to bleed a little bit for awhile after they give birth as their uterus re-adapts to not carrying and nurturing a baby. I don't know for how long but I think it's for weeks. I've been harassed-engaged by parents with newborns who appeared to be days old, a time when a woman should not be forced to be anywhere she doesn't want to be. Yet with the Russian Mafia told these parents to jump, they said 'how high.' Not by choice. Just as the Russian Mafia got recently delivered women out harassing targets the Russian Mafia wanted harassed, the Russian Mafia will get the dominants they control to influence everyone in their circle to vote in a pro-Kremlin candidate. Just telling people about this attack will help them better protect themselves but we have far more to do than that in order to survive. We must 'talk' to our ancient hard-wiring in a way it understands. The language it understands is dominant-subordinate. So we video vignettes and we ask our excellent Administration, and other noncompromised government officials, to help us talk to our people in 'modern' and in 'ancient' to cover all the bases. I discuss dominant-subordinate communication later in the book.

isn't our grandmother, grandfather, parent, doctor, mentor anymore, they are the Russian Mafia using our loved ones as camouflage. And the Russian Mafia means us harm.

That's part of the reason increasing numbers of us, including our children, are on sedatives, sleep aids, are obese, have high blood pressure, are reaching for anything to help us self-sooth – because the Russian Mafia is messing with our sense of reality, attacking our hard-wiring when most of us don't even know we have hard-wiring.

The Russian Mafia knows about our hard-wiring and is using it as a weapon to destroy us, not just to steal our country out from under us, not just to flip the world order, not just to destroy a hated rival government, but to annihilate us, the American people. Holocaust 2.0: Russian Mafia against ordinary Americans. How do I know? The Russian Mafia is deliberately destroying our families. I know because they attacked my family and destroyed it. They entrap our parents, grandparents, etc. then force those loved ones to entrap their children. After which they reveal the betrayal to the child, destroying the family. When an enemy deliberately destroys civilian families, that enemy has no intention of maintaining that culture: destroying our families is destroying our culture. Genocide.

The Russian Mafia ordered the 2023 Israel-Hamas War to unseat our President and install one of it's puppets. It's murdering and/or injuring by proxy tens of thousands of Palestinian civilians, including their babies to seize control of our country. The Russian Mafia doesn't hate Palestinians, they are elitist racists who see all people as pawns to serve them. They're slavers with a 'we're chosen by god to rule the world' inferiority complex. Their idea of nirvana is annihilating Americans. I am not making this up. They're murdering Palestinian babies and Palestinian's old people and everyone in between and they don't even hate Palestinians. They hate us. Not because of anything we've done, but because our people and our government worked hard and helped the world survive WWII, and the USSR didn't get the 'world power' bounce out of WWII that they wanted. Their government negotiated sub-optimally when they agreed to fight the Nazis, while our leaders negotiated favorable terms with England in exchange for our help. That's not our fault, but the USSR lost 20 million soldiers in WWII because of bitter winter warfare against the Germans. We had nothing to do with that but the USSR, like the Russian Mafia, were a covetous government and they resented that American leaders negotiated a beneficial post-war situation for us, while the USSR's leaders failed to do the same for their people. They decided to come after us. Jealousy, rage at not getting world domination after WWII (nobody promised them that), inferiority complex – they've got issues they're uninterested in fixing. And because they stumbled upon a blind spot in human hard-wiring last century, and because they have no problem murdering civilians, including infants, they've justified to themselves that they're chosen by god when what they actually are are fentanyl pushing, baby-killing cowards killing anybody's baby as if that infant was dirt. Any adult in the world can exit their house and see a child and murder her. That doesn't make them anything but a murderer of children. The Russian Mafia is a mass murderer of children and demands world dominance, so they can slaughter even more children. Two hundred million American children. No.

The Russian Mafia understands how our hard-wiring works and is using their knowledge and our ignorance to destroy us. Dominants in our species form part of our reality. The Russian Mafia has targeted our dominants since last century (I know because they attacked and destabilized my 29 year old mother and made me and my family homeless around 1972). When the KGB and the Russian Mafia mess with our dominants, they're messing with a fundamental core of our species, so it's easy for them to destabilize us and our children, especially since we don't know they're on the ground here, hunting. Knowing they're here and that they're attacking us will help us better defend ourselves.

Now that we know the Russian Mafia is the problem and has been the perpetrator all along, it'll be a marvelous relief for people in our country to understand we've drawn a monster working to destroy our self-image. That we're a good, decent, people, not racists. The Russian Mafia and the KGB has been pushing that lie on us since last century. Read this book, I lay out my proof.

The Russian Mafia has been on the ground since the 1960s at least, stealth attacking all our institutions, and us. The

KGB attacked me using proxies when I was around 11, and used a proxy teaching assistant to accuse me of cheating on a math test, even though he said I wasn't cheating but that another student was cheating on me (that other student was an operative, the whole thing a setup). Traumatized after one too many proxy attacks I dropped out of school. The KGB forced my high school counselor to accuse me of theft to stop me going to her office requesting advice about how to attend college when I was an impoverished, working part-time, high school student.

While the KGB, the FSB, and the Russian Mafia seize control of individuals in our institutions, our institutions aren't racist - the monster we drew is racist and is determined to destroy us. Past time to rebuild our people's self-image. We're a wonderful, hardworking people. After we get this monkey off our back (which won't be easy), we'll thrive once more.

We must educate our people, including our children, to help them recognize destabilizing attacks to help them better defend themselves, and to encourage everyone to out/report these attacks when we encounter them, even if the perpetrator is grandma, or our doctor. Currently, Russian Mafia's attacks using our trusted dominants as weapons aren't reported. That must change else the Russian Mafia will never stop weaponizing our parents, doctors, teachers, police, etc. they've enslaved to attack us.

Chapter 1.4 – Why I wrote this book – to tell my people and the world that the Russian Mafia & before it the KGB, have been committing genocide against Americans for decades, to describe what that genocide looks like camouflaged on the ground, & how we will survive

In this book I describe a series of attacks by the Russian Mafia against us and against me, and how it has deliberately deployed genocide against Americans for over 50 years, using a strategy and secret war they've named Active Measures. I describe how they've seized control of many millions of Americans, and how we'll survive their attack. I offer examples of entrapment set ups they've used against me to help people better recognize what entrapment looks like before the Russian Mafia springs the trap to help people increase their odds of evading entrapment. I've figured out how we'll evade Active Measures, with critically essential help from our Biden Administration, our security agencies, and our military. We will successfully counter Active Measures.

The Russian Mafia is deeply embedded in America. It is constantly weakening and destroying our people, from the inside out, one family at a time. They do so by entrapping authority figures like our parents, grandparents, doctors, teachers, police officers, employers, priests, landlords, etc. and by forcing those trusted people, now slaves, to entrap the rest of us. I've endured many entrapment attacks which is how I know they exist. In this book I describe them and how I evaded them. Short-answers: ethics is one of the key reasons I was able to survive unentrapped, ethics instilled in me by my young mother before she was enslaved. Had she not instilled those ethics, I wouldn't be allowed by the Russian Mafia's American law enforcement operatives to write this book. Millions of people know parts of this story worldwide but they've been enslaved by the Russian Mafia or they are the Russian Mafia.

But the primary reason I'm able to write this book outing our enemy is our nation's nuclear warheads: Active Measures is a nuclear-war-avoidance-based genocide. Because we have nuclear weapons the KGB, FSB and the Russian Mafia didn't/don't want to directly experience, they designed their weapon to accommodate their preferences, which gave us time to understand and disable their bomb.

Determined to destroy us, the KGB built and deployed Active Measures last century. It is a multi-generational (deployed 65 years ago, est.), non-overt (designed to appear nonthreatening) genocide built to exploit blind spots in our hard-wiring (our instincts), and our soft-wiring (our socialization aka how we're raised from birth).

Chapter 1.5 – Hard-wiring as a weapon against humanity in the hands of the Russian Mafia

Most people don't know about hard-wiring and soft-wiring but the Russian Mafia, and before it the USSR, knows about them and has formed them into weapons to: (1) enslave the governments the Russian Mafia wants as allies (China, India, etc.), and (2) destroy competitors, like the West-allied nations it envies and who's assets and power it covets and intends to steal. The Russian Mafia has no interest in keeping Americans alive. Only our government prioritizes the value of our lives. The Russian Mafia cares only about ascending to world dominance. No number of dead is too high a price for them as long as it's not them; no baby is too young to exploit as a blood pawn, as long as the Russian Mafia survives. They are truly evil.

Chapter 1.6 – Why I believe the Russian Mafia intends to exterminate all people living in America

A very large clue the Russian Mafia intends to destroy Americans is the fact that it is in the process of destroying families in America and has been since at least 1971, when they first attacked me.

I am directly experiencing and am directly targeted by Russian Mafia-controlled American operatives as an adult American child of two enslaved and entrapped American parents in their 80s, who are unknowingly enslaved by the Russian Mafia as of March 2024, and have been enslaved for decades. The Russian Mafia has lied to them and led them to believe their American handlers (law enforcement issuing them orders) are controlled by our legitimate government and that our legitimate government has entrapped them.

Both my parents have been forced to try to entrap me, their oldest child, their only child together, as well as been used to get information from me that the Russian Mafia has used, and/or is in the process of using, to try to destroy me, as recently as the summer of 2023 (forcing my mother to initiate the story of her experience with a palm reader in the early

1970s so the Russian Mafia could learn by my reaction if they could try to set me up using a palm reader as they've apparently successfully done to multiple family members (my mother, one of my sisters, and one of my nieces).

In addition to trying to entrap me multiple times, both my parents have been used to gas light me,[16] lie to me,

---

16 My father was used to misdirect me in many ways over the years. I told him store security and plain clothed law enforcement in stores were harassing me twenty years ago. The gas-lighting was him telling me to ignore it, to not take it personally, that stores were experiencing large spikes in shoplifting, that store security wouldn't know how much I spent at the store. I told him I'd shopped at this big box store for years, had spent thousands of dollars there on extended family, and he said the store security wouldn't know how much I spent there. That was a lie. Some of the security in that store were plain clothed police who well knew my spending habits. I'd been lured to that store by my sibling, long before I knew she was a slave, long before I knew I was being attacked in my own country by an evil enemy foreign government. Our government needs to look at how the Russian Mafia controls some of our stores. They've got their hooks deep into our food supply chains. Controlling our access to food is part of the Russian Mafia's attack on us, along with their control of our communications systems (which I learned when they've prevented my calls from going through over the last several years), their control over some of our law enforcement (which control the Russian Mafia began showing me in 2001, nearly 25 years ago, and their control of our electrical grid, which the Russian Mafia most recently demonstrated on September 29, 2023 when it attempted to destroy by laptop and this book by surging the electricity in this apartment, causing one circuit out of nine to open, the circuit I use to charge my laptop. The charger was in the plug strip, which the operative who lived in the apartment above mine could see when she lived here. She had eyes on me and the plug in the charger on my bed, and the laptop on my bed, covered by blankets, appeared to charging, but wasn't. That particular time, the laptop wasn't being charged. Had it been being charged, this laptop would've been toast and my ability to timely warn you before the November 2024 elections would've been severely compromised. Right now, 2/23/2024 at 9:08 am, the Russian Mafia has an operative in the apartment upstairs and no one lives there. The Russian Mafia controlled American operative has been noise harassing me since approximately 7:40 am or so. Mr. Putin and his Mafia are determined this book be destroyed and/or that I'm so terrorized I won't produce it. No matter my level of terror, I'm warning my people and everyone in the world.

misdirect me, browbeat me, denigrate me, and otherwise influence me under order of the Russian Mafia. The Russian Mafia has no authority in my life. To try to get me to self-harm they camouflage their evil presence in my beloved parents and force them to try to get me to cheat the government, cheat an insurance company, tried to get me to file for personal bankruptcy.

Chapter 1.7 – Dominance, hierarchy, socialization – how the lives of our ancient ancestors shaped Active Measures

People alive today are descendants of long dead ancestors who had access to dominants who were able to recognize and remember **non-overt threats**,[17] sound the alarm, and help their people evade the threat and/or develop an effective defense. Eons ago when our ancestors evolved into something approaching human-like physiology, most of our ancient ancestors couldn't identity non-overt threats, which may be part of the reason they're gone extinct.

Nearly all people can recognize **overt threats**. Someone running towards you with a knife, or if you're trapped in a burning building, are examples of overt threats. The ability to recognize and respond to overt threats is hard-wired in, meaning it's instinctive. Even newborns can recognize overt threats (a leopard sneaked into our ancestors

---

17 Non-overt threats are dangers that can kill or harm but which don't appear to be an immediate threat. Quicksand comes to mind. Our ancestors saw innumerable patches of mud and many posed no problem but some, quicksand, sucked them down and smothered them. Quicksand is an example of a non-overt threat to our ancestors and to unknowable numbers of other life forms.

camp and took hold of a baby, the baby's cry of pain alerted her clan of the immediate threat and they rushed to her defense to chase off the leopard by throwing stones at it, and waving fire-tipped logs at it they'd been burning to warm the cave). The ability to recognize overt threats is a necessity for mammals on a threat-filled planet, so that ability is hard-wired into us.

Before we evolved to be people, our distant ancestors' family's and expanded groups consisted of subordinates and dominants. Some people lived longer, were healthier and stronger, and their people noticed. Some tended to be right about a problem. For example, one ancestor recommended tribesmen not go in a particular direction because she'd seen leopard dung there when she'd been gathering berries. When those who ignored her warning were killed by a leopard, her clan noticed she'd been right. And that same ancestor was able to solve a problem about how to gather potable water from rain when she built a three foot deep pit and layered it with big waxy leaves she kept in place by placing heavy stones on them. The waxy leaves stopped the water from immediately soaking into the soil beneath the pit so her tribe had a temporary nearby source of water for drinking or cooking. Other ancestors, by their choices, tended to avoid physical injury. Our ancestors had to chase and/or trap their prey – a strong uninjured female or male hunter was highly prized. Over millennia other ancestors developed sharper knives from a particular type of stone that cut tough meat more easily, learned which wood burned longest while producing less smoke, or they were early adopters willing to try and improve upon new technology like the bow and arrow, or the sling shot. At least one of our distant ancestors created the wheel, a technological advancement upon which 21st century human society rests today. Without the wheel there would be no cars, no computers, no grocery stores, no millions of other services and products our species relies on. Over time, the more experienced, creative, and adventurous of our distant ancestors began to acquire more wealth, more food, & more helpful tools faster than other people. They excelled at surviving and they helped their people survive. Our ancestors named them dominants, or wise chiefs, females and males, who were prized for their knowledge and experience and for helping to keep their people alive longer. Over millennia our ancestors increasingly began to listen to these smarter women and men, their chiefs, and those who listened tended to live long enough to reproduce. Additionally, some of those dominants had an extra important skill: they could recognize non-overt threats and launch an effective defense. Over many thousands of years more people with chiefs who could recognize a non-overt threat and respond effectively, lived long enough to reproduce, while people with chiefs who couldn't recognize or effectively respond to non-overt threats, tended to die out before they could reproduce. People alive today are descendants of ancestors who's chiefs demonstrated an ability to recognize non-overt threats, sound an alarm, and organize/build an effective defense.

Hard-wiring is very difficult and very expensive for a life form to accurately sustain and pass along in it's genes over millennia so it's use is restricted to needs-based, immediate life-or-death threats. Our physiology (our genes and the design changes of our body over time) can embed only a small amount of important intel into our reproductive genes so a great deal of very important survival-related information that is not an immediate-life-and-death overt threat (the way that a lion chasing you is an immediate overt threat), but is only a small step from the immediacy of that threat (like the need to recognize and avoid quicksand, the need to avoid poisonous foods (like the poison found in particular types of wild mushroom), the need to know how to swim when your group lives in a flood plain. This high value information is needed and ancestors with the ability to recognize and defend against non-overt threats fared demonstrably better than those who couldn't recognize non-overt threats. In the 21st century only people descended from ancestors with access to leaders who could recognize non-overt threats survive.

Because quicksand or a flood isn't the same level of threat as a pursuing lion, and hard-wiring is severely limited, evolution off-loaded recognition of non-overt threats to: (1) our dominants. Social animals,[18] we tend to live together and are

_____

18 Social animals are life forms who tend to live together in a group of the same animal. Even when individuals leave a group they tend to reassemble into a different group of their same animal. Our distant ancestors, lacking the sharp fangs and claws of the apex predators of their day, and initially lacking the power of fire, had little chance of surviving alone. Those who survived long enough to reproduce tended to live in groups. We're descendants of those pre-humans. Our instinct

willing to invest in and teach our young); and (2) our social order (subordinates and dominants). Babies instinctively[19] assign elevated value to the dominants who take care of them.

Hard-wired into modern people is the need to listen to dominants when they warn of non-overt threats. Hard-wired means our instinctive response controls our behavior. It's a response we're born with and aren't conscious of: it's a gift of evolution. When ordinary people, aka subordinates1 see a problem and try to warn their people, our species ignores them or are irritated by the warning. Irritated targets who ignore the person trying to warn them fail to launch an effective defense so are easy prey for the Russian Mafia.

Our physiology, i.e., our body, our emotions, our instincts, calming endorphins, our increased comfort level, is our hard-wiring on display, in action, when we live in social groups which function effectively. Those hard-wired instincts control our behavior but are so much a part of our physicality, the life form that we are, that we're unaware of them on a conscious level. We don't think about our hard-wiring, and most of us don't know we are hard-wired, but instead believe our personalty and our nature are gifts from recent ancestors, our parents, our grandparents. But our inheritance spans many billions of years when earth was created and the dust from long dead stars, and planets pounded by asteroids had huge chucks flung into space, all of it recombined to form earth, a planet welcoming to life, depending on prevailing chemical

_____

to live in groups is hard-wired in.
19 Instinct means hard-wired.

conditions that waxed and waned over billions of years. No other life forms in the universe are quite like earthlings but if we become competent space travelers we may see bits and pieces of our interstellar ancestors on moons and planets who share a partial trait or a feature with us.

Humans were evolving before our ancient ancestors had the physical ability to speak, long before we'd developed language. Hard-wiring into our genes is the only way earth-bound life forms can survive. Life is tenacious on earth. If a workaround can be evolved, and if earth's chemical conditions don't prevent it, life will keep insisting on having a try. We are just one of many trillions of earthlings trying to survive.

Chapter 1.8 – The hard-wired blind spot exploited by the KGB and the Russian Mafia to destroy us is that our species doesn't "hear" subordinates' warnings of non-overt threats – we evolved and our ancestor's hard-wiring was formed and "set" millions of years ago to hear *only* dominants' warnings about *non-overt threats*. So when subordinates try to warn us, our defense systems aren't activated. No defense means the Russian Mafia will kill us

This blind-spot in our hard-wiring was noticed by the KGB (a security agency of the USSR) last century in the many wars the USSR fought. People who received accurate warnings about an impending, previously non-overt threat, ignored those warnings if they were delivered by a low status person. This behavior was noted time and again, across nationalities, cultures, ethnicity's, religions, sexes, by the wealthy or the impoverished. The KGB and the USSR decided to build a weapon around that hard-wired blind spot they saw for themselves worked nearly 100% of the time, and Active Measures was born. In 1971 when I was 11, the USSR was in America working Active Measures even though they'd signed the Geneva Conventions in 1948/1949 promising they'd not attack civilians, children under 15, or women, in war time. My family included my mom, all her children were under 15, and we're civilians, yet the USSR placed their operative into her bed and that operative lead to her involvement in a crime she said she wasn't involved in, which lead to her losing her job, our home, most of our possessions, etc.. None of that mattered to the USSR, none of it matters to the Russian Mafia. By signing the Geneva Conventions it is implied that no such attacks will take place in peace time, since governments who signed it said they wouldn't attack these groups during war.[20] I don't believe the USSR or the Russian Mafia abide by laws and treaties of other nations. They've demonstrated extreme contempt for all lives not their own.

The USSR and the Russian Mafia are liars. The Russian Mafia are criminals, best in class, world class criminals. They think people stupid enough to believe the word of a criminal

---

20 Despite the Cold War, America and the USSR had a trade relationship. In 1958 the Lacy-Zarubin Agreement saw us and the USSR (I suspect USSR operatives blackmailed our leaders into) sharing some of our technology with the USSR. And in 1975 the Jackson-Vanik Amendment, established normal trade relationships between us and the USSR (and I believe China and other nations – my theory is the USSR worked these "deals" to destroy us and benefit the USSR, and they also used them to build relationships between the USSR and China, as China got normal trade with us too). I suspect those agreements had language whereby the USSR agreed to respect our laws. The Russian Mafia respects nothing and nobody.

deserve predation. They don't care that the Russian government doesn't tell the world that it's a mafia state. They have no interest in fair play, nor do they respect other governments. To the Russian Mafia, other governments are competitors to be destroyed and/or enslaved. They have a take-no-prisoner world outlook, which they extend to our children and civilians. Most people in the world don't understand the Russian Mafia see's them as prey to be enslaved and/or destroyed. It's a big problem when ordinary people have an apex predator hunting them and they have no idea. Destabilizing democracies abound, spiking overdose rates, children stealth destabilized in their school and in their societies by Russian Mafia operatives bullies attacking children. It's like the KGB has taken aim at our kids, across all West allied nations.

The Russian Mafia believes due diligence checks are the responsibility of other people's governments and if those governments believe the Russian Mafia's lies, that's on them. They're cold blooded. Cold blooded enough, I believe, to be responsible for the murder of over 10,000 Palestinian children while blaming Americans for the 2023 Israel-Hamas War. I believe the Russian Mafia controls Hamas and individuals in the Israeli government. I believe the Russian Mafia ordered Hamas to attack Israel, and is running the war.

Our security agencies are investigating. They can't say this war is a Russian government proxy war without proof, and no matter how much they may or may not believe this is a proxy war, they're not going to say so until they have some way to prove it.

The Russian Mafia doesn't believe in leaving paper trails and our policy of not negotiating directly with terrorists, probably put into force by a Russian-Mafia controlled American operative long ago, means our outstanding professional interrogators don't have access to high value intel we'd get by negotiating directly with Hamas. The USSR and the Russian Mafia are deeply embedded here and have passed laws and instituted policies and procedures that benefit them and harm our interests.[21]

---

21 For instance, I use a mobility device and the Russian Mafia harasses me pretty much everywhere I go, and have done so for decades. I began using it when I was maybe 50 or 51 and apparently it prevented the Russian Mafia from easily anticipating my destination. So they changed bus policies in the cities where I take the bus. They had Santa Monica and Los Angeles institute new policies. Now bus drivers asks customers on mobility devices their stops when the customer enters. In a bus driver-customer relationship, the customer is the subordinate. We are from birth taught to respond to and obey dominants. If I met a bus driver on the street I'd not feel obliged to answer any question he/she put to me, but as a bus rider, in the subordinate position, I am soft-wired from birth to respond a certain way. This is an example, just one of many, of how the Russian Mafia manipulates victims, target's lives, and warfare. When they want anything at all from a target, they position that target into a subordinate position and use the dominants as a weapon against them to get from the target what they want. I have the option of telling bus drivers I'll ring the bell at my stop, and they accept that, but it took me a while, being harassed for years, to figure out how the Russian Mafia knew to have their operatives position themselves at every step of my journey, to understand that I can evade the trap by telling the driver I'll ring the bell at my stop. Another thing: the Russian Mafia uses American operatives it controls here to harass and/or attack people like me the Russian Mafia wants harassed. The apartment manager where I live harassed me on I believe it was May 31, 2023, to such an extent my heart elevated to 200 bpm, necessitating a trip to the ER. Paramedics refused to transport me to the closest ER saying it was closed but the ER they transferred me to I was harassed quite a bit and would not willingly return there. After I got out of the hospital I went to the Santa Monica Police Department to file a

I know what I know because the USSR attacked me and my family using Active Measures when I was 11, had made me homeless by the time I was around 12. They continued attacking me until their demise in 1991, whereupon the Russian government/Russian Mafia took over the attacks, expanded them, and included precision-based entrapments, and a great many more physical assaults, etc. in addition to continuing other "let's make her miserable and have our operatives cheat her" financial-based attacks the USSR had most often engaged in.

---

complaint of harassment of the apartment manager, and the nice police officer informed me that harassing someone isn't against the law in Santa Monica. So some time in the past the Russian Mafia manipulated whoever to ensure that when they wanted their operatives to harass specific targets, there's no harassment law they could be charged for breaking. At my job before last, the Russian Mafia wanted me to leave that company because the entrapment they tried failed and they had other ideas at the next law firm they manipulated me to going to. I'm African American and the Russian Mafia used two African American operatives they controlled at that company to harass me out of that company. So you see, the Russian Mafia attacks, but they make sure they don't burn their hosts. That law firm in downtown LA is in group of law firms the Russian Mafia can place victims, or their operatives, and set entrapments there. When the target evades the entrapment, which the Russian Mafia set up and sprung using that law firm's payroll department, the Russian Mafia had enough pull to position two African American operatives, force them to harass me, and get me to leave the company. To ensure I couldn't say the attacks/harassment were racially motivated, and bring harm on a company they use, an asset, they had the attackers be African American.

Chapter 1.9 – I've experienced for myself our hard-wired blind spot as I've worked unsuccessfully the last 9+ years to warn my people, and had an experience similar to Moishe the Beadle's in Elie Wiesel's excellent, haunting book *"Night"*

I've noticed as I've tried to sound the alarm about the Russian Mafia's coup here for at least the last 9+ years without getting traction that our hard-wired blind spot is real.[22] Elie Wiesel

_____

22 A mistake made out of frustration by one of the Russian-Mafia's American law enforcement operatives, allowed me to learn I was under attack when I was about 41 or so around 2001. I'd evaded entrapment attacks for years but had no clue I was doing so. Russian-Mafia American law enforcement operatives couldn't tell how I was evading them and because I didn't know that I was, and because these attacks were stealth, and because compromised American law enforcement were attacking me illegally, extra judicially, they couldn't come right out and ask me. My secret: I lived ethically, since my mid-20s, and had no idea anyone was hunting me. They'd sicced two women operatives on me and inserted into my work life, and had expanded it to occasional after work meetups. They got me to this restaurant and the ladies were talking about their sex lives. I didn't have anything to offer, so I made an off-color joke. The ladies laughed, the plains clothed compromised American law enforcement officer listening in to our conversation was disgusted. I happened to look up while I and my "friends" were laughing and noticed an extreme look of distaste on his face. I didn't know him, had never seen him, thought it weird he was listening to me and my "friends." That very night after we finished our meal I went grocery shopping alone, walking home, stopped at a grocery store, a high end one because it was closest and I was walking. I was overtly harassed by maybe five or so men in that store, uniformed and plains clothed operatives and that because the next phase of attacks by the Russian Mafia, but where they overtly attacked and harassed me. I've been kicked, stomped on, called the n word, hit by operatives entering/exiting the bus with their packages hundreds of times. On and on. It's been horrific. The upside was I'd been attacked by 30 years at that point and had the operative not lost his temper, I wouldn't have been able to put together subsequent attacks. For instance, when my family because trying to entrap me when I was 50, I'd have had no way to understand their behavior was part of an ongoing attack, had it not been for the nearly decade of experience being overtly attacked by strangers pretty much everywhere I went. Had it not been for that man's mistake, I wouldn't be writing this book, I wouldn't have been able to pull enough information together to understand the attack. When I was 58 or so and talking to my dad about writing this book,

notes it in his book *Night*[23] when Moishe the Beadle, a foreign-born impoverished Jew in the city Elie grew up in, befriended Elie when Elie was 12 or so. Moishe was one of the rare Jews to survive one of the SS's initial attacks, and tried to warn his people. Although poor he didn't beg, something his adopted village sincerely appreciated about him. So, apparently, Elie's village had other immigrants who begged. To earn an income, Moishe did odd-jobs-for hire. Think of him as a janitor chum handyman, who got paid a below living wage for sweeping up at the synagogue, or who did odd jobs around the village. A villager would pay him for light manual labor and he'd do a decent quality job. Moishe loved his adopted village and he was liked well enough in return. Not respected by the village leaders, but tolerated because he was willing to work, didn't annoy people by begging, and in general stayed out of the way. He was smart but he didn't flash. His intelligence and tenacity were demonstrated by his ability to unobtrusively rebuild his life among strangers, despite having little money, during the 1930s and WWII. Just that was an achievement, although only his friend, 12 year old Eli, truly engaged with him. Still, he was welcome in synagogue, he was part of the village.

Because he was foreign-born, he was one of the first groups of Jews rounded up in Eli's village. Villagers were lied

---

having no idea he was enslaved, he tried to talk me out of it, his main points were – (1) you don't even know who's attacking you, so how can you warn people, and (2) people will think you're off your rocker. By the time he's handlers had him say that to me, I'd had 17/18 years of overt attacks and I'd long understood the attacks weren't my imagination.

23   Elie Wiesel, Marian Wiesel (Hill and Wang; revised edition 1/16/06).

to about where Moishe and the others were being taken. They were told they were being sent to another area with better work opportunities.

Forced onto a train, the train traveled some distance, then forced off it, forced to dig a trench, witnessing the atrocities against innocent Jews, including babies, victims were forced to the trench edge and shot point blank in the neck by the SS or by their comrades wielding machine-guns against them. People fell dead or seriously injured into the trench. Moishe was shot in the leg and fell into the trench, faking death to avoid being murdered. He laid in the pile of the dead the SS hadn't hesitated to make dig their own graves, hadn't hesitated to toss crying infants into the air so comrades could shoot them, as if in target practice. Moishe witnessed this unimaginable evil.

Long after the SS left, a terrified, traumatized Moishe likely fading in and out of consciousness from blood loss was forced to witness the slow deaths of children begging for help who he could not help. Pulling himself out of the trench on top of the dead, I believe I read in an older edition that someone in a farmhouse bandaged his leg as he limped many miles home to his adopted village, his only need to warn his people, including a teenage Elie. I think these murders occurred sometime around 1941.

No one in his village believed him, including his friend Elie. The slaughter Moishe described, of infants tossed into the air and shot as if in target practice, was too horrific for the people to comprehend. Eighty plus years later it's difficult to comprehend people could commit such evil but look at the Russian Mafia's proxy-murdering of Palestinian children, and Moishe's report of the atrocities he witnessed is believable.

The villagers had yet to experience the Holocaust and could not envision such evil. They thought Moishe was unbalanced. Moishe, convinced he'd been spared by god to warn his people, was stunned that they wouldn't believe him. He tried and tried to convince them. The villagers refused to be convinced. Finally, he stopped talking about it. A few years later the SS came for the entire village. Moishe was there, with his village, loaded into cattle cars. He said to them, "I told you so."

This book is my attempt to warn you. Please don't make me another Moishe the Beadle, and force me to witness the destruction of my people by a government more evil than the Nazis.

From birth people can recognize overt threats. A newborn attacked by a predator cries and activates her/his parents (dominants) and extended clan who rally to the baby's defense. The baby recognized the overt threat because of the pain of the attack, and sounded an alarm. But there are many non-overt threats. This same baby, now a toddler, walks 10 yards behind her/his family's camp and very nearly drowns in quicksand. Babies of our ancestors who survived quickly learned to listen to the dominant's in their lives with regards to non-overt threats. Non-overt threats may be behavior, a situation, plants, animals, landscape, etc. which don't appear to be a threat, but are. Our ancestors were pre-homo sapiens and those who lived long enough to reproduce tended to have dominants they listened to who warned them of non-overt dangers.

Hard-wiring is expensive for evolution to embed into a life form's physiology, so it's use is highly needs-based. We must be able to recognize overt threats, else our distant ancestors wouldn't have survived a world filled with hungry predators with little but their wits. No hard-wired overt threat recognition – no us. On the other hand, non-overt threat recognition of say, quicksand, poisonous mushrooms, or an enemy village's scout watching our ancestors, while important, weren't as immediately essential for survival as, say, a tiger stalking them. But because ignoring non-overt threats could wipe out a village, our evolution outsourced non-overt threat recognition, piggybacked it, onto our hierarchical system. Our hierarchical system is part of our survival strategy. Top dog status evolved to help our ancestors know who to listen to eons ago when our hard-wiring evolved.

Another blind spot in our hard-wiring the Russian Mafia exploits is our species' tendency to "lean-into" the belief that high status people are more likely to be intelligent. This is part of the piggybacking where evolution couldn't afford to hard-wire an ability to spot non-overt threats, and instead linked the warnings to dominants, many who were the smartest and wisest of our ancestors, plus our newborns are hard-wired to assign elevated value to the dominants who care for them so that's how evolution did a workaround to ensure our species had access to accurate non-overt threat assessment, by hooking it to dominants. All babies who survive are reared by individuals who are dominants to the baby. Even babies somehow raised by non-human animals can be cared for by individuals in that species and if the baby is cared for by them long enough, the babe affixes dominant status and love onto the non-human caregiver because that individual keeps them alive. Many animals have enough basic knowledge to keep their babies alive. Babies must be fed on demand, kept warm, allowed to breathe, and kept safe from predators and other life-threatening events. We have fictionalized stories of non-human foster animals willing to do those essentials (Tarzan is an example).

If you look, you can see how the Russian Mafia attempts to exploit this element of our hard-wiring – to promote the idea that wealthy, high status men are intelligent. Mr. Putin has publicly spoken well of Mr. Trump and Mr. Musk. When my mother pushed me extremely hard to file for personal bankruptcy (apparently the Russian Mafia has seized control of that system in some way), she used the argument for years that Mr. Trump was president and had filed bankruptcy many times, and look "he's the U.S. President." My mother is compromised by the Russian Mafia but believes her handlers are controlled by our government. So the Russian Mafia is using the argument that "wealthy, successful" men understand our systems and are smart to exploit them.[24] I doubt Mr. Putin uses these arguments to encourage businessmen in Russia to exploit Russia's systems but he's subversively encouraging white collar crime in America to lessen the taxes our nation desperately needs, in addition to his helping to embed politicians who refuse to allow our IRS to consistently receive the amount of taxes we need to thrive and keep our promises to nations negatively impacted by the climate catastrophe, which catastrophe the Russian Mafia is largely responsible for producing by refusing to allow it's allies to work with us to effectively mitigate the catastrophe. Mr. Putin inherited Active Measures, he didn't create it nor did he discover the blind spots. Think of him as an especially evil trust fund recipient who, instead of inheriting wealth, inherited a weapon he's determined to wield to destroy a billion+ West-allied people. Think of him as an especially brutal corporate raider determined to seize control of assets he covets (like America's financial status in the world and/or Ukrainians food and fertilizer production capacity) so he can acquire even more wealth, even more power. Think of him and his Mafia as the cancer that they are. When cancer is spotted in the body it is

---

24 The Russian Mafia's American operatives forced my mom from around 2011 throughout much of former President Trump's presidency had her try to browbeat me into filing for personal bankruptcy. She used former President Trump's status as the U.S. president and as a successful businessman to try to convince me to file for bankruptcy, since she argued, he'd done so and look where he was. These types of Russian Mafia ordered parental attacks are, I'm positive, being waged across our nation and around the world and since most democracies aren't into everyone's family dynamic, the Russian Mafia is having a field day destabilizing people within the supposed safety of their home and family.

eradicated or it kills the patient. Mr. Putin is the wealthiest man in the history of our species, part of his wealth earned from forcing innocent women and children into prostitution around the world, and hooking children on fentanyl and other drugs. Every dollar or ruble the Russian Mafia earns through it's evil work, Mr. Putin gets a cut. He's got, probably at most, 30 years of life left and he's determined to go out feted by his Mafia and allies by destroying us. He's an entitled, smug, elitist, imperialist who believes everyone but him is beneath him because he's willing to murder other people's babies, and ambush and entrap innocent babies' parents.[25]

Chapter 1.10 - What entrapment looks like to the victim before the Russian Mafia springs the trap

What I noticed when compromised dominants tried to entrap me was the casual manner in which they, unprompted, offered to lie, cheat, steal and/or told/advised me to do the same, as if lying to and cheating Medicare and Social Security is normal, no big deal.

This is how the Russian Mafia sets their trap for ordinary people who have no espionage training: the dominant casually offers to "help the person out" by lying to their insurer on their behalf. The dominant doesn't use the words "lying on the patient's behalf." He explained his offer by saying he prioritizes his patients over Medicare. I'm insulin resistant and obese. The insurance I have will pay for a weight loss shot if

---

25 Mr. Putin is trying to access our hard-wiring where, piggybacked onto our hierarchy survival strategies eons ago, there sits a tie-in between intelligence and status.

I'm diabetic. I'm not diabetic so I don't qualify. The general practitioner offered, unprompted, and not for the first time,[26] to list me as diabetic, a lie. Anyone who accepts such an offer will be entrapped by the Russian Mafia, and their quality of life destroyed for the rest of their life. Far worse, the Russian Mafia will use them to enslave their entire family (including their children), blackmail, coerce and/or influence them into voting for one of the Russian Mafia's many puppets here. The reason I can write about it is I'm an adult child of entrapped parents and entrapped half-siblings, but I'm not entrapped myself. Entrapped people, from what I can gather, either accept a plea deal to stay out of prison or they go to prison for a very long time. I don't know if the type of financial crime against the state the Russian Mafia seems to prefer to entrap people which guarantees them a free public defender. Most people, especially since the Russian Mafia targets parents, accept the plea deal fearing their children will be raped in foster care, something parents cannot bear the thought of. Deliberately entrapped people have no idea they've been targeted by an enemy foreign government operating here illegally and law enforcement who entrap them are also enslaved by the Russian Mafia. The Russian Mafia won't tell victims that it will slowly

---

26 In a previous entrapment attempt, it was for another item diabetics qualify for but pre-diabetics don't. I declined then and showed the doctor I wasn't happy that he tried to lure me into breaking the law. I didn't accuse him of anything, I reacted in a way that let him know I wasn't interested in lying to acquire something I don't qualify for. Nevertheless, this past summer 2023, he was used to try to entrap me with a weight loss drug my insurer pays for if the patient is diabetic. I've never claimed to be diabetic. I'm insulin resistant. To me and my insurer, this is an important difference.

eat their entire family over years, and destroy them with relish. How do I know? I'm living it.

There are entrapments by our legitimate government. You'll recognize them by: (1) our legitimate government arrests you in front of the world: your family knows, your employer knows, your neighbors know. The Russian Mafia stealth arrests you: your family, your employer, your neighbors don't know. That's because the Russian Mafia uses entrapped people to entrap and/or harass their extended family, co-workers, neighbors. How do I know? Because I've been harassed and/or endured entrapment attacks by compromised supervisors, co-workers, neighbors, extended family, even at convenience stores, etc.; (2) our legitimate government won't force you or your children to harass other people; and (3) our legitimate government won't in their plea deal force you to assist them in perpetuity in any way law enforcement requests. From what I can tell from my entrapped family, my mother may have been entrapped when she was 29. She's in her 80s and her handlers still force her to get information from me they will try to use to entrap me. I've no idea when my other family members were compromised but most, if not all of them, are.

Because I've evaded entrapment attacks over decades, the Russian Mafia unrelentingly harass me, in retaliation, to stealth kill me using "natural causes," another word for stressing me while ensuring I avoid compromised medical care providers who they force to try to entrap me. The Russian Mafia on the ground in Southern California is all about showboating to me they control the ground here. They're convinced no one will believe me, so they gloat, laugh, smirk, even belch right in my face, as they walk closely past me and/or harass me in non-obvious ways.[27]An example is one of their operative's cut in front of me in line at the pharmacy as I sat on my mobility device some years ago, and begin an animated conversation with the pharmacist cashier. When I said nothing in response to their attempt to bait me, their conversation petered out and the operative walked away, or the Russian Mafia will have their operatives swarm me, one of their favorite attacks. They know which products I'm likely to buy as they've been on me for decades. They'll position operatives

---

27 A plain clothed police officer at a big box store exited his pickup with his maybe 12 or 13 year old son and began to harass me in front of his son. I was using a cane, not having bought a mobility device yet. I'd taken the bus to the store and was crossing the parking lot from the bus stop. As I weaved around cars to get to the store's entrance the officer weaved around the exact cars, making noise to ensure I heard him. After he'd tracked behind me for maybe four lanes, from one group of cars to a group closer to the entrance, so for maybe five minutes, I glanced at him. He smirked at me. I've never seen a child more embarrassed by his father. I was being stealth harassed by a plain clothed police officer but his child's embarrassment was so obvious I couldn't help but see it. They were Caucasian, the boy following behind his smirking father as his father stalked me. I walked so slowly that had they continued following me they would've caught up with me, so they stopped following me as we approached the entrance. I think that attack was sometime in 2010, before I got my mobility device. After I got my mobility device I was harassed by another operative, a woman, who appeared to have a plain clothed police officer with her, directing her. She wore a kerchief over the lower part of her face. Having shopped at that same big box store I was heading to the bus stop when a woman, in her 20s or 30s, blocked the sidewalk maybe 5 yards or so ahead of me, the sidewalk I needed to access to get to the bus stop, and struck a kung foo pose. The plain clothed officer ducked out of my sight, the building afforded him a nook to hide from my view. I stopped my device as I had no way to get around her, and then took an alternate parking lot to reach the bus stop.

in front of the products to block my access to food. In the belching incident which occurred about a month or so ago (I'm writing this sentence in March 2024), I was pushing my mobility device on the sidewalk because the battery had died and I had to push it if I intended to get to my apartment. A Caucasian male operative approached me and walking very close to me he belched right in my face, and kept walking. The Russian Mafia are all about showing me they can knife me at any minute, and want me to be very, very afraid. They've terrorized me for so long, for decades, I don't know if I can feel even more fear unless they imprison me here, or kidnap and ship me to one of their gulags. Regardless of my fear or my death, I'll never stop warning my people and the world about the evil and the threat that the Russian Mafia poses to innocent people worldwide. There are a thousand ways to quickly and stealthily attack ordinary people you want to stress into having a heart attack and stroke. The Russian Mafia knows them all and demonstrate how deeply embedded in our country on a near daily basis, whether I exit my apartment or not.

Chapter 1.11 - How the Russian Mafia weaponizes our socialization to entrap us

Socialized from birth to see dominants like our parents and doctors as positive, kindly, loving and to be obeyed, even when parents and doctors force shots on us, the Russian Mafia corrupts and weaponizes trusted dominants like parents and doctors and unleash them on noncriminals untrained in espionage, it's a seriously confusing gas lighting attack especially difficult on our children. Socializing is soft-wiring (how our parents' raised us: "say thank you, say please, wash your hands, brush your teeth, take a bath, use a tissue to wipe your nose, call an adult Ms. or Mr., don't interrupt an adult when they're speaking unless you ask for permission to interrupt them and they grant it, etc."). People are hard-wired to obey dominants, because our ancestors who survived to reproduce obeyed their dominants when our hard-wiring was established. Our ancestors passed their survival genes to us. People who didn't get robust hard-wiring didn't survive. There were many human species as recently as 100,000 years ago. Homo sapiens are the only species left. Our inheritance (our hard-wiring) is what helped us survive when our extended family tree didn't. So when the Russian Mafia seizes control of dominants and uses them as weapons to corrupt, groom and lure ordinary people, including our children, the Russian Mafia attacks the reality all of us have lived since birth. That's one of the reasons I believe so many of our young people are on medications now: the Russian Mafia uses their parents, their medical care providers, their teachers, anyone with access to the child to emotionally destabilize them and upend their sense of reality.

Parents, extended family, and medical professionals are part of a baby's life from birth. From birth we are taught to trust these dominants before all others.

One of the doctors I occasionally see suggests, by his repeated offers to lie on my behalf, that lying, cheating and scamming to access a medication or products from the government, is normal, acceptable behavior. When the Russian Mafia has dominants do that, it's like they're speaking a different language, which the Russian Mafia uses to confuse targets.

Most people are socialized to trust dominants so when a parent or a doctor, trusted dominant, offers a desirable service and offers and/or advises you to lie to get it, it doesn't feel nor sound like the entrapment that it actually is because the offer is made as if lying, cheating or stealing is the norm. The dominant doesn't say that their offer or recommendation, their casual presentation and who presents it, a trusted dominant, set the trap, easily, against people who've no idea a foreign enemy government is using our nation as it's hunting ground to make their bones and to destroy Americans, because they can think of nothing better to do than to ambush unarmed civilians, i.e. our children and crow to the world their dominance, while the world's species dies under their "management." No telling what they say to dominants to get them to suggest targets do illegal acts. I can't imagine dominants believe their handlers are Americans but that's what they're lead to believe because their handlers are Americans but their handlers' handlers are the Russian Mafia. I'm 100% positive American dominants aren't told their handler's are controlled by the Russian Mafia. I think compromised dominants are so terrorized by their handler's they'll do anything so long as they don't have to witness their child's death. My mother is in her 80s and has been forced to browbeat me and try to get me to do many illegal acts, or she's been used to fish to see if her handler's can try a different type of attack. She probably has given up trying to understand what's going on. Her family has been destroyed by evil men she doesn't even know. They don't care that they've destroyed her life. They only care about world domination.

Chapter 1.12 - The Russian Mafia has entrapped our dominants and use them to corrupt subordinates

Subordinates don't know whether they're coming or going because the Russian Mafia deploys compromised dominants to exploit our socialization, soft-wired from birth. Most people are socialized (raised) to believe dominants are benign, helpful, kind because from birth dominants (usually our parents) have generously served the subordinate, providing food, comfort, nurturing, and often love. The Russian Mafia flips everything the victim has been taught, exploits the emotional tag between dominant and victim but, having seized control of the dominant without informing the victim (so the victim believes they're communicating with a trusted confidante not an enemy foreign government who's deliberately and illegally accessed them by compromising their parent (or other dominant)) begins to destabilize and corrupt the subordinate. The parent (or other dominant) is prevented from warning the target and, my theory is, the parent has been lied to by the Russian Mafia, told their handlers are American law enforcement forcing them to entrap their child, while the truth is the law enforcement who has entrapped them are themselves entrapped by a foreign enemy government, probably unknowingly. I've no idea how compromised law enforcement have the attack on them justified by their compromised supervisors. The Russian Mafia works an endless circle of lies. I've been harassed for decades by police, uniformed and plain clothed, my dad is retired law enforcement who tried to entrap me and who was forced to influence, misdirect and misinform me for many years. I know some of our law enforcement have been weaponized by the Russian Mafia in their coup I know the Russian Mafia intends to destroy us. They've already started. This isn't a normal war. This is genocide. The Russian Mafia intends to annihilate us. When an enemy foreign government deliberately, over fifty years, destroys families, keeps anti-American sentiment ongoing worldwide so that when they move to lower the boom on us, they've built up so much anti-American hatred that few countries will want to our aid. We really need to destroy the Russian Mafia. I'm sorry we need to kill people but there is no upside to the Russian Mafia. They demonstrate they are bad seed in nearly everything they do. If the world knew an eighth of their covert operations , they'd be pariahs. Their government allies are their slaves and so don't have the luxury of turning away from them. That's going to be a problem for Chinese and Indian people as the Russian Mafia embeds itself ever deeper into their countries and governments. At least this book exists, to help citizens in those countries understand what's being done to them, why, and by who.

Dominants lure and corrupt subordinates in a very casual way, giving them no indication that they're leading them into a trap that will forever destroy their quality of life. When my parents and my doctor tried to entrap me their behavior told me that their advice, order, suggestion and/or offer was normal: there was no hint that they believed what they were doing was illegal or wrong. That it's normal for a doctor you barely know to offer to lie to Medicare on your behalf because "he considers his patients his top priority, not Medicare's rules," or something along those lines doesn't make sense if you consider what a person must do to become a licensed doctor in America. They study for years and likely spend hundreds of thousands of dollars in education-related expenses. Barely knowing you, and unrequested by you, they're willing to throw it all away by defrauding Medicare. That doctor knew his offer was unwelcome: I declined it. But his handlers forced him to make a similar one in the summer of 2023 after one of the Russian Mafia's owned slaves (an apartment manager) noise attacked me and frightened me so severely, my heart rate rose to 200 bpm necessitating an ER visit. Knowing I was writing this book, the Russian Mafia attempted to lure me into filing a lawsuit against the operative who attacked me, which lawsuit they would've turned into a quagmire, like they're making the 2023 Israel-Hamas proxy war a quagmire for our magnificent Biden Administration, security agencies, and military. Bait and switch. Lure. Lie. Destroy. That's what the Russian Mafia is all about. Oh and enslaving the world.

The Russian Mafia is bad. Confused, agitated and enraged people which is what people become after the Russian Mafia attacks them by flipping our socialized expectations (our history & our experiences) against us. People victimized in such a way are ripe for triggering an overthrow of our government. The Russian Mafia lusts for the overthrow of our government so it can move in for the kill against our people and our government.

The Russian Mafia is all about destabilizing and triggering us to overthrow our government so it can install one of it's many puppets to murder us. Ordinary people in America must understand despite whatever lies the Russian Mafia's American influencers tell them the Russian Mafia is coming for us, not just for our government. They use ordinary people, they understand the power of ordinary people, they hate us and ordinary Americans and ordinary immigrants in America are the us the Russian Mafia hates. I'm sorry to tell you this frightening news but you must be warned.

Our President, his Administration, our security agencies, our military and the alliances they've formed worldwide are the only protections standing between us and the Russian Mafia.

Chapter 2 – How the Russian Mafia uses dominants as weapons to destroy us

Compromised dominants groom subordinates to believe that our country and our people are corrupt. Both my parents, one a retired police officer, and one of my doctors, tried to get me to cheat our government, unprompted. We are hard-wired and soft-wired to perceive dominants are helpful, benign, not evil so when the Russian Mafia forces dominants like our parents, our police, our teachers, and our doctors, to tell us to lie, cheat and steal from our government, it creates a dissonance, confusion, because this new information works against our established reality. People become very confused, irritated, and generally unhappy, agitated, when their world makes no sense. Remember, we are hard-wired and soft-wired to trust dominants. When they do harm instead of good, or suggest we self-harm, dominants enslaved by the Russian Mafia begin the break down of our families, our society, nation, culture and government.

The Russian Mafia prioritizes entrapping parents and I've witnessed and been targeted by entrapped parents forced to use their children as beards and/or soldiers in the attack. Once children are entrapped by their parents, especially if they trusted their parents as most children do, subordinates lose faith in everything. People who've lost faith in their essential relationships are more easily triggered to overthrow their government. There is a method to the Russian Mafia's madness: push people in America to overthrow their own government so the Russian Mafia can push one of it's many American puppets who "rescue" the brave, embattled citizenry. We're embattled alright, but the Russian Mafia is the attacker, not our legitimate government. They've seized control of some of our dominants and have turned them into mini-bombs throughout our nation to destroy our people, our culture and our government. Oh, and while lying and saying our nation is sinking into self-induced corruption while the Russian Mafia knows they're secretly attacking us.

Dominants form essential societal structures to support our reality. After only a few days of life, we know that someone will come at our cry if they've done so since we've arrive. When the Russian government forces dominants to teach subordinates the opposite of what we've been taught from birth, our reality is destroyed and with no replacement, our culture and people become unmoor-ed. More people, including our children, rely on sedatives and sleep aids because the Russian Mafia is stealthily destroying our sense of reality.

Chapter 2.1 – Our children are especially vulnerable to this attack as they're hard-wired and soft-wired to obey their parents

Children are hard-wired and soft-wired to obey their parents. They have no defense against anti-socialization warfare and anti-hierarchical warfare where their parents, doctors or teachers corrupt them. Our police are dependent on chain of command. Unless warned, they have no defense against hierarchical-based psychological warfare. We must warn our people and all people worldwide, else the Russian Mafia will destroy and/or enslave the world.

Chapter 2.2 - Attacked and used as a training vehicle by the Russian Mafia who are grooming entrapped parents to do as ordered, including bringing their newborns along on attacks[28]

I've been forced to engage with entrapped parents who had newborns, in other words, I was harassed by entrapped parents

---

28 The Russian Mafia has evolved Active Measures. They now promptly begin the grooming of entrapped parents to use their children as ordered by the entrapped parent's handlers. When my mother was destabilized around 1972 she was for a time unable to function. That was 53+ years ago when the USSR ran Active Measures. Although she beat me when I was 15 (around 1974), something highly unusual for her, she wasn't deployed to overtly entrap me until I was 51 (around 2011). Nowadays, the Russian Mafia immediately begins to groom entrapped parents to destabilize the parent-child bond, to re-educate the parent to understand that their first loyalty is now to their handler, not to their child. The Russian Mafia has found that it's easier to compromise child victims than it is to compromise adult targets who've had some socializing by an unentrapped parent. Quality parenting like my young non-compromised mother gave me before we arrived in California is high value and when the Russian Mafia stops entrapped parents from ethically socializing their child, that makes the Russian Mafia's destabilization of that child victim much, much easier. Because the Russian Mafia's compromised American law enforcement operatives have attacked me for over 50 years and had witnessed no violence from me in response to their attacks, they felt I was a safe choice as groomed entrapped parents to unquestioningly bring their children on order. They didn't reassure the parents, is my guess, they simply ordered the parents. American law enforcement knew I'd hurt no baby, or parent, unless the parent physically attacked me and even then, probably not. That's partially how I've learned so much about the Russian Mafia's exploitation of the socialization of our species – they used their decades of attacks against me, and my usually non-violent response to their attackers, to groom entrapped parents to blindly follow the Russian Mafia's orders to bring their child with them on harassment attacks. For newly entrapped parents, they're much more likely to follow future orders to place their babies in harms way if the first few times they're ordered to do so the entrapped family escapes the experience physically unharmed. That's how I've experienced 1,000+ American children and their parents being groomed by the Russian Mafia-controlled American law enforcement handlers.

who were forced to bring their newborns with them into an harassment attack where they were the attacker. I've been attacked for most of my life so the Russian Mafia knows I'm not violent and present no danger to the parent or their child. Apparently, because I'm non-violent after decades of attacks, the Russian Mafia's American law enforcement operatives use me to train entrapped parents to blindly follow orders, including bringing their newborns, something almost no parent would willingly do.

My parents were used as weapons to attack, influence, entrap and/or destabilize (gas light) me, but I was adult before the worse of their attacks were unleashed. Children born into such a corruption of the parent-child relationship are severely disadvantaged by the Russian Mafia to live a normal reality. I believe the reason we've experienced a huge spike in mass murders by young people over the last three-four decades was the Russian Mafia testing out how they can control American children here, as well as children immigrants. I also believe messing with our kid's reality is why so many of our children are on medications so young: I know how the Russian Mafia attacks. I believe most adults untrained in espionage struggle to connect the dots, so no way can children do so.

Chapter 2.3 – The KGB, Watergate, Jackson-Vanik: how the Russian Mafia is trying to steal our Presidency, our Democracy, & our lives

Our government knows the KGB embedded hundreds of thousands of KGB-controlled spies when they got Jackson-

Vanik passed in 1975. Our government wanted us to know so they authorized trusted authors to inform us.[293031] Our government already knows some of those spies were deployed to destabilize our government and our nation's systems (gerrymandering is a result, and our increasingly dysfunctional government, evidenced by the House GOP, and by our President being unable to get Congress to approve desperately needed money to innocent people we're determined to help: Ukrainians and Palestinians), probably by blackmailing some of our leaders and individuals in our law enforcement.

But what our government and/or our ordinary people didn't know is exactly what the KGB did with those hundreds of thousands of seed spies, because the KGB took precautions to hide that information from our government. Ordinary people, untrained in espionage, would not understand they were experiencing a coup aimed at our hard-wiring and so couldn't inform our government of something the victims didn't recognize.

---

29 Craig Unger "*American Kompromat*" (Penguin Random House LLC 2021).
30 Craig Unger "*House of Trump House of Putin,*" (Penguin Random House UK 2018).
31 The KGB got that legislation passed probably by blackmailing an unknowable number of our people in our government, and by entrapping former President Nixon with Watergate, to force him out of office, presumably because he refused to pass Jackson-Vanik. (I've read that former President Nixon refused to allow the passage of Jackson-Vanik but I'd like to have it confirmed.) If it's true he refused to allow that Amendment to pass Congress, I'm positive the KGB destroyed him politically using Watergate. Former President Ford was acceptable to the KGB because he was willing to allow passage of Jackson-Vanik, not knowing it was a key component of the KGB's multi-decades long coup and destruction against us.

There was no way our legitimate government could anticipate the KGB's attacks on ordinary people living in America, intending not only to coerce/influence them into voting for one of the Russian Mafia's puppets, but that the KGB/Russian Mafia's intents to annihilate Americans and our government and steal what we have for themselves, similar to their end-game intention of stealing Ukraine and Ukrainian's food production and fertilizer capacity from Ukrainians.

Nuclear armed governments protect their people from most attacks but hard-wiring-based attacks on this scale, against hundreds of millions of people, is a new war for our species. New and highly effective.

Had our government and our allies governments not changed course in the 2010s in response to the Russian Mafia's attacks against Ukrainians and other innocent people, the Russian Mafia would destroy us in November 2024 by installing one of their many puppets into our Presidency. But because former President Obama, President Biden, our allies' governments, and our extended government, including our non-compromised security agencies, and our military were paying attention, and because they began sharing high value information with the public in newspaper articles, they helped us, their people, begin to connect the dots.

I wouldn't be able to write this book outing the Russian Mafia's genocide against us had not all those non-compromised, suspicious, observant people shared critically important information in newspaper articles and books with us.

Chapter 2.4 – Why the KGB and Russian Mafia's deployment of dominants has been so highly effective

The KGB, and later the Russian Mafia, destabilize person by person, attacking our hard-wiring, using entrapped proxies. To speed up the process they entrap dominants and force them to entrap subordinates who trust them. A key component of their coup, last century the KGB found that most people who are lured/groomed by trusted dominants won't report corrupting and destabilizing attacks initiated by those dominants to authorities. People are *hard-wired* to see (*frame, think of*) dominants as benign from birth because our distant ancestors survived to reproduce because they relied on the dominants who cared for them from birth), <u>and</u> (double-whammy) we're *soft-wired* (socialized from birth) to embrace the reality that dominants are loyal to us, actively work to ensure we survive, are our cheerleaders, and want what is best for us.

When the Russian Mafia deploys enslaved dominants as weapons against us, to see this attack (given that most of us have no knowledge of our hard-wiring) would require that an ordinary person ignore their hard-wiring (our hard-wiring is a very large percent of who we are), while simultaneously ignoring their soft-wiring (soft-wiring is everything they'd learned since birth (most people don't remember everything they've learned since birth)). For an untrained person to spot this type of warfare and deem their dominant unreliable would require them to become, suddenly, an entirely different life form. If hard-wiring is, to pick a number, 70% of who we are, soft-wiring is the other 29%.

It's impossible for us to disassociate in the way we'd need to to spot this type of attack. I only spotted it (1) because I have decades of experience being attacked. However, it wasn't until around 2001 (the last 24 years) when I was attacked in such a way that I understood I was under attack (due to an operative's error, and to a miscalculation by the Russian Mafia); and (2) because our government began to publicly release (in newspapers and high quality books) information they'd previously kept confidential longer, outing the Russian Mafia here, that I was able to connect the many dots the USSR/Russian Mafia attacked me with.

I was only allowed to learn I was under attack because of an operative's error. Frustrated with me after a decade of the Russian Mafia watching me evade their attacks (I didn't know – I live an ethical lifestyle which unknowingly frustrated the Russian Mafia's main strategy of entrapping people by luring them into corruption), and seeing me evade their latest attack which they'd expected to succeed (they'd finally embedded two operatives into my life as office "friends," only to watch me leave that company in a way those operatives couldn't convincingly emulate), an American law enforcement operative ordered I be overtly harassed, intimidated, and/or physically attacked nearly everywhere I went, from around 2001 to the present. It's an horrific attack. One which the Russian Mafia, convinced they had us after 9/11, allowed their

American operatives to continue and expand. They believed, (1) given my subordinate status in my species, and (2) given the Russian Mafia's mastery and knowledge of how human dominance hierarchy works, and (3) given their huge success destabilizing us using their hard-wired dominance weapon they have very high level of confidence that it works on the overwhelming majority of people, and (4) given the large numbers of compromised slaves they control in America (and probably worldwide), that no one of note would believe my claims of attacks. And, for most of the last 8-9 years, no one of note believed my claims. The Russian Mafia failed to take into consideration that their success installing pro-Kremlin governments around the world, and their theft of Crimea from Ukrainians in the 2010s would cause other targeted governments to notice their success and grow concern about the Russian Mafia's success. Which caused West-allied governments and their allies to listen harder to their citizen's seemingly unrelated complaints of harassment.

So an ordinary person targeted by the Russian Mafia is highly unlikely to report to law enforcement that their 80 year old grandmother keeps urging them to cheat on their taxes, or is browbeating them into filing for personal bankruptcy.[32]

---

32 The Russian Mafia, deploying proxies across a wide spectrum of our society, government, families, and businesses, ensured that I was coerced into ballooning their debt load when the Russian Mafia had compromised people I care about communicate a variety of financial needs to me (I'm in school - I want to attend city college but have no transportation there; my teeth are causing reoccurring, life-threatening infections; I have no food; I can't pay my rent; I can't afford car insurance, I need tutoring), were some of the many ways the Russian Mafia deployed my beloved family to weaken me financially. And by

I estimate the Russian Mafia controls, easily, tens of millions of people living in America (based on my being surrounded by Russia-Mafia controlled American operatives since I was 11). This book is me telling my government, my people, our allies, and innocent people worldwide how the KGB and now the Russian Mafia made that happen using

_____

ordering my entrapped landlord to refuse to make routine repairs, I was coerced to use credit cards if I wanted to live in a habitable apartment (my landlord wouldn't replace a very old stove because, they said, since one of the four burners still fired, proving the stove worked, they wouldn't replace a functioning stove. (I had to buy and install a stove top if I wanted an operating stove. I cook my meals. I need a functioning stove so I bought another stove top and had it installed. The landlord gave me permission to do that so long as I agreed to leave the upgrades when I moved. I agreed, having little choice if I wanted a habitable apartment). Additionally, because of faucet leaks under the kitchen sink the landlord hadn't successfully repaired for many years (I'd call to report the leak, the handyman would appear to have fixed it, but a few months later, the leak would be back, or another one would appear. I now believe the Russian Mafia had operatives in the apartment while I was at work to create a leak here and there. Because of the leaks from the kitchen faucet for many years, the water wore away the kitchen linoleum to bare wood. The landlord refused to replace the worn linoleum with a new linoleum flooring. Their best offer, they said, was to cover up the care wood with linoleum from another pattern, meaning most of the linoleum, which was original to the apartment and was old even before I moved here, would be one pattern and the landlord would buy a patch of linoleum of it's choice (meaning the cheapest, ugliest, mismatched linoleum remnant they could find on sale), and glue it over the bare wood, as unattractively as one of their handymen could manage. (I know they'd do this because a few of the kitchen tiles had to be replaced (I think due to water damage but I don't remember the exact details). The landlord agreed to replace them (the two tiles would've cost maybe $1.00 total. They have handymen employees to install the tiles). One of their handymen brought two unglazed tiles, which had a lot of variation on the surface, perfect for catching bits of food and water and which were a very similar color to the tiles needing to be replaced. I asked him not to install them and told him I'd buy the replacements myself, which I did, and which are in place to this day. The handyman did install them. When I asked permission to pay for a floor and install it myself, the landlord agreed. I replaced very old carpet in one of rooms and flooring in the kitchen. Rough estimate of the cost: $5,000 in 2008-09 dollars, racking up the credit card debt, which debt the Russian Mafia pushed

Active Measures to destroy us, so that people everywhere can better defend themselves against the evil that is the Russian Mafia.[33][34]

---

me hard (deploying my beloved mother) to file personal bankruptcy. Today that cost would more likely be double. So, thinking in 2024 finances, a renter would pay their rent, plus $10,000, for flooring if they wanted to live in a habitable apartment. I also made addition repairs at that time.

33 The KGB/Russian Mafia is best in class for not warning victims about upcoming attack, pre-attack. It's March 2024. Most Americans have zero idea that if the Russian Mafia installs one of it's puppets this election cycle, our quality of life will be destroyed by 2030. The Russian Mafia runs endless articles about President Biden's age, or about how many court cases former President Trump is juggling, or speculative articles about Mr. Trump's memory. What the Russian Mafia isn't allowing to be run in our media are articles accusing the Russian Mafia for ordering Hamas to attack Israeli, and that the attack is a proxy war to unseat our President and install one of their preferred candidates into our Presidency.

34 I also believe the KGB had a hand in assassinating President Kennedy and his brother, and other leaders, using rogue individuals and/or American mafia at the time our people were murdered. As the Russian Mafia loves demonstrating to me nearly every time I leave my apartment, or even when I'm inside it, they have endless ways to quickly assassinate a target and disappear the hit person. Hooking targets on drugs is their absolutely favorite way to assassinate people in America. It's dirt cheap (less than a penny a hit if the fentanyl shipment is large enough, it's highly effective, and because fentanyl precursors flow from China, the Russian Mafia enjoys deniability. Remember, the KGB got hundreds of thousands of Soviet spies embedded here by passing Jackson-Vanik. They already had their attack apparatus in place before then (my mother was destabilized by 1972). But what ordinary people may not understand is that even in the 1970s the KGB was taking steps to form pre-BRICS partnerships. There's a reason China and other governments are allied with the Russian Mafia: it's because the KGB delivered trade partnership to them which modernized their nation and their military, which the Russian Mafia is deploying to stalk us and our allies.

Chapter 2.5 – The Russian Mafia deploys 'racism' as a camouflage in America and uses it as a weapon against all Americans

I'm African American. The Russian Mafia built in our country the lie that most Caucasian Americans and most of our law enforcement are inherently racist so that when they wanted a person of color beaten or destroyed, the Russian Mafia had a cover story established. In Israel, the Russian Mafia has already established that the Israeli government is inherently racist against Palestinians (the Israeli government, a right-leaning government, is under the misapprehension that the Russian Mafia is a desirable ally and I believe strongly prefer to uncouple from America). I believe some in the the Israeli government think the Hamas attack is good because it gives them justification to destroy Hamas in Gaza. These people aren't asking themselves whether it is wise to ally themselves with a mafia who is knowingly facilitating the continued murder and/or injury of tens of thousands of children not at all involved in the attack on Israel. The Israeli government's naivete calls to mind the KKK's in the 1950s and 1960s, as they embraced the KGB and all things antithetical to sharing power with minorities and women, and allowing homosexual people their civil rights. Now, 70 years later, those men and women's descendants are in the fight of their lives to evade genocide, and worse, they don't even know they're under attack.)

When individuals in the Israeli government look at the Russian Mafia and see a "friendly" they've made a fundamental mistake. When a person or an entity facilitates the bombing of millions of people, not because of anything those people have done but to unseat a political rival and seize world dominance, that is not an entity to align with. Israeli government official should think this thought experiment: what if the people bombed in Gaza were their allies, what if they were Americans, or other Israelis, a group some in the Israeli government doesn't hate. When you swap out Palestinians and insert any innocent group of people being bombed, you reach the conclusion that the Israeli's must stop bombing Palestinians in Gaza until they can move them out of harm's way. Some in the Israeli government don't get that while see the Russian Mafia as a friendly, creating an opportunity for them they believe will aid in their long term safety, that very Russian Mafia is placing articles in newspapers and social media worldwide labeling the Israeli government racists. And wait, there's more. After the Russian Mafia tightens the noose around the Israeli government's neck, people of color worldwide won't come to their aid because the Russian Media has labeled the Israeli government as racists. And the Israeli government was so clueless that they ignored everyone warning them this was happening. Israeli's are a victim of the Russian Mafia. A different type of victim than say the tens of thousands of innocent Palestinians, but a victim nonetheless. What we need to do is the world come together and destroy the Russian Mafia. My people, if we don't get a clue, we're going to be chewed up and spit out.

The Russian Mafia embedded itself deeply here and has seized control of so many of our systems and businesses and media that they can establish pretty much "cover" characteristic they want. Examples of effective covers they've created here are fentanyl, mass murder, and homelessness. The Russian Mafia assassinates ordinary people here by hooking them on fentanyl. Another way they assassinate is by using mass murders suddenly rampaging. I guarantee you that the Russian Mafia has murdered people it didn't want talking by hooking them on fentanyl and/or by having one of the people they've destabilized pick up a weapon and start shooting. I've personally been harassed by Russian-Mafia controlled American operatives fronting as homeless. And these particular "homeless" don't stink. Having a stealth army of homeless means anytime the Russian Mafia wants to attack, say, our post office boxes so fewer of them will be available during a presidential election where many of us vote by mail, is they get their "homeless" to vandalize post office boxes. Before you know it, if you live near where I do, within the last several years, rough estimate, 80% of the post office boxes, including all the post office boxes that used to be in front of the post office, are gone. The Russian Mafia always has a cover story and it'll be a believable one. But it'll be a lie. Many of our post office boxes haven't been removed because of homeless vandalism, although that's the cover story produced, but because the Russian Mafia is determined to unseat President Biden this November and install one of it's many puppets and since many of us vote by mail, and since the Russian Mafia has kept our government crazy busy by stealth attacks here and against our allies, it's easier for our government to miss a blind spot, like why are significantly fewer post office boxes. For the missing mail boxes set up to work, the mail boxes had to be removed years before the election, so they were.

I'm untrained in espionage. I was a legal secretary in business law firms. I'd been directly, overtly attacked-to-my-face for decades before my government told me who was on the ground here. I thank them for telling me. The upside for me is, unlike the previous 30 years of proxy attacks by the USSR's American operatives, I know the Russian Mafia is the perpetrator. I know why they're here, and that they intend to destroy us. I know that, not because my government told me but because the Russian Mafia has shown me. We must stop them.[35]

I was around 15 when Jackson-Vanik was passed, which normalized trade relations between us and the USSR, as well as normalized trade relations between us and other nations, including China I believe. What our government, our

---

35 I was allowed around age 41 to learn that I had a seriously evil, highly motivated enemy. My entrapped father influenced me to deny the attacks for maybe 5-6 years by gas lighting me but the Russian Mafia's attacks were so in my face unrelenting, I had to accept my father was wrong, which I'd done by age 48 or 49. I had zero idea who my enemy was (until January/February 2021) but I knew the police were part of it. In California during former Governor Schwarzenegger's term there was a very brief police walkout, around 2008 or so. For one or two days the police didn't report to work, no one harassed me on the street, on the bus, etc., proving to me that my own police were extra judicially managing the operatives attacking me, when they weren't themselves harassing me. I also witnessed a police officer forcing an operative to harass me sometime in 2008-2009.

people, and the world doesn't know is that the KGB has exponentially increased those hundreds of thousands of spies into, rough estimate, many millions, having already seized control of our law enforcement's entrapment mechanism (my mother was destabilized by the KGB around 1972 and assuming she was illegally entrapped at that time, long before the hundreds of thousands of Jackson-Vanik spies were embedded here, the KGB already had in place an effective entrapment system up and running before they got Jackson-Vanik passed.

They've likely entrapped tens of millions of ordinary people here. The reason we're listing (deteriorating before our own eyes) is because the KGB and the Russian Mafia is destroying us, our families, our children, our culture and our government. Consider the 2023 Israel-Hamas proxy war. Look at the Russian Mafia's willingness to injure and/or murder tens of thousands of children. The Russian Mafia stands to benefit if they succeed in placing one of their many puppets into our Presidency. If we allow them that, it's a mistake we won't recover. We cannot continue underestimating the Russian Mafia's determination and ability to destroy us, and survive.

Chapter 2.6 – Our government knows the 2023 Israel-Hamas War is a trap, that our alliance with Israel is bait. They know the Russian Mafia sprung the trap & which governments are helping the Russian Mafia. We must start thinking like scrappy underdogs who have a lot to lose, rather than as a great nation of caring people trying to build an alliance between Middle East governments who are already controlled by the Russian Mafia

The Russian Mafia is the most effective enemy we've ever faced. They've built a weapon attacking our people which we must disable before November 2024, and we've not yet begun to disable that bomb. We're helping important allies in two wars, and we've not yet begun to build an effective defense for our own nation.

We must survive to fight another day. We cannot allow the Russian Mafia to install one of it's many American puppets into our Presidency. Active Measures is set to Close November 2024. We won't survive if we allow the Russian Mafia to take control of the White House. Just like Israeli's can't oust a deeply unpopular Mr. Netanyahu or ordinary British people can't out their unpopular current government, the Russian Mafia has embedded itself here and in our allies' nations decades ago, and using proxies, ensured that pro-Kremlin governments can't be immediately removed. We won't be able to get that person and their team out. Successfully closing Active Measures is the Russian Mafia's nirvana, the USSR set it in play sometime around 70 years ago. The Russian Mafia won't allow us to come back from Active Measures. They are mafia, they make their bones destroying us and have zero intention of allowing us to survive. Modern Americans have never been annihilated. No people like it and we don't deserve it. We're a wonderful people who drew an evil enemy. We have to go to survival mode, difficult because so many of our people are secretly enslaved, which makes it extremely hard to trust anyone.

The pro-Kremlin nations worldwide are cons (lies), meant to lure us into a false sense of security. What we can expect if we allow the Russian Mafia to succeed in their coup is Holocaust, our destruction, our genocide.

I recommend that our legitimate government re-evaluate right now. We must build an effective strategy to re-elect President Biden, that's essential for our survival. We prioritize freeing American hostages Hamas is holding, we change our policy to negotiate directly with Hamas, we adhere to the minimal requires of our alliance with the Israeli government, and we publicly criticize the Israeli government if they continue to prosecute the war ignoring the value of Palestinian civilian lives. We keep our military in that area on high alert. We beg and plead with Congress to approve aid to Ukrainians and Palestinians. If Congress refuses, we focus on re-electing our team. Anyone but the Biden-Harris ticket presents problems for us because the Russian Mafia owns so many of us, from operatives representing as "homeless" to multi-millionaires and wealthier. My prayer is that as we get the word out about this coup, that our people and our allies will believe us, and come forward. The Russian Mafia will have thought of a workaround so we can't rely on our people being able to tell us what was done to them.

The Russian Mafia's genocide against us is on the horizon, bad moon rising. I believe the Russian Mafia facilitated 9/11 and set us on a wild goose chase across the Middle East and Africa leaving our people vulnerable to the Russian Mafia's brutal, evil grip. Sometime around 9/11 is when the Russian Mafia's controlled American law enforcement operative felt emboldened enough to launch overt attacks against me, and which the Russian Mafia have not called off since. The Russian Mafia thinks they have us.

I believe the Russian Mafia ordered the 2023 Israel-Hamas War as a proxy war to install one of their puppets into our Presidency. Covid saved us before, it won't save us again. We must save ourselves. Just like the German Jews in 1930 couldn't imagine that a party would come to power and strip them of their civil rights, that's what happened, even though 1930's Germany was a government of enforced rules and laws. The Nazi party changed the rules and laws.

European Jews and other Jews have been persecuted over centuries. Nevertheless German Jews in 1930 had no concept that a Holocaust where millions of them would be murdered was about to happen. Americans have no idea how fast the Russian Mafia will steal our country from us if we allow them to install one of their many puppets into our Presidency.

People who say they won't vote for President Biden are voting for our genocide at the hands of the Russian Mafia and we must explain clearly why that is the case. I'm trying my best to explain how the Russian Mafia is attacking us, how they've destabilized us using our hard-wiring against us as a weapon, why so few of our leaders (outside our President and his Administration) are sounding the alarm.

Just as President Biden is contractually bound by our alliance with the Israeli government so that he and other government officials aren't at liberty to discuss specific details of the agreement, some of which details are confidential (which is why the Russian Mafia ordered the Israel-Hamas War to unfurl the way that it is, to exploit those confidential points), our society's other leaders have been illegally enslaved, compromised and entrapped and have been told that if they even reveal they have a plea agreement, that agreement will be rescinded and they'll be imprisoned.

It's impossible for the Russian Mafia, even though I believe they control tens of millions of us, to suddenly imprison millions of people, many who are elderly, who suddenly decide to discuss their plea deal so no matter what their handlers threaten, the Russian Mafia won't be able to hold the line against millions of victims, especially since our President will offers pardons to non-violent victims who allege entrapment.

I believe once people understand we're experiencing a coup by the Russian Mafia, the jig is up. Our President will grant immunity (clemency) and suddenly victims will be able to tell us how they've been attacked. The Russian Mafia will lie, cheat, kill, mass murder in as many ways as they can but our nation will survive the 2024 planned coup if we don't allow the Russian Mafia to install one of their puppets into our Presidency.

The Russian Mafia began Active Measures against us sometime in the 1950s or 1960s and kept it going after we'd established normal trade relations with them, where, presumably, they promised to honor our laws. The Russian Mafia lies. By the time I arrived in Southern California in 1969, the KGB was established in the KKK in the Southern California city my mother relocated to when she was offered a job here in 1969.

We need to ask ourselves why the Russian Mafia has elevated the strategic importance of the Middle East to us over the decades & why they're exploiting that elevated importance now. Our government knows Israel is bait. Our government knows post 9/11 the Middle East was a quagmire for them and they know the Russian Mafia is working to ascend. They know that but they don't understand the hard-wiring attacks and how to evade that trap.

We're experiencing an 'okey-doke.' If we hold our present course in publicly defending the Israel government's slaughter of innocents, the Russian Mafia will install one of their many puppets into our Presidency & slaughter us.

Think "scrappy underdog, not world leader." If we pretend from now through next year that we're not a world leader, but are instead a scrappy underdog working to evade the poison-tipped talons of an evil apex predator, we'll survive the 2024 attack. Scrappy underdogs are focused on their own survival, not facilitating the survival of other nations. Just as the Russian Mafia has attacked us using hard-wiring psychological-based warfare, they've used the same attacks against their allies. These attacks work across all cultures. Have you noticed when Mr. Jingping and Mr. Putin are together in a photo op or being videotaped that Mr. Jingping clings to Mr. Putin's hand? That's because Mr. Putin owns him and Mr. Jingping is a slave. That doesn't mean Mr. Jingping is stupid, it means he's a victim of the Russian Mafia's psychological warfare and that being close to Mr. Putin makes him feel safe. I noticed former President Trump's behavior when he was with Mr. Putin in Finland. To my untrained eye, former President Trump looked as if he felt elevated to be in Mr. Putin's presence. Mr. Putin is not a man to let anywhere near you.

I'm positive that the Russian Mafia has quite a few of their lies all planned out, should they be allowed anywhere near our White House. They'll have their lies all planned out to Americans who will be forced to facilitate the murder of Americans, just like the Russian Mafia has manipulated Hamas to facilitate the deaths of Palestinian babies, and Jews in the

Russian Mafia, at least some of whom had to okay Hamas attack on Israelis.

Our people don't know the Russian Mafia has us on the precipice of genocide. If people knew, they wouldn't vote in one of the Russian Mafia's puppets. Our first loyalty must be to our own survival. It's a serious problem when people don't know their survival is on the line with our next presidential election. If left to the Russian Mafia, by the time everyone knows, we'll be on the firing line being shot.

Our nation has legitimate strategic needs in the Middle East but those needs won't matter if we're destroyed. We must refocus back on our country (out of the Middle East), and call a spade a space with regard to our ally bombing Palestinian civilians. We must find a way to temporarily relocate Palestinian civilians out of harms way, as best we can, and our priority through November 2024 is to save ourselves by re-electing President Biden and V.P. Harris.

Chapter 2.7 – What the Russian Mafia's genocide looks like on the ground – they attack our families and other trusted relationships by entrapping our authority figures and setting them loose on us

The Russian Mafia is working their genocide within ordinary families and within other trusted relationships (like a virus kills a life form by infiltrating and seizing control of it 'one cell at a time'). It entraps dominants and forces them to entrap the subordinates under their influence, while simultaneously exploiting our nation's right to privacy laws which prevents our government from monitoring our private relationships.

Because ordinary people have no idea an enemy foreign government is working genocide against us by exploiting our hard-wiring and soft-wiring (our hard-wiring is our instincts - we instinctively trust the dominants who care for us from birth; our soft-wiring is our socialization (aka how we're raised in our families),[36] we don't sound the alarm to our government,

---

36 Women's bodies are flooded with pro-baby hormones pre-birth, during birth, and post-natal, so that even if a woman is forced to birth her baby alone (or she's unconscious, seriously ill, suffering from postpartum depression, was raped, doesn't want the child, or there are birth-related complications (like a breach birth, or if the baby isn't head down in the birth canal)), most women even with the intense pain of labor (which though bad rarely lasts longer than a day)), will successfully birth her baby and, if she keeps her baby warm and dry, allows the baby to breathe freely, and feeds her child, unless the baby is born with a serious health problem (most babies aren't born with a serious health problem), then her baby will survive. Babies who survive infancy do so only because dominants (usually parents) provide at least minimal care from birth including: (1) helping the baby be born – babies who aren't helped to be born or who aren't helped shortly after birth by, at minium, being kept warm, die. It's good to feed an infant as soon after birth as the infant will accept (her/his mom's breast or a bottle), but more important than immediate feeding, a baby must be wrapped in something dry to keep them warm (or born in a warm which most people would find uncomfortably warm). Newborns cannot adjust their

79

who is monitoring us in all the legal ways it can, working hard to help us survive an enemy it knows is here, but who hides it's attack methodology from our legitimate government. The Russian Mafia is 'in-it-to-win-it' meaning destroying Americans, and our government, is their intention. We can hope for the best but that didn't help victims of the Nazi

---

body temperature and are very easily chilled in a room temperature most people find comfortable. A 75-80 degree ambient (room) temperature feels uncomfortably cool to a newborn who developed and thrived inside their 99+ degree mom. Many babies worldwide die annually because those helping them be born didn't understand they must be immediately wrapped in something dry and kept warm (not hot – a baby developed inside a woman's body which probably never exceeded 102 degrees). Being held on a parent's, grandparent's, sibling's, aunt's/uncle's (or other trusted person's) bare chest, skin-to-skin contact with a baby's face and body laying on someone's bare chest or bosom helps sooth the baby – and as long as the baby has covering on them holding in their body warmth (a baby blanket or other soft material that's dry and warm), and as long as the baby can breathe easily, they'll calm after the unexpected drama of being born. Their umbilical cord must be cut, within five or ten minutes of their birth is fine - severing it doesn't hurt the baby. That cord is largely how they were able to access their mother' body's nutrients before and during birth. A study found that keeping umbilical cords on the longer side, I don't remember the length but say by leaving 5 inches attached to the baby (a longer length than most people would ordinarily leave attached to the baby), allows the newborn to continue accessing the nutrient-dense, and calming hormones packed into the gift bequeathed them from their mother's body. Babies left with a longer umbilical cord remnant fare significantly better, all things being equal, than a newborn left with a shorter umbilical cord. The remnant dries and over time falls off on it's own but before it does, the baby benefits physically and emotionally from those nutrients and hormones, helping them to more calmly assess and adapt to their new environment. Birth is a big event in a baby's life. Without warning, usually within a few days, they transition from receiving oxygen via their umbilical cord, which is all they've ever known, to becoming an air breathing mammal. That's a shock to anyone. Babies benefit from the calming hormones and nutrients in their umbilical cord remnant and when left longer, their body accepts the gifts, which helps them start life on the right foot, like colostrum (pre-milk their mother's body produces before her milk arrives) helps them, like skin-to-skin contact helps them, like being kept warm and dry, and being able to breathe freely helps them. So, a baby must be helped into life by a dominant (even a young sibling 4, 5,

government nor is it helping innocent Ukrainians and Palestinians, who, despite the Russian Mafia's many lies and their operative's many lies, we and our government are working steadfastly to help.

Holocaust 2.0, Russian Mafia-against innocent Americans.

Two clues I'm right that the Russian Mafia started the 2023 Israel-Hamas War as a proxy war to destroy us: (1) the war directly benefits the Russian Mafia's goal of seizing world domination as they're working it to blame our President for the war they themselves ordered, yet no one is publicly saying so in the articles I have seen. Whenever a war directly benefits the Russian Mafia, the Russian Mafia has made that war happen. It's not big on transparency and it doesn't care that our security agencies know – it's put in play policies and procedures in our country and government, has been embedded here for so many decades, that people who inherit procedures the Russian Mafia set in place don't question established policies, even when those policies are detrimental to our self-interests (like the policy where American negotiators don't negotiate directly with terrorists, which the Russian Mafia set into place and is currently using to stop our interrogators directly engaging with Hamas leaders). We're in a life or death struggle with a best in

_____

or 6 years old is a dominant in their relationship with their newborn brother or sister and can help the baby survive by knowing and supplying what the baby needs). When dominants provide life-sustaining care the baby is taught through that interaction (socialization (soft-wiring)) that dominants are pro-baby. Because babies instinctively want to live, anyone helping them live is good. To a baby, dominants mean good, someone helping them survive.

class, world class enemy who hates us and who has been stealth destroying us, including our children, for decades. We're on the ropes. We must be able to jettison policies and procedures that are getting us killed or are harmful to us.

The timing of the war, the way it was unfurled to negatively impact our President's chances of political survival, and the fact that it hit key points in our defense alliance with the Israeli government, insuring we are contractually obligated to assist the right-leaning government, that the Russian Mafia knew the agreement prevents government officials from publicly discussing confidential aspects of our agreement with Israel, means our President and his Administration are by law and by contract prevented from publicly discussing these details which would fundamentally alter our understanding of the war (and show in no uncertain terms that the Russian Mafia is, yet again, using innocent babies to seize world dominance). The last fact the Russian Mafia wants the world to know, as it prepares to install one of it's puppets into the U.S. Presidency, is that it is the party responsible for the murder and injury of tens of thousands of Palestinian and Israeli civilians.

Even the fact that a deal between us, the Saudis and Israel is in play, very strongly suggests that the Russian Mafia ordered Hamas to attack Israel, the deal an added incentive to our government to back Israel, despite it's killing of Palestinian civilians. The Israeli government has the right to defend itself but as Biden Administration officials have repeatedly said, how they prosecute this war matters. The 2023 Israel-Hamas

War has been precisely deployed (by the Russian Mafia) and if you know anything about wars, you know they don't precisely do anything. For instance, in a more naturally occurring war, it would've made sense, given their clear intent to steal the West Bank from Palestinians, for Israeli settlers to attacked Palestinians and the war to have grown out of that. But the Russian Mafia needed Israelis to be attacked, and at this time the Russian Mafia can only get some of the Soviet Jews in the Bratva to okay an attack against Israeli Jews, they couldn't get Israeli Jews to mass attack Israeli Jews, plus to activate the clauses of the U.S.-Israeli agreement the Russian Mafia needed another government or entity to attack Israelis. That's one of the reasons I think the Russian Mafia formed an alliance with the Iranian government, to acquire the terrorists the Iranian government supports. The war unfurled, hitting specific beats that are to the Russian Mafia's advantage, but to our disadvantage. That doesn't happen by accident.

I published an earlier version of this book on 10/5/2023, which was a surprise to the Russian Mafia.[37]

---

37 I published an earlier version of this book on 10/5/2023, which book the Russian Mafia and/or it's allies tried to destroy on 9/29/2023 by causing a power outage in Santa Monica (the city where I live) which power outage they used as a cover, when the actual weapon was the power surge meant to fry my laptop, which they aimed at the one circuit (of the nine total in my apartment) I usually use to charge my computer. Had it been charging the Russian Mafia and/or it's allies would've fried it. The test strip my charging brick was plugged into would've kicked open from the surge, thus preventing a fire, but the elevated amount of electricity that got through, the leading edge of the surge, would've destroyed my computer's delicate circuits. The Russian Mafia had an operative in the apartment above mine monitor me at around 3 am as she stood over my bed in her apartment watching me, apparently reporting to her handlers that there was a charger brick

Although they knew I was writing a book and had prevented me from getting technical help for the last several years, the rough draft I published contained information the Russian Mafia didn't want Americans to have in the lead-up to the 2024 election. Accessing my book shortly after it was published the Russian Mafia didn't want the conversation in

---

plugged into the outlet strip on my bed and that my computer was also on my bed. Because of the blankets on my bed the operative couldn't tell that the computer wasn't plugged in. It was normal for me to use that test strip to charge my computer. Since she could see that the charging brick was plugged into the test strip, and since she could see the computer was also on my bed but partially covered by blankets, it made sense to believe the computer was being charged since that's what I had done in the past. The fact that I didn't charge it that morning as I tried to go to sleep, and the fact that, waking up to go to the bathroom a few hours later, and seeing that there was a power outage in the apartment I automatically removed the charging brick, even though it wasn't charging anything, was how by god's grace I exited that attack with a functioning computer. (I don't know if the brick would've suffered from an electrical surge.) Because the upstairs operative had gone to bed by the time of my bathroom run, and because the Russian Mafia and/or it's operatives caused a power outage and so couldn't see that I'd unplugged the power brick during my bathroom run, when they later turned the electricity back on and surged excess electricity into the one circuit they'd targeted (the one my power strip was plugged into) the circuit breaker in my circuit box tripped open, thus preventing a fire as it's designed to do. The Russian Mafia had every confidence that they'd destroyed my computer. Upon waking I saw that some outlets worked while other's didn't. I saw from the circuit box that only one of the circuits in my apartment had clicked open. Only then did I understand that my computer had been targeted by the Russian Mafia with the help of the upstairs operative. I wasn't shocked because since 2010, even before I knew my enemy is the Russian Mafia, their operatives amused themselves by destroying my small kitchen appliance, a vacuum, etc. On a small income I couldn't afford to replace the good quality long ago purchased blenders, and food processors they destroyed. It took me years to buy cheaper replacements. I experienced the Russian Mafia's operatives as extremely cruel for the pleasure of being cruel. And what I witnessed them do to our children is beyond evil. Amber's story in this book is true. The upstairs operative the Russian Mafia used in it's attack no longer lives there and I suspect they killed her, making it appear to be natural causes. She was in her 80s and had served the Russian Mafia by unrelentingly harassing and/or torturing me for over a decade. If she's

84

America to focus on their coup, but wanted the world blaming Americans for the murder of the innocent Palestinians the Russian Mafia is responsible for killing: Palestinian civilians.[38]

The second clue the war was ordered by the Russian Mafia is the strong anti-American tone of the war. The Russian Mafia intends to destroy us, all four hundred plus million people living in America and to facilitate that, has been attacking our nation's self-image since, easily, the1960s, so that when they go loud here the world won't come to our defense. If you notice, articles blame America for the war even more than they blame the Israeli government. That's not an accident.

Two through lines the Russian Mafia is the aggressor in a war are (1) the war directly benefits the Russian Mafia, and

_____

dead I believe they killed her because they didn't want her available to answer any questions if American interrogators questioned her. I listed some of her attacks against me in my book. The last thing the Russian Mafia wants is an operative who can corroborate a victim's account of the Russian Mafia's attacks in America, especially before the November 2024 election.

38 I suspect they planned to unfurl the war in January 2024, to give the Biden Administration little time to effectively respond and get out of the Middle East to win re-election, but they pushed the attack forward to ensure no one would be talking about my book. I published that book because the Russian Mafia has increasingly threatened me with death and I believed them, so I was determined to get something out to the world before I was assassinated. Because the book's sudden appearance surprised them, they attacked Israeli's earlier which has given our people more time to save ourselves.

they don't publicly say so; and (2) Americans are blamed for the harm to any victims.[3940]

Another goal of the Russian Mafia is to shove Israeli civilians right-leaning, make them embrace their pro-Kremlin government, a government most Israelis before the war were resisting, hard. Although the Russian Mafia will slaughter Israelis like they will slaughter Americans if we allow them to seize control of our Presidency, because there are a significant number of Soviet Jews in the Russian Mafia, they prefer to not annihilate the sole Jewish state. They will if they can't shove Israelis civilians right but it's not their first choice. American Jews will be slaughtered along with all Americans. The Russian Mafia loathes us. And it's not because of anything we've done, but only because our government was more powerful than their's post-WWII. Clues of the Russian Mafia's mastery over the people they own (and that we need to understand that they intend to destroy us) is reflected in their ability and willingness to (1) get Soviet Jews in the Bratva to

---

39 Our security agencies don't publicly accuse entities without evidence and the Russian Mafia doesn't leave paper trails for enemy security agencies to find, especially security agencies and their citizens the Russian Mafia is working to destroy. The Russian Mafia had their American operatives install some time ago the "we don't negotiate with terrorists" policy to protect the Russian Mafia from the problem of their terrorists negotiating with professional American interrogators.

40 General Valery Gerasimov, Chief of the General Staff of the Russian Federation,"The Value of Science in Prediction," *Military-Industrial Kurier* (February 27, 2013). General Valery Gerasimov states: *"... wars are no longer declared..." "...The very "rules of war" have changed...." "...The role of nonmilitary means of achieving political and strategic goals has grown, and, in many cases, they have exceeded the power of force of weapons in their effectiveness...." "...applied in coordination with the **protest potential of the population....**"*

sign off on the attack against Israeli Jews; and (2) to get Hamas to continue a war where they've seen for themselves is resulting in the deaths and injuries of many tens of thousands of innocent Palestinian civilians, including children. Most of Hamas' members are radicalized Palestinians who were once Palestinian children. When in pre-war the enemy demonstrates total control over all parts of their weaponry, that's not something to be ignored, especially in a hard-wired based war. The Russian Mafia is demonstrating this control in Gaza, but I'm telling you the Russian Mafia, and before it the USSR, entered our nation sometime in the 1950s or 1960s and began destabilizing us, that as I child I directly experienced their effectiveness and that over the majority of my life they've successfully attacked me in my own country, it would be naive and foolish to believe that the Russian Mafia can't call to arms the tens of millions they control here. Their first choice is to avoid nuclear war, hence they are determined to install one of their many puppets. But of course they have a plan B. Russian Mafia Jews prefer to not destroy Israel because it is the only Jewish state and even though Russian Mafia Soviet Jews' first loyalty is to their Mafia, they are Jewish, and prefer not to destroy the only Jewish state in the modern world. But they will if Mr. Putin orders it, or if whoever else in their circle who controls them tells them they must. It's not their first choice but they will if they believe they can't enslave Israeli Jews by other means.

Chapter 2.8 - The Russian Mafia camouflages their coup by stealth seizing control of our parents and/or other authority figures (via entrapment[41]). Most of the war is being waged in our families (and in other protected spaces), where our legitimate government is prevented from monitoring us

Most of the coup is being waged in families and in other trusted relationships i.e. parent/child, doctor/patient, priest/parishioner, spouse/spouse, teacher/student, etc. The Russian Mafia camouflages itself within the family (or other trust relationship) by entrapping the dominants (parent, grandparent, doctor, other authority figure) and forcing them (by lying to them) to entrap the subordinates within their circle of influence.

Our government doesn't ordinarily monitor or infiltrate relationships between family members, doctor/patient, spouse/spouse, priest/parishioner, landlord/renter, supervisor/employee, teacher/student, grocery store/customer, etc. because it is prevented by law from doing so, which laws guarantee our right to privacy. That's why the Russian Mafia is exploiting these specific categories of relationships to destroy us.

---

41 Installing an operative into a target's bed is one of the Russian Mafia's and before it the USSR's favorite ways to entrap and destabilize a woman. These attacks are cheap and highly effective. The Russian Mafia already owns the operatives involved: only the victim is free in this assault. As a target who unknowingly evaded the KGB's "romance/mentor" attack, I can attest that these attacks are a highly effective emotion-based ambush. The only reason I evaded the trap was because the man was married. I refused to steal something precious from his wife: her belief in him.

Ordinary people have no idea the Russian Mafia is working genocide against us by entrapping our parents (or other authority figures) and forcing them to entrap, groom, lie, misdirect, lure or obtain information from us we'd only reveal to those trusted people (parents, doctors, spouses, priests, etc.) which information the Russian Mafia then exploits to destroy the subordinates that dominant can influence. Because ordinary people don't know they're under stealth attack, they can't report the hidden attacks our government. When the Russian Mafia deploys a beloved grandmother, or a trusted doctor, to lure a victim to lie and thus entrap them, from birth we're taught to "see" dominants as pro-us. We cannot accept that a dominant is actively working to harm us when one of them offers to "help" us, as one of my doctor's did when he offered to lie to Medicare so I could get a product I don't qualify for. One of my doctor's twice, my multiple occasions, and my father once, tried to entrap me. My father was a police officer, he retired as a law enforcement officer. I've loved my mother my whole life. And from birth my mother taught me to trust doctors. People have re-frame ready to actually understand whats being done to us when our grandmother tells us to lie on our taxes. The Russian Mafia knows 99.99% of us will not out our parents and authority figures. We must rethink that hard-wiring and soft-wiring. But first we must be told we're hard-wired in a way where we understand what that means. Because our government isn't privy to these attacks, but know there's a problem, they're trying to see how we're being attacked. We must help them. Our legitimate government is a critical component of our defense. We won't survive the Russian Mafia unless we tell our government what's going on in our families, in our doctor's office, with our teachers, at the grocery store, etc.

The Russian Mafia can do these types of attacks because they've seized control of our law enforcement's entrapment system,[42] who have seized control of our trusted family leaders (and other authority figures), the dominants I referenced above. Likely in different ways, under different conditions, but there's zero reason a law enforcement officer should be telling his adult daughter to take what she can get from one of our social safety net programs unless he's forced to tell her. The same goes for my mother, and for my doctor.

Exploiting our socialization (we're raised to obey our parents, doctors, and other dominants), and by manipulating hard-wiring most of us don't know we have, the Russian Mafia is destroying us and our nation one family at a time, one doctor's visit at a time. Having seized control of the dominants

---

42 Possibly by forming partnerships with the KKK, who had many law enforcement officer participants in the 1950s and 1960s. It's highly likely that when the KGB was here in the 1950s deciding on how they'd destroy us they chose to embed into the KKK because so many law enforcement in that era were part of the KKK. KKK were exploited via their racism, homophobia, sexism, etc. to let the monster in the door. They weren't the only Americans conned. Decades later, we're on the very edge of being annihilated and most of us don't even know, by the most evil government our species has ever produced. There are many hundreds of books about the Nazis. The Russian Mafia leaves them in the shade when it comes to evil deeds, done cheap. They're murdering our people using fentanyl, killing us for pennies per death, daily. They've stolen our parents from us and lie to the world saying we're corrupt. They are monsters.

in the family and forcing them to entrap, groom, lure and/or influence those in their circle of influence, the Russian Mafia is set to Close Active Measures this November.

Some years before I learned the Russian government is on the ground here, when I was 58 or so (in 2018-2019), I'd been overtly attacked by strangers for nearly twenty years and I told my dad that I was going to write a book detailing the attacks. I didn't know he is enslaved, he was used far less often than my mother after I was sickened and was used to influence and misdirect me in my 30s and 40s.[43] He's a retired police officer and my father. Most people in America don't think their parents are entrapped by the Russian Mafia or entrapped by anyone.

And if any non-compromised government official heard any part of my father's and my conversations (except for where he told me flat out, unprompted by me, to "get what you can from the government...[it] didn't care about me" (I still remember much of the conversation because that's the first and only time my father has told me to lie. And I also remember it because he didn't make sense. I wasn't asking my dad whether I should lie to the government, it didn't occur to me to lie to the government, so him raising the subject of lying to the government was … strange. I listened politely and not long after hung up. I talked to my dad to tell him I'd been advised

---

43 Even as I type these words an operative upstairs is dropping things loudly in the hallway upstairs to stop me from typing. 2/14/2024 – 9:56 a.m. (How the Russian Mafia uses American proxies to harass targets stealthily so no one but the target and the aggressor knows the person is under attack).

by a legal agency who helped people at the time and I was telling him what the attorney had told me (I believe that attorney was an operative controlled by the Russian Mafia). I hadn't asked my father should I lie. I have health problems, built by the Russian Mafia but I didn't know that in 2010. I was telling my dad what I'd been advised to do and he introduced criminality into the conversation, which was bizarre. I didn't call my father's lane-shift bizarre. I was socialized to listen courteously to a parent or authority figure so I listened respectfully. I didn't' agree to do as he suggested, I didn't challenge him, I thought he made no sense but I'd never say that to my father. I evaded his entrapment attempt because I didn't agree to steal from the government. Had I agreed with him in any way, I would be enslaved by the Russian Mafia and prevented from writing this book, like all the other American slaves are prevented form coming forward. The Russian Mafia is mafia, the most evil, best-in-class mafia. It and the KGB/FSB can keep ordinary people silent. Unless they're warned, then all bets are off. Rough estimate, I believe the Russian Mafia has entrapped tens of millions of people here and those people aren't imprisoned. There is zero way compromised law enforcement has the legal authority to imprison tens of millions of illegally entrapped Americans. We just need to get this story out and most people, those who where entrapped with non-violent crimes, can request pardon after our legitimate government reviews the trumped up charges against them. President Biden will have no problem pardoning illegally entrapped, innocent people. Many people have been further criminalized and groomed by their Russian Mafia controlled American handlers. I know, for sure, some compromised victims are groomed to embrace cheating on their taxes and to encourage extended family to do the same. Many people now "see" themselves as criminals because the Russian Mafia needs all of us they can sucker into believing we're corrupt to drink the kool-aide. We're not corrupt. We've been attacked by an evil enemy mafia, that's what's happened to us. To survive, we must free ourselves. If we don't free ourselves, the Russian Mafia will destroy us.

Anyone listening to my dad's and my conversation (and people were listening, compromised law enforcement needed to be listening because this was an entrapment attempt (I didn't any of this at the time, I had to figure it out later). But if noncompromised law enforcement untrained in recognizing the Russian Mafia's stealth coup were listening, they might label our conversations "family issues" instead of "genocide against Americans," because they wouldn't understand, because my father isn't free to save so, that he's a slave, that he was forced to try to entrap me. I doubt he knows he's a victim of a foreign enemy mafia working a coup.

The Russian Mafia moved their genocide to "family interactions" because our legitimate government doesn't monitor that space, and if someone glanced casually in a family this war would appear to fall under the category of "family" when it's actually genocide, warfare, a coup against us, using hard-wired, dominance-specific warfare, a different kind of warfare, a reconfiguration of war to make it bespoke because it's using our hard-wiring and our soft-wiring against hundreds of millions of people, and is largely deployed in our families, after dominants are entrapped by compromised law enforcement. By seizing control of the dominants in our lives, the Russian Mafia seizes control of our families, our culture, and our nation, in order to destroy us.

Dad strongly discouraged me from writing this book, told me people would think me crazy. He asked me if I knew who was attacking me. I said I didn't know. He used that in his argument, that no one would believe me if I didn't even know who was behind the attacks. I'd already told him years ago that I was being harassed by store security. My theory is he has no idea he's a slave of the Russian Mafia. I don't believe many if any of our people know their handlers are controlled by the Russian Mafia. I believe entrapped Americans and entrapped immigrants believe their handlers[44] are owned by our legitimate government because that's what the Russian Mafia has told them. They lie so smoothly and ensure victims believe it's our legitimate government who's entrapped them.

---

44 Handlers are what I call people who tell entrapped people what to do after they accept a plea deal. The Russian Mafia's plea deals force victims to assist law enforcement whenever requested for the rest of their life. When any plea deal you're offered makes you assist law enforcement forever, that's a red flag the Russian Mafia is involved.

Chapter 2.9 – The importance of dominants[45] in our species is hard-wired in. The Russian Mafia exploits our hard-wiring in order to destroy us

Our species consists of two subgroups: dominants and subordinates. We are born subordinates. Newborns are subordinates dependent on the dominants in our lives for survival. Those dominants are usually our parents, grandparents or other care givers. Subordinates are group members. At birth we are served by dominants because it's necessary else babies wouldn't survive. Mothers are hard-wired to care about their babies via pregnancy-related hormones that flood their body during pregnancy and post natal. We enter life as subordinates, that's our default position, so our species has built our reality to reflect that. We are taught to obey dominant teachers, given medical advice and forced to accept shots by dominant doctors, our groceries are stocked by dominant grocery store managers who order foods for us, we pay our rent to dominant landlords who have the authority to make us homeless if we don't pay rent. In most people's lives, we are subordinates to somebody, usually to many somebodies. Wealthier people are subordinate to fewer people but generally, even wealthy or powerful people are subordinate to someone in their lives. In fact, powerful people probably enjoy

---

45 I tried to tell one of my doctor's and one of my teachers about the attacks I endure. My doctor didn't want to know and the teacher shut me down with 'I'm not interested in politics.' The Russian Mafia controls movements by controlling dominants. When dominants are compromised they're only open to what the Russian Mafia is open to and the Russian Mafia is about destroying anything and everything that doesn't directly benefit them.

being subordinate to a trusted person as a break from the responsibility of constantly wielding power - being in charge is hard work. Sometimes a powerful person is responsible for many people's lives. That's not easy.

Chapter 2.10 – The Russian Mafia prefers to attack targets when the victim is in the subordinate position

Based on how the Russian Mafia attacks me they prefer to attack targets when the victim is in a subordinate position. To achieve that, the Russian Mafia watches the target, ambushes and entraps the individual they're subordinate to, and uses that dominant as a weapon, as camouflage, to gather intel, entrap or groom the target. We're socialized to perceive dominants as positive, helpful, benign – not as an enemy, not as someone who is setting us up for harm. The Russian Mafia exploits our socialization, i.e. our soft-wiring (aka how we're raised), and our hard-wiring. We are descendants of (1) ancestors who obeyed unknown generations of parents, and (2) dominants who could recognize non-overt threats and warn their people to raise an effective defense.

Chapter 2.11 – Genocide is the endgame

The Russian Mafia's endgame is to use us to overthrow our own excellent government, install one of their many puppets into the U.S. Presidency, and exhaust and destroy us as a people. This is a genocide attack, a war of hatred against Americans by the Russian Mafia. *If we allow them to take us, we won't be coming back from that. There is zero upside to allowing them to destroy us.* We're an amazing, hardworking, ethical, kind, generous, non-racist, intelligent people who had the misfortune to draw a monster mafia's covetous eye. We must understand the national self-loathing the KGB, FSB and Russian Mafia has built in us since the 1960s is a lie, they had to convince us that we're immoral, corrupt,[46] but the immoral, corrupt government is the Russian Mafia.

Active Measures isn't a normal war, like a dispute over an asset or territory. They intend to destroy us by installing one of their puppets into the White House November 2024. Begging them for mercy won't work, although they'd enjoy that. These are people who've knowingly and willingly proxy murdered and/or injured tens of thousands of innocent people over the last six months, who have proxy bombed two million people as a pretext to destroy Americans (by unseating our duly elected President and installing one of their puppets), and as a pretext to show Israeli civilians right. The Russian Mafia cares not at all about Israeli but Soviet Jews in the Bratva prefer not to destroy the only Jewish state, so the Russian Mafia is working to shove Israeli civilians right. If Israeli civilians allow themselves to be shoved right and to embrace a pro-Kremlin government, they will long for death. Just as the the Russian Mafia seized Hamas and forced it to start and continue a war that murders/injures Palestinian children (children Palestinian men love and value), the Russian Mafia

---

46 Janis Berzins "The New Generation of Russian Warfare" "...the Russian view of modern warfare is based on the idea that the main battle space is in the mind." Aspen Institute | Prague, *Aspen Review* (March 2014). https://en.m.wikipedia.org/wiki/new-generation-warfare.

will force Israeli Jews to harm other Caucasian Jews, where ever the Russian Mafia orders them so to do. It is far better for Israeli civilians to not go there. I had a beloved boss, an attorney – I was his legal secretary, working for him and two other attorneys. I don't know how the Russian Mafia entrapped him but I believe they did. They, I believe, were forcing him to attack me, and probably do other evils as well. He committed suicide instead. At that time, I had no idea he was under attack. Before he died, the last Christmas he was alive, he gave me a Christmas present. He knew I was unentrapped because, I believe, his handlers were forcing him to attack me and he understood that since someone wanted me entrapped, that meant I wasn't entrapped. Entrapped people are prevented from reaching out to non-compromised government officials but unentrapped people can sound the alarm. He was prevented from telling me anything. After he died I was deposed, his handlers wanting to know what I knew. He'd told me nothing, thus saving my life. He gifted me with a education package, he'd been a teacher before he became an attorney. He gave me an education package which encouraged me to keep learning. I liked his gift but at the time I probably would've liked cash. Nevertheless, since he'd given me a present which pushed me to keep educating myself, I did so. I watched the videos he gifted me with teaching me all kinds of history and things I hadn't known before. Entranced by the huge variety of subjects, I continued to order more educational videos on my own, for years after his death.

He couldn't tell me what had been done to him. Yet he knew I was unentrapped so he knew I had the potential to stop the people who'd drove him to suicide. He was a brilliant man. He knew, since he couldn't escape the trap that ensnared him, other entrapped people wouldn't be able to either. I knew none of that then. What I knew then was that I loved him, liked him, respected him, and was devastated and stunned when he killed himself. I've always loved learning, and he encouraged me, by his final gift to me, to keep learning. I don't know how I've learned all the facts in this book, but I have, despite being overtly attacked for 23+ years, and covertly attacked using proxies for 30 years before that. Without the educational videos he gifted me, and my continuing to learn the 20+ years since his death; without my young mother's outstanding parenting, years before the KGB destabilized our family; without my government's development of nuclear weapons which forced the KGB and the Russian Mafia to take the long route in attacking our nation, giving me time[47] to connect the dots and figure out how to disable their bomb, without all those people, I couldn't have written this book.

Because I remained unentrapped, I haven't been directly prevented, as I believe all other entrapped people are prevented, from alerting our government about the genocide we're experiencing. If this book goes loud, one of the reasons is because that man, that honorable, ethical, smart, handsome, wonderful man who couldn't free himself from those who

---

47 With the critically important information from quickly declassified reports our security agencies now report in our newspapers and books.

entrapped him, but who figured out how to throw a 'Hail Mary' pass to his legal secretary that could get past his Russian Mafia-controlled American handlers without raising their suspicion, to save his people - he'd be very pleased with this book. This book is for him, and for the world's children, and for all of us victimized by the Russian Mafia.

The Russian Mafia does not care when good people kill themselves. To the Russian Mafia, that's fewer people for them to have to murder outright. Don't dance with the devil Israeli civilians. You will regret it.

They've worked Active Measures against us (a hard-wiring based genocide), despite the fact that we've had a trade partnership with them since 1975, and a technology agreement with them since 1958, despite the fact that they stole our nuclear weapon's intellectual property. They've been on the ground in America deliberately destroying our families for well over 50 years. They've destroyed our children.

Chapter 2.12 – Most people's default world view are as subordinates, which is our first position after our birth

Over time, through experience, employment, education, soft-wiring (cultural pressures), personality, nurturing, etc., many of us become dominants. But because we are born subordinates, dominance is always our secondary position. Most of us look out at the world as subordinates. Subordinates become dominants in their child's life when they become parents. Siblings can become dominant if they are first born. Subordinates can become dominants through their work: our

President is dominant because he is commander in chief of our military. Subordinates become law enforcement officers and as part of their job order dominants or subordinate individuals in certain situations. Subordinates become teachers and become dominants in relationship to students they teach; bus drivers are subordinates who, when they drive a bus, are dominant over passengers. Subordinates become doctors and become health advisors, dominants to patients. Employees in grocery stores are subordinates but become dominants in relationship to customers when they order and stock essential foods. In these examples, everyone listed is both a subordinate, and if they're a dominant it's because of their job or because they're a parent or an elder sibling. Most people's default is subordinate, most of us see ourselves as part of a group, not the leader of the group.

Chapter 3 - What a dominant is – why the Russian Mafia targets them, & the essential 'secret sauce' dominant's are in our species

Dominants are ordinary people with societal-approved authority over others. The Russian Mafia targeted them for multiple reasons: (1) to force them to entrap subordinates, like our children or friends. When you're a foreign enemy government stealth attacking another country it's faster and easier to entrap dominants and sic them on others who control others than it is to entrap subordinates and dominants separately by enslaving our dominants they speed up their destruction of us. An example is when they entrap a landlord, then use that property owner to destabilize all the tenants the Russian Mafia wants attacked under that landlord's circle of influence. I don't know how many properties the owners of this building own/manage but they are many. When destroying a people as the Russian Mafia is actively working to destroy us, it's more efficient to entrap a property owner who has access to many people's lives than it is to approach individuals separately. 'Attack-via-dominant' is faster, more efficient and highly effective when the targeted nation's people don't know they're being attacked in a secret war; (2) dominants function as our species' alarm system for non-overt threats. Eons ago when our hard-wiring evolved people who had dominants who could recognize a non-overt threat and launch an effective defense survived to reproduce, so when the USSR and Russian Mafia entrap dominants, and don't allow them to sound the alarm, the Russian Mafia disables our species' alarm system; and (3) the Russian Mafia forces dominants to corrupt the target nation thus destabilizing it's structures in order to commit genocide against us, for example when my parents, one who was a law enforcement officer, and my doctor, encouraged me, unprompted, to lie and cheat the government, and I said no. When targets refuse to lie and cheat, the Russian Mafia harasses and/or kills them. I know because I'm one they couldn't entrap and they've just kept the harassment ongoing. Today (2/14/2024) I typed this book in my apartment and I was noise harassed multiple times. No one witnessed the harassment but me and the people who noise attacked me but that's what the Russian Mafia does: it uses proxies to attack targets. If we make the really bad mistake of voting in a pro-Kremlin candidate, the Russian Mafia will dispense with proxies and simply order our police to break down our door just as the Nazi's stripped German Jew's civil rights from law abiding citizens. Things won't end well for us if we let that happen, but the Russian Mafia will be ecstatic. If you care about our nation and our nation's children I beg you to not vote in any pro-Kremlin candidate, ever.

Chapter 3.1 – The Russian Mafia is gas-lighting & grooming us to force us to overthrow our government

Many of our people no longer believe in anything much because of the gas lighting the Russian Mafia has stealthily deployed over decades, the reason behind the gas lighting is to trigger us to demand the over throw of our legitimate government so the Russian Mafia can install one of it's puppets. It's part of the Russian Mafia's coup, to have us lose belief in our government so when the Russian Mafia triggers us, so we'll all storm the Capitol, and do the Russian Mafia a favor by overthrowing our government and installing one of their many puppets, after which the Russian Mafia will enslave and annihilate us. To destabilize us they have dominants or authority figures casually advise us to break the law, and they offer this unsolicited advice to "help" the target. Then they stealth arrest the victim and offer a plea deal that is slavery. Parents who refuse to abandon their children to foster care are one of the favorite people to entrap because they're the easiest.

The corruption and gas-lighting the Russian Mafia is doing to our people by having dominants lead subordinates to corruption is a part of their destabilizing attack part of their genocide against us.

Parents are dominants in a parent-child relationship, even after the child is an adult the parent remains the dominant, out of the adult child's courtesy and respect. Dominants function as our species' T cells,[48] and are responsible for (1) recognizing non-overt threats, (2) sounding an alarm when they spot such a threat, and (3) launching an effective defense in response to that threat. People have little problem recognizing overt threats. If someone hits you, or you're trapped in a fire, you immediately know there's a problem and seek to escape. But there are many threats that are not overt, and do not appear dangerous but are. The Russian Mafia's coup was built and deployed by the USSR in such a way as to not trigger our species' alarm system. It is a slow moving, multi-generational, secret, stealth, psychological warfare-based, hierarchical warfare-based, socialization warfare-based, non-overt coup keyed precisely to our species blind-spot of not recognizing an attack unless it is overt, not recognizing a non-overt attack unless a dominant warns us, as well as our difficulty recognizing a grooming or luring attack from a loved one or a trusted friend (most of us have no idea what grooming, luring, or a psychological warfare-based coup looks like), the Russian Mafia is evil to camouflage itself in the skin

_____

48 Specialized cells that launch an effective defense

of trusted parents to force those parents to entrap their own children.[49] The Nazi's didn't entrap Jewish parents and force them to lead their children into the gas chambers. The reason the Russian Mafia gets away with forcing parents to entrap their own children is because they lie and say that they're our legitimate government, and handlers of entrapped victims probably are uniformed and/or identification bearing American law enforcement. We're all being conned. All to destroy us, our families, our society, our culture, and to steal our nation out from under us. The Russian Mafia is highly covetous and is willing to destroy anyone to get what it wants.

---

49 Try, although I know it's hard, but try to see people as what they do, not for what they mean to you. If you can learn to put aside your love for them and see their behavior separately from your love and affection for them, you'll have a better chance of evading some types of entrapment.

Chapter 3.2 – The Russian Mafia exploits our loyalty, gratitude, & respect towards our loved ones to shield it's genocide, knowing people are unlikely to rat out 94 year old grandma who keeps urging them to cheat on their taxes. _**And because the Russian Mafia tells victims our legitimate government is running these attacks, voters loathe our government**_ **(the Russian Mafia considers lies a legitimate weapon of war).**[50] **The Russian Mafia has enslaved millions of our people. When we out dominants who've tried to entrap us, we'll free them from the Russian Mafia's enslavement. After that, they'll have quite a bit to tell us, & it's intel we need to know**

The Russian Mafia exploits our protectiveness towards our loved and trusted dominants, and towards our loved ones whatever their status. They know we are hard-wired in such a way that we protect dominants, because we still perceive them, as we did in our infancy, as crucially necessary for our survival, even after we become adults. This intense protectiveness is a left-over from our infancy, when we physically couldn't survive without our parents or care givers. The Russian Mafia knows that reporting dominants who attempt to lure/groom us into corruption is anathema (feels instinctively wrong, because it goes against our hard-wiring and our soft-wiring (socialization)). Emotionally it feels akin to reporting ourselves to authorities for a minor infraction, something like jaywalking on a quiet street. Most people have no intention, initially, of allowing their dominants to

---

50 Janis Berzins "The New Generation of Russian Warfare" "...the Russian view of modern warfare is based on the idea that the main battle space is in the mind." Aspen Institute | Prague, _Aspen Review_ (March 2014). https://en.m.wikipedia.org/wiki/new-generation-warfare.

lure/groom them, but because the Russian Mafia seizes our people and won't let them stop pestering us, or, more stealthily, everyone in their circle of influence, what the Russian Mafia has done is place bombs in our families. The Russian Mafia is camouflaged in the skin, face, body, and (in a patriarchy-based hierarchy) high status men they use as a wrecking ball to destroy people's families, to the victim's rage and despair. And because the Russian Mafia tells victims that our legitimate government is working these entrapments, people grow to loathe our government and blame them for destroying their families, when it's actually the Russian Mafia destroying our families, while laughing their heads off. Most cultures world-wide are patriarchy-based so all the Russian Mafia has to do is entrap the dominant male and deploy him to entrap the extended family. Women are seized as well, as is my mom in my family. Once entrapped, the family is used in any way the Russian Mafia chooses. Current examples of how the Russian Mafia exploits entrapped men to destroy people in their culture those males ordinarily protect, is the Israel-Hamas War. The Russian Mafia is forcing Hamas to continue a war that's seen tens of thousands of Palestinian children killed and/or injured, which is contrary to what Palestinian males are hard-wired and soft-wired to do. And in the same war, the Russian Mafia coerced at least some Soviet Jews in the Bratva to okay Hamas' attack on Israeli Jews, something Soviet Jews wouldn't ordinarily agree to do. These examples demonstrate the power of the hard-wiring/soft-wiring weapon in the hands of an evil mafia who cares nothing about the people they control, but only about acquiring world dominance.

The Russian Mafia exploits our past relationship with them, when we were newborns and depended on them 100% for survival. Although we no longer depend on them 100% for our survival, that soft-wiring is part of our socialization, so they know we won't to others

Chapter 3.3 - Socialized to protect our family, which the Russian Mafia exploits against us

People are socialized to protect their families so the majority of victims of these attacks are children, adult children, extended family and/or friends, who wouldn't dream of reporting these attacks to authorities, and have no idea how to process them. Who do you report your parent to when they urge you to break the law or do other bad things? I suspect the Russian Mafia has flooded our nation with compromised parents, doctors and other authority figures who lure victims into entrapment. I wouldn't be at all surprised if at least one of the mass murders we've endured the last several decades was orchestrated by the Russian Mafia.

Chapter 3.4 – The Russian Mafia exploits a parent's love for their child & a child's love for their parent to seize control over the family & then destroys the family (genocide)

The Russian Mafia destroys our families by forcing entrapped parent(s) to entrap their own children then shows the (adult) child (and probably minors) their parents' betrayal. Overseeing and forcing this familial destruction are compromised law enforcement the Russian Mafia use as camouflage to lie to the family that they're representatives of our legitimate government, making everyone hate our innocent government who has no idea that their name is being exploited in vain by an evil enemy government.

The Russian Mafia exploits a parent's love for their child and gets them to accept a plea deal to save the child from foster care and/or having their life turned upside down. The Russian Mafia piles on the guilt to parents, conning them into doing whatever is necessary to protect their children, never telling the victimized parent that they've been deliberately, evilly targeted by an enemy foreign government working a coup and genocide against our people.

The Russian Mafia exploits a child's love for their parent(s)/siblings, extended family who grooms, corrupts and compromises them, our socialization (how we were raised) preventing even adult children from reporting these attacks to to legitimate law enforcement and preventing our government from getting to the bottom of how the Russian Mafia is destabilizing our nation and our people.

The attacks aren't presented to the target as attacks. They're presented as 'I'll help you out...,' 'I think you should...,' 'you'd be smart to do ....' Or they're offers. One of my doctor's offered to lie on my behalf to cheat my insurer, twice, unprompted. I'm interested in the weight loss jabs but can't afford them. I'm insulin resistant, One operative got me to ask my doctor (the Russian Mafia owns so many people here, they've quadruple-teamed me at least two times to influence me to act or make choices. They're bad seed). My doctor offered to lie to Medicare. A doctor isn't an ordinary professional in a person's life. Doctors/nurses are present at our births. We're taken to doctor visits from the youngest infancy. We're forced to accept shots from these professionals. They aren't our parents but our parents force us to engage with them, and in a child's life, parent's are top dog. When a parent or a doctor lures a victim to corruption, it hits hard-wired and soft-wired beats, which is why the Russian Mafia has entrapped our medical professionals. Authority figures like parents, grandparents, doctors are key dominants and when they groom victims into corruption, that's a fast way to destabilize a society.

Most adult children have been socialized to respect and value their parents so would never consider ratting them out to law enforcement when their parent tries to 'help them out' by suggesting they cheat on their taxes or commit other finance-related crimes. Adult children don't know that their parent is entrapped by an evil enemy foreign government who is forcing them to order and/or lure their own children into breaking the law. They've no idea their parent is enslaved - their parent isn't allowed to tell them and the Russian Mafia has zero intention of warning their victims. This is a coup. This is genocide. The Russian Mafia facilitates the murder of children. A government like that could care less about destroying innocent families.

The Russian Mafia exploits sexual need and love by installing their operatives into a target's bed. The couples I've seen where the Russian Mafia placed an operative into an innocent person's bed is always entrapped. I believe "romance/mentor" sex entrapments are invariably successful. I've never seen an innocent person escape such an attack once they begin the 'romance.'

The Russian Mafia destroys our families to shove us, hard, into overthrowing our government so that they can install one of their many puppets, and really get the genocide started. The puppets they'd install would commit mass murder against most of us over the next decade(s) – if you think it can't happen, talk to Hamas. Before the Russian Mafia got a hold of them they'd have never considered continuing a war that killed/injured nearly 30,000 children in six months. Or talk to Soviet Jews in the Bratva. Although it's hard to know for sure with mafia members, it's hard to believe they'd have approved an attack against Caucasian Israeli Jews unless forced by their group of strategists.

The Russian Mafia exploits a parent's love for their children by coercing the parent into accepting a plea deal rather than placing their children into foster care. Later, after the Russian Mafia-controlled American law enforcement handlers have groomed and destabilized the parent(s), they force the parent to entrap their own children, and show the adult children this betrayal, thus destroying our families as the Russian Mafia commits genocide against us.

When the Russian Mafia wanted to entrap me they forced my entrapped parents to urge me to lie, cheat and steal from various government programs. I was over 50 when this attack happened and I was so stunned by their separate suggestions, I didn't know it was an attack even though by that time I'd been overtly attacked in the street by strangers for over a decade. Had it not been for that decade of overt attacks by strangers, I'd have not been able to understand that my mother and half-siblings had somehow been co-opted by some hostile entity; it took me awhile to understand my loved ones were trying to entrap me. I had no framework to help me switch them emotionally/psychologically from love/support to hostiles working to harm me. It was a new reality for me as I worked to evade homelessness, having been worked to exhaustion by the Russian Mafia. I'd unknowingly successfully evaded entrapment since my 30s but in my 50s the Russian Mafia deployed my loved ones against me in a new and usual way, telling me to my face, flat out to lie and cheat government entities like that was totally normal. I was flabbergasted. My father was used to try to entrap me I believe when I was 50 but he was a trained police officer, was deployed in a different way than my other relatives, and it wasn't until I understood the Russian Mafia is on the ground here (which I learned in 2021), that I understood my father is enslaved by our evil enemy government, the Russian Mafia.

And it wasn't until a decade later that I learned the name of my enemy, the Russian Mafia (they're not big on transparency). I refused my compromised loved ones orders/offers/unsolicited advice, which is how I remained unentrapped. I also refused when my doctor twice tried to entrap me using Medicare fraud. These attacks are insidious because the Russian Mafia uses a known and trusted dominant in the victim's life, and *people are socialized to obey the dominants in our lives from birth*. So this is a socialization attack, a hierarchy attack, it messes people up on so many levels, it's not even funny. While the Russian Mafia laugh their heads off.

Chapter 3.5 – How we evolved to listen to dominants

Before we evolved to be people, our ancestors family's and expanded groups consisted of subordinates and dominants. Before we could talk, before we developed language, or writing, our distant ancestors noticed that some among them lived longer. Those were the ones who tended to be right about a problem. They were dominants, people who excelled at surviving and who helped their tribe survive.[51] Their improvements, technologies, accurate observations, and knowledge we use to this day, and our dominant and subordinate ancestors discovered and/or created the prototypes.

---

51 Discovering which wood burned longer and cleaner helped our dominant ancestors skin their food animals and prepare meals faster. Their people stayed warmer longer and their babies tended to survive because they were kept warm and dry. One of our dominant ancestors created the wheel, which serves our species to this day. We'd have no cars without wheels. We'd have no transportation system. No computers. Wheels are used in tens of millions of applications. These smarter ancestor dominants were early adapters who used flaked stone knives, and built wells that actually held water until needed.

In their time, dominants were female and male chiefs, prized because they kept their people alive.

Over millennia our distant ancestors began to increasingly listen to these smarter women and men, their chiefs, and those who listened tended to live long enough to reproduce. Additionally, our ancestors whose women and men chiefs (dominants) who could recognize non-overt threats and launch an effective defense, lived long enough to reproduce, while other people's chiefs who were unable to recognize non-overt threats, tended to die out before they could reproduce. Over thousands and thousands of years, people alive today are descendants of ancestors who were either dominants or who had access to dominants with a demonstrated ability to recognize non-overt threats, sound an alarm, and build an effective defense.

Non-overt attacks were less immediate in our distant ancestor's lives than overt threats. Although deadly, they weren't deadly frequently enough to warrant investment by our hard-wiring in the same way we can recognize overt attacks. Yet non-overt threats occurred often enough to devastate a tribe when our ancestors ignored them. So our hard-wiring evolved the strategy of off-loading the job of recognizing non-overt threats to our hierarchical system. Hierarchy became part of our ancestor's survival strategy. Dominants, who tend to be higher status, got the hard-wired job of recognizing non-overt threats, sounding an alarm, and launching an effective defense. Because the USSR stumbled on that intel, it built a secret, undeclared, psychologically-based war model to ambush, enslave, and/or entrap our dominants. Unless dominants raise the alarm, subordinates don't "see" non-overt threats. The Russian Mafia's coup is a non-overt threat deployed to exploit our species' hard-wired blind spot.

We're hardwired to listen to dominants. From birth, we're hard-wired and soft-wired to listen to our parents and other dominants in our lives, doctors, grandparents, babysitters, aunts, uncles, teachers, police, grandparents, etc. When subordinates raise the alarm warning of a non-overt threat, people ignore them because eons ago, before we were people, subordinates of our ancestors didn't effectively protect our ancestors from non-overt threats enough to be hard-wired in. Today subordinates have access to high value information (their experiences being harassed/attacked by their family and larger community) that the Russian Mafia hides from our dominants. The Russian Mafia itself relies upon our hard-wiring, believing it will ensure that our dominants and other subordinates, and our allies' dominants and subordinates will continue to ignore a subordinate trying to sound the alarm. But this isn't two million years ago. And the Russian Mafia's success, especially over the last couple decades, has been noticeable. It is well known as an evil entity targeting children (Syrian War, Ukrainian War). I'm 100% positive it ordered Hamas to start the 2023 Israel-Hamas War, where it, I strongly believe, is forcing Hamas to continue a war disproportionately killing and/or injuring Palestinian children against Hamas' will.

Chapter 3.6 – Dominants, our hierarchy & our hard-wiring are our species' survival strategy the Russian Mafia is exploiting; but we can effectively counter their attack to save ourselves

Dominants function as our species T cells[52] to help us survive non-overt threats. Our species is hard-wired and soft-wired (socialized) to hear only our dominant's warnings of non-overt threats. When subordinates try to warn our species, they are ignored[53] – it's a hard-wiring blind spot the Russian Mafia is

---

52 In our bodies T cells recognize viruses and launch an effective defense against a viral attack. When T cells aren't available and a deadly virus attacks a person, that person dies.

53 I've tried to warn my people for over eight years we have a problem but got no traction. I'm a dominant to my childhood siblings because I'm the elder child, but to the world I'm a subordinate, so my species ignored my warnings. In Elie Wiesel's book *Night* (Hill and Wang; revised edition (1/16/06)), Moishe the Beadle, an impoverished Jew, accepted by the villagers but foreign born, survived an SS attack years before Elie's village is attacked, and, shot in the leg, limped his way back to his adopted village, only to be disbelieved by the villagers. He knew he'd been spared by God to warn his adopted villagers and he was dismayed and embittered that they ignored him. Our species can't hear non-overt threats from subordinates. We're hard-wired to hear warnings of non-overt threats from dominants because when the hard-wiring was established in our distant ancestors eons ago, it was the dominants of that time who had the ability to understand non-overt threats who survived and reproduced. People alive today had ancestors who were dominants who could perceive non-overt threats and launch an effective defense. Now our species is hard-wired this way and the Russian Mafia is exploiting it to destroy Americans and everyone living in America. I also believe I read an operative discussing Active Measures and he said that people ignored accurate information. He didn't say why. Based on my personal experience trying to warn my nation, and based on Moishe the Beadle's experience as reported by Mr. Wiesel, we've got a blind-spot in our hard-wiring. That's what the USSR built their coup around to destroy us. There's no way we can beat hard-wiring unless we know it exists and build our defense on how subordinates identity dominants – we're all subordinates under the

exploiting. <u>The Russian Mafia's coup is a non-overt threat</u>. By entrapping dominants and preventing them from sounding the alarm, no alarm is sounded. To counter this attack, we must free our entrapped dominants and/or build a huge team of non-compromised dominants and flood our media with them warning our nation of this type of attack. We identify the team member to the audience in a recognizable way with their dominant status, paired with their subordinate, and the dominant warns the audience.

Many, not all, of our dominants are slaves of the Russian Mafia. That's why Mr. Putin believes he'll succeed in placing his puppets in the White House, because he already controls so slaves in America.

Most people see overt danger and evade it but warning our species of covert danger we're hard-wired to look to our dominants.

Chapter 3.7 – The Russian Mafia's attack under the hood – ordering their enslaved dominants to tell everyone in their 'circle of influence' to vote for a pro-Kremlin candidate. To survive we'll tell our people the truth, that they've been conned by an enemy foreign government determined to annihilate us

We must disable this bomb or life as we know it is over for everyone living in America – we must go loud. Deploying our dominants-as-bombs has the potential to install one of the Russian Mafia's puppets into the White House. Most people don't know the dominants they trust are slaves of the Russian

---

hood. It's our first position. We must tell the world or the Russian Mafia will destroy and murder it's way to world dominance.

Mafia. Even the slaves don't know, nearly everyone has been lied to and believe their handler's are legitimate American law enforcement. We can disrupt this bomb in part by telling our people what's been done to us. We'll have to blow the entire scam, offer everyone immunity else people won't come forward. The Russian Mafia will tell their slaves the pro-offered immunity from further prosecution is a lie, so we'll have to show we're telling the truth. And we've got to move fast, we're nearly out of time. My father, when he was trying to entrap me, indicated deep distrust in our government and he's a police officer. He's a slave and must say what his handlers tell him but we need to be ready for the Russian Mafia deploying dominants like police, parents, doctors, teachers countering our nation's attempt to defend ourselves.

Unless dominants raise an alarm, subordinates don't see covert attacks like stealth coups, until the attack has been sprung, hot, right in our face. The KGB and the Russian Mafia prioritized attacking our dominants to disable our alarm system to prevent us from launching an effective self-defense, allowing the Russian Mafia to destroy us. I think using our dominants as camouflage may have been an added bonus: to destroy our families and to commit genocide against us.

Chapter 3.8 – How we fight back using our dominant teams – move this to the dominant chapter

Since our dominants have been largely compromised, President Biden's Administration and his re-election committee must be informed of this component of the Russian Mafia's attacking our democracy, so that we and our hardworking, best-in-class Biden Administration, security agencies and military can launch an effective defense. All ordinary dominants aren't yet owned and we need many teams of them flooding our media, newspaper, etc., educating people about this type of warfare. We've got an excellent President who non-compromised dominants will want to help.

Chapter 3.9 – The Russian Mafia uses our dominants to quash my ability to sound the alarm, to quash my nation's ability to raise our defenses, &/or to try to entrap me – five examples of how the Russian Mafia deployed two teachers, two doctors, and my police officer father to try to entrap me or try to stop me sounding the alarm about their genocide-coup set to Close in November 2024

This strategy probably works to quell rebellions as well, deploying dominants against subordinates, but in the following examples the Russian Mafia used five people in my life who I turned to for help to prevent me from warning my people about the Russian Mafia's genocide-coup or to try to entrap me. In this genocide-war, the Russian Mafia has prioritized owning our dominants and you'll understand why after I recount my interactions with them. In these attempts to warn my people I'm subordinate and the compromised dominants (one doctor isn't compromised but was uninterested in helping me) were/are deployed to misdirect me, quash my warnings and/or entrap me.[54] The Russian Mafia likely uses this attack-strategy in any country they target.

With my teachers I am a student, and in a teacher-student relationship, the teacher is the dominant. The Russian Mafia understands the power of dominants in our species and works them as weapons to destroy the populace. In my case by having two of my teachers misdirect and/or refuse to listen to my complaints, the Russian Mafia ensured that I would not be able to harness the power of those educational institutions to amplify public knowledge about the Russian Mafia's genocide-coup. Compare that to how the Russian Mafia has deployed it's protesters to attack our higher education so even dominants in that system who believe the Russian Mafia is running a proxy war in Gaza dare not speak of it publicly. Do you see? The Russian Mafia is manipulating our people, institutions and systems, deploying the GOP so that no one is publicly willing to call out the Russian Mafia. To achieve this 'silence of our higher educational system,' psychological warfare in play on

---

54 I also wrote and/or emailed the last three Presidents, contacted the American Civil Liberties Union, emailed the FBI several times, emailed an international human rights agency years ago (I forget it's name), emailed an government oversight official website (I forget the name of it but it's on my phone), wrote Norway seeking political asylum (police were extra-judicially attacking me and I had no idea why), and may have contacted other agencies/institutions I don't remember years later, offhand. The Russian Mafia had uniformed and plain clothed police increase their harassment's against me, especially after I wrote former President Obama at the end of his presidency. I wrote former President Trump before his Finland meeting with Mr. Putin. Seeing that I didn't write him again. I believe the Russian Mafia deployed the shoplifting entrapment when they understood Mr. Biden might get elected. They didn't want me alive to write him. Unless we effectively defend ourselves President Biden will be our last president not owned by the Russian Mafia. So we need to handle this.

the ground, the Russian Mafia deployed: (1) GOP operatives, (2) pro-Palestinian protester operatives, and (3) antisemitism protester operatives. As General Gerasimov said in 2013, and I paraphrase 'war is undeclared, it is deployed in nontraditional ways that far outperform military success.'[55] When you look at the Russian Mafia's success in preventing anyone from publicly outing them about their proxy war, you see what General Gerasimov is saying.

1. To get me to mislabel the Russian Mafia's war on American civilians as racism, the Russian Mafia had one of their African American teacher operatives deploy the Russian Mafia's 'racism in America' camouflage in order to hide their genocide-coup. I told one of my city college teachers, who is African American, when I attempted a beginning tech class maybe ten years ago that I was being harassed, including by law enforcement. She said that was normal for African Americans. She told me Tom Bradley's daughter was harassed for many years. What this did was tell me the attacks I was experiencing were race-motivated, business as usual in America, but I know that's a lie because I'm African American and I was alive when my country wasn't like this before the KGB seized control of it. The Russian Mafia has built racism

---

55 General Valery Gerasimov, Chief of the General Staff of the Russian Federation,"The Value of Science in Prediction," *Military-Industrial Kurier* (February 27, 2013). General Valery Gerasimov states: *"...wars are no longer declared…" "...The very "rules of war" have changed…." "...The role of nonmilitary means of achieving political and strategic goals has grown, and, in many cases, they have exceeded the power of force of weapons in their effectiveness…." "...applied in coordination with the **protest potential of the population….**"*

here as a 'hide,' like they've built homelessness as a hide to use when they need property damage done (say to our post office boxes so we'll significantly reduce the number of post office boxes available to people in the November 2024 election), and they've entrapped our parents to use as hides from which to destroy the subordinates in their circle of influence, like they use fentanyl, mass murder, and police brutality as hides in order to assassinate people they want killed. The Russian Mafia uses poison against targets in England, Germany, and elsewhere they want security agencies to recognize poison as their 'signature.' But in America, the Russian Mafia uses racism, police brutality, homelessness, fentanyl, mass murder, and our dominants and other loved ones (parents, grandparents, siblings, doctors, teachers, etc.) as hides (camouflage) from which to do their evil. The Russian Mafia deployed an African American teacher to tell an African American student who sought her help to, in essence, what the attacks I was experiencing. This is to the Russian Mafia's advantage. Anyone they can get to mislabel genocide perpetrated by a foreign enemy government as racism, allows the Russian Mafia to continue their genocide uninterrupted;

2. The Russian Mafia deployed another teacher to shut down critical information I tried to give her about the Russian Mafia being on the ground here – she told me she doesn't discuss politics. I told another one of my city college teachers within the last two years, a woman who taught one of the programming-related classes I took. I tried to tell her about the

problems I'm experiencing. She shut me down by saying she doesn't discuss politics. I was trying to tell her my experiences. I may have mentioned Mr. Trump but I don't remember my comments being Trump-centered. My communications with her is saved somewhere on the city college system, I give our security agencies permission to look at it and at any of my communications with my teachers and staff there.[56] (They'll notice one of the Financial Aid officers was rude to me; a tactic I now understand the KGB/FSB has long deployed when they're trying to discourage me from pursuing something.

---

56 Even before I began taking programming classes over the last couple years, the Russian Mafia deployed entrapped extended family members to discourage me from taking programming classes at city college, and the Russian Mafia has deployed unknown numbers of proxies in, rough estimate, a couple hundred stealth noise attacks against me when I'm studying, taking tests, Finals, doing homework, etc.; or they've had a family member call me with emotionally devastating news, as when they deployed my beloved mother to tell me my much loved and respected mentor had died, that news told to me during Finals, after I'd spent the nigh in ER, after the Russian Mafia had deployed their operatives to noise harass me so excessively that my heart rate spiked to 200bpm, necessitating trips to the ER. At the ER I was further attacked. I began taking programming classes in order to (1) get my computer up and running without it being hacked so I could write this book – I'd tried for years to get tech help but the Russian Mafia is all over me and no store employee would help me and the Russian Mafia stopped me from getting through to Ubuntu when I tried, multiple times, to reach them (the Russian Mafia played elevator 'musack' on the line when I tried to reach Ubuntu, in order to taunt me, after I tried reaching Ubuntu's American offices and a few international ones); (2) having not yet written this book and so not having connected as many dots as I have by writing it, I'd hoped to learn enough programming to work part-time from my apartment as a programmer; and (3) I'd hoped I could help my country as I understand we need capable programmers and I like programming and I like helping my nation be stronger so working as a programmer was a win/win to me. My family member suggested, unprompted by me, that I take online classes at an online school. My compromised sister backed her husband. They were deployed as influencers. The Russian Mafia has sought to contain me, as if I'm a virus, and 'natural causes' kill me, for decades.

They don't want me to have money. I didn't understand how they'd pushed me from job to job until I wrote this book, when I had to remember why I left each job. On nearly every job I was harassed out by suddenly inexplicably hostile supervisors/co-workers, or I quit when they used a 'romance/mentor' attack against me at a law firm and I left because my supervisor was married, or they worked me to exhaustion after I evaded their embezzlement entrapment).

The Russian Mafia had already prepared the programming teacher, students must pre-register so the Russian Mafia knows which courses I'm taking months before the class starts. This particular teacher lied multiple times: before our first quiz she told us we'd have a certain amount of time to take the quiz, and that the quiz would have a certain number of questions. Her timed quiz had more questions than she'd told us, and she gave us less time than she'd initially told the class. I mentioned that to her and she said she'd adjust my grade accordingly, but she lied again, and didn't do that. I earned a B and should've earned an A. It's extremely important to the Russian Mafia that I'm not perceived as an A student, not only because of how they've had teachers run most of the programming courses I've taken but also because of how unrelentingly the Russian Mafia deploys it's operatives to harass me when I'm taking quizzes, Finals or doing programming homework assignments. The class I took from her wasn't a tech heavy class and anyone who applied themselves would've earned an A. Since I earned a B in a not challenging class, the Russian Mafia is setting up my grades to appear as if I'm not programmer material because they don't want me to believe I can be a programmer, and they want anyone looking at my grades to think I'm not a programmer. After writing this book and better understanding how the Russian Mafia is working to destroy us, my focus is on helping my nation, our allies, other innocent people worldwide, and our species survive.

3. <u>The Russian Mafia deployed one of the doctors they own to try to get me on psychotropic drugs after I told her I was being harassed by police for no reason</u>. In my 50s I told one of my doctor's at a world-renowned university hospital that I was being attacked by American law enforcement and asked for help. She is owned. Instead of helping me she recommended I go to the county and begin taking psychotropic medications, this despite the fact that I have medical insurance and that none of my other doctors recommended I get on psychotropic drugs. She also lied in her doctor's notes about what I told her. When I asked her to correct her notes, she refused. When I complained to the university, they refused to make her change her notes, saying it wouldn't change her diagnosis (sleep apnea). I complained to a state agency and no one would make the doctor correct her doctor's notes, even though I'm a customer and bought a service and the doctor didn't accurately report in her notes what I told her during my appointment. I knew because patients are given a copy of the doctor's notes and I read her notes. Reading them, her notes

were so inaccurate that I told her so, citing specific inaccuracies, asked her to change her notes and was surprised when she refused. She was a sleep specialist and I believed I had sleep apnea so I'd made an to be tested. The Russian Mafia compromised my first sleep test at the university, had me placed in a sleep study room and then pumped sleep inducing, odor-less gas into the room to make me sleepy. Because the Russian Mafia had openly terrorized me for about 15 years by that point, and I was unable to sleep naturally, when I inexplicably became very sleepy in the sleep room before the study even began, I grew very alarmed because I never felt that way, and it was so sudden and so different that it frightened me deeply. I knew, but had no way to prove, that someone was compromising the sleep study. In my 50s I had no idea who my enemy was but I knew I had one, and that it was a powerful enemy who controlled police, doctors, some of my family, my landlord, and many hundreds of strangers on the street. Because I'd been attacked overtly for years, I didn't spend my time denying that someone was trying to throw the sleep apnea test, I just accepted that that was the case. I was too terrified to sleep, even with the sleep-inducing gas, so even though I stayed in the room and remained attached to the leads for the prescribed time, the study was a fail because I couldn't sleep. I was told I'd have to return and I did. The next time I was placed in a different room and was able to sleep a little. The test showed I have sleep apnea, a mild to moderate case rather than a moderate to severe case. The Russian Mafia didn't want me diagnosed with sleep apnea. I've no idea why.

By having a doctor refuse to help me, the Russian Mafia prevented my ability to harness the power of that world-renowned medical and educational facility to warn the world about their genocide-coup here. Had that doctor not been compromised she would've advised me on a course of action, or at least asked her institution and the Ph.D's who work there how to counter the attacks I told her I was experiencing. So the Russian Mafia, by compromising medical staff, prevents subordinates from sounding the alarm because, in my case, the dominants threw out the subordinate's observations and experiences. The doctor's hospital and the State of California backed her up, preventing our nation from learning that an enemy foreign government is on the ground destroying us. America has plenty of powerful think tanks and many world-class, best-in-class resources, but the Russian Mafia controls access to them by controlling doctors and other gatekeeper dominants.

Another doctor, my GP a few years ago, 'didn't want to hear' my information that I was being attacked, and her staff stopped filling my prescriptions and stopped taking my phone calls to schedule appointments. I had to find another doctor just to get prescription refills. I chose the only one I saw in the city where I live, but he's compromised and has tried multiple times to entrap me, while he strongly recommended a medication that swelled my legs. When I told him my legs were swelling and he saw that they were, he, supposedly, didn't connect the dots between the medication's side effects and my swollen legs for a few years from which I began telling him. As far as I can tell most of the doctors I have access to appear to be compromised or they've been warned by their Administrations about me and simply aren't interested in prescribing quality medical care. I went recently to the ER for a medical emergency and that's the first time in a long time that I believe I got a correct diagnosis about the problem I experienced. I'm grateful. I have a book to publish.

My GP before my current entrapment-springing one took a 'head in the sand' approach when I told her at least twice I was being attacked. I saw that doctor maybe 25 times over the years and it took me awhile to rebuild my courage to tell her the attacks I was experiencing after the debacle with the sleep study doctor some years earlier. Her staff stopped taking my calls. The doctor was uninterested in helping me. I don't think she was entrapped. If she were entrapped she'd be unable to help me or, provide supbar care, or like my present GP, be actively working to destroy me. In her case, my sense was she wanted to work in her lane, and not get involved with her patient's other problems. And like Moishe the Beadle when he tried to tell the villagers about the horrors he'd witnessed, and those villagers had no framework in which to place such information and so ignored it. My previous GP may have had no framework in which to place 'police harassment and strangers attacking me on the street for years.' We want our doctors to be caregivers but some of them, stressed, sick and obese themselves (this doctor was obese and middle age) can only give what they can give.

This doctor's unwillingness to continue having me as a patient, or, and far more likely, since she wasn't owned, and since the Russian Mafia wanted me with one of their compromised GP's, and since the Russian Mafia has a history of controlling who I can reach on the phone, it's very likely that they prevented my calls from going through to her office, thus encouraging me to seek medical care elsewhere. The GP I'm with now, who is as far as I know the only GP from my medical system in this city, has been deployed twice to try to entrap me by offering to defraud Medicare on my behalf, unprompted by me. He's trying to entrap me. I only go to him when I absolutely must.

4. The Russian Mafia deploys my current GP to encourage me to commit Medicare fraud. Because my previous GP from maybe five years ago (now apparently retired) became unavailable to me probably because the Russian Mafia prevented my calls from reaching her office and/or prevented my messages from being accessed by non-compromised staff, after I wrote a polite letter to one of their doctors about ten years ago complaining about his subpar care, and he was fired (apparently I wasn't the only patient who complained), the Russian Mafia made compromised medical office staff nonresponsive to my requests for prescription refills and scheduling appointments – meaning I'd have to take the bus

across town in order to make a doctor's appointment, then return for the appointment). The Russian Mafia ensured that the staff didn't show me how to engage with their online patient app, so I wasn't tech literate enough to go online and make appointments and request refills. I had to switch to another doctor who was available. While it's true he is available he's also been used to try to entrap me twice over the last three years with Medicare fraud, his most recent entrapment attempt the summer 2023. If I had to chose between a doctor who didn't want to know about the attacks on me and a doctor who actually attacks me, I'd choose my previous doctor. At least she didn't promote a medication that swelled my legs and then ignore the systems for two years when I reported it. She didn't help me, but she didn't harm me. My current GP harmed me and continues to be deployed by the Russian Mafia to harm me.

In a doctor-patient relationship the doctor is the dominant. When my doctor's refused to help me, and actively work to hurt me by (1) advising me to get on psychotropic medications unnecessarily, (2) running a bogus sleep study test so I'd not be properly diagnosed, and (3) trying, twice, to entrap me using Medicare fraud, the Russian Mafia has demonstrated that it's flipped our medical care system to destroy Americans, instead of our systems intended aim which is to help us stay healthier, and to help us survive medical emergencies.

5. <u>The Russian Mafia deployed my father, a retired police officer, to tell me not to write this book, told me people would say I was crazy, his argument being, when I first decided to write it, I didn't know my enemy's name, and that since that was the case, that I shouldn't write it</u>. I knew I had an enemy, I knew their behavior. I knew their many attacks. But not their name. My father used that as his reason that I not write this book – because I couldn't name the perpetrator. I finally learned the name of the perpetrator when was 61. I initially told my father I would write a book when I was 58 or so, maybe 59. Had I done what my father told me, there'd be no book and the Russian Mafia would destroy me, my nation of 400 million, and our allies beginning in earnest this November 2024.

I didn't know that when I decided to write this book. My initial motivation was to protect children that I saw being used as beards and co-soldiers by their entrapped parents as their entrapped parent(s) harassed me. By 58 I knew my mother and some of my siblings were part of the attacks against me. I'd been surprised by the attacks, didn't know enough to label them as the entrapment attempts they were. Children blind-sighted by compromised parents and siblings would be devastated. My intention was to warn them to help them survive. At that time I knew nothing about the genocide. I saw the evidence and followed it because I was being attacked and couldn't not see the attacks, unless I was willing to take the psychotropic drugs and street drugs the Russian Mafia urged on me. I refused to self-destruct or emotionally destabilize. Emotional destabilizing warfare is another type of warfare the Russian Mafia deploys against me by stealth murdering my loved ones. They've murdered my favorite aunt, my beloved grandfather, and my beloved mentor. The Russian Mafia seeks to stealth murder me using 'natural causes' but they will knife me if they can get one of their mentally unstable people near me, or they'll have one of their operatives make an error on a repair and cause an explosion. Just days ago one of their operatives was banging on something in the laundry room that didn't require banging.

For years my dad misdirected me, specifically when I mentioned being harassed in stores and he gave me a cover story about stores experiencing increased shoplifting. He told me misdirection stories nearly 20 years ago and like any of the Russian Mafia's cover stories, they had a veneer of truth but were meant to hide the truth that Russian Mafia is running a genocide-coup here.

In a father-daughter relationship, the father is the dominant. And when your father is a police officer, many daughters would do as he ordered. Yet by the time of his order, I'd experienced nearly 20 years of overt harassment, and physical, psychological and/or sexual assaults. My mother and some of my siblings appeared to be a part of it, and thousands of strangers, including children, harassed me nearly everywhere I went. Police were involved but instead of protecting me they oversaw the attacks (I witnessed one force a young man to harass me), and/or police stealth harassed me themselves. My mother had tried to get me to lie and cheat and steal when she'd raised me to do the exact opposite during my young childhood. All these children harassing me, most with their parents, why in the world would parents harass a stranger and bring their children with them, and in some cases, have their children assist them. I had to write this book to warn those children (since whatever this was ensnared the best parents, like my mother had been). I had to warn the world's children, so that they might better defend themselves against whatever this was. I didn't know when I decided to write this book that I was witnessing genocide against us by the Russian Mafia.

Chapter 3.10 – The USSR[57] used children they owned to threaten me when I was 11

The first bully to directly threaten me was at least a grade or two ahead of me, at least two or three years older, and significantly larger. After I picked up a brick to defend myself in response to his threat, an adult operative (the man the USSR installed in my mother's bed to destabilize her and our family) called off the child-bully's attack.

The parent of the that child, likely a KKK member who's local chapter, along with the KGB, controlled the neighborhood where we lived, and which local chapter the

---

57 I strongly suspect the Russian Mafia continues to use children to attack targeted children in America similarly as to how the USSR attacked me when I was child starting in 1971, using child operatives the USSR controlled.

KGB had formed a partnership with to remove people of color from the neighborhood who displayed strength,[58] did not sign off on their son being hit by a brick, but had apparently approved their son beating up a smaller, younger African American child.

My mother was ordered by the USSR to beat me when I was 15, as the three teenage girls the USSR sicced on me in high school were unable to get me alone to attack me. I watched them multiple times from windows as I studied in the library but although they waited outside a long time to attack me they wouldn't attack me in the library in front of the high school library staff who would've outed them as the aggressors.

---

58 A child defending herself and her siblings the KGB interpreted as a sign the child and her family were strong and required immediate destabilization. My poor mother. She was only 29 and solely responsible for four children. She'd survived an horrifically abusive childhood, the murder of her beloved grandmother when she was 12, endured several failed marriages by her mid-20s, but had pulled herself up by her own bootstraps, thought over her options, devised an effective Plan B and became a surgical technician. She bought a house with her own earnings as a surgical technician, was the hero of her own life, only to be attacked by the KGB in her late 20s. The KGB was the most evil security agency in the history of our species, far worse than the Nazi's experimenting security agencies because the Nazi's in the 1930s and 1940s largely restricted themselves to Germany and the surrounding nations while the KGB was worldwide. The KGB illegally stole not only my mother's house from her and made her children homeless but they stole her career, her children, her grandchildren, and her great grandchildren. She's a civilian, bothered no one, was a full time mom and a full time hospital employee. In about a year after they'd placed their operative into her bed, my mother had lost her home, her career, she was sick, and the KGB, fronting as American law enforcement, my guess is they threatened to take he children from her, put us into foster care. Using lies, using stealth, the KGB was here illegally, destroying families as early as 1971-72.

The Russian Mafia weaponized my father to denigrate me into not writing this book and my mother, for several years, to denigrate me into filing personal bankruptcy, which system, apparently the Russian Mafia now controls.

My parents have been unable to entrap me but my half siblings born by my mother, are entrapped and where my mother was the weapon who entrapped them, those adult children loathe her: the Russian Mafia destroyed my family. Traditionally, our species doesn't destroy the enemy's family structure in a coup, and usually doesn't destroy families unless they are involved in running the government. Even in genocide, enemy's don't usually destroy the targeted people's family and social structure. The Nazi's didn't insist Jewish grandparents, parents and religious leaders betray their loved ones by having them enter the gas chamber under false pretenses. As evil as the Nazi's were, even they didn't try to destroy Jewish families before murdering them. Betrayal before murder is worse than just plain murder. Both are horrific but betrayal by your mother, father, grandparents, your priest, is next level evil. You destroy families when you intend to destroy the people and their nation.

The entrapped adult children in my family let their children know, my mother's grandchildren, that she is not to be trusted. One of my mother's grandson's is an artist and she's never done nothing to harm him, but he won't even speak to her because his father loathes her. My mother is not at liberty to explain the cause of the break to me. Her plea deal prevents her from discussing any of this. She's not the only American great grandmother tricked into slavery by Russian Mafia-controlled American law enforcement who has had her family stolen from her.

The Russian Mafia forced my mother to entrap at least some of her children and then showed my siblings that betrayal, breaking our family unit, stealing my mother's children from her, her grandchildren, her great grandchildren is evil and the Russian Mafia just doesn't care. I believe my mother took a plea deal when she was entrapped in 1972 or so to keep her children out of foster care but the Russian Mafia destroyed her family anyway because they're working genocide here, not just a coup. They intend to murder us. They're proxy murdering/injuring tens of thousands of Palestinian children to install one of their puppets here and they could care less about innocent Palestinians. Us, they hate. Us they've schemed to destroy for maybe 70 years, have had the bomb up and running probably 60+ years. They destabilized my family 53+ years ago. All these attack against civilians and women and children living in America after the USSR signed the Geneva Conventions and anti-genocide international agreements in 1948-1949. They flat out lie.

In our species families form the cornerstone of our culture, our society, and our government. Destroying our families is genocide against us and only an enemy who intents to destroy us would deploy it. The Russian Mafia is that enemy.

Our species' consists of dominant and subordinate subgroups and the Russian Mafia's bomb targets those subgroups.

Chapter 3.11 – Most of my supervisors in the businesses where I worked threw me under the bus when the Russian Mafia, fronting as American law enforcement, spread their lies. One supervisor, however, refused to attack me. He helped me learn enough to warn you

One of my favorite bosses, an attorney, was entrapped and committed suicide, I believe because he refused to live his life destabilizing innocent people. He was very smart and he'd been a teacher before he became an attorney. He knew I wasn't entrapped because his handlers were trying to force him to attack me. He knew they'd kill me if he tried to warn me so he told me nothing about his entrapment. Instead, he gave me a Christmas present of educational videos his last Christmas alive. Everyone knew he'd been a teacher so his gift made sense. I'd been hoping for a monetary Christmas gift ($100 was what some secretaries received from attorneys they worked for) but I love learning so was genuinely appreciative of his gift. What he was saying to me with his gift was to keep learning. I liked the videos so much that I bought more on other subjects. Because I am unentrapped I am free to warn you, although the Russian Mafia has threatened me, deployed proxies to physically attack me, and has psychologically attacked me for most of my life. Still, because I am not entrapped, I am free to complain. Entrapped people are slaves.

They are not free to complain. They cannot sound the alarm. I believe there are tens of millions of entrapped people in America alone. When I was his secretary aged around 41-42, I didn't know enough to understand what was going on. It was right before I became his secretary that the Russian Mafia went loud on me (2001-2002), attacking me using strangers and plain clothed law enforcement and security in grocery stores, on the bus, on the sidewalk, etc. but it was decades before I learned the perpetrator was the Russian Mafia. Only unentrapped people are able to speak. This brilliant attorney, my friend, who refused to attack me when other supervisors had at it, would've warned the world if he could have. My theory is he was told that if he talked he'd be imprisoned on whatever they'd entrapped him with, where he'd be raped and murdered, and that his loved ones would be murdered. He was married and I think his parents and at least one sibling were alive. He didn't discuss his family with me. The Russian Mafia controls our jails and prisons. They've worked unrelentingly for decades to put me in the criminal justice system by springing entrapment attacks against me. Failing that, they tried to get me to file for personal bankruptcy, where, apparently they control some aspect of that system. Two of my immediate family members, and an HR worker at my previous job, encouraged me to file for personal bankruptcy. The Russian Mafia had me deposed after his death. They knew he'd refused to harass me and/or try to entrap me (unlike several other supervisors still in my future), and the Mafia wanted to know if he'd told me anything. I knew nothing about their attacks on him so I could tell them nothing except that he was a nice man and a hard worker. I was devastated by his death. I kept on learning, his 'Hail Mary' videos a pass to help me warn our nation and the world's peoples about the evil that is the Russian Mafia. This book is dedicated to him, and to all people worldwide attacked by evil people.

Chapter 4 - <u>Who modern humans are under the hood</u>

Modern humans are a hard-wired species, meaning we're descendants of long dead ancestors who were not human as we are today, but who's societal organization we still share because both we and our ancestors build our societies around our families (not all mammals do), and modern humans inherited bits of our ancestor's genes which predispose us, under certain conditions, to make broadly similar decisions in response to certain stressors. This predisposition instinct or preference is our "hard-wired" respond to our environment and shapes our societies in ways not that different in the fundamentals from our distant ancestors. We receive the hard-wiring in our genes, the reconfiguration of genetic material we inherit from both our parents when they mate and the female becomes pregnant.

An extended family-based species, our ancestors survived long enough to reproduce and some of their preferences and instincts were passed along to us from those bits of their genes over millions of years and many trillions of sex gene reconfigurations.

Babies born to our ancestors were cared for by their parents, extended family, and/or members of the group. Then as now babies were born helpless and dependent upon their caregivers for survival.

Our ancestors' societies and our modern human societies consist of two subgroups: dominants and subordinates. All people are born subordinates, by that I mean we're born dependent on others for survival. I'll call the people newborns rely upon for care, dominants. Dominants in a newborn's life are usually their parents, grandparents, etc. and in relation to newborns dominants are far larger, have significantly more life experience, and much more survival knowledge, which benefits a newborn relies upon from birth. Dominants in our ancestors time might have been the baby's parent, and so was primarily responsible for keeping the baby safe, or might have been a dominant in the larger group, a person who was a little more skilled and/or adept at keeping themselves and their group members alive.

In modern times dominants are ordinary people who have societal-approved authority, power, or control over others under certain circumstances, or who provide an essential service in society. Examples include parents in a parent-child relationship, teachers in a teacher-student relationship, landlords in the landlord-tenant relationship, generals in a general-soldier relationship, doctors in a doctor-patient relationship, police in a police-citizen relationship, employers in a employer-employee relationship, bus driver in a bus driver-passenger relationship, etc. Essential service providers include grocery stores, utilities, water and sewage departments, etc.

Dominants and subordinates then and now are members of the group. At birth subordinates are served by dominants because babies wouldn't survive without them. Mothers are hard-wired to care about their babies via pregnancy-related hormones that flood their body during pregnancy and post-natal. Fathers are socialized (soft-wired) to care for their babies pre-birth and after they're born.

Because we enter life as subordinates, that is our first and default position and we build our reality from that point of view. We're socialized (raised) to obey dominants: parents, teachers, medical care providers. We're forced from infancy to accept shots from doctors, our grocery stores are provisioned by managers who ensure the store has food for us to buy, we pay our rent to landlords who have the authority to make us homeless if we don't pay. In most people's lives, we are subordinates to somebody, usually to many somebodies.

From birth dominants provide essential services, food, physical comfort, often love. Because they respond to our cries and respond with appropriate remedies we cannot get for ourselves, babies attach high value to dominants who care for them and instinctively assign dominants an elevated value from the earliest days of life. This is called socialization or soft-wiring and is how a baby begins to make sense of the world. The instinct to highly value dominants is part of our hard-wiring, a gift from bits of our ancestors' genes passed to us (in utero) from distant and recent ancestors, who survived long enough to reproduce. Those ancestors, too, assigned elevated value to the dominants in their lives. So think of hard-wiring as an inherited boon, something evolution gifted us because it helped our ancestors survive to reproduce. Surviving to reproduce is like a lottery win in evolutionary terms.

In a functioning family dominants become trusted by a baby within their first days of life. The baby associates bigness with strength, intelligence, comfort, safety, knowledge and food, all good things in a baby's life. Babies associate dominants with goodness. To a baby, dominants are what good means. In a functioning family, that basic orientation, dominant=good, lasts through the baby's childhood and into their adulthood.

The dominant=good orientation is found across all ethnicity's, cultures, religions, nations, income levels, etc. the KGB discovered some time last century, and they decided to build a psychological-based weapon to exploit it our socialization (the similar way babies are raised worldwide).

Since nearly everyone in the world is raised to believe the dominants in their lives, i.e., parents, grandparents, doctors, etc. are good, and because subordinates are taught to obey and trust them without question, and are taught that those dominants intend the best for them, the KGB theorized that a relatively fast way to seize control of a nation was to compromise it's dominants and unleash them against the subordinates.

The Russian Mafia used that attack against me as recently as the summer of 2023, when they had the doctor I saw at a medical appointment, after I asked him if I qualified for one of the injectible weight loss drugs as I'm obese and insulin resistant, but not diabetic. He told me that no, pre-diabetics don't qualify for these weight loss drugs under my insurer, but he offered to tell my insurer, his offer unprompted by me, that I'm diabetic. He said the insurer doesn't question doctor's labeling of patients, and that I could get the medicine that way. I declined his offer.

Many people would assume, because they've been taught from birth that doctors are helpful and benign, and would never offer to do something that would land the patient in prison, that accepting the doctor's offer to 'help' me get a weight loss medicine was not a trap, and yet that's what it was. An attack by the Russian Mafia entrapping the innocent as it works to destroy us via coup. The Russian Mafia exploits victim's socialization, knowing that trust in doctors from birth is how many children are raised, and since a doctor's dominance in a baby's life means he's good to the baby's mind, trusting patients are unlikely to see an medical advisor since birth as suddenly out to get them, yet the Russian Mafia is a world class, best in class mafia, and can get most people, including doctors, to do anything.

Babies are raised to believe that doctors and other care givers are trustworthy, and that's one way the Russian Mafia is destabilizing us and has us on the brink of a coup. Mr. Putin is determined to Close Active Measures before 2025 and we dare not allow him to succeed. If he installs one of his puppets, Mr. Putin will destroy us. The Russian Mafia has been working this coup since, rough estimate 65 years. They know this is it, given that they strongly prefer to avoid nuclear war.

The 'help' the doctor offered is a crime that would've allowed me to benefit from defrauding an insurer. Had I accepted, I would've been charged with defrauding Medicare.

Listening in to the doctor make his pitch were compromised law enforcement who had ears on the room because otherwise how would they know what my response was. Compromised American law enforcement owned by the Russian Mafia have worked to entrap me for, easily, thirty years.

Had I accepted the doctor's offer I would've been arrested, jailed and after a few weeks in jail, perhaps a plea deal would've been presented to me. Conspiring to defraud Medicare I've no idea how long that prison sentence is but say conspiracy is 10 years (conspiracy is if you agree to commit a crime. The Russian Mafia likes it because victims don't have to do the crime to be charged with conspiracy, they only must agree to do it. I learned that after I'd unknowingly evaded multiple entrapment attacks and finally understood why people were ordering, recommending and/or telling me to do all kinds of illegal acts. They have to raise the subject, invite me. The invitation is the conspiracy: they invite, if I accept, I'm toast.

The plea deal would've meant no more jail/prison for me under certain conditions. Based on how other entrapped victims are used to try to entrap and/or harass me, the plea deal means I'm owned by the Russian-Mafia's owned American handlers for the rest of my life. I've no idea if that means the Russian Mafia can rape me but I've been sexually assaulted by at least three of their operatives so no evil is beyond them.

I wouldn't have the option to avoid their phone calls, or discuss my plea deal with anyone but them, including other law enforcement at the highest levels in our country. If I break any of these terms I would be immediately sent to prison where I'll serve time for the original crime plus added time for breaking the terms of the deal. I'm not at liberty to say I took a plea deal. When law enforcement say I need to be some place, I get to that place. When he or any other law enforcement officer asks me to help in any future law enforcement matter for the rest of my life, including entrapping and/or lying to other people, including grooming children, I agree to do so. Because I've been harassed and/or attacked for over 53 years I know American children are victimized by the Russian Mafia, and before them, the KGB.

As you can see from this example which happened except I refused the doctor's offer, so was offered no deal, the KGB combined the socialization of how people are raised to trust doctors with entrapment. Entrapment is another component of the Russian Mafia's coup, where they compromise/entrap/manipulate blindly trusted dominants (parents/doctors, etc.) to lure/groom/entrap innocent subordinates. When people take the deal versus prison they lose their quality of life forever, and if they have children (and these attacks are aimed to target parents, which I discovered as a child), the family is broken when the Russian Mafia forces the parents to entrap their own children, after which the child blames the parent for ruining their life.

The Russian Mafia controlled compromised American law enforcement's insertion into the victim's life means they never get their life back and they're forever destabilized. The handler is inserted into the victim's life to keep them agitated, miserable, unhappy, enraged, and traumatized because frustrated, miserable Americans are who the Russian Mafia intends to use to overthrow our legitimate government. One of the Russian Mafia's presidential puppets will say she/he was just responding to the demands of the citizenry when the puppet president striped us of our democracy.

Victims are criminalized by their Russia-controlled American handlers, lured, destabilized. From what I've witnessed in my family, victims are encouraged to cheat on their taxes to give the Russian Mafia more leverage on them. The Russian Mafia aren't people you want to have leverage on you. Over time, the victim is influenced into the GOP Base in any number of ways where they vote for whichever candidate the Russian Mafia tells them to.

This is how the Russian Mafia manipulates our voting system: rather than hack voting machines, they hack our people. Then throw them away in the trash heap.

Another attack parts of which I've endured but because I refused, when told, to agree to break the law when my parents ordered it, I wasn't entrapped, but some of my siblings are entrapped, is when the Russian Mafia forces compromised parents to entrap their children. After the child is entrapped they blame their enslaved parent and the family is broken. Deliberate destruction of family, the foundation of American culture, and the foundation of most all human cultures, is genocide. Genocide is what the Russian Mafia is perpetuating against me and my nation. Genocide means the Russian Mafia is using Active Measures to remove Americans from life.

To encourage targets to lose trust in their family, our government, anything/everything, the Russian Mafia has respected dominants casually offer to cheat and/or lie on the subordinates behalf, which, because we're raised from birth to see dominants=good, when dominants are corrupt and encourage subordinates to be corrupt, it creates dissonance and confusion in the targeted subordinate, making them feel the foundations on which they've built their life are crumbling. This again is to push subordinates to demand the overthrow of our legitimate government in the hope that a different government will be more ethical. Our legitimate government is magnificent. We had the very bad luck to draw the eye of the evil, highly acquisitive Russian Mafia.

The Russian Mafia reveals it's presence in none of this. It attacks using proxies and does it's best to make people blame our legitimate government who often has no idea the Russian Mafia is working these attacks on the ground here. As you can imagine, what the Russian Mafia is doing is illegal. The last thing they intend to do is flash their destabilizing attacks before our legitimate government.

Another type of psychological warfare attack is when good people who work their guts out get a kick in the teeth instead of a positive reward. The USSR leaned in heavy with this type of attack against me in my 20s, to enrage me. I call these types of attacks the "life isn't fair Russian-Mafia coup attacks" because the Russian Mafia uses them to enrage targets and then trigger them to rampage. I believe the Russian Mafia ordered the 2023 Israel-Hamas War[59] to unseat our excellent

---

59 In this book I argue that the Russian Mafia ordered the 2023 Israel-Hamas War and I state that the Russian Mafia started the war as a proxy war to unseat our President and to seize control of us and of our nation by installing one of their many puppets here, but I have no conclusive proof to offer other than that the war benefits them directly, which in my readings on the Russian Mafia doesn't happen by accident. Russian General Gerasimov stated in his 2013 paper communicating war strategies and tactics that the Russian Mafia deploys protesters to accuse, harass, and pressure enemy governments, which has certainly been done in this war. (General Valery Gerasimov, Chief of the General Staff of the Russian Federation,"The Value of Science in Prediction," *Military-Industrial Kurier* (February 27, 2013).) Additionally, the Russian Mafia has access to Hamas as they have formed an alliance with the Iranian government, which government is Hamas' benefactor, and the Russian Mafia has a history of involving itself in proxy wars against us. It also has a genocide called Active Measures in play on the ground against Americans I believe they intend to close this November 2024. For these reasons, and because the KGB and later the Russian Mafia has attacked me since I was 11 years old and continue to do so to the present date via proxies (my age is 64), I am 100% sure the Russian Mafia ordered this war to destroy me, my nation, my people, my government, and to shove Israeli civilians right

President and seat one of it's many puppets. Hamas are largely radicalized Palestinians, radicalized means the Russian Mafia frustrated them so much those men can be triggered and aimed at anything and they'll destroy everything in their path. As you see when you allow the Russian Mafia to enrage you, things don't end well for people you care about. Hamas never would've attacked Israel had they known that less than six months later 15,000 Palestinian children would be dead. I've experienced the Russian Mafia try their enraging attack against me. They had my supervisor suddenly, for no reason, begin to treat me derogatorily. That went on for some months. Then the Russian Mafia had a co-worker, a woman of color like me, approach and ask me if I wanted to join a complaint against the supervisor charging racism. I declined. There was a problem but I didn't' think it was racism. The Russian Mafia wanted me to strike out at my supervisor, gave me a reason (abused by her), and a method (lawsuit, money). Ethics saved me. I had no idea it was entrapment at the time. I assessed what I knew: there was a problem, I didn't think it was racially motivated. So I refused to join a complaint alleging racism.

Most people don't know the Russian Mafia is working these scams here and so innocent people are lured into messes day in and day out. People must be warned.

Dominants are the glue that holds our species together. By messing with dominants, flipping them from kind, helpful, benign to corrupting, the Russian Mafia demonstrates it intends

---

to embrace their right-leaning government.

to destroy us, our culture, our society and our government. It doesn't care that it's working against the norms we were raised with.

When the Russian Mafia deploys dominants to compromise subordinates in ways the subordinate perceives as wrong, but framed by a "good" dominant, this kind of anti-societal attack attacks the foundation of who we are. The Russian Mafia doesn't care.

In our species dominants are the glue holding our species together. One of the essential roles dominants play is warning subordinates about non-overt threats. By targeting, enslaving and entrapping American dominants, the Russian Mafia has ensured that we have many fewer non-compromised dominants available to warn us about non-overt threats, which the Russian Mafia's coup is a non-overt threat. Because we're hard-wired to "hear" dominant's warnings of non-overt threats, when dominant's don't sound the alarm, no alarm is raised and the people are destroyed. I refuse to let my people be destroyed.

When I see no one accusing the Russian Mafia with committing yet another proxy war murdering over 10,000 Palestinian children in order to unseat a political rival, our President, it suggests to me the Russian Mafia has seized control of people we'd traditionally rely upon to warn us about the Russian Mafia's coup attempt.

The Russian Mafia knows we're hard-wired to hear dominants sound the alarm about non-overt threats which is part of the reason they've enslaved so many of our dominants. To evade this trap we must not only free our dominants, which will take quite a bit of time so, the fastest thing to do is hire non-compromised dominants and produce a series of vignettes and play and present them, flood our media spaces with them to disable the dominant-based bomb the Russian Mafia has built to destroy us.

The vignettes will tell our people what they need to know about this election cycle but it must be done by dominants being dominant, with subordinates believably responding to the dominant in the vignette. This isn't hard to do, we just need to know a dominant/subordinate vignette is required to disable the Russian Mafia's bomb.

Chapter 4.1 – The Russian Mafia now says it doesn't declare war, leaving it to the innocent to somehow know they're under attack, including children[60]

Caveat: Although I refer to compromised American law enforcement throughout this book, the majority of American law enforcement, and the overwhelming majority of our security agencies, and our military, are not compromised. If those groups were compromised, they'd stop me from writing this book where I out the Russian Mafia's attacks against ordinary Americans and immigrants. While the Russian Mafia is working to stop me writing/publishing this book, and upon publication will do anything/everything to undercut my credibility (which is why I've included my flaws and every mistake I've ever made), this book contains no lie or exaggeration about the several thousand attacks the USSR, Russian government, &/or Russian Mafia have made against me using operatives they control here, and/or that I've witnessed. I was around 12 when the USSR destabilized my mom. She's never been able to fully explain what happened to her, and I stopped asking when I was a 12 or 13 because it appeared to me to cause her great pain. I swear on the bible that this book tells the truth to the best of my knowledge. Stacy Hackney, 3/6/2024.

People living in America, and likely our allies worldwide, are under stealth attack by the Russian Mafia. As I type these words, our President and his Administration are being manipulated by the Russian Mafia. I believe the Russian Mafia started and is running the 2023 Israel-Hamas War. Knowing and exploiting the confidential language of our alliance agreement with Israel, the Russian Mafia set the Israel-Hamas War proxy war in play in such a way that our government

---

60 Gen. Gerasimov, Chief of the General Staff of the Russian Federation,"The Value of Science in Prediction," *Military-Industrial Kurier* (2/27/2013). *"...wars are no longer declared...." "...The very "rules of war" have changed...."*

cannot publicly call for a cease-fire, unless certain conditions are met, which conditions the Russian Mafia has ensured that neither Hamas nor the Israel government meet. If we call for a cease-fire outside contractually agreed parameters in our agreement, we end our alliance with the Israeli government, which the Russian Mafia knows we don't want to do because that alliance is important to our military which, again, the Russian Mafia knows. The 2023 Israel-Hamas War is a proxy war to destroy Americans yet no one is shouting it. That no one is shouting it demonstrates how effective Active Measures has worked for the Russian Mafia. If they succeed, Americans will be destroyed. We're pre-Holocaust:

I refuse to let my people be destroyed. If we let the Russian Mafia steal our freedom and our nation, life will be a nightmare for us, and to be frank, most of us won't live long. I know there are plenty of examples of pro-Kremlin governments that appear to be somewhat functional. Italy, Hungary, Israel, but these are cons (lies) to get us to take the bait that pro-Kremlin governments aren't so bad. Lies, if you know anything about the Russian Mafia you know they lie straight-faced, without blushing.

Chapter 4.2 – Our nation is 'listing' because the Russian Mafia is secretly attacking us

This is what psychological warfare looks like to people not trained in espionage. The Russian Mafia eventually flips victims into the GOP Base via influences in a variety of ways:

online, prepper chats, church influencers,[61] installing compromised lovers into a target's bed,[62] having a compromised authority figure introduce the target to an operative of the Russian Mafia, etc.

My best tip right now is to not vote for pro-Kremlin candidates. You will, of course, do what you feel is right but if you vote in a pro-Kremlin candidate, you'll be helping the Russian Mafia destroy us, our culture, and our country. They'll

---

61 The Russian Mafia and before it the USSR, pushed entrapped victims into embracing religion. The type of religion the Russian Mafia pushes people into prioritizes faith, meaning the Russian Mafia grooms victims to obedience. Victims will, no doubt, be prepared by their ministers to expect they'll be attacked because of their faith when in reality the Russian Mafia is working to steal their nation from them. Our security agencies, our unentrapped citizens who have a clue, our unentrapped allies, are trying to help victims keep their freedom, which, the Russian Mafia has stolen from them when it deliberately destroyed their life and steered them to religion. Ordinary people simply have no idea the Russian Mafia and it's predecessor government have been stealth attacking us for about 65 years and that until recently, our legitimate government didn't know. Our government didn't know because the Russian Mafia deliberately hid the attacks from our legitimate government.

62 Shaun Walker, Pjotr Sauer and Tom Phillips "Panic and Emotional Pain as Alleged Deep-Cover Russian Spies Vanish," *The Guardian* (April 3, 2023). "Campos Wittich was a Russian spy living under an assumed name in Brazil. At least six such suspected spies have been unmasked… in various locations over the past year, suggesting there could be….more. He was a deep cover spy working for an elite intelligence programme…." "She (the victimized girlfriend) suspected nothing at all," said the friend of Campos Wittich's partner." "She is just a lovely woman looking to create a family and get married to a man she thought was the man of her life." "She is really scared of this situation and hurt by all the pain of having an abrupt cut-off in a relationship that was perfect in her eyes," said one of friends. The victimized girlfriend of the spy is a veterinarian. She had no suspicion she was involved with a Russian spy. The romantic partners of both spies have been left devastated, said their friends and acquaintances in Athens and Rio."...Campos Wittich may have been in an earlier 'embedding' stage of their missions, which could have gone on for decades if not detected," reporters state.

never tell you because they're not known for their truth telling; your education will be a nasty surprise down the road. To them, nirvana is world domination and before they can achieve it, they believe Americans must be destroyed, not just our government, but us, our families, our children, all our systems. They're extremely acquisitive and they're thieves; they see something they want and they take it. Their bombing of Ukrainians is Exhibit A anyone can see.

I've seen the Russian Mafia deliberately psychologically destroy a child and make her into a weapon. To do that they forced her to endure years of unimaginable physical pain, as well as many years of psychological pain. I saw them do that. That's in addition to their attacks on me and other loved ones.

The Russian Mafia is a group of men who hate women, all women.

I've witnessed, experienced, endured and/or read about their attacks against nearly all ethnicity's of women but because I'm African American I've had an up-close and personal view of their attacks against women of color.

I've witnessed the Russian Mafia's American operatives force American parents to groom their children and force them to use their children in harassing attacks against me. I've witnessed American parents forced to use their own children as co-operatives, as soldiers. I'm talking American children, age 10 or so, being shown by their entrapped parent that it's perfectly fine to harass a woman on a mobility device, under order of the Russian Mafia in America.

The Russian Mafia intends to destroy us. They have nothing to offer us but lies, destruction, and death. If you vote them in, you will regret it. Any descendants you have, any friends not already enslaved by them, will be. Our descendants will all wish for death and the Russian Mafia will happily accommodate them.

Chapter 4.3 – Setting the trap – the Russian Mafia is working to lull us into a false sense of security with how they interact with pre-Kremlin governments – first thing to know about the Russian Mafia: they lie

The Russian Mafia is working to not spook us with how they're running other pro-Kremlin leaning governments but really, a proxy war where (I believe) they're murdering thousands of babies to unseat our President is not supposed to spook us? Apparently they don't care what our security agencies, our military and our President think because those professionals are restricted by policies and laws, which the Russian Mafia knows and is manipulating, and has baked into their attack on innocent Palestinians. It's extremely interesting that the Russian Mafia controls our media to such an extent I've read no one outing them about the proxy war I'm convinced they're orchestrating and running. If we let the Russian Mafia near the White House, it's over for us. We won't survive.

Chapter 4.4 – Genocide deployed

At present and for the past fifty+ years, entrapment was/is the USSR's/Russian Mafia's go-to to embed themselves into noncriminal victim's lives. Once entrapped, the target is a slave and never gets their life back. The Russian Mafia, and before it the USSR, worked for decades to entrap me and to place me into the criminal justice system, which apparently it rules – apparently it believed it could more easily attack and destabilize me. I managed to evade entrapment but I believe millions of innocent victims didn't, I know some who didn't, and I know they form a percentage of our GOP Base. I know two GOP Base who are entrapped because they told me they're in the GOP Base.

The Russian Mafia prioritizes entrapping parents and other dominants because they've found it easier and faster to enslave a larger number of people when they exploit our hierarchy and hard wiring.[63] Children are socialized from birth to obey their parents. Entrapping parents and forcing parents to entrap their children speeds up the Russian Mafia's destruction of us, and facilitates it's genocide against us. Family's are the cornerstone of Americans, and the cornerstone of our species. Destroying our families destroys humans and our culture. That

---

63 Using psychological warfare (aka lying with a thin veneer of truth), and exploiting our species' hierarchical hard wiring (subordinates like children are socialized to obey parents, who are usually dominants in the parent-child relationship, so ambush/entrap the parent & force the parent to entrap their own child (both my parents were used against me in that way). The Russian Mafia is working as fast as possible to first enslave and destroy Americans, and to replace our government with the Russian Mafia as world ruler in Mr. Putin's lifetime. He is 71 in 2024.

is why I'm charging the Russian Mafia and the Russian government with genocide in this book: that's what they're doing to us. My book is an attempt to stop them. I also offer survival tips which I hope will better protect you against the Russian Mafia.

The Russian Mafia is, arguably, the most powerful mafia our species has ever produced, is a hybrid of the Russian government and the Russian Mafia.[64] At present it is demanding world dominance over all governments. If Americans allow that, the Russian Mafia will destroy us. If we try to form a peace deal with the Russian Mafia, they will destroy us. The Russian Mafia has shrugged off the Geneva conventions.[65][66][67] The USSR and the Russian government have a long history of reneging on agreements. I know because, despite promising to honor the Geneva conventions in 1948, and despite their promises to obey our laws as trade partners in the 1975 Jackson-Vanik Amendment, the USSR's operatives were on the ground in America attacking me via proxies in 1971 when I was 11. They didn't cease attacking me until their demise in 1991, whereupon it's successor government, the Russian Mafia, continued and expanded the attacks. The KGB in the form of the USSR, and the Russian Mafia have attacked me since approximately 1971.

---

64 Karen Dawisha, *Putin's Kleptocracy*, Simon & Schuster (2015).
65 Gen. Gerasimov, Chief of the General Staff of the Russian Federation,"The Value of Science in Prediction," *Military-Industrial Kurier* (2/27/2013).
66 Throughout the book I'll usually refer to the Russian government as the Russian Mafia because they are entwined entities who work together.
67 Karen Dawisha, *Putin's Kleptocracy*, Simon & Schuster (2015).

The Russian Mafia benefits from old KGB psychological warfare attack strategies and the mafia's blackmail and coercion strategies. Mr. Putin, the Don, is the wealthiest man in the history of our species,[68] taking his cut from every crime the Russian Mafia makes, including drug running, money laundering (aka dark money), slavery, gun running, etc.[69]

Active Measures has been a boon, first for the USSR, and later the Russian Mafia because the USSR embedded unknown numbers of spies, operatives and/or slaves here after it got the Jackson-Vanik Amendment passed in 1975. That Amendment, which former President Nixon refused to authorize and subsequently found himself removed from office via the Watergate,[70] was how the USSR seeded hundreds of thousands of spies here. Over 600,000 Soviet immigrants were granted permission to live here, and those immigrants were

---

68 The Russian government and it's Mafia are one of the most successful illegal drug dealer systems in the world; targeting anyone, including children in West aligned nations is part of the Russian government's & it's allies' strategy to destroy us. According to Craig Unger in his book *House of of Trump House of Putin*, "...Putin and his cronies [...] were allowed to make billions through drug trafficking, extortion, elaborate financial schemes, the sex trade, arms deals, and the like...." Mr. Unger also states "...made Vladimir Putin the richest man in the world, with wealth, according to Hermitage Capital Management CEO Bill Browder, approaching $200 billion." Craig Unger, "*House of Trump House of Putin*," (Penguin Random House UK 2018).

69 According to Karen Dawisha in her book *Putin's Kleptocracy* (Simon & Schuster, 2015) "the FSB (Federal Security Services) has 'absorbed' organized crime ..." She also states "....Russia under Putin had become a virtual 'mafia state' in which state structures operate hand in glove with criminal structures to their mutual benefit, ..."

70 I strongly suspect Watergate was a setup by the KGB to oust former President Nixon after he refused to authorize the passage of the Jackson-Vanik Amendment.

first approved by the USSR,[71] and approved by our government, second.

Active Measures is the Russian Mafia's ongoing war against us and is responsible for the deterioration of families, our government, and our nation. It's impossible for ordinary people without espionage and psychological training and other training to suss it out. The only reason I did was because one of the Russian Mafia's American law enforcement operatives grew frustrated as I slipped the noose, leaving the company where they'd laid a trap for me and getting a job elsewhere, largely because the operatives they'd assigned to me were too 'sticky,' were all over me, not sexually, but unnaturally clingy. Had I been a teenager their inexplicable attachment to me wouldn't have worked, I was too busy helping my family, working and trying to get to college. But when I thought through leaving that company my thinking was "it's like I'm head cheerleader in high school, and these grown women are junior cheerleaders, fascinated by everything about me. I'm out." I guess the Russian Mafia was so desperate to get any operative into my life, they had their people focus unnaturally on me. I quit that job to get away from them. And because I had no idea I was being hunted, I enraged the law enforcement operative orchestrating and managing the attack. When he went loud (ordered strangers to attack me), the Russian Mafia

---

71 Craig Unger, *House of Trump House of Putin,*" (Penguin Random House UK 2018) regarding the USSR's exploitation of the Jackson-Vanik Amendment passed in1975 to seed our nation with 600,000 Soviet immigrants, spies, slaves, an unknown number of whom were sworn to aid the USSR.

didn't reign him in. He made a mistake, the Russian Mafia compounded it. They'd been so successful with their 9/11 attack (I believe they facilitated that attack) maybe they were giddy, thinking they own the street here now. That's actually how their American operatives behave with me: they are determined I know they own the street here. When you have an inferiority complex, no matter how successfully you ambush innocent people, you're going to have a problem maintaining self-control because ego gets in the way. The Russian Mafia started celebrating in 2001. Now they're working to Close Active Measures. They're not going to make it. When they spike mass murders here and fentanyl overdoses spike we'll tell our people, warn them, better protect them. The Russian Mafia has laid a lot of evil here and when they don't win in 2024, they'll go spiteful because that's who they are. They're men who are proxy murdering Palestinian babies to seize world dominance. But Americans will now know who's behind proxy attacks against us and why. We'll decide how to respond.

The USSR, and it's successor government, allied with the world class criminal mafia[72] is responsible for the deterioration of our families, our schools, our education systems, our housing system, the drug flows into our country, our medical system, our media, etc.

---

72 Karen Dawisha, *Putin's Kleptocracy,* Simon & Schuster (2015).

Chapter 4.5 - The Russian Mafia takes over from the USSR in the 1990s

The USSR ended in 1991 but it got far along in it's goal of destabilizing us. The subsequent Russian government, working with the Russian Mafia (I call that mafia-government hybrid the Russian Mafia), continued and expanded Active Measures with the intention of being the Closer government of that war. Mr. Putin, President of the Russian government and Don of the Russian Mafia,[73] intends to destroy us on his watch. He is 71 in 2024 and the richest man in the history of our species.[74][75] Given his crime-wealth he'll be able to fight off death, unless he is murdered, let's assume to the age of 100, which gives him another 20+ years to destroy and/or enslave us, and install the Russian Mafia as top dog in the world which is the legacy he intends and is working to leave the Russian Mafia. A world ruled by the Russian Mafia is bad because it is a criminal

---

73 The Russian government and it's Mafia are one of the most successful illegal drug dealer systems in the world; targeting anyone, including children in West aligned nations is part of the Russian government's & it's allies' strategy to destroy us. According to Craig Unger in his book *House of of Trump House of Putin*, "...Putin and his cronies [...] were allowed to make billions through drug trafficking, extortion, elaborate financial schemes, the sex trade, arms deals, and the like…." Mr. Unger also states "...made Vladimir Putin the richest man in the world, with wealth, according to Hermitage Capital Management CEO Bill Browder, approaching $200 billion." Craig Unger, "*House of Trump House of Putin*," (Penguin Random House UK 2018).

74 According to Karen Dawisha in her book *Putin's Kleptocracy* (Simon & Schuster, 2015) "the FSB (Federal Security Services) has 'absorbed' organized crime …" She also states "….Russia under Putin had become a virtual 'mafia state' in which state structures operate hand in glove with criminal structures to their mutual benefit, …"

75 According to Craig Unger in his book *House of of Trump House of Putin* (Penguin Random House UK 2018) As Oleg Kalugin, former head of counterintelligence for the KGB told Mr. Unger…. "the Mafia is one of the branches of the Russian government today."

organization that doesn't value life and will do anything and everything to it's advantage. I believe but can't prove, they ordered the 2023 Israel-Hamas War to suppress President Biden's approval rating and remove him from the White House, to install one of their puppets in 2024. If I'm right, the Russian Mafia is knowingly responsible for the deaths, injuries, and rapes of approximately 40,000 innocent Gaza Palestinian civilians and Israeli civilians in under six months: is another one of their many evil crimes against humanity.

Mr. Putin is very confident that if his government can install a puppet in the White House the Russian Mafia will seize control of the world. They see no down side to trying. They believe they'll be rewarded if they install one of their puppets, and they believe it unlikely we'll risk nuclear war in retaliation against them if they fail. Neither do they believe we'll drop nuclear weapons on them if they succeed as they'll rely on their many puppets to prevent a retaliatory bombing from us on them.

Although they could walk away from this, they won't: they believe they've got us. Mr. Putin and his Mafia are determined to prove to their allies, who are also their slaves, that they're smarter than Americans. Mr. Putin's ego won't let him stop. The Russian Mafia hate us, contrary to their propaganda not because of anything we've done (they hated us long before we put sanctions on them in response to their unprovoked bombing of Ukrainians), but because sometimes some males need a "kingdom on the hill" to hate, scheme against, target, work to destroy, and measure themselves against. The Russian Mafia actually needs someone to hate – that way, they don't have to do the hard work of strengthening their people and building up their nation into the powerhouse it could be. But the Russian Mafia embraces cheating, lying, baby-killing, and destroying it's way to the top. They will murder or try to anyone who calls them out. They are cowards. When grown men murder babies to ascend to world dominance, they are cowards. Because they see people as weapons and pawns, the Russian Mafia won't invest sufficiently in ordinary Russians. They can't see the value of their own people, or the value of any people but themselves. They think because they've ambushed and killed children and families that makes them worthy to rule the world.

The Russian Mafia is not worthy. They lust for dominance over our species while they loathe our species, unable to see the beauty and strength in us. Instead of marveling that our hard-wiring and soft-wiring has kept our species alive when all other human species died out, the Russian Mafia is actively working to murder about half a billion West-allied people, not because we've harmed them but because our government is top dog and the Russian Mafia can't deal. How can a government of the last human species on this planet plan to murder half a billion people, and willingly proxy murder and/or injure 30,000 children and call themselves men. Instead of strengthening their species, they're about destroying our species. They are anti-life. They need to go. They are troubled. Off the rails.

Jealousy and envy runs strong in the Russian Mafia. Instead of the Russian government investing in it's people, and it's country, it developed a psychological weapon to destroy it's enemies, and entrap and blackmail it's allies. A hybrid of the old KGB, the Russian Mafia, and a corrupt Russian government, the Russian Mafia believes they can take the world because they've blackmailed so many of the world's wealthy, and so many ordinary people. I bet most all their victims don't know the Russian Mafia owns them but have been led to believe the legitimate government in whichever country they're attacking does.

The Russian Mafia are responsible for endless suffering, death, and destabilizations of ordinary people around the world. The people coming to America are victims of the Russian Mafia and it's Mafia associates and cartels, deliberately destabilizing and corrupting their nations so ordinary people can't survive. Instead of remaining in their home country with no hope of building a life, immigrants risk their lives to immigrate.

The loss and despair on the faces of Gaza Palestinian civilians cradling their loved ones must not be endlessly perpetuated by this evil group of men daring to demand world domination using the blood of innocent babes as pawns, while blaming an innocent country (America) for this horror. Past time to end the Russian Mafia.

Mr. Putin likes the narrative that he entered the world the child of a factory worker, who leads his country to world domination against overwhelming odds. The Russian Mafia are gamblers, are highly acquisitive, and have a covetous eye on assets of value in other nations, which is currently on display by their attack against Ukrainians. They covet our government's financial power and that the world's currency is the dollar. To the Russian Mafia the world is their oyster – there to serve them. Their sense of entitled is off the charts, even for humans (humans are a young species so we have an excuse for being immature but excuses don't matter on an overheating planet).

The Russian Mafia believes everyone but them is stupid, that they are ordained by god to rule our species, that women are less than men, that people of color are less than the Russian Mafia, and that everyone on this planet is meant to be the Russian Mafia's slave. In the 19th century the Russian government exhibited an inferiority complex, abusing/murdering Soviet Jews and other minorities in and around Russia. Their government has believed they were the master race, just as the Nazi's did, since the 19th century ordinary Russians are just trying to survive and avoid the demon eye.

The Russian Mafia is working very hard to install their puppets into the presidency and into cabinet positions in 2024. We can expect they will continue working very hard to seize control of our government over the next several decades, minimum. This book details many of the attacks I've experienced at the hands of USSR and Russian Mafia controlled American operatives from 11 to my present age of 64. Most of my life I've been attacked by the USSR and later the Russian Mafia. As you can imagine, begin stealth attacked by an enemy foreign government determined to seize control of our country is extremely terrifying. I believe I'll be murdered by them but it is my intention to publish this book in the hope that it will help you, my people worldwide, survive and rid yourselves of the Russian Mafia.

Chapter 4.6 – A coup to the Russian Mafia is them installing one of their many puppets into our Presidency. They deploy denigrating psychological attacks, forcing entrapped dominants to sneer at their children - both my parents were deployed thus

The Russian Mafia's coups are when they install a pro-Kremlin candidate[76] into the targeted nation's top governing positions, and if possible, throughout the targeted nation.[77] You'd never vote in a pro-Kremlin candidate if the Russian Mafia showed you, for example, how they're hooking our teenagers on fentanyl. The following is an example, based in part on my personal experiences being offered drugs by the Russian Mafia,[78] and other types of attacks I've endured over decades,

---

76 Julie Hirschfeld Davis "Trump, at Putin's Side, Questions U.S. Intelligence on 2016 Election," *New York Times* (7/16/2018).
77 Gen. Gerasimov, Chief of the General Staff of the Russian Federation,"The Value of Science in Prediction," *Military-Industrial Kurier* (2/27/2013).
78 The Russian Mafia has enlarged the number of operatives bequeathed to them from the USSR via the Jackson-Vanik Amendment. I've never indicated interest yet Russian Mafia operatives have offered me drugs

and in part based on newspaper articles I've read and video discussions of parents/families destroyed by their children's/sibling's overdose deaths: The Russian Mafia destabilize our children by forcing a child's entrapped parent(s) to denigrate him, make him feel stupid (both my entrapped parents were forced by their Russian Mafia controlled American handlers to denigrate me. That's how I learned this type of attack exists). The Russian Mafia forces compromised American teachers to undermine children's ability to learn by attacking their confidence. For example, when you give students a quiz without the teacher introducing the quiz material beforehand and/or by not teaching the material and giving students the opportunity to complete and turn in the homework to be graded (in other words, testing a student on

_____

and/or alcohol multiple times. In the last five years, drugs of my choice were offered for free (I was on my mobility device, scooting on it using my feet because Russian Mafia's American operatives destroyed my batteries and I couldn't afford to replace them). The second offer was during my aunt and her operative 'friend's' visit. They left a drug behind, instead of throwing it away. I allowed them to leave it not understanding at the time it was yet another setup, thinking that if my aunt returned she might want it. After I understood it was a setup, meant to tempt me to experiment, I threw it away. Never trust the Russian Mafia. Don't take drugs. My aunt's friend operative sexually assaulted me during their visit in the middle of the night. I later told my mother and my aunt. My mother, a victim of child sexual assault, said she didn't believe me, although I've never lied to her in my life, meaning not only have I never lied to her and she's never known me to lie to her, she's never caught me in a lie. She was forced to gas light me because the Russian Mafia are masters of psychology. When victims tell dominants in their life and the dominant refuse to believe them, that's a ding on the subordinate's emotional psyche. The Russian Mafia has used multiple family members to gaslight me. For victims of abuse, part of the healing is when dominants believe them. When dominants believe, victims feel empowerment. The Russian Mafia is pro-slavery and loathes women. The last thing it wants any victim, but especially female victims to feel is empowerment.

unlearned, untaught material.) (I experienced this strategy in two separate classes within the last few years as an adult student, that's how I know this type of attack exists) and students fail, students believe they don't have the intelligence to learn the subject. The "battlefield is the mind," the Russian government says.[79] The Russian Mafia considers children and civilians fair game.[80] Scholastic failure depresses a student's self-confidence, makes him or her more vulnerable to peer pressure and more willing to seek out something to help them feel better. When the Russian Mafia stealth attacks our people, they make sure there are plenty of self-destruction options at hand people can reach to self-sooth.

They will insert a "white knight" into a targeted child's life: a school friend, say, who comes to the child's defense when he or her is being bullied at school. The Russian Mafia never shows the victim that the children-bullies and the child-white-knight are soldiers owned in the Russian Mafia. How do I know? Because I've been attacked since I was 11 by children controlled by the USSR, and later, I was and am harassed by people, including children, controlled by the Russian Mafia. I've been the victim of a white knight romance attack. White Knight romance attacks are brutal because they're keyed to our hard-wiring and soft-wiring both. And, the Russian Mafia

---

79 Janis Berzins "The New Generation of Russian Warfare" "...the Russian view of modern warfare is based on the idea that the main battle space is in the mind." Aspen Institute | Prague, *Aspen Review* (March 2014). https://en.m.wikipedia.org/wiki/new-generation-warfare.

80 Gen. Gerasimov, Chief of the General Staff of the Russian Federation,"The Value of Science in Prediction," *Military-Industrial Kurier* (2/27/2013).

doesn't tell victims the Russian Mafia is running an attack against them. The Russian Mafia is destabilizing us because they've mastered some components of human psychology but their attack has a weakness: to work people must not know the Russian Mafia is attacking then or know how the Russian Mafia is attacking them. Once people are told, in a way they can hear, people better defend themselves. Their attacks are based on our hard-wiring and/or the socializing we received from birth. However, information telling/showing us how we're being attacked will help us launch an effective self-defense, which the Russian Mafia knows and is why they hide their war against us from us.

The Russian Mafia is running a functioning child army in America comprised of American and/or migrant children. The children I've seen in this army are overwhelmingly American children in number, not overwhelmingly immigrant). In my above example, the white knight "friend" gets the victim to try fentanyl, hook him/her, and destroying his/her life for decades to come, if not outright killing the victim with a few too many grams of fentanyl. Overdose deaths are part of the Russian Mafia's destabilization (i.e. destruction of us). They are meant to get grieving families and communities to lose faith in our government so the Russian Mafia can use us to overthrow our government. The Russian Mafia is all about the overthrow of our government and the installation of one of their puppets. I've been 'shown' by decades of harassing attacks by compromised law enforcement on the street that somebody powerful had a problem with me. That powerful entity is the Russian Mafia. Blind, naked ambition for world dominance, and they will kill anyone and destroy anything to achieve it. I am not exaggerating.

If the Russian Mafia showed Americans any part of this example attack and publicly accepted responsibility for it, they'd never get pro-Kremlin, right-leaning candidates in office, so the Russian Mafia doesn't show what they're doing here, but instead show the deterioration, which they then 'American deterioration' rather than as 'an attack by an enemy foreign government.'

We're experiencing a coup attempt. It's been a work in progress for decades. I know because the USSR first attacked me decades ago and continued it's attacks against me until it's demise in 1991, after which the successor government, the Russian Mafia (a hybrid of the Russian government and the Russian Mafia), continued and expanded the attacks.[81]

It's illegal for the Russian Mafia to work a coup against us, our Constitution prohibits it. They're here illegally and know they're here illegally but don't care. They are criminals. Criminals don't care about laws. As far as they're concerned, we have nothing to stop them. We'll see about that. First, we warn our people, all our people, and all the world's people about this attack. Then we establish a shared, accurate reality.

---

81 General Valery Gerasimov, Chief of the General Staff of the Russian Federation,"The Value of Science in Prediction," *Military-Industrial Kurier* (February 27, 2013). General Valery Gerasimov states: "...*Wars are no longer declared....*" "*The very "rules of war" have changed..*"

The Russian Mafia will fight us on this. Then we eliminate the threat.

The Russian Mafia is wrong. We have plenty to stop them but we've had a slow start because (1) the Russian Mafia are flat out liars: they said they wouldn't behave as they're behaving when they agreed to the Geneva Conventions – they lied,[82] and they don't give warning shots, meaning they attack victims without warning them, in other words, they ambush children, women, civilians and families in enemy countries who have no idea they're under attack. People who know they're under attack behave defensively, better protect themselves, which is why the Russian Mafia doesn't warn them;[83] (2) a secret war exploiting our hard-wiring and soft-wiring (socialization) to use as weapons against us are tough to spot, especially since (a) nobody's talking, (b) ordinary people and families are the target, and (c) we have no espionage training, or even (d) know we have hard-wiring, much less the (d) blind spots in our hard-wiring, or (e) we're hard-wired and soft-wired to protect our dominants, making it unlikely that we'll report their corrupting influence since we (f) have no idea they've been enslaved; and (3) our legitimate government officials are targeted in a different way. My entrapped father is retired law enforcement. Presumably he was entrapped in a different way than my mother, his attacker's would've studied

---

82 Geneva conventions signed 1948 & 1949.
83 General Valery Gerasimov, Chief of the General Staff of the Russian Federation, "The Value of Science in Prediction," *Military-Industrial Kurier* (February 27, 2013). General Valery Gerasimov states: *"...Wars are no longer declared...." "The very "rules of war" have changed.."*

him. Neither are free to discuss their entrapment. Neither have told me they're entrapped. I know they are because they've tried to entrap me. Entrapment isn't normal behavior or normal parental behavior unless the perpetrator is raised from childhood to believe lying and cheating is normal. *One of the Russian Mafia's adaptations* of the entrapment system they inherited from the USSR is they now force entrapped parents who accept a plea deal to immediately use their children, at minimum, as beards, when the entrapped parent is forced to harass targets the Russian Mafia wants attacked. I know because I've experienced, rough estimate, 1,000+ of child-inclusive entrapment attack starting when I was 11 and continuing into March 2024. I've endured a great many attacks by children: lots of swarming at bus stops, skateboarding nearly into me, etc. When the child is accompanied by an adult, depending on the age of the child (younger children are more often used as beards to cover the harassment of their parent or to engage me; older children are often used to aid their parent in harassing attacks. As the target I've witnessed both). The attack is altered, perhaps depending on age, or other factors.[84]

The USSR seeded operatives in the hundreds of thousands by getting passed then exploiting the Jackson-Vanik

---

84 General Valery Gerasimov, Chief of the General Staff of the Russian Federation, "The Value of Science in Prediction," *Military-Industrial Kurier* (February 27, 2013). "*... wars are no longer declared...*" "*...The very "rules of war" have changed....*" "*...The role of nonmilitary means of achieving political and strategic goals has grown, and, in many cases, they have exceeded the power of force of weapons in their effectiveness....*" "...Among such actions are *the use of special-operations forces and internal opposition to create a permanently operating front through the entire territory of the enemy state....*"

Amendment,[85] and because they prioritize the use of psychological warfare against civilians, including our children, while refusing to declare war, neither civilians nor our government know the extent to which the Russian Mafia is wrecking havoc, but there's widespread Russian Mafia orchestrated damage throughout our nation.[86] Using my life being attacked by the Russian Mafia as an example.

Given the difficulty millions of us face finding affordable housing, the dysfunction in parts of our government, and our overdose death rates, it's becoming increasingly clear we have a serious problem. This book is to help Americans and people worldwide better understand how the Russian

---

85 Craig Unger, *"House of Trump House of Putin,"* (Penguin Random House UK 2018) regarding the USSR's exploitation of the Jackson-Vanik Amendment passed in 1975 to seed our nation with 600,000 Soviet immigrants, spies, slaves, an unknown number of whom were sworn to aid the USSR.
86 Gen. Gerasimov, Chief of the General Staff of the Russian Federation,"The Value of Science in Prediction," *Military-Industrial Kurier* (2/27/2013).

Mafia's stealth attacks civilians without declaring war.[87888990] This book is to inform the world how the Russian Mafia runs it's coups from the point of view of a citizen victim.

Our legitimate government feels badly that the Russian Mafia got the drop on us but the KGB did the brunt of that last century, probably blackmailing us into Lacy-Zarubin and Jackson-Vanik, long before our present leaders were in office.

---

87 Not long ago, in the last two years, I tried telling one of my teachers about the attacks here and she said, and I quote "I don't follow politics." Teachers are dominants in a teacher-student system. The Russian Mafia prioritizes entrapping dominants: control the dominant, you control the subordinates the dominant controls. Teachers have the ability to impact the students they control. There's a reason the Russian Mafia has acquired China and India's governments: those two leaders control over 3 billion weapons the Russian Mafia now has access to. Just as they had my parents refuse to believe me in the dominant-subordinate parent-adult child relationship, the Russian Mafia weaponize teachers to shut down any high quality intel a student knows by having the teacher refuse to listen to the student. Here's an analogy: storm is far out to sea, in colder waters. As that storm moves over warmer waters, it acquires more energy, more power. By having dominants shut down subordinates, the Russian Mafia keeps the storm always in cold waters, so that it doesn't build up energy and become powerful. Dominants are extremely important in the Russian Mafia's coup. Police, parents, supervisors, landlords, grocery stores, doctors, etc. - all these groups and many I haven't named, are entrapped and used as weapons to entrap and/or control subordinates.

88 Janis Berzins "The New Generation of Russian Warfare" "...the Russian view of modern warfare is based on the idea that the main battle space is in the mind." Aspen Institute | Prague, *Aspen Review* (March 2014). https://en.m.wikipedia.org/wiki/new-generation-warfare.

89 Shaun Walker, Pjotr Sauer and Tom Phillips "Panic and Emotional Pain as Alleged Deep-Cover Russian Spies Vanish," *The Guardian* (April 3, 2023).

90 General Valery Gerasimov, Chief of the General Staff of the Russian Federation,"The Value of Science in Prediction," *Military-Industrial Kurier* (February 27, 2013). *"wars are no longer declared…." "…The very "rules of war" have changed…." "…The role of nonmilitary means of achieving political and strategic goals has grown, and, in many cases, they have exceeded the power of force of weapons in their effectiveness…." "…Among such actions are the use of special-operations forces and internal opposition to create a permanently operating front through the entire territory of the enemy state…."*

Active Measures is an undeclared war our leaders didn't know was aimed at our entire population, including our children. While we must be told how this attack has succeeded against us for so long (that shared knowledge is essential to shore up our defenses), right now we're in survival mode that too many of us are unaware of. This book intends to tell our people and everyone in the world about the Russian Mafia's attack strategies against innocent ordinary people and their governments.

There's no upside for us in being totally ignorant that we're under attack by the Russian Mafia and it's all upside for the Russian Mafia that we not know they're a player on the ground here.

It's not only what we say to warn our people, but who we use to say it. Non-compromised dominants are an essential component. This war is dominant-keyed so our defense will be dominant-keyed.

We must inform our population about this attack and about how the Russian Mafia destroys our families, destabilizes our society, while they declare that we're corrupt [91] without admitting they're forcing our dominants to corrupt us.[92] They are successfully attacking our government, all to

---

91 Coming from the Russian Mafia, that's bizarre. I believe they are one of our species most prolific mass murderers, far worse than the Nazi's.
92 Julie Bosman "Fentanyl Cuts a Bitter Swath Through Milwaukee," *New York Times* (12/12/22). Glenda O. Hampton, Executive Director of Gateway to Change, a drug treatment center in Milwaukee said "I've seen a lot of terrible drugs. This is the worst."

achieve their concept of nirvana, top dog status in the world by destroying us and enslaving everyone else.

Short term solution we must not vote in their puppets – long term solution our people must discuss our options and decide how to respond to the ongoing threat of the Russian Mafia.

Chapter 4.7 – Our defense will save us – it'll be quick and dirty, but effective, even inexpensive. Dominant-subordinate communique are cheap dates. We only needed to know the solution & that's not hard to figure now that we understand the attack

Sometime after WWII, in the 1950s or 1960s, the USSR decided to to destroy Americans. Not just our government, but ordinary Americans, including immigrants living here too. They called their secret war Active Measures.[93][94][95][96] The core idea is build around subversion and ambush, destabilizing, controlling, and destroying individuals, families, entities, and businesses. There is an emphasis on secrecy and stealth, with the desired result being the destruction and/or enslavement of Americans, and the destruction of our government. We've endured many unexplained suicides of ordinary and noted people over the last several decades. One of my beloved bosses committed suicide. I now suspect he was being pressured to harass me. I believe many of these suicides were by our people who refused to act as destabilizing dominants in their families

---

93 https://en.m.wikipedia.org/wiki/Active_measures.
94 https://en-m.wikipedia.org/wiki/unconventional_warfare.
95 https://en.m.wikipedia.org/wiki/Russia_mafia.
96 https://en-m.wikipedia.org/wiki/Cold_War.

and in our country and so chose suicide instead of betrayal. I believe these people are our heroes – refusing to cause harm to their loved ones or co-workers by entrapping them.

Active Measures is in play[97] and is the reason we and our nation have deteriorated the last 50+ years. To begin the destabilizing and criminalization process, the Russian Mafia prioritizes the entrapment of dominants. By weaponizing our dominants the Russian Mafia (1) forces our beloved and/or trusted authority figures to corrupt and entrap us; (2) commits genocide against us by destroying our families, using our trusted authority figures as beards to camouflage the Russian Mafia's stealth infiltration into the lives of noncriminals, including our families and our children, in order to exploit and manipulate subordinates socialized from birth to obey people they've trusted since birth into corruption, drug addiction, and to coerce/influence them to vote into office puppets the Russian Mafia wants installed here, so they can destroy us, while the Russian Mafia publicly crows about the deterioration

---

97 Active measures – the Russian Mafia (1) use our dominants to disable our species' alarm system (eons ago dominants were the pre-people and people more likely to notice and launch an effective defense against non-overt attacks). They reproduced and dominants who didn't recognize non-overt attacks didn't, so we're descendants of a particular intelligence that was hard-wired into us over time – listen to dominants when non-overt threats arise; (2) corrupts and destabilizes a targeted nation making the people cynical and doubtful of their government. The Russian Mafia puts that in play when they force authority figures to corrupt subordinates. I didn't experience this until my 50s and 60s but I still found it shocking; and (3) speeds up entrapment which the Russian Mafia coerce voters they've entrapped and/or influenced to vote for the puppets the Russian Mafia wants installed into power positions. If we let the Russian Mafia steal our country and our lives from us, we won't be coming back from that.

of Americans, which deterioration the Russian Mafia is facilitating here, in the epitome of evil gas lighting. They entrap parents and force them to entrap their own children, and/or the Russian Mafia entraps siblings and force them to entrap non-compromised siblings. I'm a survivor of both these types of attacks which is the only reason I know the Russian Mafia is working them. Nobody told me, I didn't read about them, and I'm not making this up. Forcing family members to entrap each other destroys our families, which are an essential feature of our species, our society, our nation and our government. By destroying our families the Russian Mafia commits genocide against Americans and everyone living in America. Genocide is a war crime. The USSR, the predecessor government to the current Russian government, signed an agreement not to commit genocide in 1949; [98] they lied; and (3) by targeting and enslaving our dominants the Russian Mafia has significantly reduced the number of leaders not in our government, security agencies and/or military, who are not slaves. Where are our noted writers, philosophers, editorialists, and journalists? Why am I the only person I've read accusing the Russian Mafia of yet another proxy war, this one with the intention of unseating our President and installing one of their many puppets into the White House.

---

98 Convention of the Prevention and Punishment of the Crime of Genocide art. 2, 78 (UN.T.56 277, 9 December 1948).

Chapter 4.8 - The Russian Mafia is actively, hence the title Active Measures, committing genocide against us

The Russian Mafia are our species, but they're off the rails. Long before we issued sanctions on them, and remember President Biden tried to talk Mr. Putin out of attacking Ukrainians, warned him of the severity of the sanctions, but Mr. Putin blew him off, going for the brass ring, world dominance. The Russian Mafia believes that what they want trumps what everybody else in the world wants. They've convinced themselves that the Russian Mafia are the chosen people, and that only their wants and needs matter. In a world of 8+ billion people the Russian Mafiosi, a group of under 5 million men, believe they have what it takes to annihilate 400+ million Americans, destroy the economies of our allies who they expect to cower in fear and awe of them, and that we'll just lie down, with no upside for us to do so. We already know the Russian Mafia intends to destroy us. No one expects them to show mercy because they are murdering by proxy, before our eyes, tens of thousands of Palestinian civilians, many thousands of them children. They are showing us who and what they are.

We won't attack Russian civilians, which the Russian Mafia will exploit, but the Russian Mafia, who've been stealth attacking our children, our senior citizens, our families and our government for decades, can expect to receive what they're giving. No matter how they think this will unfold, as Russian General Gerasimov's says, and I paraphrase, 'war evolves in unpredictable ways.' I'm not magical thinking or using bravado. I'm not psychic. The Russian Mafia's success the last couple decades installing pro-Kremlin governments worldwide has drawn attention. People begin to put two and two together and before the Russian Mafia knows it, they've got a world of 4+ billion people aware of their stealth destruction, and with the knowledge to defend themselves against genocide. I call the knowledge and ability to defend yourself a win.

This is the group of people determined to rule our species. If Americans think that the Russian Mafia, who are partially comprised of Soviet Jews, some of whom had to sign off on murdering Israeli Jews, and that that Mafia won't destroy Americans in a long lusted-&-schemed-for-fruition of a secret war called Active Measures the KGB set in play about 65+ years ago, those Americans are "magical thinking." This is our Holocaust. Fortunately for us, we know how not to deal with the Nazi's aka the Russian Mafia. We know what we can expect from them as we watch them proxy-slaughter Palestinians babies in order to seize world dominance. This is a black-versus-white war. The Russian Mafia has nothing to say except "I could've been a contender." It's up to us to accept their "leadership." They've got nothing on offer. Baby killers have nothing to offer.

The Russian Mafia has been working Active Measures since the 1990s, when they came to power in Russia, and before them the KGB. Active Measure's isn't in response to sanctions we placed on them in response to their unprovoked attacks against Ukrainians. The Russian Mafia is bad seed, an example of the worse our species can be.

Far more evil than the Nazi's. The Nazi's didn't enslave Jewish parents and force them to lead their children into the gas chambers. In America, the Russian Mafia entraps American parents and force those parents to enslave their own children. That's why when you see photos of Mr. Putin he's so often smirking. He's slave owner of millions of Americans and that's his idea of happiness.

The Russian Mafia intends to destroy us and will do so unless we launch an effective defense against Active Measures. Then our people must decide how to deal with the quality of threat they are to us, and to the rest of the world. If I'm right about them starting the Israel-Hamas War to unseat our President, that they've aimed their protesters at us to blame us, knowing the Russian Mafia is responsible, they will spend the next decades destroying innocent civilians in their tantrum and insistence on being top dog, despite it being obvious that they aren't worthy to lead any species.

Non-dominants, which most people are, can try to warn the tribe from now until eternity about non-overt threats, but the tribe won't see the problem until the threat is overt because our hard-wired blind spots were created eons ago. Dominants in pre-homo sapiens societies were able to decern non-overt threats and helped their people survive so often that we're descendants of those people – they are our ancestors who survived long enough to reproduce. Evolution linked dominants to non-overt threat alarms and effective defense because in our pre-human ancestors' societies, dominants functioned as our de facto alarm system for handling non-overt threats and we're their descendants meaning that people who survived long enough to reproduce tended to listen to the advice from dominants (parents, wise women/men, grandparents). That tendency to listen to dominants is in us, dormant until needed.

We didn't have language or writing to pass this information to every generation so evolution tacked it to hierarchy in our species. We're hard-wired to listen to dominants. Evolution piggybacked non-overt threats onto our hierarchical system because non-overt threats happened often enough that we needed help to survive them. Dominants are top dog via our hard-wiring so evolution used our hierarchy to help us survive, and packed the additional punch that we listen to our dominant's warnings, particularly with regard to non-overt threats.

Chapter 4.9 – Active Measures is designed to ensure we're unprepared for attack & blame our outstanding Administration when this attack is nobody's fault but the Russian Mafia

Throughout our species history, and throughout our long-dead ancestors' histories, dominants were relied upon to sound the alarm against non-overt threats. The Russian Mafia is working a multi-generational, non-overt coup, the very kind dominants in our species and in our ancestors histories were designed to spot, raise an alarm and help us defend against. So. Where are our dominants sounding the alarm? The Russian Mafia has enslaved many of them. I know that because my parents, some of my teachers, a doctor, landlord, the dominants in my life are compromised. For decades the Russian Mafia tried but failed to entrap me, but it's succeeded in an unknowable number of victims. Dominants are targeted and attacked in a secret war our people have no idea is occurring. The Russian Mafia has flipped many American dominants into destroying their own nation, like one of those autoimmune diseases that attacks the host's body.

Not only is the Russian Mafia actively working to destroy us, our culture, and our families, it is determined to disable our species alarm system so there are many fewer dominants who are free to raise the alarm about their attack. Dominants help our species sustain our culture – long before our ancestors created language and writing, dominants held our culture and communicated our history to us – the oral tradition of telling stories was a way our dominants (elders) passed life strategies to us and practical knowledge. The Russian Mafia has worked unrelentingly for decades to try to entrap me. The last overt entrapment attempt against me was this past summer 2023 when a general practitioner attempted to get me to lie to Medicare to get access to one of the weight loss drugs. The weight loss drugs are only available to diabetics on Medicare and I'm pre-diabetic. The doctor offered, unprompted, to tell Medicare I'm diabetic to help me receive the weight loss drug.

I declined his offer. When a professional in the course of doing their job offers to lie for you in order to "help you out," red flag, that professional is entrapped by the Russian Mafia and the Russian Mafia is working to entrap you. You do not want to be entrapped by the Russian Mafia, not by anyone but especially not by them. They force people to throw their own children and their own country under the bus, by order of the Russian Mafia doing business as our legitimate government by camouflaging itself in some of our compromised American law enforcement's uniforms or identifications. Endless lies, and endless cons.

Another nurse at a different hospital, where I was taken against my will when I suffered a medical emergency in 2023, encouraged me to ask the doctor if I qualified for the weight loss drug. And the ER doctor at that hospital urged me to make a followup appointment with a general practitioner. The Russian Mafia has quadruple teamed me before. It is widely underestimated how many slaves the Russian Mafia controls here, all entrapped, all believing they're entrapped by our legitimate government, who has no idea the Russian Mafia is running these scams in their name.

The problem was getting people to talk to our legitimate government, impossible to do because people are entrapped. It so helps that they're out here, defending us against the Russian Mafia. Without them we'd be toast. They see their mistakes, but the Russian Mafia can play nearly anyone. I only know what I know because they embraced dominance warfare so completely, & feel so entitled, they thought my status so low that nobody would believe me. Fortunately we're an adaptive species. Low status people can be vetted just like anybody else. All I had to do was tell the truth. I guess the Russian Mafia didn't consider that we'd notice their success installing pro-Kremlin governments and try to adapt.

Chapter 4.10 – Anatomy of a psychological warfare-based coup: <u>The Problem</u> - the Russian Mafia has planted a secret, highly effective, psychological warfare-based bomb in Americans called Active Measures, they've set to detonate November 2024. <u>The Solution</u>: launch an effective defense which requires we understand the intricacies of their attack, and deploy our non-compromised dominants in an information-sharing blitz to help our people evade this non-overt, multi-generational bomb

The Russian Mafia, the most powerful mafia in the history of our species, loathes us and intends to destroy us. Their preferred method of annihilating us is via coup because they strongly prefer to not be bombed – a hot war with us would be a bombing war, probably nuclear. Although the Russian Mafia is determined to destroy us,[99] and despite their infinite number of lies are the unprovoked aggressor in this undeclared war, they don't believe they can survive a nuclear bomb dropped on them. They're determined to destroy us and they don't consider

---

99 The Russian Mafia are the drivers of this world domination flip. Because Active Measures has been extremely successful in their destabilization of us, they believe they can destroy us with minimal risk of bombing to them. They will not stop working to install puppets into positions throughout our country unless we destroy them.

it a win if they're dead, so a hot war between us and them is out from their point of view at this time. No matter their 'nuclear referencing' comments, they're very heavily invested in our 2024 Presidential election and are betting on Active Measures paying off for them in November 2024. After which they expect to install one of their candidates into the U.S. Presidency. We have no choice but to stop them installing one of their puppets into our Presidency but we also must understand they'll throw a huge tantrum if they fail, where they'll kill a great many of us using their proxies, drug flows, mass murders, mass shootings and by ordering their American operatives to kill as many people here as they can. As long as the Russian Mafia has deniability that's all they'll care about when our President remains in office. They're big on deniability. Or they were: they might not care so much after this book blows their cover.

If we allow them to insert one of their puppets into the Presidency, they will annihilate us. That is their intention. I am 100% positive after over 53 years being attacked by the KGB, FSB, and the Russian Mafia that we have zero chance of survival if we let them steal our Presidency and our nuclear launch capability. They believe they have us. They are 95% confident. My book has rattled them but other than killing me, which they've told me they'll do, and which I accept as the price for helping my people survive, and helping innocent people worldwide not see their babies bombed by this evilest of Mafia's, I look forward to helping my nation build our effective defense and am 100% confident we'll disable their bomb.

The Russian Mafia could continue to stealth control our people but Active Measures was never about anything but the Russian Mafia destroying us and flipping the world order, with them controlling everyone. They *need* the world to see them as the big cheese even though they hate all people, and would kill a person just as soon as look at them. The Russian Mafia has an inferiority complex that's fundamental to who they are. Their sense of inferiority is what is driving them and we can't fix that problem.

I refuse to let them slaughter us because they feel inferior to us. Negotiating with them will only get us killed if we believe anything they say.

They inherited Active Measures from the USSR and seeing that the USSR had gotten far along in destabilizing us, decided to continue and expand the war to include: (1) Fentanyl drug flows into our nation,[100] (2) getting us hooked on pain medications and prescription opioids, (3) entrapping our

_____

100 They consider it a win when they stealth murder one of our people who they prevent from becoming a soldier their soldiers would have to face when the Russian Mafia expects to land their military here. Murdering us using fentanyl cost them pennies per dead American, and involve no risk to them of being bombed by us. They consider fentanyl a highly successful strategic success, a win-win for them. Because the precursors flow from China they have deniability and at this stage of their coup, they don't want to have obviously dirty hands before Americans because they expect Americans to welcome them, just as they expected Ukrainians to welcome them. They've lied so much, spread so much propaganda, and own so many slaves in America, they've convinced themselves (and their American slaves in the know tell them what they want to hear), that they'll be rescuing us. It's easy even for professional liars to get lost in the lies.

dominants (parents, teachers, doctors, police, government and military leaders, landlords, supervisors, etc. to unleashing reconfigured Russian Mafia-controlled "leaders" into our society to corrupt and destabilize our people), (4) destabilizing our families, children, and students, (5) pushing entrapped people into religion or sex relationships where they're encouraged to blindly follow a compromised influencer/spouse/parent, (6) eliminating affordable housing to foment societal unrest and prepare us to overthrow our legitimate government, (7) promoting gerrymandering nationwide to install pro-Kremlin candidates, and (8) destabilizing our government – easily seen in our dysfunctional House GOP. These are just a few of the Russian Mafia's many stealth attacks against us from 1971 to the present. I suspect the USSR was here wrecking havoc before 1971 but they attacked me in approximately 1971 when I was 11, so that's the starting date I'm using.

Chapter 4.11 – The Russian Mafia works their coup

The Russian Mafia is in the process of installing a series of pro-Kremlin American puppets into the White House. If successful they will force their puppets to destroy us and our government. Because such an attack has never happened to modern Americans most of us have no idea such an attack is possible. In 1932 German Jews had no idea that Nazi's would take office the next year and attempt to annihilate them. Why we should be concerned is I believe the Russian Mafia ordered the 2023 Israel-Hamas War to unseat President Biden, uncaring that thousands of innocent Palestinian civilians, innocent Israeli civilians, and/or some number of innocent foreign born workers in Israel, would be killed, injured and/or raped. If I'm right, and I believe I am, the Russian Mafia could care less about murdering Palestinian children as long as they can use the blood of children to unseat a political rival. The Russian Mafia doesn't hate Palestinians or Israelis but they hate us. If they're willing to facilitate the slaughter of babies to ascend to world domination, that's a mafia who is a threat to our species.

A short term solution for us is we refuse to vote for their puppets. To achieve that we must (1) explain to our people how this type of weapon works, and why it's such a serious threat that our lives and our freedom are at stake; and (2) because this bomb is keyed to our hard-wiring, dominants must do the warning.

Chapter 4.12 – Our ancient ancestors were hard-wired to hear dominants warning of non-overt threats eons ago

Over many thousands of years, some dominants tended to become the wealthiest/more knowledgeable of our ancestors, and our infants became hard-wired to assign elevated value to the dominants who cared for them in their infancy and/or to people who were introduced to them as part of their family and trusted circle during their childhood.

Modern humans are hard-wired to hear only dominants when people warn of non-overt threats, but this changes when the threat becomes overt - then nearly everybody can see the problem. In the case of the Russian Mafia, they already have their ducks in a row when their attack is overt to everyone. We have to fight them before they've destroyed us. Waiting until they're overt, in our face, means we'll have a very serious problem because they'll control the weapons and those wielding the weapons. They already know the law enforcement who will cause them a problem and they've already eliminated them in a variety of ways, or, after they go loud so everyone see's them, they'll eliminate leaders they know will cause them problems.

Most of us don't understand they will install one of their many puppets in the White House if we don't stop them and what that actually means for us: death. Just as the Nazi's coming to power in 1933 meant death for German Jews, other Jews, gypsies, homosexuals, and disabled people, any pro-Kremlin candidate in our country, but especially one in the White House means death for us.

This is a zero-sum game that's not a game.

Chapter 4.13 - We're in a re-imagining of WWII. This time Americans are German Jews and the Russian Mafia are the Nazis. We're pre-Holocaust

Like the Nazis, the Russian Mafia isn't broadcasting their plans for us but they're giving plenty of clues. One clue they mean us harm: they're deeply embedded here, which our House GOP dysfunction displays, but mostly they're destabilizing ordinary people and our people and our government must be warned.

Another clue: their protester operatives blame us and our government for the Palestinian's suffering. If I'm correct that the Russian Mafia ordered the Israel-Hamas War, the Russian Mafia is, apparently, willing to continue the murder of innocent children and their families until November 2024. If we let them. We'll have to find a way to stop them. If I'm right about the Russian Mafia's hand in this war, and I believe I am, the Russian Mafia needs killing.

Before German and Polish and other Jews in the area, as well as disabled people and homosexuals were gassed, many Jews, despite their centuries of oppression, had no idea the Nazi's would attack them in the way that they did. Victims never see the Russian Mafia coming. The Russian Mafia lives to ambush, loves to attack civilians. They need world dominance. They can't bear life without world dominance. When they don't get it, they create problems for their species. Nazi's, too, prioritized stealth against their innocent victims. Modern Americans can't conceive they can suddenly find themselves targets in their own nations, but German Jews attacked in WWII were German natives and suddenly a hostile anti-Jewish party took power and stripped them of their rights. Even General Gerasimov in his 2013 paper discussing the Russian Mafia's war strategies tells his audience (I'm paraphrasing him) 'a seemingly stable nation can be flipped into chaos in weeks, days.'[101] I'm sure he was talking about us.

---

101 General Valery Gerasimov, Chief of the General Staff of the Russian Federation,"The Value of Science in Prediction," *Military-Industrial Kurier* (February 27, 2013).

All the Russian Mafia has to do is install one of it's many puppets here and coordinating with their other pro-Kremlin individuals and power players, figure a way to overthrow our government. With regards to the Russian Mafia, hoping for the best from them is not an option. They're mass murderers and mafia. Expecting mercy from people who, if I'm correct, knowingly slaughter Palestinian children is stick-your-head-in-the-sand magical thinking. I'm positive my President, our security agencies, our military and many of our allies have few illusions about the Russian Mafia.

Something was going on in that part of the world in 1930s Germany, and in 19th to 21st century Russia, where when a certain type of man gets into power, he thinks he's all that and feels the need to destroy millions of people. Mr. Putin and the Russian Mafia have an unquenchable, unrelenting need, not a want, a need, to dominate the world and destroy West allied people and our governments.

They need to enslave, punish and destroy Americans who barely pay attention to them because we're busy trying to survive in a country the Russian Mafia has stealthily increasingly made hostile. They're a jealousy and hatred-based group of men who see something they want that other people have, and attempt to steal it or destroy it. When they can't steal it, they kill or destroy the people who had it, then they steal it. They are Mafia. The hatred is all on the Russian Mafia's side. They have a serious inferiority complex going back a century, at least. No sane group feels the need to destroy hundreds of millions of people not bothering them,[102] in order to seize world dominance because the Russian Mafia won't bother to retrain 400 million people to accept slavery. The Russian Mafia does not value life, not our lives, not anyone's life but their own. To them, ordinary people living in America, are the enemy. Our government is the enemy and the people who've benefit from our government are the enemy.

The Russian Mafia is an entitled slaver mafia. It enslaves and traffics women, children and civilians even now. Most of it's current victims outside of America (in America the Russian Mafia is stealth destabilizing everyone), are people of color. By flipping the world order, it intends to enslave and destroy even more people. If you don't believe me, and if we don't act, and you live another twenty years, the Russian Mafia themselves will bear out my predictions. I accurately predict them because they've attacked me over 53 years of my 64 years of life. While our security agencies know more about the Russian Mafia than I'll ever know, I know the Russian Mafia in a way our security agencies don't: I know their cruelty to children on the ground. I know, 100%, they're willing to slaughter us and will do so unless we defend ourselves.

Would you do that, if you could? Would you start a proxy war using Palestinian babies as blood pawns, watch

---

102 President Biden pleaded with Mr. Putin to not attack Ukrainians, warned him about the damage of the sanctions. Our President didn't know the Russian Mafia expects to seize control of us this presidential election cycle. But. We have to allow them to do that. If we refuse, the Russian Mafia has a different scenario than the slaughter of us they're planning.

thousands upon thousands of them be murdered in order to unseat a political rival and flip the world order? I think 99.99% of people wouldn't do it, and .01% would, and that number includes the Russian Mafia and some of their allies.

Their lust for power at any cost is unacceptable.

The KGB put Active Measures in play after they stole our nation's nuclear weapon technology to build their our nuclear weapons. Continued it after we'd agreed to share much of our technology with them in the Lacy-Zarubin Agreement in 1958, continued it after we normalized trade relations with them in the Jackson-Vanik Amendment in 1975. In these agreements, they agreed, presumably, to obey our laws, but they lied straight-faced without blushing, already dreaming of the day they'd have annihilated us. There's a reason one of China's spokespeople publicly called us ghosts in 2023 after we accused China's government of illegal spying on us using one of their balloons: ghosts is what the Russian Mafia has told their closest allies that we are. By the time the USSR had signed onto Jackson-Vanik in 1975, they'd made me homeless and had my mother beat me, something she'd never done, because the KGB wanted me beaten and mom's handlers forced her to, probably under threat of imprisonment if she refused to honor the plea deal I believe she was coerced into accepting, and probably by threatening her that they'd place her children into foster care because she wasn't following the KGB's orders. In foster care mom's children would've most assuredly have been raped by compromised children and adults in the foster care system. So she beat me, and by doing so saved me and my siblings from being raped. But I left home two years later, traumatized by the unearned beating, my leaving allowing the KGB to destroy one of my loved ones. I did not know it until decades later, as I didn't learn the Russian Mafia was my, my nation's, and innocent people everywhere enemy until 2021, less than 3 years ago. To the best of my ability I began a company and did what I could to acquire the knowledge and assets to write this book. The Russian Mafia has stopped me every step of the way. We're blessed that my government and some of our allies' governments are working to help us survive. They're working largely blind because they don't know exactly how the Russian Mafia's attacks ordinary people and families on the ground. The Russian Mafia hides these attacks from our legitimate governments.

The Russian Mafia wants all children, all women, all people of color, all Caucasians, all Americans, and all immigrants living in America beaten and destroyed. The Russian Mafia is insane. They've got a 19th century imperialist's mindset in a 21st century world, and, to repeat myself, they think they're smarter than everyone else because they inherited a secret war already in play that weaponizes our hard-wiring against us. All they pretty much had to do was expand it and be evil enough to kill anybody not them. They have no problem killing anybody not them.

The only reason they've nearly destroyed us is because they lie by default, ambush everyone, and exploit our hard-wiring. Most of us know nothing about our hard-wiring or about our ancient ancestor's evolution, so we don't know enough about the life form we are. That must change or the Russian Mafia will slaughter us. They are the worse of our species, unworthy to be called men. Babykillers.

Their extreme inferiority complex is part of the reason they spent the 19th century attacking Soviet Jews and other minorities. Something about not being seen in the world as the top dog government has driven an already psychologically unstable government, insane. Their lust for world dominance is unquenchable. The Russian Mafia is a threat to earth habitability because they'll force all governments capable of self-defense to never give up the military advantage of petroleum-based products since the Russian Mafia won't. An insane Mafia with nuclear weapons. Fortunately, they don't want to be bombed but they've stealth enslaved many of our people and, I suspect, plenty of their allies' government officials, and have many evil tricks up their sleeve, trick which they will deploy when they find we've prevented them from installing one of their puppets. I'm 100% positive our people will refuse slavery and the Russian Mafia is 100% positive we will too, so they've already begun to destroy us by stealth destroying our families.

But it will help our people and the world's people psychologically and emotionally to understand that it is the Russian Mafia who has destroyed our families, who has ordered the murder and/or injury of many thousands of

Palestinian civilians and Israelis, and who is in the process of destroying our nations.

We must find ways to kick the Russian Mafia out of our nation and we must understand the nature of their multi-generational, multi-decades long non-overt attacks. Telling our people about how the Russian Mafia has run stealth entrapment attacks against noncriminals here for decades is a start. We must find a way to become a manufacturing powerhouse again, help other nations who need our help build their inner manufacturer to strengthen their people and their economies, and develop better defenses to protect our people and our infrastructure. There's a reason the Russian Mafia has stealth attacked us. They live to ambush. If we experience a hot war with them, we can expect they'll attack our water systems, our grid (so no cooling in summer heat and humidity and no warmth in the winter), our food production systems, etc., in ways similar but worse than what they've displayed against Ukrainians.

The only reason they've not bombed us is because we've got bombs and we'd retaliate. Because they're determined to destroy us, they're attacking innocent families a world away who've done nothing to them. They made my mother cry, made her give up her first dog, made her lose her only home. They took an amazing young mother and incapacitated her and then made her beat her child for no reason. As recently as this summer they used my mother who is in her 80s, to try to fish for information to see if I'd go for the palm-reader okey-doke they used to entrap my mother nearly 55 years ago. My mother believes her handlers issuing her orders are American law enforcement because the Russian Mafia flat out lies and couldn't care less about lying to women, or anyone, and use compromised American law enforcement as camouflage their attacks. To them, my mother is a pawn and a slave. That's it. Alive only because they intend to use her to compromise me.

The Russian Mafia doesn't see people. They see pawns, people as weapons. If I'm right about their proxy war in Gaza, they don't see 15,000 children dead because of them, they couldn't care less. All they care about is world domination and to them, any number of deaths of people of color's babies is worth it to get them what they want.

They're not worthy of the title men, or even the title of people. They're not men or people as we define them. I'm not saying they're subhuman. They are human. I'm saying they've drunk the "we are the master race" kool-aide and because of their success continuing Active Measures, a secret genocide war deployed by the USSR attacking our hard-wiring, the Russian Mafia believes they've found a way to flip the world order. They don't care how many people they murder as long as those people aren't them. The Russian Mafia simply does not care that they'll kill half a billon people and that those people will be us.

They'll always deny this is their intention but one of the only truths they've ever told us is 'we want to flip the world order.' Their lies will sound like the truth. They'll have a veneer of the truth, but if won't be the truth as far as Americans' future and the Russian Mafia are concerned. The truth is they believe they can and will flip the world order, and they see no downside to trying. They don't believe we'll bomb them for trying to install one of their puppets into our Presidency.

We're going to have to rethink under what conditions we're willing to bomb the Russian Mafia. Currently, the Russian Mafia is rewarded for continuing to try to install one of their puppets into the U.S. Presidency and throughout power positions and ordinary businesses and entities in our country, because they receive no negative feedback (bombing) when they try, and continuing to try offers the promise of success. World dominance is their nirvana, it's all they care about, and being alive to enjoy world dominance, so continuing to try to install puppets throughout our nation promises nirvana. If your nirvana is world dominance and no one stops you from trying for it, and you don't care how many people you murder and there's no downside for how many attempts you make, you'd try for it as long as you lived, right? That's the Russian Mafia's mindset. Criminals don't care that they're breaking the law.

The Russian Mafia will say that the loser always has a case of sour grapes and American's slaughtered would, obviously, devastate people living in America but this isn't a traditional war where people are fighting over resources. This is a genocide, the deliberate destruction of all things American.

No matter what lies the Russian Mafia tells Americans they currently own, it's just more lies to get them to acquiesce to the initial stages of the Russian Mafia's genocide against us.

Mr. Putin is Don of the Russian Mafia, head guy and/or part of a group of head guys, the group of other head guys you don't hear about publicly. I don't know the head guy's names but I know they exist. The Russian Mafia consists of some Jews in Mr. Putin's inner circle and I believe that at least some of those Jews had to sign off on Hamas' attack against Israeli civilians on 10/7/2023. Mr. Putin is not Jewish, and is extremely racist, sexist, and entitled; he'll never admit it publicly unless he seizes world dominance. I don't know how many powerful Soviet Jews are in the Russian Mafia's inner circle, but we can assume anyone in Mr. Putin's inner circle fears him. I'm positive he ordered the 2023 Israel-Hamas War as a proxy war to unseat our President. Thousands of innocent Israeli Jews were injured or killed, some of them raped. I believe at least some Soviet Jews signed off on the attack. When Jews across nationalities can be manipulated to kill other Jews, we have a problem. Because Jews have a long, long history of oppression in much of the world, and Jews are educated to know about their oppression. When Caucasian Soviet Jews are willing to authorize the attack, injury, murder, rape and/or oppression of Caucasian Israeli Jews, there is a problem. Any individual can be manipulated to betray their group and I believe the Russian Mafia has got psychological warfare so dialed in that Jewish criminals chose the Bratva

(Russian Mafia) over their culture. What this means is, the Russian Mafia will have little problem manipulating Americans to work with them as the Russian Mafia continues to stealth destabilize us. Those who work with them will be promised survival and a good quality of life.

Caucasian Jews authorizing attacks on Jews who are people of color is not new. Our species has racism issues perpetuated by, you guessed it, the KGB and later the Russian Mafia who stealth seized control of parts of our media, movies, etc., and by perpetuating racism on the ground in the lives of ordinary people. Caucasian Jews' racism against Jewish people of color is nothing new. Caucasian Jews authorizing the murder of other Caucasian Jews is a big flashing red light.

There's nothing we can say to convince the Russian Mafia to stop them attacking us. We must effectively defend ourselves against Active Measures.

Chapter 5 - A dominant's work in our species and why the USSR and the Russian Mafia has worked since last century to enslave them

As a refresher, modern dominants are ordinary people who have societal-approved authority, power, or control over others under certain circumstances, or they're workers who help provide an essential service in society. Examples include parents in a parent-child relationship, teachers in a teacher-student relationship, landlords in the landlord-tenant relationship, generals in a general-soldier relationship, doctors in a doctor-patient relationship, police in a police-citizen relationship, employers in a employer-employee relationship, bus driver in a bus driver-passenger relationship, etc. Essential service providers include grocery stores, utilities, water and sewage departments, etc. Dominants and subordinates are us, members of our group.

When a non-overt threat occurs and dominants don't raise an alarm, the people are destroyed. That's why the Russian Mafia, and the KGB before it targeted our dominants for enslavement for decades. I can't tell you exactly when my mother was enslaved but she was destabilized when she was 29. She lost her job and her home. I was around 12. When I was 15 she beat me, something she'd never done, and she did it for no reason, I'd done nothing wrong. She created a scenario and I was confused, and protested the beating but certainly didn't retaliate against my beloved mother. Apparently the KGB was hoping I'd retaliate. My theory is her handler's forced her to attack me, and probably threatened her with prison for reneging on her plea agreement and threatened foster care for her children, which threat I believe worked in 1972 and in 1975. Around that time the KGB had been trying to get me beaten up by a group of three girls in high school they controlled but because I was often studying, and/or I believe I may have had a part time job around that time, and when I wasn't studying or working, I had chores at home to complete, and I helped my family in other ways. The KGB was unable to get me beaten up by the teen operative proxies they owned in high school, so they had my mother beat me. Her beating me for no reason when I was 15 was the reason I left home as soon as I could when I was 17. When I left home, that left one of my loved ones vulnerable and unprotected. I didn't know the KGB was an aggressor in my life so I had no way of countering their attacks against me and my family. By writing this book, I hope to prevent other innocent families from enduring the attacks the Russian Mafia waged against my innocent family.

Chapter 5.1 – The KGB built Active Measures to exploit the chink in our hard-wiring: we're hard-wired to hear _only_ dominants' warning of non-overt threats because when this part of our hard-wiring was established millennia ago, the then-dominants were the part of our group most likely able to spot the non-overt threats

This blind spot where we ignore subordinate's warnings of non-overt threats is the reason the Russian Mafia and before it the KGB began to attack and destabilize dominants. To ensure that there are many fewer dominants publicly sounding the alarm about the Russian Mafia's non-overt coup they're working against us. Subordinates like me sounding the alarm doesn't work to activate our species' defense mechanism: to our hard-wired "hearing," it's like water off a duck's back. The Russian Mafia has enslaved or stealth destroyed as many dominants as they could since 1971 (probably even earlier), the year they entered my life and stealth attacked my family.

Fewer non-compromised dominants mean fewer leaders Americans have to rely on to warn us of this threat. So far, after months of the Israel-Hamas War, I've read no one but me saying this is a proxy war by the Russian Mafia to unseat our President and install one of their many puppets here. That's so obviously the case, is evident even in the war's design, aka how the Russian Mafia unfurled the war to hit key components of our alliance's legal agreement with the Israeli government.[103]

The silence is ominous. I'm sure I'm right that this is a proxy war but apparently the Russian Mafia controls so many of our leaders (dominants) here that, outside of the Biden Administration, our security agencies, and our military (who are contractually controlled by our agreement with the Israeli government), most ordinary dominants are prevented from publicly accusing the Russian Mafia. Just as Soviet Jews authorizing Hamas to kill Israeli Jews is a flashing red light

---

103 I haven't read the U.S.-Israeli defense agreement, parts of it, the parts President Biden and the Biden Administration aren't publicly discussing, are confidential. But I was a legal secretary for 20+ years and I've been attacked by the Russian Mafia for decades. None of the people they've used to attack me are allowed to tell me they're entrapped, and have accepted a plea deal. I know 100% President Biden would never willingly allow the Israeli government to dumb bomb babies unless he absolutely had no choice. He's genuinely empathetic. The Russian Mafia loves torturing him, his Administration and our legitimate government with their murders of innocent Palestinians. The Russian Mafia are masters of psychological warfare. This war is keyed to our hard-wiring, so we have many fewer dominants sounding the alarm, but it's also keyed to our wonderful President and his Administration's hatred of innocent people being attacked. Gaza Palestinians are being used a cannon fodder, blood pawns by the evil Russian Mafia, and they couldn't care less who knows it as long as those people aren't publicly declaring it. In this book, I'm publicly declaring it. I'm willing to die to publicly declare that the Russian Mafia is murdering innocent babies, understanding that the Russian Mafia has no problem killing me. I'm not a martyr, I just refuse to be bullied. By anyone. There are millions of people like me. Journalists worldwide have been attacked by the Russian Mafia and their allies to shut them up.

that we have a problem, no American dominants I've read sounding the alarm that the Russian Mafia ordered this proxy war to destroy us is another flashing red light that something is seriously wrong. Where are our noted writers? Philosophers? Ethicists? Patriots?

Chapter 5.2 – From manipulating Hamas (who are mostly Palestinians) into continuing a war that's seen tens of thousands of innocent Palestinian children killed and/or injured, to forcing at least some Caucasian Soviet Jews in the Russian Mafia into approving Hamas' 10/7/2023 attack against Caucasian Israeli Jews, the Russian Mafia demonstrates it's mastery of psychology as a weapon to coerce individuals into doing what it wants – we'd be very foolish to ignore this mastery as the Russian Mafia works to Close their coup against Americans in November 2024

This war bears so many of the hallmarks of how the Russian Mafia's stealth attacks the innocent: (1) large numbers of innocent people are killed (Ukrainians/Palestinians/Israelis); (2) no one blames the Russian Mafia; (3) Americans are blamed for the war (in Ukraine we're blamed for daring to support Ukrainians; in Gaza for the excessive deaths of Palestinians despite the fact that our Administration has fought since the Israeli military began using dumb bombs to indiscriminately bomb Gaza Palestinians to stop them from doing so. Before Israel entered Gaza we did our best to slow them down and make them consider their end-game, but the Israeli government is right-leaning and were hot to kill as many Hamas as possible, irrespective of the deaths of innocent

Palestinian civilians. I believe that had the right-leaning Israeli government wanted only to target Hamas, they would've moved Gaza Palestinian civilians out of harm's way in whatever way they could manage).

The Israeli government doesn't seem to get that the Russian Mafia is determined to add them to their coterie of weapons to destroy America, that they're being used by the Russian Mafia to try to destroy us. If an allied government is this clueless, this uncaring about the fate of one of their allies, and if they keep being this easily played by the Russian Mafia, the Russian Mafia will eat them for lunch. They're not stupid, but too many of the right-leaning government apparently believe the Russian Mafia is their "friend," and is "helping" them exit the U.S.-Israeli alliance some right-leaning Israeli's want destroyed, like the Russian Mafia wants our alliance with the Israeli government destroyed. When the Russian Mafia wants something, it's always to the disadvantage of the other party and always to the Russian Mafia's advantage. If I were Israeli, I'd push as hard as possible to prevent Israeli civilians from going pro-Kremlin but Israeli's must understand, the Russian Mafia is deeply embedded there and has no intention of letting them slip away. To escape, the Israeli people must work around Russian Mafia controlled Israelis and I can tell you from personal, direct experience, that's extremely hard to do if you don't know the Russian Mafia is on the ground in your country working a coup.

Yes the Israeli people can go pro-Kremlin but remember: to unseat their rival, the Russian Mafia orchestrated the murder and/or injury of tens of thousands of Palestinian children who had nothing to do with Hamas' attack, and is continuing to murder those children. Some Israeli people may think that's okay, enraged as they are by Hamas' murder of their loved ones, but they need to remember that at least some Soviet Jews in the Russian Mafia signed off on Hamas murdering Israeli Jews, and they need to understand that Palestinian civilian's aren't the problem, despite Hamas being largely comprised of radicalized Palestinians. The Russian Mafia is who is stalking them. If Israeli's go pro-Kremlin, I can tell them from direct personal experience of the Russian Mafia and the KGB, that they will regret it. And they need to know something else: the Russian Mafia actually are mafia, probably the world's most effective mafia. Once you take a deal with the mafia your hands are as dirty as their hands. Meaning that you won't be allowed to sit on the sidelines when obviously immoral choices are the ones on offer. For example, I doubt most Israeli's would've agreed to dumb bomb, kill and/or injure over 30,000 children who had nothing to do with Hamas' attack on Israelis. Yet, if the Israeli's "dance with the devil" they, like Hamas, won't be allowed to bow out of nasty, ugly choices involving the mass murder of innocent children.

The 2023 Israel-Hamas War is a proxy war by the Russian Mafia to shove Israeli's pro-Kremlin. Don't fall for it Israelis. If you do, the Russian Mafia will destroy you.

Remember, the Russian Mafia are masters of psychological warfare. This war was designed to steal something precious from the Israeli people, their sense of safety, and something far more important, their freedom. Hamas members are now slaves of the Russian Mafia. If they could, they'd call a cease fire but they're owned so they can't. They're not even allowed to negotiate directly with American negotiators, something the Russian Mafia set in place long ago by some of their American operatives. Israeli's may not know it but the 2023 Israel-Hamas War is a variation of the Russian Mafia's power grab of Israeli court's power. I am praying for the Israeli people. If they can't get their people to not right-lean, if I were Israeli I'd leave Israel. I've seen that the Russian Mafia forced Soviet Jews to authorize Hamas' attack against Israeli Jews, and I've seen that the Russian Mafia forced Hamas to continue a war that's murdering and injuring tens of thousands of Palestinian children, when Hamas is largely Palestinians who were once children. If Israeli's look at what the Russian Mafia is stealth doing they'd understand that aligning themselves with the Russian Mafia is a bad idea they won't be able to come back from. Although the Russian Mafia is determined to destroy Americans, and although the Russian Mafia think they have us this November, I won't let the evil Russian Mafia destroy my people, my nation, my government or our allies. This book is my attempt to help innocent people worldwide survive and understand which government is the actual threat. The Russian Mafia is hands-down, the bad seed.

Look at Hamas, the Russian Mafia has manipulated them so they're involved in a war that's killed well over 10,000 Palestinian children with an unknown number of innocent children injured. Hamas are mostly Palestinians, radicalized, but still Palestinians. You must know they're heartbroken about the deaths of all those Palestinian babies, and the other innocent Gaza's deaths. If they were in control of this war a cease fire would've been negotiated by them months ago. Because we're not negotiating directly with them, we can't access their love for their people to stop this war.

The Russian Mafia (the Bratva) has a sizable number of Soviet Jews, and yet at least some of them approved Hamas' injury, rape and/or murder of thousands of innocent Israeli Jews (and other ethnicity's working in Israel).[104] When a mafia gets Jews to destroy other Jews, that's a seriously evil mafia who is demonstrating it's ability to get people to kill their own

---

104 I've not read where the Israeli government has listed the number of Israelis and foreign born who were injured but not killed by Hamas but since 1,200 Israelis were killed, and I've no idea how many innocent Israeli women were raped, I'm assuming that at least a thousand people survived the attacks. From accounts of war attacks I've read over the years, there are at least an equal, usually a greater number of people who are injured but survived. Gaza's are rough estimate, nearly 30,000 killed and an unknown number injured and buried beneath collapsed buildings. Gaza's are enduring war, there's no way to track the missing and injured but rough guess there are at least as many injured/missing as confirmed dead. That's how I came up with my guestimation of the thousands of innocent Israeli's killed, injured, or raped in Israel. Given that 1,200 were murdered, I added an estimate of a 1,000 injured survivors, to come up with my comment that thousands of innocent Israelis were injured, raped and/or killed by Hamas, and that at least some of the Russian Jews in the Bratva had to sign off on the attack of Jews. Jew on Jew violence, after all they've endured, is especially reprehensible in my opinion. I feel the same about Hamas on Gaza Palestinian violence: Palestinians hurting Palestinians: that's a seriously evil mafia to get groups of people hurting their own people.

group. The Russian Mafia won't blink at destroying Americans if they're willing to harm Israeli Jews.

The Russian Mafia is working genocide here to annihilate our people. All dominants are under attack because they form essential structures for our culture. Dominants in a family are parents, grandparents, sometimes aunts and uncles, sometimes an elder sibling. Dominants because of their job includes President Biden who is Commander in Chief of our military; generals, captains, supervisors at work, doctors, teachers, etc. Dominant businesses provide essential services, i.e. grocery stores, post office, pharmacists, water, power (electricity/gas), waste management, medical care, builders of affordable housing, etc. The Russian Mafia is attacking all these dominants, and more.

Although pretty much everyone is a subordinate, as that is our first position in life, there are a great many dominants and we will listen to them when they warn about non-overt threats but we must be shown they are dominants as we perceive dominants to be.

Chapter 5.3 – The Russian Mafia targeted dominants since the last century so that when they made their move, we'd have many fewer leaders outside of the our political leaders sounding the alarm about their coup

I read the newspaper routinely and I've seen no dominants outside President Biden, V.P. Harris, other cabinet members, our security agency leaders who occasionally give interviews and our military, but outside from them, our other society dominants have largely been silent. This is not accident. Active Measures is keyed to our hard-wiring, the way our species' use of dominants and subordinates has been circumvented by the evil Russian Mafia who means us harm, so it's eerily quiet in terms of no one saying the 2023 Israel-Hamas War is a proxy war. Right now, we're not adequately defending ourselves against this coup. That will change moving forward.

Chapter 5.4 – The Russian Mafia is running an Exhaustion Warfare play against our Administration

The Russian Mafia is running an exhaustion warfare attack against President Biden, his Administration, our security agencies and our military. Our people are not complaining but it's been one crazy emergency after another and they're exhausted. I recognize the attack because the Russian Mafia targeted me with it from around 2008 until I collapsed in 2010.

The attack is meant to make them collapse, to make them hate their job, yet we need them, so they're working their guts out for us. We must help them. We will help them. Vote for them. Tell them they're doing an outstanding job. Tell them you appreciate their sacrifice. What they're experiencing isn't normal – it's an attack by the Russian Mafia, is a stealth war, part of their genocide against us, so that our Administration will be so exhausted, they won't have time to assess what the polls mean because they're being kept super swamped with the endless emergency-abuse authored by the Russian Mafia. Without the Biden Administration, without our security agencies, and without our military, the Russian Mafia would've stomped all over us long ago. What we're witnessing is an Exhaustion Warfare attack that's been in play since the Biden Administration took office. This is not a normal presidency, this is our Administration, our security agencies and our military under attack by the Russian Mafia. The Russian Mafia goes on and on about President Biden's age. You put a 40 year old into the presidency under an Exhaustion Warfare attack and watch that person just collapse. Only years of experience, determination and loyalty to us has that team on their feet fighting for us. We're not going to get better than what we're getting from our dedicated, determined public servants. If we let the Russian Mafia seize our Presidency, they'll just laugh at our demands and shoot us, or knowing them, they'll poison us with colon causing toilet paper and speed up fentanyl shipments. If you think I'm making this up, you're wrong.

Chapter 5.5 – The Russian Mafia uses compromised dominants to push Americans into overthrowing our government

When the Russian Mafia secretly enslaves a dominant important to us, say a parent, grandparent or mentor, and forces them to try to corrupt us, the Russian Mafia force us to "perceive" the authority figure we may have trusted our entire life as corrupt. This causes a subconscious unresolved conflict, when targets are repeatedly forced to engage with a compromised dominant. Our valuation of the dominants in our lives is hard-wired into us, is part of our instincts, is the foundation of who we are, embedded into our very genes.

And, our dominants hold part of our self-image. When we look into a mirror, our reflection is not all we are, we are also the cause of our grandfather's delighted smile as he witnesses our pleasure in our first taste of an ice cream cone when we are 5. We are hard-wired to perceive our dominants as good, as we are hard-wired to perceive ourselves as good. When the Russian Mafia flips our dominant to corrupt, it attacks our own self-image, flips our self-image to corrupt.[105] Psychologically the initial attacks are akin to rubbing fur the wrong way, or dragging fingernails across a chalkboard, but instead of one of those sensory assaults, this attack targets our self-image and our sense of reality.

The Russian Mafia uses dominants like parents, grandparents, doctors, etc., to destroy and destabilize our nation and our people. I doubt the Russian Mafia explains to it's enslaved dominants why it's forcing them to encourage corruption in their child, if the exploited dominant a parent. My guess is the general practitioner is told to offer the crime as one of our government's spot-checks on fraud prevention, the lie being that if a patient who wants a product won't agree to cheat Medicare when their doctor offers it, then it's unlikely that

---

105 The Russian Mafia is all about making Americans and their other enemies believe their enemy is corrupt, while they have long been the villain. No one wants to see themselves as corrupt, as evil, but the Russian Mafia really is evil and doesn't care. The Russian Mafia is in-it-to-win-it and doesn't care how many innocents it has to murder to be top dog of our species.

person is lying about feeling too unwell to work. That sounds like a reasonable argument against Disability fraud but the Russian Mafia's has mastered creating plausible lies. I am insulin resistant, I'm obese, I have high blood pressure, all which the Russian Mafia made happen by refusing to allow the doctors they own to accurately assist me when I sought medical help decades ago. The truth is the Russian Mafia is forcing dominants to offer corruption in order to destabilize targets psychologically, to get people into the streets demanding the overthrow of our government.

The Russian Mafia is determined to use Americans to overthrow our government, and vote in one of their puppets, to manipulate us into demanding our government be overthrown. What a crock. Militarily, the Russian Mafia's soldiers would be used as closers, the heavy lifting is destabilizing us using our dominants. To achieve that the Russian Mafia must create an emotional impetus as to why people would be willing to vote in the Russian Mafia.

First off, the Russian Mafia isn't going to tell anyone that their puppets are their puppets. The last article I read quoting Mr. Putin about his preferred candidate for November 2024, he said he preferred President Biden. He lies. If that were true he wouldn't have ordered the 2023 Israel-Hamas proxy war to unseat our President. Mr. Putin is a walking lie. He doesn't care that people know he's lying. The only thing he cares about is seizing world dominance, destroying everyone living in America, and being feted by his Mafia and slave-allies, who he anticipates will lavish praise on him for, wait for it, outsmarting Americans and murdering babies. No one will mention his slaughter of innocents to him.

Second, the Russian Mafia will have their candidates push traditional values, which will appeal to the person who had their parent/doctor try to corrupt them, and who is wandering around with a subconsciously, unwanted, self-labeled corruption. Third, after the Russian Mafia shows the subordinate that their trusted dominant is corrupt, and after the subordinate talks to their dominant, trying to convince them of the error of their ways and since this is a setup, the subordinate's concerns will be dismissed by the enslaved dominant.

Influencers will be deployed to groom and lure the target into overthrowing our government, to invite them to protests and invite them to cute dates if the Russian Mafia thinks the target would be open to that. According to General Gerasimov, the Russian Mafia deploys protesters as part of their nontraditional warfare model.[106] I know you're thinking the Russian Mafia can't possibly have time for all this but that's where Jackson-Vanik comes in. The USSR got hundreds of thousands of operatives embedded here, probably by

---

106 General Valery Gerasimov, Chief of the General Staff of the Russian Federation,"The Value of Science in Prediction," *Military-Industrial Kurier* (February 27, 2013). General Valery Gerasimov states: *"... wars are no longer declared..." "...The very "rules of war" have changed...." "...The role of nonmilitary means of achieving political and strategic goals has grown, and, in many cases, they have exceeded the power of force of weapons in their effectiveness...." "...applied in coordination with the* **protest potential of the population....**"

blackmailing Jackson-Vanik through. I believe they outed former President Nixon when he refused to approve Jackson-Vanik. Those hundreds of thousands of destabilizing spies have been built into probably tens of millions of Russian Mafia controlled operatives here. There's a reason Mr. Putin is constantly smirking when he's asked about America. We're in extremely deep trouble because the Russian Mafia has stealth run rampant for so long, but we'll save ourselves. If you like the Biden Administration, say so. He's an outstanding President, the Biden Team (President Biden, V.P. Harris, their Cabinet, our security agencies, and our military) are best in class.

Regarding getting targets to overthrow the government, their compromised supervisor and/or co-workers will be used to influence them and compromised law enforcement will entrap them. Once the victim is entrapped their quality of life is over and they're never allowed to re-stabilize. They will be a slave of the Russian Mafia just as their parent, grandparent and mentor are. An additional reason targets are allowed to see that their dominants are corrupt is to make them believe that breaking the law is "normal." It's not. If anyone offers lawbreaking to you, that's the Russian Mafia working to ruin your life. Decline and evade that person.

As you can see, the Russian Mafia is working a patchwork of reasons to get voters here to install one of their puppets: to get Palestinian Americans to vote for Mr. Trump, the Russian Mafia started the 2023 Israel-Hamas War. It has all the earmarks of one of the Russian Mafia's proxy wars: (1) the Russian Mafia is not publicly blamed; (2) the American people and our government are publicly blamed; and (3) as if by magic, the war facilitates the Russian Mafia's intention of seizing world dominance.

By showing targets their trusted authority figures are corrupt, victims become much more willing to take to the streets to demand the overthrow of their legitimate government, especially since the Russian Mafia will deeply influence targets and the Russian Mafia has it's puppets offer a variation of law and order, promising ethics and tradition.

Our hard-wiring is largely who and what we are but 99% of people don't know it. The Russian Mafia, and before it the USSR, know it and have keyed their weapon to vulnerabilities most of us don't know we have. Fish in a barrel. But our government and allies are sharing what was once confidential information with ordinary people, so we're better able to connect the dots.

As the Russian Mafia says, the battlefield is in the mind. The Russian Mafia has zero problem: (1) lying to us to get us to overthrow our legitimate government so the Russian Mafia can annihilate us, (2) proxy murdering and/or injuring tens of thousands of Palestinian civilians including babies, to seize world dominance, (3) stealth destabilizing Israeli civilians to force them to embrace their right-leaning government by ordering Hamas to attack Israelis.

The Russian Mafia is a reliable enemy, never a reliable ally.

Victims unaware they've been attacked by the Russian Mafia using our hard-wiring against us seek to resolve the conflict, not understanding that our instincts (aka our hard-wiring) are powerful drivers on our behavior and when the Russian Mafia deliberately attacks us using our hard-wiring we're at a serious disadvantage because we know almost noting about our hard-wiring.

The Russian Mafia doesn't tell victims: (1) they're being attacked by a hostile enemy foreign government running an illegal undeclared war targeting innocent civilians, including our children. (2) that the Russian Mafia is working genocide against us and has no intention of ever stopping attacking us; (3) the Russian Mafia is exploiting hard-wiring most of us don't even know we have, (4) that the Russian Mafia has enslaved our parents and is extorting them in true mafia fashion, under pain of prison and/or death. Our dominants are led to believe their American law enforcement handlers are controlled by our legitimate government; (5) the Russian Mafia has worked decades to annihilate us, through our trade partnership in the 1970s and beyond, all premeditated; (6) that the Russian Mafia has effectively off-lined (compromised/blackmailed) many of the dominants (leaders) we'd ordinarily rely upon to recognize this non-overt coup which is based upon their exploitation of our hard-wiring and our ignorance that we are hard-wired and what that means when an enemy attacks us exploiting vulnerabilities we don't know we have. None of these truths will the Russian Mafia tell us. This November 2024 Mr. Putin intends to be the Closer of Active Measures by installing one of their many puppets into the White House. If we allow that, they will destroy us.

Chapter 5.6 – The Russian Mafia is big on 'logistics warfare' to install new 'procedures & policies' that facilitate their future attacks. One reason they downed General Austin was to put into place the cabinet list where all cabinet level employees must report anesthesia-requiring procedures, which list makes it much easier for the Russian Mafia to stealth attack our Cabinet (& their families)

The Russian Mafia worked a 'logistics attack' against me, changing procedures on buses I take to make it much easier for them to harass me (which is where I learned about 'logistics warfare'). They've also removed most of the post office boxes in my area by deploying a 'homeless vandals' cover story attack as they gear up to try to unseat our President.

The Russian Mafia is working hard to seize control of our nuclear launch sequence, but not only that. I'm suspicious about the new list the Russian Mafia put in place which makes it easier for them to track (and stealth attack) all cabinet members undergoing procedures requiring anesthesia. That list isn't a good idea because it makes it even easier for them to hunt and destroy our cabinet members the year they expect to Close Active Measures.

The Russian Mafia made General Austin's complications happen. Downing a member of our nuclear launch team is flashy and designed to get our security agencies' attention. The Russian Mafia had their operatives in play (people in our government our government trusts) who reported back to them our reaction, what we did, how we did it. They'll work our reactions to their advantage in their next attack. Possible short term goals: (1) they want us to change our nuclear weapon launch configuration in order to include more people, at least one of whom will be compromised by the Russian Mafia, but will have been cleared by our security agencies (our security agencies don't trust, especially with an evil Russian Mafia ascending (who is deploying an effective weapon (dominance/hard-wiring)), but despite their diligence the Russian Mafia embeds operatives into our government (and elsewhere, i.e., our families). We must come up with workarounds as best we can; (2) the Russian Mafia wants us to question whether General Austin was compromised while anesthetized or during his hospital stay since, I'm guessing, he had no Secret Service detail. The Russian Mafia loves deploying racism, it's the gift that keeps on giving. After all, the KGB got one of it's major footholds into our country by courting the KKK. Racism has helped them nearly destroy us. They go with what works); (3) the attack on General Austin was intel gathering to learn how the team protecting cabinet members responds to stressors. For instance, if our legitimate government believes General Austin was compromised would they promptly remove him, be so concerned about the African American vote during an election year they'd keep him in place (I'm positive my government wouldn't do that), or to avoid the appearance of racism would they install another African American male to replace General Austin. If our government did, that person is more likely to be owned by the Russian Mafia. The Russian Mafia sickened General Austin for a combination of reasons. They've already established a new policy where cabinet members must report when they'll be anesthetized, which gives the many Russian Mafia-controlled American operatives an easy way to target individuals. I've experienced a successful logistics-intel-attack by the Russian Mafia.

I use a mobility device, not because I can't walk but because the Russian Mafia has destroyed my ability to sleep normally which has negatively impacted my stamina. When I walk blocks in the city where I live, there aren't seats along the sidewalk for a walker to rest. I prefer to not sit on the ground when I tire, getting up would be difficult since I'm obese, so I use a mobility device to get from point A to B. I think I bought a mobility device in 2010 and I take the bus. The Russian Mafia's American operatives found it difficult predicting where I was heading (it's not like strangers on the bus can ask me and expect I'll answer, especially since the Russian Mafia has allowed their American operatives to overtly harass and/or attack me since around 2001). To get the information from me, they used their power players in bus agencies management to change their policies in the two bus agencies I usually take. They have the driver ask me my destination exit. In a bus driver-passenger interaction the driver is the dominant and I've been socialized from birth (as we all have) to accommodate dominants in routine interactions. It took me awhile to come up with a workaround. This occurred years ago, before I began this book so I was slower making connections and figuring out how to better protect myself. By writing this book I've learned so much more, I've connected the dots with so many of the attacks and seen how the Russian Mafia operates here. I'm hoping that my readers will also benefit from my decades of trauma so they're more timely, and effectively defend themselves against the Russian Mafia.

The Russian Mafia either has operatives listening on the bus or they have some other way of hearing that information. With it, they position operatives at my exit stop who block my exiting the bus, block the sidewalk I try to use to reach my destination, have operatives harass me all along the route, back and the return, and they position plenty of operatives at my destination. The Russian Mafia got this change enacted when I was maybe in my mid-50s. I knew by then I had a powerful enemy but it wasn't until I was approximately 61 (in 2021) that I learned the Russian Mafia was the perpetrator. This is a policy the Russian Mafia stealth introduced and has weaponized against me and is one that will remain in place which they will use to attack anyone else using a mobility device. Policies remain in place unless someone high in the hierarchy complains and works to remove them. I'm 100% positive the Russian Mafia will weaponize this new policy where cabinet members must report when they're being anesthetized. Having someone they may have already compromised suggest this new policy in response to General Austin's hospitalization may be how the Russian Mafia worked this attack. While we're focused on 'did the Russian Mafia compromise General Austin when he was unconscious' is a perfect cover for the Russian Mafia to set in place the policy where all cabinet members must report serious procedures, giving the Russian Mafia a centralized way to attack our cabinet. I'm basing my conclusion on direct, personal experience.[107] We must reconfigure our new cabinet reporting

---

107 The Russian Mafia deployed homeless operatives as vandals to provide cover for the significant number of mail boxes they've had removed as the Russian Mafia prepares to Close Active Measures this November. They couldn't have someone in the Post Office Administration say, for no reason, that post office boxes needed to be removed. To provide scaffolding for their people(s) in our Post Office the Russian Mafia deployed vandals, fronting as homeless. Homeless are one of the groups the Russian Mafia have deployed to harass me for well over a decade. A giveaway is that 'homeless' the Russian Mafia owns don't stink. I was a homeless child (a church lady took me and my family in for nearly a year until we found a decrepit building to rent). I've been near many homeless people, sat next to many on the bus and/or at the bus stop to know that most stink. This is because the Russian Mafia ensures that our agencies don't provide them enough bathrooms and showing facilities for any poor person to access. This is on purpose, it is another prong of the Russian Mafia's coup against us, the part where they push Americans into the streets demanding that we overthrown our legitimate government. The Russian Mafia deploy homelessness as a visual cue, a visual lie to our society that our government doesn't serve our impoverished people. This is a lie in my own personal experience: when the USSR destabilized my mother and made me homeless when I was around 12, it was my government, not my mother, not my father, who fed me and my impoverished family and who put a room over our head. The amount of food stamps we received carried us 3 weeks out of the 4 - we didn't know about food pantries, my mother had worked as a surgical technician where she'd earned a living wage, so she'd not needed to learn about food pantries. Since our food ran out every month despite my mother's buying the most affordable ingredients she could find, we planted a garden next to our decrepit rental, but that garden

policy to effectively defend our cabinet, else the Russian Mafia will attack them when they're most vulnerable. Just as the Russian Mafia got a germ infected scalpel (or other germ-compromised device) into the hands of General Austin's doctors, they'll do the same, and worse, to our other cabinet members. The Russian Mafia is 'in-it-to-win-it.' They expect to Close Active Measures this November. Let's not make it easy for them.

---

refused to grow anything but zucchini so that's what I often ate the last week of the month: zucchini, potatoes and eggs fried in bacon grease mom had saved over from the package of bacon she'd bought at the start of the month. The Russian Mafia deliberately worked to make me believe that our legitimate government forced a thousand parents to entrap their own children, was forcing those parents to bring their children along with them on stealth harassment attacks against me. My father told me, when he tried to get to me cheat my government by lying to Social Security he said to me "the government doesn't care about you. You're just a number to them." When I'd been 12-15 and my mother was unable to work and provide for me, it was my government who fed and housed me and my immediate family. I know, from direct, personal experience that my government cares about it's poor people. And I know from direct, personal experience that the Russian Mafia loathes us and is determined to destroy us. When the Russian Mafia came with it's many lies, using my parents to lie to me, my government had ensured I had food to eat for most of the month, while it was the KGB who robbed me and my family of our home and stole my mother's career from her. It is the Russian Mafia who is working genocide against me and my nation. The Russian Mafia is a world class, best-in-class liar. They're mass murderers of children. They force women into prostitution. They hook innocent children on fentanyl. Believing anything they say other than "we want world dominance and we're willing to slaughter anyone to get it," will get the listener and their loved ones killed.

Chapter 5.7 – The Russian Mafia deploys ubiquitous 'workmen' operatives, 'homeless' operatives, 'mail carrier' operatives, & 'delivery' operatives to harass me, to vandalize American property (specifically post office boxes as they gear up for the 2024 election), & to open my mail & other items before I receive them

Sometimes the Russian Mafia needs vandals to attack specific property here, like our post office boxes. The Russian Mafia is confident it will embed one of it's puppets into our Presidency this November but to cover as many bases as possible, they began removing post office boxes in my area over the last two years, rough estimate. On the main street near where I live there were, as recently as three years ago, 11 post office boxes on a major street within three miles of my apartment. I know because I used most of them within the last three years. Now there are only 2 post office boxes, approximately three miles apart. Even the post office boxes that used to be in front of the post office have been removed. When I asked why all the post office boxes in front of the post office were removed, the employee said repeated vandalism.

To ensure that there are fewer post office boxes up and available for voters this November, I believe the Russian Mafia began removing post office boxes a few years ago, with 'repeated vandalism' providing a cover story for their removal. Removing them long in advance of November is down to the Russian Mafia's penchant for deniability, as in 'no, of course the many fewer post office boxes have nothing to do with us working to install one of our puppets into the presidency.'

The Russian Mafia has deployed workmen to noise harass me over the last approximately five years. Before I understood they were operatives I explained I have sleep difficulty and asked them to be quiet, I asked them multiple times. They ignored me. They were doing a renovation, apparently, next door which apparently ended a year or so ago. Now I believe some of those 'workmen' got themselves hired on by the management company of this building and have had a field day, with, I believe the other workmen operatives employed by the owner of the building, to go to town noise harassing me. I've thought I wouldn't be surprised if the items heavy items and appliance they apparently throw on the floor upstairs will come crashing down into this apartment. On several occasions it was very obvious the workmen were attacking. It's been pretty bad.

Workmen operatives, harassing me in a small building where other operatives also live and who I'm guessing those operatives aren't here when the workmen operatives are free to noise harass and startle me as loudly as possible, is how I know that 'workmen' operatives are a thing. The Russian Mafia also deploy 'homeless' operatives to harass me and have done for years. I know they're operatives because although they're dressed in raggedy clothes that have seen better days, they don't stink. The Russian Mafia ensures our country has an insufficient number of bathroom, laundry, housing, and showering facilities for impoverished people: they're trying to drum into us the lie that our government doesn't care about impoverished Americans. I was once a homeless child. My government made sure I had food for most of the month, and a place to live. The rental was cheap because it was decrepit but the roof didn't leak and there were no rats. There was the occasional water bug but compared to living on a sidewalk, which I never had to experience like too many of our people are experiencing, a place where a family can close the door and lock it against wandering threats means a lot. My family and I certainly appreciated it. Our government cares deeply about impoverished people, but the Russian Mafia is working to convince us otherwise.

For over a decade the Russian Mafia has deployed operatives in the mail and delivery services to open my mail and products ordered online before I receive them, so I'll get mail from my mailbox and it's been opened, or, far more frequently, a small part of the flap is lifted, as if by accident, but done many dozens of times, it's not an accident. Apparently, operatives slip a camera into the small tear in the envelope or packaging to read my mail. Also, most of my packages are open and/or compromised before I receive them. Usually the packaging has been breached, so the Russian Mafia's American operatives can see what I've received or access food items (you never want the Russian Mafia near your food so I tend not to eat food delivered opened). I've complained to the Santa Monica Police (my mail was stolen out of my mailbox on more than one occasion), I've complained to the management company who owns this building, I've complained to the ordering service where I placed the order. The breaches continue.

Chapter 5.8 – This is what genocide looks like on the ground before the Russian Mafia springs it's trap

This is what genocide looks like sans the Nazi's gas chambers. And because the Russian Mafia knows we won't report these attacks by our parents, doctors, other mentors, the attacks never stop. There's no way our government can know these attacks are occurring, so the Russian Mafia continues them until they destabilize the victim, driving them to medications (our young are unusually dependent on prescription medications to help them stay centered, a response to the Russian Mafia's unrelenting and stealth attacks against them), to street drugs and/or into other high risk behaviors where the Russian Mafia awaits to stealth assassinate them.

The Russian Mafia destabilizes our culture, our nation and our government when it forces dominants to lure/groom/entrap subordinates in subordinate's face.

For example, when my entrapped parents and my compromised doctor casually told me to lie to obtain a desirable but illegally obtained advantage from my government,

When the dominant's we're hard-wired from birth to "see" as good, attempt to get us to lie (cheat and/or steal), what the Russian Mafia is actually doing is beginning the process of social unrest to destabilize targets to get them to eventually take to the streets demanding the overthrow of our government.

We're hard-wired to *want* dominants to maintain our reality that dominants are "good."

We're far more willing to overthrow a government where dominants are corrupt, being set up to hope that a new government will re-establish our hard-wired view of an acceptable world order. This is a drip, drip, drip process over time, months, maybe years, not an immediate collapse of the target, and it's done with enough lead time to get the victim onboard.

The Russian Mafia doesn't tell targets it's forcing trusted authority figures to try to corrupt them as part of the Russian Mafia's multi-generational, non-overt coup against us, as part of their push to get us to take to the streets demanding our legitimate government stand down, and a new government take it's place, one the Russian Mafia controls, where one of their puppets will promise an ethical society. The Russian Mafia are world class, best in class liars. What we'll get if we're foolish enough to believe more of their lies is closer to annihilation. I am not exaggerating.

Because modern Americans have never been annihilated we've no established steps to recognize a a bespoke Holocaust aimed at our hard-wiring. And the Russian Mafia has continued to deploy and expand the Active Measures genocide the KGB built and deployed last century, where it has greatly reduced the number of non-compromised dominants available to warn us of this non-overt attack. We're facing a trap, a psychological-based, hard-wired based trap. And most of the scouts we're hard-wired to rely upon, our dominants, have been compromised, off-lined, silenced. This warfare attacks our sense of reality, and weakens our self-confidence, which is one of the things the Russian Mafia is going for, but the main thing they're after in the short term is to agitate us into overthrowing our government. A version of their attack is in play against Americans in America and I suspect may be in play against Israelis in Israel.[108]

---

108 If we allow the Russian Mafia to install one of their many puppets into our Presidency, they'll use that puppet to hire other puppets, and the Russian Mafia will force them to destroy us, our government and our nation. The Russian Mafia is working a 21st century Holocaust without the gas chambers. They've ensured there are plenty of poisons victims case take to kill themselves. If you think the Russian Mafia can't manipulate Americans into killing other Americans, existing evidence to the contrary include Hamas, who they are forcing to continue a war against their will that is facilitating the murder/injury of tens of thousands of Palestinian civilians, including Palestinian children. Hamas is largely made up of radicalized Palestinians who didn't sign up to help slaughter/injury 50,000 innocent Palestinians in under 6 months. Another example of the Russian Mafia's success in weaponizing a group it controls against that group's own people are the Soviet Jews in the Bratva, at least some of whom had to sign off on Hamas' attack against innocent Israeli citizens. When Caucasian Jews sign off on the murder, rape and/or injury of thousands of other Caucasian Jews, that's big. The Russian Mafia keeps racism going in every country it's working a coup against and in it's own nation and neighboring nations meaning Caucasian Soviet Jews signing off on murdering Jewish people of color, given the Russian Mafia's centuries of racism wouldn't be a surprise. But Caucasian Soviet Jews signing off on an attack against Caucasian Israeli Jews must not be ignored. **The Russian Mafia can get Jews to attack Jews, and Hamas (which is largely made up of radicalized Palestinians) to facilitate the murder of Palestinian babies. That means the Russian Mafia will be able to get Americans to facilitate the murder of other Americans if we allow the Russian Mafia to get anywhere near our Presidency.**

When the Russian Mafia, camouflaged in the trusted persona's of our entrapped parent, doctor, or other trusted authority figure (dominant), shows us that dominants we trust are corrupt (by enslaving our trusted authority figures and forcing them to tell us to lie, cheat, steal, etc.) that sets up an unresolved conflict in our subconscious. Our hard-wiring (aka instinct, the very being we are), and most of our experiences since birth (our socialization, soft-wiring), tell us dominants are good, while before us is a trusted dominant showing us dominants are corrupt. This dissonance is something people are hard-wired to avoid.

Chapter 5.9 - An example of how the Russian Mafia quadruple-teamed an American on the ground

This type of attack is intended to destroy us, our society, our culture, our families and our government. It is highly dangerous but on the surface doesn't appear to be so. It is easily framed by Mafia-controlled law enforcement as "fraud prevention," when, for example, they forced my general practitioner doctor, unprompted, to offer to help me cheat Medicare to get a weight loss medication, when that same general practitioner had offered to cheat Medicare on my behalf a couple years previously. The Russian Mafia knows I won't steal from anyone, that's not the point of this attack. By forcing my doctor to casually offer me corruption, the Russian Mafia pushes dissonance into my life to destabilize me, and to stealth assassinate me by elevating my blood pressure, by destroying my ability to sleep naturally, by destroying my sense of safety. Forcing compromised dominants to stealth agitate the target.

One of the reasons the Russian Mafia forces parents and other authority figures to try to corrupt us is because the Russian Mafia knows our hard-wiring doesn't like it. Psychologically it's akin to dragging fingernails on a chalkboard. It makes us uncomfortable but because we don't know about our hard-wiring, we don't know why it bothers us, only that we don't like it. It agitates us emotionally, psychologically, and sets us up to *look for* a different system where dominants are "good," to stop the dissonance between our hard-wiring and the lied reality the Russian Mafia establishes. The Russian Mafia will have it's candidates offer variations on law and order to get targets to vote them in.

Chapter 5.10 – The Russian Mafia is destroying Israeli civilian's sense of safety to drive them to support their right-leaning government

Because of these attacks I know how the Russian Mafia got Americans to attack the Capitol, and I know how the Russian Mafia is stealth attacking innocent Israeli civilians, doing everything to make them feel unsafe in their own country, forcing them towards the monster who is, in fact, destroying them: the Russian Mafia.

Chapter 5.11 - An example of how the Russian Mafia quadruple-teamed an American on the ground; what psychological warfare looks like on the ground

Attacking me on 5/31/2023, the Russian Mafia quadruple-teamed me: (1) they forced one of their operatives (the manager of this apartment) to noise harass me to such an extent while I was working on my Final that his attack caused my heart rate to spike to 200bpm, necessitating my first trip to the ER that day (5/31/2023); (2) the paramedics and other 911 personnel who arrived in response to my call for help refused to take me to the closest hospital to my apartment's ER when I asked to be taken there, telling me that that hospital's ER was closed to ambulance deliveries, forcing me, during a medical emergency, to take whatever medical help was available; (3) the Russian Mafia then forced a compromised ER nurse to be "super friendly" to me, forcing me, because of her status, and mine, and my hard-wiring, to try to respond appropriately despite my experiencing an ongoing medical emergency. She stayed with me for maybe 15 to 30 minutes or so after I'd arrived at the ER, to ensure my heart rate continued to stabilize (the Russian Mafia's cover, the Russian Mafia <u>always</u> has a believable cover). The Russian Mafia forced that same ER nurse, who they actually deployed as a weapon against me in ways that didn't, on the surface, appear to be an attack, to initiate conversation with me (in a nurse-patient relationship with the patient experiencing a medical emergency, the nurse who is providing essential medical assistant is the dominant, the patient is the subordinate receiving assistance. People are hard-wired and soft-wired to accommodate dominants). Talking to me, the nurse required a response and I did the best I could. Forcing her to be unnaturally cheerful and animated was part of the Russian Mafia's attack. If you've ever been sick you know how difficult it is to be cheerful when you feel unwell. The Russian Mafia forced even more stress on me than their initial attack by ordering the ER nurse to be unnaturally cheerful and to focus her specifically on me. To anyone observing the nurse would've appeared merely friendly. If you've ever been to the ER, one where the ER nurses aren't controlled by the Russian Mafia, ER nurses are very busy once the patient gets back to the ER. ER nurses are trying to fix your problem or trying to stabilize you, they're not trying to be your bestie. The compromised ER nurse was ordered to behave as if she was trying to be my bestie. Based on how the compromised paramedic kept trying to draw attention away from my medical emergency while I was in the ambulance (repeatedly saying "this is the third elevated heart rate case I've had this week and it's only Wednesday"), and based on the ER nurse's determined attempt to make jokes and to get me to laugh, the Russian Mafia absolutely didn't want my government to understand that the Russian Mafia had attacked me, although the Russian Mafia had attacked me and continued their attack from the morning of 5/31/2023 through the next ER visit, after I was discharged from the ER, while I waited for the ride service, where 'homeless' operatives harassed me repeatedly when the ride service took me to the ATM, after I collected my prescriptions, by having an operative steal my sleep aid medications in the ER, by having my mother call me after I got out of the ER to tell me that one of my most beloved mentors had recently died (the Russian Mafia is determined to emotionally destroy me but I refuse to be emotionally destroyed by those evil people). After I nabbed and wiped off a dusty sleep aid that I'd dropped on the floor a while ago, I was able to get enough sleep to complete my Final, my quiz and other necessities. While I did those things the Russian Mafia had one of the neighbor operatives begin to loudly play a movie right as I began to take my last quiz for the class.

The difference between me and other victims is that I know the Russian Mafia is attacking me using proxies while most other people don't know the Russian Mafia is working a multi-generational genocide against us. Because I know about it I'm not in denial – I can warn the world. For years the Russian Mafia had my father gas light me into denial. Those days are over. Denial, when dealing with the highly proactive Russian Mafia, will get you killed. The irony is for decades the Russian Mafia's American operatives have demonstrated to me, on a near daily basis, that they own the ground here. I believe them. But they don't own all the ground and our security agencies, our Administration, and our military are pushing back. We get this book out and start telling people about it, we'll see some positive movement in our polls. The Russian Mafia has been threatening to murder me for many years and I believe them. They will kill me if they get half a chance. Until that day, I'm outing them and warning the world about the evil they do and the evil they are.

The ER nurse and I discussed multiple subjects and one of the subjects was my obesity and the new weight loss drugs and my inability to qualify because I'm insulin resistant, not diabetic. She told me to ask my doctor, said that I might still qualify for the shots because of additional related health issues, like obesity. When I left the ER for my first ER visit that day (I returned later due to a different though related problem), the ER doctor, who couldn't find a physical reason for why my heart rate elevated as it did, told me to make an appointment with my primary care doctor to follow up. This is standard advice and isn't indicative that the ER doctor is owned but it is indicative that the Russian Mafia knows all the standard practices in America and can predict to a high degree of accuracy, how a person is likely to respond to the information they receive, given hard-wiring and socialization; (4) they forced my primary care doctor to offer to lie on my behalf to Medicare when he told me he'd designate me as diabetic when I asked him if I qualified for the shots, given my insulin-resistance, not diabetic designation.

Most people would never believe (and our government doesn't track a law-abiding person's every interaction), that the the Russian Mafia is so deeply and widely embedded here that it was able to create and exploit a medical emergency, which attack by the manager it initially launched to get me to withdraw from the programming class I was enrolled in that semester.[109]

Most people wouldn't believe the Russian Mafia was able to deploy four American dominants it controls, (1) my landlord, (2) the paramedics assisting me during a medical

---

109 The Russian Mafia, and before it the USSR, has worked for nearly 50 years to stop me learning advanced subjects. When I was in high school earning mostly As and Bs, (while working part-time jobs because my family needed the financial assistance) and doing family chores like ironing some of my family member's clothes, helping prepare family meals, and helping the youngest get to school because our mother was at work at that time of morning), I sought out my school's guidance counselor for advice about how to attend college. I didn't know how I could afford it since our family was working poor. (The USSR had deliberately made me and my family working poor when it destabilized my mother and made us homeless when I was around 12.) I knew I needed to attend college or my family would never earn enough income. I made sure I earned good grades so I could help myself and them by making sure I had an opportunity to earn more income. As a fast food worker earning minimum wage, a library assistant in the school library, and other low-wage part time jobs, I knew I wouldn't be able to afford college without scholarships and grants so I made sure I took the necessary courses to be accepted to college. I didn't know this subject, I was 15 and 16 but I kept trying to learn how I could attend college. My high school guidance counselor was the only resource I was aware to tell me how apply for college. I sought her out having no idea the KGB was on the ground in America working stealthily and using proxies to stop me from getting ahead. Showing my guidance counselor my completed classes and grades, I asked her how to apply to college. She discouraged me, told me not to apply for scholarships/grants I could read for myself that I qualified for. An American owned by the USSR in 1975, 1976, when the USSR didn't want me to attend college, they forced her to accuse me of stealing $20 from her purse, to stop me going to her office to ask her for advice. I don't remember what my high school GPA was, but for at least two years it wasn't far from 4.0, while I took college preparatory classes, not basic classes. When she accused me of theft I was devastated, heartbroken, confused as to why she didn't want to help me go to college. I may have been 16, maybe 15 when she accused me, and would graduate at 17. I'd never been accused of theft before and I expected to be arrested because of her accusation, having no idea how our laws worked. The day she accused me, after her accusation I walked to the front of the school and sat on the school lawn to wait for

emergency, (3) the ER nurse, and (4) my primary care doctor, to attack me but that's what they did.

I can tell you that we're in great danger from the Russian Mafia. Whether we do or not will depend on a national conversation where, hopefully, we'll be able to hear each other over the Russian Mafia's lies and noise.

---

the school bus. I'd never steal from anyone but because she was an authority figure and I was a teenager, her accusation hurt me in a way the USSR intended. This attack is classic anti-America Active Measures: aiming a psychological weapon, an authority figure, at a child, to destabilize her. At their heart, these types of attacks are meant to initiate self-loathing and self-doubt in children and adults but, fortunately for me, the USSR didn't enter my life until I was 11 and my mother and grandparents had had 11 years to socialize me in ways that helped make me strong. I knew I was a good person because my people let me know that I was. It's crucially important that a child is shown that they are loved by their family. Were it not for my mother's and grandparents and my extended family socialization of me during my first years of life, their love and acceptance of me, I'd not be writing this book. Love and acceptance of children by their families, helps the child understand they are of worth. At the time the USSR deployed my guidance counselor to attack me, I thought all dominants (police) would automatically believe her. At 15, 16, I didn't understand racism, I think I thought more in terms of authority figures and was confused as to why my guidance counselor didn't, for some reason, like me when I was polite to everyone. I had no idea there was a foreign enemy government stalking me and weaponizing proxies against me. It was excruciatingly painful to be accused of theft. I remember that I felt very sad, devastated, and confused. Because the USSR successfully stopped me asking my guidance counselor for college guidance when I was 15-16, they knew that attack worked and used a variation of it when they got a teacher's aid they owned when I was in engineering school to accuse me of cheating on a math test. I dropped out of engineering school, that attack by one of their proxies just one of many in my 20s (operative supervisors at my first full time job after college (so around 1984-85), physically assaulted me, sexually assaulted me, psychologically assaulted me, so many psychological attacks over most of my life. I endure them but the pain has been excruciating. I need you, my people, my nation, my species to survive. I need the habitability of earth to be maintained. I'm convinced Americans nor our species will survive unless we destroy the Russian Mafia.)
In high school, because the USSR refused to let the guidance counselor help me despite my excellent grades, I received, I think, one grant (that paid for much of my college expenses), and I think I won a $200 or

It'll do no good to introduce someone as a dominant and have them make a statement. We're dealing with human hard-wiring. Dominance, their work as a dominant, as well as a subordinate must be included in whatever communique we produce.

_____

$300 scholarship. It's been 50 years but I think that's all I got with a nearly 4.0 GPA for two years taking college prep classes, a good score on my SATs, and probably a 3.5 GPA average for the two other years. And these grades were with me working part-time, participating in high school extra-curricular activities because colleges preferred students who participated in extra-curricular activities; this was despite the USSR forcing three high school to threaten to beat me up and those three girls stalking me (I barely knew their names so there was nothing I'd done to them); this was despite my mother beating me for no reason when I was around 15. This was with the USSR having made me homeless about three years earlier, and despite my mother's low income, and me helping prepare my family's meals and me and my sister ironing our family's clothes. We lived in subpar housing because that's what my mother could afford. Because the USSR stopped me from learning about scholarships/grants (this was before the internet was widely available, so it was 1975, 1976), they stopped me from accessing thousands of dollars in grants/scholarships I qualified for, so I worked during some of my college years, when I would've otherwise been able to focus completely on earning high grades in pursuit of a career. My freshman year GPA was subpar, I had no idea how to access help since my high school counselor had basically run me off from seeking help, this was by design of the USSR. I rallied and improved my grades in my sophomore year and beyond but I never achieved the college scholastic achievement I would have had the USSR not stealth attacked me from 11. I am one woman in America but I'm 100% positive that variations of the attacks against me were done against unknown numbers of other people, targeting people of color and women, and now Caucasian Americans. By forcing their compromised teaching assistant at a world renowned university to accuse me of cheating when I was around 26, the USSR got me to stop working to achieve a professional career, using a great many proxies it owned to stealth attack me. The USSR cheated me out of that career, cheated my government and wider society out of the taxes I would've contributed as a high wage earner, cheated my family out of a mentor who would've acted as scaffolding for the career aspirations of my extended family (even recently one of nephews has one of his academic advisors refuse to allow him to graduate from engineering school, because, as I understand it, my nephew didn't pass one especially challenging class

For example, parents are dominants in their children's live, so you can have a parent wrangling a wiggling toddler, controlling the toddler in a believable, simple scenario that most parents and children would recognize and find believable, and have that dominant, as they're doing their dominant work, talk about the Russian Mafia and it's threat against our people. As the dominant works in the shot, the subordinate must too.

This war is keyed to our hard-wired dominant-subordinate hierarchy. It does no good to have a talking head with their position floated underneath, dominants must be shown doing dominant's work, so viewers accept them as dominants, then the dominant and make their argument. The argument won't matter unless the dominant is demonstrated .

_____

and apparently isn't allowed to continue to try. As far as I know my nephew now works as a tutor, not allowed to become an engineer. Just one Russian Mafia or KGB compromised American operative in a victim's life, like my high school guidance counselor, and my teaching assistant in a university math class, are deployed to destroy American families and to destabilize our society. All stealthily, all with proxies. And only the Russian Mafia knows, victims have no idea they're victims because the Russian Mafia is committing genocide against us.

Despite the KGB (who I didn't know was the puppet-master in my life), I found ways to help my extended family attend school by providing vehicles for three of them. I paid for tutoring for my nephew. I paid for thousands of dollars in dental care for another loved one, I bought computers for my immediate family and their children so they could learn how to use computers, bought them when computers were overpriced, so I was in my late 20s, maybe early 30s because I was determined we wouldn't be homeless again, even though most of my siblings had married and/or become parents. I helped feed my extended family over many years, bought them necessities like furniture, bought my nieces/nephews clothes they needed, etc. I did all that and more despite the USSR, who has covertly worked to destroy my family and my nation since I was 11.

Chapter 5.12 – We need our non-compromised dominants to launch a specific type of appeal – a dominant is recognized by their work and their control and/or authority of others. Dominance is established by physically controlling someone else, ordering a subordinate to do as the dominant demands/requests, or a dominant requesting something from a subordinate that requires the subordinate to agree to adjust their behavior

Dominants control others. A demonstration of a dominant controlling a subordinate's behavior must be filmed along with the dominant's warning. This shows our people that dominants are warning us about a non-overt threat so we need to pay attention.

We'll tell our people the reason behind the vignettes (they're like a vaccine to activate our species' ancient defense system). We wouldn't need the vignettes if the Russian Mafia hadn't attacked and enslaved our dominants to disable our species alarm system. Since they did we need a booster to activate our hard-wired (instinctive) self-defense mechanism to survive genocide.

Chapter 5.13 – We must make many videos and ask our people to watch at least one of them. We can ask our non-compromised dominants to make a presentation as well

In many small vignettes we need three components: (1) a dominant demonstrating dominance in a believable way, (2) a subordinate responding to the dominant's dominance in a believable and age-appropriate way, & (3) the dominant telling the audience about how the Russian Mafia and before it the

KGB worked genocide against us for decades and that's how our country got on the ropes. The dominant will tell the audience an example of Active Measures on the ground in play hurting ordinary people, after she/he is established as a dominant. This will work because our hard-wiring requires that a dominant tell us and the vignette will show the person talking to us is a dominant and the audience will known why the information is delivered in this way. The Russian Mafia has built a bomb keyed to our hard-wiring. We're taking no chances and will dismantle the bomb using hard-wiring. We'll tell the audience what we're doing, we're not hiding anything from them. Our government officials will tell our people why we're countering the Russian Mafia this way and our officials will also make statements, but because the bomb the Russian Mafia is deploying is based off our hard-wiring we must disable the bomb in a way our ancient hard-wiring understands. Unless our hard-wiring gets the memo, we won't survive.[110]

Fortunately it's not at all difficult to demonstrate someone is a dominant in a brief amount of time. The interaction between the dominant and subordinate is just as important to our ancient hard-wiring as our warning about the Russian Mafia. We're all experts on dominant-subordinate interaction because we've interacted with dominants from our

---

110 Star Wars: A New Hope is a great example of what we're looking for: not the production values, the basic information showing dominant, subordinate, subordinate's response to the dominant's dominance (1) Leia's capture by Vader; (2) R2D2's capture by the jawas; and (3) Luke's introduction with his aunt and uncle demonstrating dominance in different, sex-appropriate ways, and his response.

first moments of life. We see how to disable the bomb from the design of the KGB attack: the KGB and the Russian Mafia has targeted dominants, we know we rely on dominants to warn us about non-overt threats, so our defense must deploy dominants warning our people about this threat, but tuned to our ancient hard-wiring. Think of it as a vaccine to activate our 'we must organize to fight or we won't survive' genes. We create hundreds of cheap brief videos, a couple minutes long at most, and ask every voter to watch at least one, explain that our national security is at stake, explain what we're doing and why. Tragically, that's the truth. The videos will establish a dominant and a subordinate relationship and then we have the dominant tell the audience about an attack the Russian Mafia has committed. Our ancient hard-wiring "hears" the dominant, it doesn't matter that the dominant is an actor in a video, we accept dominants as "real" in films all the time. We just needed to understand we're facing a dominant-keyed bomb so we need to disable it using dominants. When people watch the videos they'll be surprised because they'll be very simple, and brief. Dominant-subordinate language is extremely simple. We just needed to know that we need to use it to disable this bomb. Watch the first 15 minutes of Star Wars: a New Hope. Leia/Vader, R2D2/Jawas, Luke/Aunt Beru, and Luke/Uncle Owen are all excellent and clear examples of dominant/subordinate interactions. Brief, to the point. In our videos we'll have the established dominant tell us about the non-overt threat. Our ancient hard-wiring will "hear" the actor because we'd have established the dominant-subordinate relationship before giving the warning. Because these videos will be cheap to produce, we don't need famous actors, it'll be an inexpensive, effective inoculation. Plus we'll have our excellent Biden Administration officials, security agencies, our military, in their roles as President and his Administration, taking questions from reporters. In a dominant-subordinate interaction, the subordinates are critically important – as long as they behave normally, which in the case of our country, the reporters tend to ask challenging questions. So just having a Cabinet official, Presidential spokesperson and/or our President, security agencies heads and/or military leaders take questions from reporters establishes a dominant-subordinate relationship after which our leaders can warn us about the non-overt coup from the Russian Mafia.

We're dealing with hard-wiring so we must key our defense to our hard-wiring. Telling the audience the speaker is a dominant won't work. The presentation must include dominant work subordinates (all of us) recognize as within the range of behaviors a dominant does. Our first interactions after birth are with dominants. We're all experts on recognizing dominants.

A parent giving her little one a bath and discussing the Russian Mafia's attacks on us over decades, saying she doesn't want her son enslaved by the Russian Mafia, while she prevents the child from running and slipping on the wet floor, might work (it is essential to catch the toddler's believable response to his mom's save – his believable reaction is as important to our hard-wiring as his mom's save (which is the dominant's work in the vignette). The opening sequences in Star Wars is a great place to see what I mean about dominant-subordinate interactions. The subordinate's believable reaction makes the sequence real, "locks it in" as believable). A teacher teaching his class about how the Russian Mafia has stealth attacked our nation for decades while he hands out yesterday's graded homework might work, we watch a student flip through his returned assignment and slightly grimace at the red marks and the C-. Dominants must be shown in the environment where they are dominant and doing something a dominant would do in that setting. If we weren't dealing with hard-wiring we could have talking heads warn people but by themselves talking heads won't disable the bomb keyed to our hard-wiring.

We will deploy our officials to help us survive, they will warn us. In their official capacity, answering questions from the press, they participate in a dominant-subordinate interaction our hard-wiring will accept, but we need many more dominants than we have non-compromised government officials, and we need them deployed in multiple languages, the languages spoken across our country, so vignettes keyed to our subcultures will help us survive. To deactivate the bomb requires the warning be told clearly and simply, while the dominant is doing a dominant's work and while the subordinate engages with the dominant. So we need to provide three types of information nearly simultaneously, first that the person warning us is a dominant, which requires dominance-on-display, and that requires a subordinate responding to the dominant: that takes seconds. Then the dominant must warn us about the non-overt threat so we'll know to pay attention. The action, where a dominant is filmed doing what dominants do, depending on their work as a dominant, must be shown, as well as the subordinate responding. We could have a nurse in the NICU feeding and changing a baby, show the baby is not happy, and then the nurse feeding the newborn and the baby's contented and increasingly drowsy drinking, as the nurse gently strokes his/her tiny fingers. Then have the nurse discuss the Russian Mafia's Active Measures against us and the nurse's refusal to let this baby grow up in a world where families have been destroyed by the Russian Mafia. The nurse must discuss a specific attack and how that attack is a threat to our society and to the baby she/he is caring for.

We can spot a dominant-subordinate by their work at 1,000 paces. This isn't hard to do, we just need to understand that we must do it to dismantle this bomb.

Chapter 5.14 – Examples of effective dominant/subordinate vignettes from a film most of us know and love – Star Wars: A New Hope

Here are some examples of dominants being dominant and subordinates being subordinate in Star Wars: a New Hope. (1) Darth Vader and his army attack and capture Princess Leia and her ship. We see the attack and their successful capture of the

Princess. This is a routine example, set in space, of what dominants working to capture an enemy government envoy, do. Darth Vader is successful, Princess Leia is captured, she doesn't like it but she's handcuffed and surrounded by the enemy so she walks with her captures to the brig, as you'd expect of a princess and professional envoy. If she collapsed, shrinking in terror, her title of princess wouldn't match with her defending herself, firing her weapon and downing of the enemy earlier. Titles matter in video set ups, behavior matters, the dominant and subordinate are a package, our vignettes must reflect that.

Another example is when the jawas ambush R2D2. We witness the ambush and the capture. This is an example of how a dominant predator ambushes a less powerful prey. The prey is shot, he goes down, he's carried off by the hunters. Another example is Luke on the farm with his Aunt and Uncle. His Aunt calls him over and asks him to do something for her and he agrees, an example of an Aunt-nephew relationship where the Aunt is the dominant demonstrated by her request of Luke and his immediate agreement to fulfill her request, but this relationship takes into consideration sex roles: Luke's Aunt requests his assistance and his immediate acquiescence gives the audience information about the quality of young man he is: he respects women, something girls and women love to see in a character. This relationship is further aided by the ages of the actors and their appearance. Another demonstration of dominance, after Luke's Uncle buys the droids, he orders Luke to clean them up before dinner. Luke protests but his Uncle holds sway. This example of dominance contrasts with Luke's Aunt's request and is something people generally know about men and women. In many societies men demand, women request, because in many societies women have much less financial power and status, especially in the 20th century when Star Wars was released. The Uncle's demand is an example of him being a dominant when he prevents Luke from visiting his friends and Luke protests, but acquiesces (which is a classic dominant/subordinate teenage interaction; as is when Uncle Luke promises Luke something at the dinner table, and Luke, having heard that before (as many children have), isn't buying - that too is classic parent/subordinate interaction). Dominants control subordinates physically and behaviorally, as these examples demonstrate. In our presentations to our people, we must not only have dominant speak their warnings about the non-overt threat, we must create a scenario where the dominant/subordinate is part of the presentation, as well as the subordinate's response. The warning about the non-overt attack, the demonstration of dominance and the subordinate's reaction, are all required.

Dominants expect immediate acquiescence from the subordinate, and the audience will too. The first 15 minutes of Star Wars, a New Hope show multiple dominant/subordinate interactions: Leia/Darth Vader, R2D2/Jawas, Luke/Aunt Beru, Luke/Uncle Owen.

Unless our non-compromised dominants raise the alarm, the Russian Mafia will install one of their puppets here and destroy us. We won't be coming back from that if they get their way.

Their dominant status must be shown, along with their quality, clear description of the non-overt threat. Both are required. A parent chasing a laughing toddler and catching him or her up in her mother's or father's arms, as the child attempts to chase a ball nearly into the street and a car drives by, demonstrates dominant status and the believable work of physical control over another that a parent does in a parent-child relationship, in a way our species interprets as dominance in that scenario. The parents save and the toddler's believable reaction are required to be filmed. The toddler's realistic reaction is critically important as it's necessary to shoot both the dominant and the subordinate they're controlling, and the subordinates response. Establishing dominance/subordinate requires we understand what we need to do. The Russian Mafia won't believe we'll have disabled the bomb and rolled out our defense within the time needed to save ourselves, but we will. (We aren't experiencing what those poor Nazi victims did.) After establishing dominant/subordinate behavior to satisfy our ancient hard-wiring (and free our inner ninja), which will only require only 20 seconds of video time, the dominant will then briefly, clearly talk to the audience referencing specific attacks by the Russian Mafia against innocent people in America to warn us about this secret, non-overt threat. Our species will be able to 'hear' the warning now and launch an effective defense.

Another layer of this attack against us by the Russian Mafia is their use of dominants to lead subordinates to corruption. This is done to entrap victims so compromised law enforcement have entry into their lives and forever destabilize them. This means that whatever mistake the target made by trusting the operative, because they've taken a deal (which is illegal under the Geneva Conventions, people can't be forced to give up their rights,[111] but, of course, targets are told this by the Russian Mafia), the Russian Mafia prevents them from ever righting their ship, psychologically or emotionally taking their life back. With Russian Mafia compromised American law enforcement operatives continually destabilizing noncriminals slowly, unrelentingly criminalizing them, and fronting that they're doing so under order of our legitimate American law enforcement, ordinary people have no idea of what's going on, what reality is, or how they we're trapped. Destabilized, traumatized by unknowingly compromised and uniformed American law enforcement, victimized Americans can't think. If they could think they'd understand there's no upside for our legitimate government to enslave us, but there's plenty of upside for the Russian Mafia if they can get Americans to believe it. They then use enslaved American's despair to push us to overthrow our government. The Russian Mafia is forever working a scam here. They love the idea of conning vulnerable

---

111 Geneva conventions, people can't be forced to give up their rights

Americans into buying the lie that our government is so corrupt it's still stealth enslaving Americans. Whatever lie they can float to overthrow our government, they'll float and they'll call us suckers for believing them.

Chapter 5.15 – How the Russian Mafia is using the Middle East to destroy us and how to save ourselves

The Israeli government is right-leaning and appears to be pro-Kremlin. We don't abandon Israelis, we honor our deal, but neither do we ignore reality. Can we renegotiate our alliance with them? The 2023 Israel-Hamas War is clearly demonstrating that the deal in place is being exploited by our enemies to destroy us. Like any government and people we have the right to defend ourselves. Do we have a way to renegotiate our alliance with the Israeli government? I know some American Jews won't like it that we're renegotiating but we don't owe it to any ally to self-destruct because they've been entrapped by the Russian Mafia. We ourselves are being successfully attacked by the Russian Mafia. We wouldn't expect the Israeli government to self-destruct along with us should, god forbid, the Russian Mafia installs one of their puppets. Looking at this war, you can see how the Russian Mafia has exploited our alliance with Israel to Close Active Measures. What we're seeing done to the Israeli government, their isolation, setting them up to embrace the dark side that any alliance with the Russian Mafia brings. But the horror I pray the Israeli people survive is nothing compared to what the Russian Mafia would do to Americans. I am not exaggerating when I say Holocaust against Americans. This kind of stealth attack works because there's no recent history of modern Americans being destroyed. Again, the Russian Mafia is doing nothing to spook us, even ensuring that no one in the newspapers I've read is outing the Russian Mafia for starting the 2023 Israel-Hamas proxy war when it's really clear that the Russian Mafia did so.

The loud silence of no one outing the Russian Mafia publicly about this proxy war demonstrates the Russian Mafia's control over our media and over many of our leaders.

What do we need from our Middle East allies? We must get a clear-eyed view of what we want from them. I'd like to better understand why the Middle East is an essentially important region to us. Our government believes it is and I'm positive the Russian Mafia has manipulated politicians currently out of office so our current leaders inherited conditions on the ground.

The Russian Mafia, I'm positive, facilitated 9/11 to get us big into the Middle East in a way we hadn't been for. I'm positive the Russian Mafia is the reason our exit from Afghanistan was messy; and now the 2023 Israel-Hamas proxy war, designed to keep us in the Middle East so the Russian Mafia can install one of it's puppets in the White House, with the Saudi Arabia deal in play so we'd display our best alliance-building self to tempt Saudi Arabia into forming an alliance between us, Israel and Saudi Arabia.

The Middle East territory is the gift that keeps on giving to the Russian Mafia, where they, time and again, attack us there. I strongly suspect that the Russian Mafia elevated the Middle East's importance in a way the Russian Mafia is exploiting to destroy us.

Chapter 5.16 – Examples of attacks by the USSR & how they deployed compromised Americans against me

Attacked for decades by the USSR's and the Russian Mafia's American proxies (examples include my car insurance agent who talked me into getting hit and run insurance, only to have my insurer deny my claim when another operative hit and ran me, in front of at least two police units, one whom took off after the perpetrator and learned who he was. When I asked for his identity in order to file a police report, Santa Monica Police refused to give me his name, even though at least two police witnessed the hit and run. To repair my car cost me $3,000 out of pocket). Operatives at a law firm cheated me out of a week's salary. The KGB was the attacker and at that time they were working hard to enrage me (as they enraged then triggered citizens who stormed the Capitol in 2021, and I suspect how they enraged and triggered Hamas over decades and then unleashed them upon innocent Israelis in 2023. I understand Hamas is a terrorist organization and the Houthis are too, but the Russian Mafia is a malignant mafia many steps ahead, agitating, enraging and destabilizing. We need to re-instate negotiating directly with terrorists because the Russian Mafia owns these people and they have intel we need. I'm 100% positive the Russian Mafia made 9/11 happen. No proof except for how they went loud attacking me around that time. They knew they'd focused our security agencies and military to the Middle East and even now, they're playing us with the Middle East, having us bend over backwards to maintain a relationship with the Israeli government, which government they obviously control. We must understand what is being done to us, and by who. We're on the very precipice of our own Holocaust and despite the horrific accounts we've read about the innocent Jews and other ethnicities tortured and murdered, it's far worse when those attacks happen to you. Denying what the Russian Mafia is, while we see photos daily evidence showing us, will get us killed, not help us survive. We dare not allow one of their puppets into the White House. Covid saved us in 2020. It won't save us in 2024. We must save ourselves.

I waited at a stop light with other people. Police witnessed the attack and chased the driver, only to later refuse to give me his name so I didn't know how to file a complaint against him, a hit and run operative one, my insurance agent) e they had advise to destroy as they used proxies to destroy me and career aspirations) it wasn't until around 2001 when strangers started attacking me overtly in grocery stores, on buses (I've experienced many operatives on buses and on sidewalks who would drop coins (to get me to look at them and engage with them) as harassing attacks

Chapter 5.17 - The Russian Mafia is working to shut down our economy, which is one of the reasons they're working to shut down our border and starve our IRS of the great deal of funding our nation needs to thrive, expand our nation-wide electric grid, re-commit to becoming a manufacturing powerhouse and keeping our promise of gifting financial aid to countries threatened by the climate catastrophe

I believe one of the reasons the Russian Mafia is working so hard to close down our borders so they can get a better handle on destroying our families. The constant influx of immigrants makes it harder for their (I believe) millions of operatives to destabilize us. They're doing all they can to sicken and destabilize our people but the constant influx of younger, willing workers, makes it more difficult for the Russian Mafia to destroy our economy, which is one of their many goals. You can see that goal when they use their operatives to ensure our IRS is chronically underfunded. And it's not just well-off people who're encouraged to under-pay their taxes: I've experienced two tax-related entrapment attempts. A compromised couple in my family are used to offer tax services to the more impoverished of us. I suspect their handlers encourage them to cheat on their own taxes and help at least one of my relatives cheat on their taxes, to the detriment of the IRS.

When covid happened and there were programs and tax breaks for low income people negatively impacted, which group included me and my mom, I urged my mother to apply. That's when I was told that she was listed as a dependent on one of her children's taxes and so was unable to benefit from the tax breaks and covid-related financial programs, which she would've qualified for. A program that included helping people who earned/received under $60,000 or so to pay their rent, and/or utility bills. And I believe the federal government sent people two checks. She didn't receive either because she was a dependent. I shared from my checks with her. I'm not aware that any of her other children shared their covid-related federal checks with her and she resisted when I told her I believe she qualified for a (specific covid housing program – name of program deliberately omitted here). I don't apply for programs I don't believe I qualify for, which my family knows so my mom knew I'd never recommend a program she didn't qualify for, one which would've helped her pay three months rent, and utilities. I qualified for the program but received only two months rent and financial assistance for my gas bill was approved. The Russian Mafia worked and set up and entrapment against me weaponizing that program but I spotted the trap (just barely) and was able to evade them. They're evil, exploited a federally authorized grant program, one of the few programs I qualified for, to try to entrap me (they changed the rules of the program mid-stream and failed to tell me, while slow-walking and bounced my completed application out of their web page, then lied about it, lied, lied. It was bad, but I evaded it but it was yet another horrific attack by the Russian Mafia, leaving me no way to know which American operative set it in play. I filed a complaint online with some oversight government system but never heard anything back.) I also received two covid grants for students, which I greatly appreciated. As you see by the Russian Mafia refusing to allow the IRS to raise taxes, the Russian Mafia is all about weakening our finances, which is key to our ability to thrive. We've never had an enemy this accomplished and this embedded we've never had a situation where our people were so unaware of the attacks against us by a foreign enemy government. This must change or things won't end well for us.

Chapter 5.18 – Why you should believe me & the many attacks the Russian Mafia's operatives have made against me: within the last few years one operative tried to kick me in the face as I sat on the bus – I'd had no interaction with him previous to the assault

I am a 64 year old African American woman[112] who was first attacked by the USSR when I was around 11, living in Southern California with my mom and half siblings in 1971. I have over 53+ years experience being overtly and covertly attacked by the USSR, the Russian government, and the Russian Mafia while living in America. I've never lived outside America or traveled outside America. I'm not an operative nor have I been trained in espionage. I'm not a government employee. I worked for 20 years as a legal secretary/legal word processor in law firms. No person or government asked me to write this book. These experiences that I state are my own, I directly experienced first hand, or I

_____
112 In 2024.

witnessed the attacks on someone else when I say they weren't done to me.

When I was around the age of 12 the USSR made me and my family homeless. The USSR continued stealth harassing me using proxies until the USSR's demise in 1991. The subsequent Russian government merged with the Russian Mafia[113] and intensified their attacks against me to include more physical assaults and entrapment attacks. When I unknowingly evaded multiple entrapment attacks in my 30s by living an ethical lifestyle (having been raised that way by my young mother), the Russian Mafia-controlled American law enforcement team tasked with entrapping me at that time (2001) grew frustrated. Apparently, the Russian Mafia pushed them to entrap me, and being unable to do so, operatives began overtly harassing and/or attacking me on the street when I was around 41, in 2001 and beyond. Unknown numbers of skateboarders and/or joggers (who appeared to be plain clothed law enforcement) swerved nearly into me, a plain clothed law enforcement officer walked nearly into me, forcing me to step down into the street to avoid him when I worked at a law firm in downtown Los Angeles. Unknown numbers of police in their cars stared at me as I waited at various bus stops over twenty years. Once on the bus at night after work I was the only passenger and the bus stopped and a uniformed officer entered and walked past me, looking at me, and walked down and back up the aisle and exited the bus. I guess but don't

113 Karen Dawisha, *Putin's Kleptocracy*, Simon & Schuster (2015).

really know, law enforcement watching from the bus cameras wanted to see how I'd react to a uniformed officer physically close to me and staring at me.

I wrote previous presidents and other organizations asking for help, and in retaliation police harassment spiked as multiple police did either drive-by's staring at me as I waited at the bus stop, or were already at the bus stop when I arrived on my mobility device and stared at me from their police cars, daring me to stare back at them. Of course, I didn't participate in a stare down with armed police officers stealth harassing/attacking me extra-judicially for nothing I'd done.

Unknown numbers of operatives jaywalked across major streets to walk nearly into me. In grocery stores or big box stores I've often been swarmed; in pharmacies, on the street, on the bus, pretty much everywhere I go. The woman who used to live in the apartment above mine harassed me unrelentingly for well over a decade. She no longer lives here but workmen who are refurbishing the apartment noise harass or stealth monitor me routinely, even today. (March 2024)

I've lost count of the number of physical assaults I've endured by operatives who enter and/or exit the bus and stealthily, deliberately hit me with their bags or purses as they pass me on my mobility device. I lean as far away from the aisle as possible. I've been kicked, stomped on, nearly kicked in the face by an operative on the bus (the face kick I only avoided by leaning far back on my mobility device – the Russian Mafia had one of their male operatives try to kick me in the face while I sat on my mobility device on the bus, minding my own business – I think this attempted assaulted occurred the last five years (bolded because I might remove it. Although it's true I don't want to alarm my legitimate government), called the n word, sexually assaulted by Russian Mafia American operatives. I've been threatened by Russian Mafia-controlled operatives using their vehicles, have had multiple operatives rev their vehicles as if they don't see me and peel out in front of me only to stop as if in surprise as they spot me on my mobility device, as I'd paused to let them turn the corner and they, obviously, refused to turn until they could fake not seeing me and rev their engine at me to threaten me, as I prepared to cross in the intersection in front of them. I experienced two attacks like this, and even worse. This is how the Russian Mafia threatens me. I've been hit by two separate Russian Mafia-controlled operatives in their separate vehicles in two separate incidents. Years ago when I had my car, an operative hit and ran and two police in two separate vehicles as if they'd been told to be there to witness the attack: one chased the perpetrator while the other officer checked to see if I'd been seriously hurt. Although I wasn't bleeding, I was stunned by the sudden impact. The operative hurt my back when he slammed his car into mine. I was at a stop light with several other cars in Santa Monica, then wham!, I and my vehicle were hit out of the blue with no warning. The Santa Monica Police refused to tell me the operative's name after they learned his identity. No way I could sue him without knowing his name or anything about him. I did have car insurance, and two police offices were there so maybe they filed a report. I ended up having to pay $3,000 to repair my car. Maybe this was around 1985. One of my supervisors at work had physically assaulted me and I'd left her department, not filed a lawsuit. I was attacked by proxies routinely but there was no way I could know the USSR was the perpetrator. That was when the USSR was working to enrage me, like it and/or the Russian Mafia enraged the Americans who attacked the Capitol in 2021. The Russian Mafia stealth harass and enrage ordinary people, then trigger them and aim them at our government, to overthrow our government. If we let them con us out of our democracy, they're slavers, all they have on offer is slavery.

Our socialization is who we are, who we're trained to be from birth. It's nearly impossible for someone untrained in psychological warfare to spot these kinds of attacks unless they know an enemy is on the ground wielding it.

After the Russian Mafia tried for decades to entrap me and failed, their American operatives decided to work me to exhaustion until I collapsed. During that process, they used my supervisors and coworkers to harass me as well as many hundreds of strangers on the street. They used various family members to financially stress me in multiple ways, as well as my landlord, who refused to make routine maintenance repairs. The Russian Mafia flooded my mail box with credit card offers. I'd gotten out of credit card debt several years earlier, but the Russian Mafia wanted that attack option because they figured a way to entrap me using personal bankruptcy filing. I never found out what that trap was because I refused to file for bankruptcy although they forced my mother to browbeat me for years to do so, using my mother's body as camouflage to hide their identity and their evil intent.

One of the worse things about the Russian Mafia, and there are so many evil things they do, but their willingness to attack victims using the victim's loved one's as hides and camouflage is next level evil. The parent is a slave and has no ability to warn their child. Imagine the damage the Russian Mafia forces entrapped parents to do to their five year old and what that five year old becomes when 18 and able to legally buy a gun.

Compromised parents, I theorize, face very long prison sentences. One suggested they'd be killed, if they talked. I'm also certain that entrapped American parents have been told or led to believe their handlers are American law enforcement. Their handlers are American law enforcement but they're controlled by the Russian Mafia. It's possible entrapped American law enforcement don't know this but you'd have to ask them and just like other entrapped victims, they're prohibited from discussing their plea agreement or even admitting they have one. I know at least six entrapped people, none of whom have told me they're entrapped or accepted a plea deal. None have admitted they have a plea deal. I believe confidentiality is one of the terms of the plea deals. If people break their plea deals they believe they'll be imprisoned, where the Russian Mafia rules. For now.

I borrowed thousands to help my family buy food, pay dental bills for loved ones, bought a needed car for a loved one in graduate school, made repairs to my apartment the owner should've made like putting in new flooring when the linoleum was worn down to bare wood, and replacing the stove when the owner refused to because one of the four burners still operated, so they said the stove still worked, etc. The Russian Mafia removed me from the desk I worked on and had me working for, at times, 6 or 7 attorneys. At the same time the Russian Mafia had strangers harassing me on the bus to and from work. Then they had a partner at the last law firm I worked at yell at me for no reason as he overworked me. I collapsed from overwork and stress.[114] While I was sickening (the Russian Mafia had operatives on the bus spray a mace like mist repeatedly near me to cause repeated sinus infections, and the Russian Mafia had operatives harass me, often multiple times a day to and from work. All my vacation time was used for sick time as the Russian Mafia had it's operatives harass me

---

114 Usually a legal secretary works for 2-3 attorneys/paralegals – to sicken me the Russian Mafia had the law firm where I worked pile on up to six attorneys, so twice the work load. I had about 20 years of legal secretarial experience which is why I was work for so many attorneys for so long before collapse. To add insult to injury, this was during the 2008 collapse, the law firm I worked asked staff to take a 10% pay cut. Workers were laid off and I couldn't afford to be laid off so I continued at that job while I prepared my resume to find an additional part time job. Before I could, I collapsed from overwork. Then the Russian Mafia began the next phase of their attack – they used four members of my immediate family to try to entrap me. I never want anyone to go through something like that. It was horrific.

unrelentingly), the doctors on my plan were controlled by the Russian Mafia and weren't allowed to diagnose me. Operatives finally stopped misting me with the mace-like substance after, I suspect, operatives deploying it began to sicken as well. Because the Russian Mafia controlled employees in most of the law firms I worked at, when I called in sick citing yet another sinus infection at my last legal secretarial position (I believe the mace-like mist was one of the reasons I kept getting re-occurring sinus infections which ate up all my vacation time as I forced to use it as sick time (one of the reasons I was unable to attend my beloved grandfather's funeral), the Russian Mafia's American operatives got that intel where I reported another sinus infection. And if Russian Mafia-controlled American operatives misting me with the irritant got reoccurring sinus infections too, that's probably why they stopped attacking me with it, for they finally did stop, I believe the last full year I worked there, 2009, although they may have stopped in 2008, one of those two years. So for at least a year or two American operatives misted with a mace-like product, harassed me unrelentingly to and from work, had the law firm I worked at remove me from one desk (I'd been hired for (the okey-doke)) and had me working for sometimes as many as 6 or 7 attorneys, assigned to 5 but available to help other attorneys who's secretaries hadn't come to work that day, and the Russian Mafia had one of the partner's yell at me for no reason.[115] In the exhaustion attack against me by the Russian

---

115 This work load, and the yelling, and the attacks to and from work by
    strangers, and my family's need for money with my other loved ones

Mafia I see reflections in the exhaustion warfare they're waging against the Biden Administration, probably in our security agencies and military as well. Exhaustion warfare is a type of nontraditional warfare. I read about how the Russian Mafia tortured Mr. Navalny (the Russian citizen they recently murdered), how the short box was used to destroy his health and body. I read about how the Russian Mafia is trying to run our Administration into the ground with overwork deploying Exhaustion Warfare. I read how innocent Palestinians are starving and innocent Ukrainians are being bombed. I hope this book gives people worldwide a better idea about who and what the Russian Mafia is, so people can make an informed decision. If we allow them to seize control of our species, we will be a dead species sooner rather than later. It would be as if the Nazi government won in WWII and has seized control of the world.

After I collapsed, the Russian Mafia had both my parents and two of my siblings, and other extended family, try to entrap me. I have no training in espionage and I had no idea the Russian Mafia was in America until 2021 when our

---

refusing to help when I asked if we could share the load, and the Russian Mafia cutting my salary (under the pretext that the 2008 downturn required the law firm to ask employees to take a 10% pay cut, and by refusing me off a desk I'd been hired to work at (that was the set up desk where I evaded the business credit card entrapment), and the landlord where I live refusing to make routine repairs but giving me permission to pay for them myself, twice of what a legal secretary normally carries, was the Russian Mafia's pushing me to exhaustion after they saw that I wouldn't steal from the company (vis-a-via the business credit card entrapment attack I evaded).

government allowed the information to be reported in the media and I happened to read the article.

Chapter 5.19 - How the Russian Mafia weaponizes our dominants to destabilize us and set us up to overthrow our government, while the Russian Mafia destroys our family's to commit genocide against us

The Russian Mafia weaponizes dominants to mess with subordinates' sense of reality and to deliberately destabilize our society. When dominants try to corrupt subordinates it shocks us and makes us doubt our family, our society, our government, and our reality. To set the trap, the Russian Mafia forces compromised dominants offer corruption to noncriminals as the norm. The dominant doesn't flat out say cheating is normal. Instead, they offer to lie, and offer to help the victim lie, for the victim's gain. In my experiences, the target the Russian Mafia uses to be scammed is the state and/or federal government or one of the social safety net programs. But the Russian Mafia also tried to set me up as a drug mule, and they worked another line of attack to get me to file for personal bankruptcy. A drug mule is obviously illegal. I don't know how they've compromised our personal bankruptcy system but apparently they have: they had my mother push me for years to file for personal bankruptcy after the Russian Mafia deliberately exhausted me and prevented me from getting helpful medical attention, and they had one of the Human Resource operatives recommend at my last job, unprompted, that I should file for personal bankruptcy. One of my compromised siblings was allowed to let slip that that they've filed for bankruptcy and had a positive experience, this information being shared with me within the last twelve months. The Russian Mafia's push to get me to file for personal bankruptcy has lasted for over 15 years, from my late 40s, maybe age 47 or so they began their campaign of stealthily getting me to increase my credit card load after I'd gotten it down to zero.[116]

Ch 5.4 – Examples of how the Russian Mafia deploys dominants to attack victims: doctor, bus driver, store cashier – we can no longer afford to blindly trust – the Russian Mafia attacks us when we're in subordinate position, using the dominant as a weapon to destroy us

On four occasions the Russian Mafia had cashiers not scan some of the items I'd presented for purchase. In the cashier-customer relationship, the cashier is the dominant. Customers automatically assume when they place products into their cart and present the cart to the cashier that the cashier will ring up everything in the cart. Not always.

The Russian Mafia tried to entrap me four times within the last several years deploying a shoplifting-cum-hard-wiring-

---

116 Weaponizing multiple individuals in my family to have urgent financial needs, and other family operatives unwilling to assist when I asked them, and weaponizing my landlord to refuse to make routine repairs. I borrowed to financially assist my family and make routine repairs the owners refused to make and I also bought needed household items, and a few wanted items, like carpet as my old carpet had been laid, estimate, over 20 years ago, and I installed a new electric box with the owner's permission to try to ground the electrical system in this apartment, where most of the plugs are not ground. The Russian Mafia has used that weakness to destroy multiple small home appliances I've struggled to replace.

cum-soft-wiring combo attack. In a cashier-customer interaction, the customer 'labels' the cashier as the dominant. From birth we're taught to trust dominants and, unless our family environment was completely dysfunctional, we are hard-wired and soft-wired (socialized from birth) to trust dominants. As a customer presenting my cart to the cashier at a big box store, I automatically assumed he'd scan all the groceries in my cart and charge me. I had maybe 10 items, basics like peanut butter, I think peanuts, etc. I didn't have much money so the items were lower priced foods. Unknown to me the cashier scanned maybe six of the products. I wasn't paying attention as I sat on my mobility device and he scanned the items directly from my shopping cart, so I didn't notice that he'd missed items. Had I loaded the items onto the conveyor belt it would've been clear he didn't scan them all but since this was an entrapment attempt, I suspect he told me (as other cashiers have done now and again) to leave my items in the cart and he'd scan them from there. I believed him. Like hundreds of millions of people I've shopped for groceries thousands of times over my life and until that first attack, I'd never had a cashier deliberately not scan my chosen items while indicating that he had.

I paid what he said I owed, activated my mobility device with one hand while pulling my shopping cart behind me. I presented my receipt to the exit employee. I rolled a few feet out of the store and glanced at my receipt. I stopped rolling. I saw that several of my items hadn't been scanned. I immediately turned around, re-entered the big box store against the stream of exiting customers, and approached the cashier. He appeared flustered when I told him he hadn't charged me for everything. He asked me if the exit employee had told me. I said no, I'd noticed myself. He scanned a few of the items. I saw that he still hadn't scanned everything. I pointed out what he'd missed. He finally scanned everything and I paid him and left the store.

That store for years has used plain clothed security to harass me and attack me in so many ways I actually mentioned it to family members (who, unknown to me at the time I told them, are entrapped). Had I not immediately noticed and returned to the store, store security would've swarmed and arrested me. I estimate I had a minute before I'd have begun experiencing the horror the Russian Mafia had planned for me. At the time of that first shoplifting attack, the KGB and the Russian Mafia had worked unsuccessfully to destabilize and entrap me for approximately 51 years. Here's what I believe would've happened: after I'd exited the store security would've asked to see my receipt, I would've handed it to them. They would've seen from my cart that all the items weren't on the receipt. I would've been surprised, I had no idea the cashier hadn't scanned everything until I glanced down at the receipt as I exited the store. I would've said that's not possible and protested I hadn't done anything but go shopping after I'd come to collect my prescription. Security would've contacted law enforcement (who I later saw were already there after I evaded the trap). I believe law enforcement works security at this particular store. Police would've charged me with suspicion of shoplifting, I would've denied it. Compromised law enforcement would've charged me with resisting arrest (declaring your innocent more than once while you're being arrested can be interpreted by police as resisting arrest). After I repeated that I was innocent, compromised law enforcement would've removed me from my mobility device and slammed me to the ground where five or so men would've sat on or knee'd me. My attempt to evade the pain from their attack those officers would've reframed at my post-death inquiry as their attempt to restrain me since I continued to resist arrest. I would've been seriously injured before being taken to jail, and/or additional operatives would've assaulted me in jail. I would've died, one more innocent American assassinated by compromised law enforcement under orders from the Russian Mafia. (Again, most law enforcement officers are not compromised by the Russian Mafia, but a significant number of them are.) Just as the Russian Mafia has seized control of our parents, medical staff, other authority figures, they've seized control of some of our law enforcement, who they use as camouflage to destabilize and murder targets, not much different from how the Russian Mafia deploys 'homeless' operatives to vandalize and do other destruction on people or property here.

The Russian Mafia has embraced the concept of using proxies as camouflage to hide their evil attacks against innocent people.

How did I unknowingly evade that entrapment, the first one of it's kind I'd experienced, knowing nothing about hard-wiring, soft-wiring or cashier-customer dominance hierarchy? First, god had me glance down at my receipt. I didn't know it was god, I just glanced at my receipt like I've glanced at receipts for decades. Had I not glanced at the receipt and noticed the discrepancy within the minute I exited the store, my life as I'd known it would've ended.[117]

Another reason I unknowingly evaded the trap was because I immediately returned to the store upon spotting the discrepancy. That was down to my young mother's training. She'd prioritized developing ethics in her oldest children in our

---

117 I'm positive the Russian Mafia would've made sure I was killed while in police custody. I'm convinced some (not all) of our police are entrapped slaves just like many of our other dominants are (parents, grandparents, doctors, teachers, etc.). I know the Russian Mafia exploits them as camouflage, just as the Russian Mafia has entrapped my parents and uses them as camouflage and as weapons against me. The Russian Mafia has worked for decades to put me into the criminal justice system. They own the ground there. Not because our legitimate government isn't there but because the Russian Mafia entraps victims and those people are led to believe that those entrapping them are our legitimate government, when in reality, it's the Russian Mafia who has entrapped our people and is using them to attack and destroy us. I'm not sure exactly when the big box shoplifting entrapment occurred. Maybe sometime in October-December 2020. The presidential election may not yet have happened but because of this attack I now believe that the Russian Mafia was concerned Mr. Biden might be elected and didn't want me alive to try to tell him about the overt attacks I've endured for decades. I'd written letters to the two previous presidents describing the attacks against me and the Russian Mafia had retaliated by increasing harassment against me by uniformed police, and other operatives.

young childhood. I was raised not to steal. While some people might've interpreted the free groceries as a gift from god, I would've felt guilty for not paying for them. That guilt for wrongdoing is a gift from my beautiful young mother which has saved my life more than once, and which is how I've remained unentrapped to write and publish this book.

Finally, this store had used it's plain clothed security to harass me for decades. One operative cut in front of me in the pharmacy line and began an animated conversation with the pharmacy cashier to bait me into an altercation. Hundreds of them over many years blocked my access to usually bought products, or stealth me in many other ways over the years. The Russian Mafia has made all stores/pharmacies I visit a weapon to attack me. I can't grow my own food so I must grocery shop. This is one of the grocery stores that first began overtly harassing me, I think even before the Russian Mafia went loud on me around 2001. They just had too many people all over me. When I returned an item to the store, a customer-operative behind me would question why I was returning it. Not an employee, a stranger. This store has been especially hostile to me for years, while most of the other stores have been hostile. So, although I had no idea about the shoplifting entrapment, nor did I know the Russian Mafia was concerned that Mr. Biden might be elected, I did know that all grocery stores and this store in particular, treated me as an enemy, on the low down, and that I was in enemy territory when I was in that store. Although I didn't anticipate that specific attack I'd been stealth attacked many hundreds of times previously. Although I had no idea the unscanned foods were an entrapment when I evaded the attack, I did know the store is enemy territory. There's a reason the Russian Mafia uses ambush against us. People who have no idea they're under attack have no defense. Although I couldn't anticipate this specific attack I'd been shown by overt attacks from strangers since around 2001 that I was under attack under orders from someone, so I'm always on guard.

Chapter 5.20 – The Russian Mafia uses our socialization, our default trust in dominants (how we're raised from birth) to destroy us

The Russian Mafia exploits our socialization to use as a weapon against us – we're socialized from birth to acquiesce to dominants requests/demands. The Russian Mafia enslavement and weaponization of dominants has devastated our people. To recover and out the Russian Mafia we must report authority figures trying to corrupt us.

Some subordinates will assume what they're being told to do must not be wrong if their parent or doctor or grandparent or coach or priest tells them to do it. When the targeted person is entrapped, the victim blames the parent which destroys the family: stealth genocide committed against us by the Russian Mafia. Because the Russian Mafia is working to annihilate us, they care not at all that they've destroyed our family structure before they murder us.

Chapter 5.21 – the KGB built a weapon to exploit our hard-wiring

The KGB noticed our species' blind spot last century and decided to exploit it by building their coup around it, and to exploit the fact that we're all born subordinates and socialized from birth to obey the dominants in our lives what happens when the Russian Mafia entraps them and unleashes them against an innocent society?

The Russian Mafia ambushes, entraps, and enslaves dominants not only to sic them on subordinates to flip us into an enslaved nation, but also because dominants were, on average, the individuals who eons ago were most likely to recognize a non-overt threat, raise an alarm and help their people launch an effective defense. Our long ago ancestors included dominants who successfully recognized non-overt attacks. Dominants did that so well so often throughout our evolution that our species became hardwired to pay attention when dominants warned us of non-overt threats.

Because of our species' hardwired reliance on dominants to warn us of non-overt threats (overt threats most of us have no problem identifying and trying to evade), the USSR and the Russian Mafia ambushed, targeted and enslaved as many of our dominants as they could, and their attack is ongoing. They then sic these entrapped dominants (parents, etc.) on subordinates, working to enslave our entire nation, because that's who the Russian Mafia is. And because they've prioritized enslaving dominants, significantly reducing the numbers able to raise the alarm, that's significantly fewer dominants to raise the alarm. When no alarm is raised the people are destroyed.

Our species isn't hard-wired to respond to subordinates' warnings of non-overt attacks because eons ago, when this area of our hard-wiring was established, dominants were part of the scouts who saw the danger from a stealthily approaching raiding party, or were the people who sounded the alarm and built an effective defense. We are the descendants of those dominants, or the people who had access to those dominants' ability to recognize non-overt threats. We don't respond to subordinates' non-overt threat warnings because we're not hard-wired to. This is a flaw, a blind spot in our hard-wiring.

In the Russian Mafia's coup against us, non-compromised dominants sounding the alarm is required for our survival. We must locate and educate these people, and mass release them into our media, Youtube, books, newscasts, newspapers, everywhere, to warn our people about the nature of the attack against us by the Russian Mafia.

Chapter 5.22 - What being in a dominant position gives a subordinate – a little more confidence because often a person in a dominant position is responsible for others and their success in protecting the more vulnerable of their people leads to, or just doing their job competently adds, a little bit more self-confidence

Just a little genuine self-confidence is required for leadership. Subordinates depend on leaders to give them a sense of safety, like newborns prefer swaddling because that's what they're used to. Perfection or flawlessness isn't required. Being a "strongwoman," "strongman," or fearless isn't necessary. It helps being able to think logically on your feet, but if you're a little slower in processing, that's okay. Most people have leadership genes in them, they just don't know it, haven't been given the opportunity, or leadership hasn't been forced on them. I'm a quiet, introverted person but my young mother needed a lieutenant to help her care for my half-siblings and I was her first born. She shaped me into a protector during my infancy, and she and my half siblings taught me what they needed from a leader, so I was one by the time I was five or six. I had two beloved younger siblings who were raised with me by that age and another would arrive when I was 7. I was so young I don't remember my mother's pregnancies. She was just mom – always there, steadfast, our rock. The Russian Mafia destroyed that woman out of evil ambition to rule the world. She's still amazing but she's been destabilized for decades and has no idea who's destroyed her quality of life. I hope she reads this book.

A leader see's in three parts, or at least I did when I was a child, and juggles the needs (not wants) of three parties: themselves, the captain they report to, and the high valuables (in my case sibling-children) they are responsible for protecting. It's possible that another reason the Russian Mafia was unable to entrap me is because I view the world from a lieutenant's point of view, because I was trained from infancy to do so. My mother didn't know she was making a lieutenant; parents worldwide raise their eldest children as helpers, as my mother did. I was raised by a parent who prioritized ethics because her beloved grandmother had been an ethicist and that's where my mother had been the happiest during her childhood. I am my mother's daughter: I didn't spring forth from nothing. She was raised to embrace ethics so she raised her children to embrace them, or at least her oldest children. The KGB destabilized her by the time she was 29. They put her into survival mode which negatively impacted her parenting skills. But I can attest that she has outstanding parenting skills when she's not being hunted and her family and her quality of life is being destroyed. Every nation will not allow the Russian Mafia to kill every oldest child in the world (even their enslaved government allies might protest) but I know, 100%, that the Russian Mafia would commit that level of carnage if it would get them the world domination they believe is their divine right.

When our job requires it, or our birth order, or when we become a parent, we become dominants in relation to other people, usually to people who depend on us. Dominants become carers, people responsible for others. They have more expectations placed on them, increased responsibility, more trust, and as a result, some dominants may have their intelligence expand. They become more informed, often to take better care of the subordinates they're responsible for, and often they become more careful of those they serve, so they keep their people alive, a highly valued characteristic in a leader. Dominants learn that their actions have the power to negatively and positively impact other people so many learn to think before they speak. As a result some dominants stretch. Often, because of the load placed on them, they're forced to learn more, and grow smarter, understand more. Like when you exercise a muscle, that muscle can become stronger, serve you longer.

Chapter 5.23 – Dominants and subordinates adapt to the threat of the 21st century Russian Mafia

Our dominants and subordinates responsible for gathering intel, and our allies' dominants and subordinates responsible for gathering intel, increasingly listen to anyone trying to sound an alarm about the Russian Mafia, and they're sharing information that helps ordinary people better recognize the threat the Russian Mafia poses. And some subordinates are persistent. Because the KGB/Russian Mafia's nuclear avoiding coup takes time, and because we have nuclear weapons the Russian Mafia doesn't want to experience, if a subordinate, me for example, has the good-though-excruciatingly-painful-luck to annoy an American Russian-Mafia controlled team into overtly attacking me nearly everywhere, because I'd unknowingly evaded the KGB's and the Russian Mafia's grooming and/or entrapment attacks by 2001, for approximately two decades, and if that subordinate stumbles on an article her government has released telling people in America the Russian government is on the ground here, and if my same government and/or our allies governments publish books informing their public about how the Russian Mafia operates on the ground here, well that person is able to finally understand, communicate and frame all the attacks she's endured are part of the Russian Mafia's coup, and is able to report this genocide in time to prevent our nation and our allies' nations, and other innocent nations, from being destroyed by this malignant, evil mafia. That's a huge win for our species. I understand the Russian Mafia, in it's evil and in it's access to nearly all our institutions knows just how much our government knows, so when my government acquires a copy of my book, so does the Russian Mafia. That's a negative. The Russian Mafia is moving all they can to destroy me even harder now. I understand them. Whatever they have to do to take me out, they'll do, no matter what protection list my government puts me on. Former President Kennedy had Secret Service and I'm convinced the KGB assassinated him using American mafia.

Chapter 5.24 – We're hard-wired to rely on dominants to sound the alarm

We rely on dominants to sound the alarm about non-overt threats. The Russian Mafia has enslaved many of our dominants to prevent them from sounding the alarm. No alarm, no defense, the Russian Mafia installs a puppet, and we'll be destroyed. This is our pre-Holocaust moment. We must free our dominants and/or marshal our unentrapped dominants to warn our people and launch an effective defense.

Chapter 5.25 – Non-overt threats and our hard-wired blind spot

We heed only the warnings from dominants with regards to non-overt threats. We depend upon dominants to sound the alarm, hardwired into our ancestors eons ago. If dominants don't sound the alarm about a non-overt threat, no alarm is raised and we're destroyed. When subordinates try to warn us about a non-overt threat, their warnings are ignored and/or illicit annoyance. I've tried since 2016 to sound the alarm and I've been ignored. I didn't know who the perpetrator was but I knew there was one. I began writing to presidents as early as 2016. I've written to every president since President Obama late in his term, even former President Trump, early in his term. After I saw his meeting with Mr. Putin in Finland, I refused to write him again.[118] What usually happened when I wrote previous Presidents was an uptick in police harassment. They didn't care that I knew they monitored my mail. Most of my mail or purchased items are open before I receive them. The Russian Mafia doesn't care that I know it, they like frightening me and causing me anxiety. They open many if not most of my packages before I receive them, and quite a bit of my mail. I see the opened items but no one else see's the attacks.

---

118 Julie Hirschfeld Davis "Trump, at Putin's Side, Questions U.S. Intelligence on 2016 Election," *New York Times* (7/16/2018).

Chapter 5.26 - To exploit our blind-spot the Russian Mafia and before it the USSR spent the last 50+ years stealth attacking & enslaving our dominants, in order to greatly reduce the number of dominants we have available to sound a species alarm putting us on notice

So far, I'm the only person I've read or heard of who've called the 2023 Israel-Hamas War a proxy war. Our security agencies investigate and don't make allegations they can't back. The Russian Mafia knows that. I'm a dominant only in my relationships with my siblings because I'm first born and my family needed me to be a dominant when I was a child. I was a child-dominant to assist my family. The Russian Mafia has been on me for decades to entrap me. By attacking and enslaving our dominants, the Russian Mafia has greatly reduced the number of dominants we have available to us to raise the alarm. We must organize large groups of noncompromised dominants to get the word out about this type of attack. Without dominants raising the alarm, the Russian Mafia will install one of their puppets and destroy us.

We must change our unwillingness to launch complaints against dominants trying to entrap us, like our parents, doctor, or any other previously trusted dominant in our lives. The Russian Mafia really screws with people's head's by exploiting dominants. They've used dominants to try to entrap me for years. Dominants include grocery store cashiers. In the cashier/customer relationship, the cashier is the dominant. The Russian Mafia exploits every dominant/subordinate relationship their target is part of to destroy the target. When I take the bus I'm a subordinate to the bus driver. The Russian Mafia exploits my subordinate position to get the driver to ask me what stop I'm getting off. Based on that intel, which they couldn't themselves get from me, they have harassing operatives all along the route to my destination and have their operatives ready at my destination to harass me. I've lived this, is how I know.[119]

---

119 The Russian Mafia set into place new policies for the two bus service entities I most often take, in order to more easily harass me. It is in their best interests to murder me and natural causes is their first choice since shooting me would draw too much attention, and I declined their operatives offer of free drugs of choice. I use a mobility device. To harass me more easily the Russian Mafia had two bus services in two different cities change policies so that drivers ask passengers on mobility devices where they'll be exiting. The only reason the Russian Mafia wanted that information was so they could more easily and efficiently position operatives along every point of my trip. They already had operatives on the bus harassing me and listening to my answer. With this new policy, the Russian Mafia positioned operatives at my exit and at every point along my trip. When I went to pick up my prescriptions, the Russian Mafia had operatives at my exit point, blocking my ability to exit the bus many times. They had their operatives on the sidewalk blocking my access to the sidewalk, deliberately space harassing me by having groups of operatives slow walk in front of me (I'm writing these words March 2024. I believe the Russian Mafia worked this stealth attack as recently as January 2024 and/or February 2024). Russian Mafia-controlled American law enforcement operatives love to order senior citizen operatives, disabled operatives, and children operatives to harass me because no one thinks those groups are owned and deployed by the Russian Mafia so American operatives who manage these attacks against me believes no one, not even anyone observing me being blocked, would believe it is a deliberate attack because senior citizens, disabled people and children are considered non-threats. On their own individuals in those groups are non-threats but in the hands of the Russian Mafia, who use them to stealth attack me, they are transformed into effective weapons. The Russian Mafia harasses me in ways I recognize as harassment from their decades of attacks but which no one else will. The Russian Mafia attacks me because they're working to stealth assassinate me and while they're 100% sure no one will believe my complaints about being attacked, they'd just as soon get me gone. I on the other hand want to get them gone. Because they couldn't entrap me, their operatives do their best to harass me to death with stealth attacks. By using operatives

Based on how the Russian Mafia exploits dominants to get information from me, entrap me, launch entrapment attacks at me, I'm estimating that hundreds of millions of people in America are having the dominants in their lives try to entrap, groom, lure them and/or influence them.

Hierarchy is a key survival strategy for our species and the Russian Mafia's weaponization of it to destroy and confuse us is a big problem for us. *We must flood our nation with information about this type of attack*. Almost no one is going to file a police report against their parents or doctors or other dominants when those people try to entrap and/or groom them.

The dominant (parent, doctor) is told to offer their services/advice. Entrapment is achieved when the target agrees to cheat something or someone, usually the government. Apparently there's a mechanism in place where entrapped victims are reported somewhere to our government. The Russian Mafia must hit certain beats to get people entrapped; the Russian Mafia doesn't tell our government that they've compromised our law enforcement running these entrapment scams. Nor do they tell victims they've been illegally entrapped by an enemy foreign government. The Russian Mafia isn't big on transparency.

From birth, people are taught by real world experience that dominants 'help' us so even though the entrapment is meant to harm the target, the dominant saying "I'm trying to

in power positions in bus management the Russian Mafia got two separate cities to change their policies so that the Russian Mafia can more easily harass me to death. No lie, no exaggeration. In fact, understated.

help you," their words, and the dominant saying them, overrides direct evidence that the dominant is trying to harm them. I know I keep saying this but the Russian Mafia is evil. From birth dominants help us. Before we know words or can do anything but cry, sleep, poop and drink milk, dominants are essential for our survival. And we know it from our first days of life. By framing the attack positively as 'help,' the 'help' offer from a dominant, confuses people's hard-wiring and soft-wiring, and they accept the offer having no wish or interest in cheating and/or stealing from anyone. The item is desired but stealing to acquire it is not. How to decline a kind doctor's offer without seeming rude, when from birth we're raised to accommodate dominants, to obey them. When a dominant controlled by the Russian Mafia offers to 'help' us acquire a desired item by lying/cheating, the dominant-operative doesn't say "I'm lying and cheating to help you get this." They say "I'll help you out by helping you get this item you've expressed interest in. I'll label you as diabetic. Your insurer won't question my labeling you that." People scammed in this way have no interest or desire is cheating anyone.

This is an illegal attack against innocent people by an enemy foreign government working to annihilate us by exploiting the fact that we're raised to be polite to dominants.

Psychological warfare in the hands of a monster is destroying us. The cashier at a big box store tried to entrap me, tried to get me arrested by refusing to scan many of the few products I had in my cart. Had I not noticed and immediately turned around, the store's security, who's hunted and stealth harassed me for many years, would've arrested me and the Russian Mafia would've finally got me into the criminal justice system where they could destroy me as they're obviously are hot to do. Despite their endless lies the Russian Mafia isn't working to destroy us because of anything we've done, but because they covet the financial might our government enjoys, which our government achieved for our sacrifices in WWII when we joined the war effort and helped destroy the Nazi government. We don't have a beef with the KGB/FSB/the Russian Mafia, the Russian Mafia has a beef with us. They've developed an effective weapon and they're on the cusp of destroying us. For over 53 years they've inserted themselves into my life to my harm and to my family's harm since I was 11. The KGB attacking an 11 year old a world away sounds bizarre doesn't it, but it's true. No one could image our security agencies targeting children a world away to stop them from learning an honorable trade, acquiring a career, helping their family, helping their country thrive. Our security agencies are the gold standard – the Russian Mafia forces slavery, death and destruction. And baby killing. Yet they demand world dominance as if they're entitled to it because they're willing to kill everyone's babies. The Russian Mafia knows we won't accept their evil so they hide it, and lie like a rug.

I see their evil hand in our overdose rates and our homelessness. I see them. They have no upside. I push for their annihilation as a defensive move not because I enjoy saying we need to kill a mafia. I don't see them stopping their attacks against us. They've had a great deal of success doing what they do.

We're in trouble. The Russian Mafia is messing with people's sense of reality. When you start messing with human beings hard-wiring and use dominants to urge us into corruption, it really confuses people, especially our children, who're increasingly turning to drugs to self-sooth. This is an advanced type of warfare and all of us must be warned.

The Russian Mafia has forced my parents to try to entrap and/or influence or lure me many times. They've been doing it for years but I didn't know because I didn't know I had a world class, best in class enemy until the Russian Mafia's American operatives grew so annoyed with me for unknowingly evading their many entrapment attacks (I evaded them because I live ethically, it had nothing to do with knowing I had enemies who ordered strangers to attack and/or harass me nearly everywhere I went, which the Russian Mafia has done since approximately 2001.

Years later I witnessed a young man being forced to harass me by a uniformed police officer. He'd had no intention of bothering me although apparently he'd been ordered to do so because when he saw the police car tracking slowly behind him, monitoring him, the young man harassed me. I shifted away from him. But before he saw the police officer he'd had no intention of bothering me.

The Russian Mafia had been on me for decades and I'd had no clue because they hadn't wanted me to have a clue, just like they don't want other victims to have a clue. It was because I had a clue (after decades of overt attacks) that when they tried their shoplifting entrapment I understood, after the fact, that I'd evaded an another entrapment attack. Later I understood the reason the Russian Mafia deployed the cashier-as-weapon is because the cashier is dominant in a cashier-customer exchange and from birth we're socialized to trust dominants. I assumed the cashier would ring up all the items I had in my cart but he deliberately failed to scan about four of the 10 items. It was a lightly loaded cart. We can't go on auto pilot with the Russian Mafia hunting us using our socialization against us. I didn't recognize the shoplifting incident as an entrapment until after the fact but two decades being harassed meant I'm never left alone, and I understand I'm being hunted. When a person knows she/he are under attack they're just that little bit more aware on a subconscious level. Being that little bit more aware (and my mother-instilled-ethics), even without recognizing the entrapment before it was sprung, helped me immediately turn around as soon as I saw the discrepancy between my receipt and my groceries.

All these attacks against an ordinary person. I was an ordinary office worker before the Russian Mafia worked their exhaustion/unrelenting harassment warfare model against me. I read books for fun, all kinds of books. I often couldn't afford vacations, just one now and then to visit my extended family, so I read. I like romance novels but I read anything. I love math and learning new things. I worked and helped my family as best I could and took classes that interested me. I didn't even know I was evading the Russian Mafia, but they knew and they're committing genocide against us. Whatever they do to entrap people, it's bad: the only way out is suicide or victims are forced to attack their loved ones. When I read about ordinary and noted people committing suicide, I believe that person refused to destabilize their loved ones and friends. They can't tell their loved ones for fear it'll put their loved ones at risk. A boss I loved committed suicide and he couldn't tell me he was entrapped. He knew if he did whoever had him would murder me. After he died, I was deposed. His compromised handlers knew they'd tried to force him to attack me and that he wouldn't so they questioned me under oath after he died to see what I knew. But I was heartbroken and stunned by his death, having no idea there'd been a problem. It was obvious to those deposing me that I knew nothing about why he committed suicide. He saved my life by not telling me how he'd been entrapped. Plus he was smart: he couldn't free himself (and he was brilliant: the Russian Mafia are world class evil). But he knew I wasn't entrapped because his handler's tried to sic him on me (I've now deduced). Compromised people cannot speak (I've no idea the threats against them, but they're bad, probably they and their extended family are threatened with murder and/or imprisonment). He knew I wasn't compromised so I was free to speak but I needed to know more. So he gave me a great educational package as his Christmas present to me, to encourage me to keep learning (he used to be a teacher). Of course I watched the entire series of videos and liking them, bought even more. They didn't directly help me understand what was going on but they did widen my world view and help me better understand my species and the world, which is why I've been able to figure out the Russian Mafia's attack and how to disable their bomb. All the learning helped me understand more. I was better able to deduce, to make connections, to draw conclusions based on my experience and learning. His 'Hail Mary' play worked: I'm now telling the world what I suspect was done to him and what I know for sure has been done to me. *He was a hero*. Anyone can be ambushed if they have no idea an evil enemy foreign government is stalking them.

At the job before that they'd finally gotten to the point of inserting two female operatives as office friends into my work life, after they'd harassed me from the word processing center which I'd loved before my supervisor and coworkers began harassing me. So, to give the Russian Mafia another bite at the apple (another entrapment attack), they made the word processing center a hostile environment so that I transferred to a secretarial desk. The Russian Mafia had two operatives ready and I interacted with them, but I'd gotten another job and the operatives who'd organized the attack were frustrated they went loud on me. I wouldn't wish that experience on anyone but the upside is I finally was told I had an enemy, where before I'd had all kinds of awful things happen and/or done to me but had no idea why. To reward my decency with a kick-in-the-butt attacks, using proxies. That's how I know they enraged the Capitol attackers. They've been on me for decades and they absolutely love rewarding a decent, kind, hardworking, self-sacrificing person with hard smack downs to convince them their country is 'rigged.' I saved up money so my mother could buy another house and her handlers had her spend it and then used my siblings to drip suggestions of lies and cheating by my mom. No one in my family but me knows we're victims of a coup by the evil Russian Mafia. I hope my family reads my book so they'll finally know who crashed and burned our family and why so many of my loved ones are miserable. We really are victims and tragically for our people and nation, we're far from the only ones.

The Russian Mafia entrapped my parents and forced them to try to entrap me. The Russian Mafia also entrapped most if not all my siblings and forced them to try to entrap me and forced them to try to entrap me, thereby destroying my family. They attacked me when I was a child, they attacked my young mother, my siblings, siblings in my extended family. They are responsible for hooking some of my loved ones on drugs and causing them to overdose, and they're responsible for trying to hook me and other loved ones on drugs. I'm 64 and for decades I've directly experienced the Russian Mafia forcing entrapped parents, teachers, field trip minders, supervisors, landlord, bus drivers, store security, police, and many, many other people in America to covertly and/or overtly attack me. I've been hit by two vehicles by driven by Russian Mafia operatives and threatened by many more operatives who revved their engines when they saw me, and used their vehicles in many ways as weapons to intimidate me. The Russian Mafia has attempted to blackmail with regards to bestiality and encouraged me to become interested in that lifestyle. They also an operative encourage me to get a dog. They introduced bestiality based porn into my internet thread, left a disgusting message on my phone. I'm not into bestiality so it took me awhile to understand what they were going for. At the time of the blackmail attempt I didn't know the Russian Mafia was my enemy so I was stumped as to what was going on.

Chapter 5.27 – We must rethink our unwillingness to out trusted dominants who try to corrupt us

Another problem we have the Russian Mafia exploited is our unwillingness to out loved ones (like our parents and/or family), or other dominants (like our doctors) in our lives who the Russian Mafia has flipped and use as weapons against us. I can't remember all the entrapment attacks from dominants in my life but here are a few: both my parents, combined maybe ten separate entrapment attacks, my doctor was used in at least two separate occasions, cashiers at one big box grocery store tried to entrap me so store security could charge me with shoplifting. He refused, twice, to scan all my items. I believe I was minutes way from being arrested for shoplifting. At least two supervisors were deployed to entrap me.

The Russian Mafia force dominants to stealth harass me using children in their care as weapons. The overwhelming majority of these children were under 15. The Russian Mafia is a mass murderer. They are, I believe, responsible for the Israel-Hamas War and complicit in the murder and injury of over 30,000 innocent people in less than six months: Gaza Palestinians, Israeli's, minority workers in Israel when Hamas attacked. The Russian Mafia facilitated such evil to unseat their political rival, President Biden. Not only that, the Russian Mafia has weaponized protesters (which they've admitted to incorporating into their strategies in a 2013 paper[120] by Russian General Gerasimov to falsely accuse Americans and our President for mass death in Gaza when the Russian Mafia is the perpetrator, while Mr. Putin publicly calls for cease fire. This is the man and the Mafia/government who demand rule over our species. They don't think what they're doing is wrong because no one is outing them publicly but me. Where are our noted writers? Philosophers? Patriots? Ethicists? Is everyone owned but me? Can no one see that the Russian Mafia ordered this war to turn the world against us? People think President Biden is being blamed but that's not how the Russian Mafia is playing it. America appears as hapless because the Russian

---

120 General Valery Gerasimov, Chief of the General Staff of the Russian Federation,"The Value of Science in Prediction," *Military-Industrial Kurier* (February 27, 2013). General Valery Gerasimov states: *"... wars are no longer declared..."* *"...The very "rules of war" have changed...."* *"...The role of nonmilitary means of achieving political and strategic goals has grown, and, in many cases, they have exceeded the power of force of weapons in their effectiveness...."* *"...applied in coordination with the **protest potential of the population....**"*

Mafia is constantly creating messes for us to step into, stealthily. We're not at all evil. We've drawn an evil enemy determined to flip the world order. If we allow them that, it'll be Holocaust 2.0 but this time, Americans will be the slaughtered victims. Anyone serving as president not owned by the Russian Mafia will be made to appear not bright. Our President is smart but there's no way he can anticipate attacks the Russian Mafia set into play five years ago. The Russian Mafia could make Jesus look stupid.

We don't understand our enemy and that's just the way the Russian Mafia likes it.

Our species is hard-wired & soft-wired to hear dominants' warnings about non-overt threats. If dominants don't raise the alarm, no alarm is raised and the tribe is destroyed. When subordinates try to warn of non-overt threats their warning triggers annoyance in our species, a blind spot in our hard-wiring the Russian Mafia exploits to destroy us. Short-term solution – prevent the Russian Mafia from installing a pro-Kremlin candidate into the White House. Long-term solution – launch an effective defense which include informing our people.

Our security agencies investigate before they publicly accuse a government of wrongdoing. I've been attacked by the USSR and Russian Mafia operatives for 53 of my 64 years of life. I recognize their signature: (1) the attack benefits the Russian Mafia's publicly stated goal of removing our government and replacing us with themselves; (2) massive numbers of innocent people are murdered; (3) the Russian Mafia use it's proxies to accuse Americans or our government.

Chapter 5.28 – We'll hire non-compromised dominants for the vignettes we'll make to save ourselves

To evade this dominant-specific bomb we not only deploy our officials who warn the us and the world in their official capacity as our dominant leaders, but we hire many non-compromised dominants and deploy them in videotaped vignettes to warn all our people in all our languages about this weapon. We hire and educate them about the attack & what's at stake, & release them into our mass information systems, i.e. media, school, newspaper, books, etc. Fortunately the hired dominants can be ordinary people, but must introduce themselves as dominants in a way our species will recognize. Parents, grandparents, teachers, lawyers, presidents, police, vice presidents, property owners, bus drivers, etc. Dominants are ordinary people who have societal approval under certain circumstances to control others. Our species will only hear dominants warning's of non-overt threats.

Chapter 5.29 – Dominants & their families attacked decades ago in America – attack keyed to our hard-wiring

For decades the Russian Mafia (& before it the USSR), targeted & enslaved our dominants to prevent them from sounding the alarm to warn our species we're hard-wired to hear. To unravel this trap, we must launch an effective defense deploying non-compromised dominants to sound the alarm, flood our media with many hundreds of dominants warning people, from now through the election.

The Russian Mafia's coup is a precision weapon keyed to our dominant-subordinate hard-wiring, which was developed over millions of years, with contributions from ancestors who lived long before our ape-like ancestors. Some of our far distant ancestors were fish – later a small shrew-like mammal added it's gifts to our genes. We, and all life on our planet, are marvels of determination, tenacity, resiliency, and luck.

We are the last of our kind, we must survive. The many human species cousins we had as recently as 100,000 years ago have gone extinct, not because of anything we did but because of the environment they lived in and their abilities and limitations at the time. We are not meant to go extinct in our infancy, we're only a couple hundred thousand years old. We are meant to grow beyond our cradle to become competent space explorers, embrace the boon of life, figure out how to self-manage so our entire species thrives, become capable planetary stewards who equitably manage earth's bounty with our co-inheritors.

Our hard-wiring is ancient so we must talk to it in the language it understands. It recognizes dominant-subordinate because that survival strategy existed when our distant ancestors lived. Modern people know dominant-subordinate, we just don't know that we know it. But we've known it from birth, born subordinates and dependent for existence on the dominants in our lives. Dominant-subordinate is a survival strategy our species created to stay alive.

To talk to our hard-wiring we must create and deploy many dominant-subordinate vignettes in all the languages spoken by our people in our land. After the dominant and subordinate are established (which takes seconds), the dominant warns us of the non-overt threat. We'll ask our excellent Biden Administration and other government officials to issue public statements warning of this non-overt threat. We need to flood our nation with this information, we don't have enough government officials for what we need, so we hire non-compromised dominants, videotape a short, believable dominant-subordinate vignette and have the dominant warn our nation.

Subordinates who try to warn of non-overt threats will be ignored because our hard-wiring didn't evolve to hear them – when our hard-wiring was "locked in" with regards to non-overt threat warnings, the dominants in those times did the successful warning. They survived to reproduce, we are their descendants, we inherited part of their genes. Talking heads with their title floated beneath will be ignored by our hard-wiring because it can't understand the language. *Our hard-wiring understands only the dominant-subordinate language with regards to non-overt threats*. This blind spot in our hard-wiring is why the USSR designed Active Measures to enslave our dominants, i.e. our parents, police, teachers, etc., to prevent them from sounding the alarm. Unless dominants sound the alarm, no effective defense will be raised, and we will be destroyed.

Chapter 5.30 – The Russian Mafia deliberately use dominants as corrupting, destabilizers and entrapping agents – to make citizens lose hope in our democracy, our government, our families and ourselves

I was stunned when my parents, my siblings & my doctor encouraged me to lie, cheat & steal from the government – the Russian Mafia well knows most people won't report their parents, siblings, or doctor for suggesting illegal acts when they're just trying to "help" the target. The entrapment is framed as "help" but is actually an attack. If you accept it your life, your family's life and everyone you care about will be destroyed, through you. The Russian Mafia will force you to entrap, coerce and/or influence everyone you care about and force you to advocate to everyone in your circle to vote for one of the candidates the Russian Mafia wants installed here. This is a coup but it's also genocide because the Russian Mafia is deliberately destroying our families, the cornerstone of our species and our culture. The Russian Mafia is evil in so many ways but forcing parents and doctors they've taken control of to use as camouflage to entrap children & patients is beyond vile.

Seizing control of our dominants, via entrapment, blackmail, influencing, coercing, etc., the Russian Mafia use dominants as proxies and camouflage to influence, lure, groom, corrupt and entrap subordinates. Many subordinates (ordinary people) believe in nothing because the Russian Mafia has sent our dominants out to suggest and lure people into corruption. Entrapped dominants aren't free to reveal that they're being forced to suggest illegal and/or immoral behavior. They're slaves of the Russian Mafia and the Russian Mafia wants more slaves and as many dead and disheartened Americans as they can destroy.

When my parents engage with me they aren't allowed to tell me they're speaking on behalf of the Russian Mafia: the Russian Mafia exploits them as camouflage and use them to extract information from me because they're using my parent's shell. If the Russian Mafia stood before me they'd have no authority to question me, they're here illegally. The Russian Mafia had bus services in my area have bus drivers now request disabled people when they'll exit. I experienced this attack and it was because the Russian Mafia wanted to be able to more easily attack me where I was heading. Since they had no authority to ask me directly, they installed a policy in Santa Monica and Los Angeles buses that I ride on where drivers ask me my exit stop. I'm not obligated to tell them but it's amazing how hierarchy and dominants work: I am socialized to respond to a dominant when I am in a subordinate position, as I am when I'm a passenger on the bus, and the driver asks something me. From birth we're conditioned to accommodate dominants and it takes a conscious effort to come up with a workaround to protect myself on the bus given I know the

Russian Mafia attacks me with whatever I say. The Russian Mafia exploits our socialization how we're raised from birth. They are truly evil.

They exploit my parent's authority and access to me and use them as weapons against me. They use my parent's and my half-siblings to access me. I doubt my parents know the Russian Mafia controls the American law enforcement who control my parents. Forcing parents to entrap their own children and extended community destroys that family and community - genocide. I charge the Russian Mafia with genocide against my people because they are in the process of deliberately destroying my people and community and attempting to seize our government's assets.

People are bewildered by our nation's deterioration and alarmed they can't see the reason for it. The Russian Mafia is the reason behind it but they don't want it known because they don't want us launching an effective defense. The Russian Mafia respects none of our laws nor the treaties the USSR and/or the Russian government signed/signs. In 1971 the USSR was on the ground demonstrably breaking the Geneva conventions they'd signed in 1948, and successfully working to destroy my quality of life until 1991, it's demise. Looking back, I can see a clear difference between the USSR's attacks and the Russian Mafia's attacks. The Russian Mafia's attacks are sharper, more precise. They include entrapment attempts, more physical assaults, far more psychological assaults, and the deployment of more individuals in my family as weapons against me, including my both parents. The USSR prioritized subversively deploying proxies to financially and emotionally destabilize me, especially in my 20s, until the USSR's demise in 1991.

Because of the USSR's attacks I know how the Russian Mafia has built and attached victimized Americans to the GOP Base. Short answer: the seed the USSR planted using the Jackson-Vanik Amendment, the USSR and the Mafia grew into enough destabilizers to enter, analyze and destabilize ordinary people's lives. Decent, hardworking, self-sacrificing people are deliberately made miserable by the Russian Mafia when their best efforts financially and/or emotionally ricochet to destroy them (ordinary people have no idea the Russian Mafia is here waging secret war on them). Nothing they do turns out as it should. An example of this type of psychological warfare is the first few years of the Biden Administration when President Biden and his Administration worked like dogs and their best efforts resulted in extremely low poll approval ratings. This mismatch: the Administration working flat out to help the American people, and the American people seemingly turning their noses up in disdain, is one of the way's the Russian Mafia builds rage in ordinary people.

The USSR's and the Russian Mafia's attacks didn't work because (1) my mother made me corruption-resistant during my early childhood, long before the USSR's initial attacks on me when I was 11; and (2) during the ongoing phase of the coup they inherited from the USSR and have expanded, the Russian Mafia is prioritizing entrapment because it allows them entry into most people's lives. To the Russian Mafia access equals control because their war is undeclared. Once entrapped, victims can never psychologically or emotionally re-stabilize. Their handlers won't allow it. When I demonstrated by evading entrapment (which I did unknowingly[121] until the age of 51), Russia-controlled

---

121 With one exception – at my final job (I was 47-48?), the partner I worked for gave me a company business card and asked me to make purchases on his behalf. He told me no one but me would see the statement and that I was responsible for approving the purchases. I'd worked for partners before, in permanent and/or temp positions. None had given me a business credit card, asked me to make purchases for them, and told me I'd be the only person reviewing and approving the statement. By that age I'd been harassed by strangers on the street, the bus, in grocery stores, etc. for at least 7 or 8 years, so I knew something was going on. Quite a few of the people stealth harassing/attacking me were uniformed police. Because I was being overtly harassed I knew I had a problem. I didn't know why I was being attacked or who was behind it, but I knew there was a problem, whereas before the Russian Mafia's attack-by-stranger-overt attacks I hadn't known I had an enemy. When I was 11 the child the USSR had attacked my 10 year old brother in front of me, and another child threatened me on the school bus, and three teenagers I barely knew threatened and tried to beat me up in high school. Those attacks were overt but could be explained away as childhood bullying, KKK-related, and the teenage girls, I had no idea. Because of 7-8 years of strangers' harassment, my boss' pitch made me suspicious because he said "you'll be the only one to review and approve the statement," was so strange, I noticed. There was absolutely no reason why a legal secretary would be specifically told that only she'd review and approve a credit card statement unless it was some kind of setup. He could've simply asked me to make occasional purchases for him and it would make sense that he'd review the statement. I'd worked in legal for nearly 20 years by then and (1) no other partner had made such a request, and (2) I was being harassed, nearly daily, to and/or from work. I wouldn't have stolen anything had I not been overtly harassed/attacked, but you see the advantages to the Russian Mafia of stealth attacks, using this one example. Because their American operatives told me, by overtly harassing me, that I had an enemy, when the Russian Mafia tried to deploy an embezzlement entrapment I didn't have to rely entirely on ethics, I was able to make the connection between their operative's attacks nearly everywhere I went, and my supervisor's offer. The overt attacks/harassment,

American law enforcement operatives went loud on me out of frustration and began overtly attacking in the street, bus, and in grocery stores, etc. They'd already stealth attacked me on most of my full time job, post-university. Because they did so, I know why we experienced extremely high turnover rates as covid was ending: ordinary people are harassed at work, and many other places, by Russian Mafia's proxies, to make as many of us as miserable as possible, as they attempt to push us into overthrowing our excellent government.[122] Their coup relies on them installing one of their many puppets and agitating and frustrating us into the streets, demanding change. That's rich. The Russian Mafia would destroy our government and install their own and begin to slaughter us, stealthily, quietly. If you think excess mortality would be reported in our newspapers if the Russian Mafia seizes control you're very much mistaken. Please don't vote for any pro-Kremlin candidates. If Israelis could get a do-over you better believe

although extremely psychologically painful and terrifying (I was made homeless at around 12 and witnessed my mother's collapse -- both terrifying to most children), still, the attacks from age 40-41 were high value intel because they served to warn me. Most people are ambushed, set up, having no idea they're being hunted. When their American operatives decided to go loud (overtly attack me to-my-face when I was 40-41), the Russian Mafia lost it's ambush weapon. That's a key component of the Russian Mafia's coup against us and why so many innocent people are entrapped. The Russian Mafia's mistake starting when I was 40-41 is how I was able to eventually understand that my parents' and my siblings' encouraging me to lie/cheat were authored by the same entity who'd been attacking me since I was 40-41. So, that entity had destroyed my family, for no reason that I understood.

122 Janis Berzins "The New Generation of Russian Warfare" "...the Russian view of modern warfare is based on the idea that the main battle space is in the mind." Aspen Institute | Prague, *Aspen Review* (March 2014). https://en.m.wikipedia.org/wiki/new-generation-warfare.

Mr. Netanyahu and his government wouldn't be in office. The problem for Israelis and for all of us is the Russian Mafia. We better deal with them.

The Russian Mafia deploy entrapped dominants to entrap and/or influence those subordinate to them, using the dominant as camouflage, because that's the fastest and most efficient way to enslave millions of my people.

Subordinates aren't willfully blind, we are hard-wired and soft-wired blind: we're socialized from birth to ignore non-overt threats, we rely on our parents to warn us if there's a cover problem (soft-wiring & socialization) we don't see. When a subordinate warns other subordinates about a non-overt threat, the intel doesn't stimulate an effective response, they're ignored and/or they irritate the people they're trying to save. I'm not a dominant (except by birth order), and I've tried for over eight years to warn my people about the problems I've experienced to no avail. The warnings must come from dominants or the warnings don't register with our species. It doesn't help the Russian Mafia floods media with disinformation, making it harder for subordinates to sound the alarm in media that's compromised.

Unless dominants warn people, we won't survive. An example of what we can look forward to at the hands of the Russian Mafia is what Ukrainians and Gaza Palestinians are enduring, except the Russian Mafia hate us, so they'll treat us far, far worse.

To keep our democracy this November 2024, we must re-establish the use of dominants in our society, replace and free the ones the Russian Mafia have entrapped and/or otherwise commandeered. We need to hire a team of dominants, they can be people of note or not. We educate them and train them to point out the threat of the Russian Mafia's coup to our people. And keep saying it. Political leaders may believe this is overkill but we're dealing with hard-wiring that requires dominants, so dominants we must provide. Tragically, the Russian Mafia has provided plenty of examples of their brutality, so no need at all to exaggerate their cruelty.

Our society's dominants would naturally provide this service for us were it not for the fact that many are entrapped, blackmailed, etc. We must survive and to survive we need dominants to talk to our people about the threat the Russian Mafia is to us. *When dominants do not report the danger, subordinates do not see it*. I request the Biden Administration hire a team and train them, and send them to media outlets, etc. so people are told what's at stake, from now until the election. We've not a day to lose.

Chapter 5.31 – Friendlies – the KKK & the USSR

In 1950s and 1960s America the KKK[123] attracted citizens and law enforcement interested in destabilizing minorities, women, homosexuals, and any other group requesting a government more responsive to their needs.

As an "in" across our country,[124] the USSR had it's operatives offer the KKK and/or local police it's services discouraging, destabilizing and/or weakening minorities, which the USSR operatives did quite effectively given that they were trained in KGB methods of state control, and respected none of our laws. Using the blood and pain of minorities and women (similarly to how the Russian Mafia is using the innocent blood of Gaza Palestinian children as pawns to unseat President Biden. African Americans (also unknowing victims of the Russian Mafia) – attacked over decades in America, from I believe, but cannot prove, at least the 1950s to the present, in order to keep us in a simmering rage to use us as a weapon to destroy our nation.

We, and other people of color, and now Caucasian Americans, were first attacked by the USSR's operatives, and now by the Russian Mafia. Sometime around the mid-20th

---

123 General Valery Gerasimov, Chief of the General Staff of the Russian Federation,"The Value of Science in Prediction," *Military-Industrial Kurier* (February 27, 2013). *"...wars are no longer declared...." "...The very "rules of war" have changed...." "...The role of nonmilitary means of achieving political and strategic goals has grown, and, in many cases, they have exceeded the power of force of weapons in their effectiveness...."* "...Among such actions are *the use of special-operations forces and internal opposition to create a permanently operating front through the entire territory of the enemy state...."*
124 Ibid.

century I theorize the KGB began their attacks against people and families of color because we had fewer legal protections at that time. I believe the KGB made their bones (established themselves) in American partially by their alliance with the KKK and offering destabilization services targeting people and/or groups the KKK didn't want to share power with. Even though the KKK the KGB is on order of magnitude many times worse and are now working steal our country right out from under us, and destroying the descendants of the people who once hired them. The KKK was conned, like many other ordinary Americans. The KKK certainly wasn't innocent but they had zero idea the KGB intended to steal our country. The KGB and the Russian Mafia don't share their true intentions with their targets. The KKK didn't know they were just another group on the hit list, only further down, after the people of color, homosexuals and others groups were destabilized. A shared skin tone means nothing to the KGB or the Russian Mafia. They used it to gain entrance to the KKK while spouting racist and homophobic tropes but to them, the KGB/FSB and the Russian Mafia are the master race: everybody else, irrespective of skin tone, is nothing. Worse than their hatred of others is their intention to do something about it. Think about it like this: most people in the world with an opinion don't go out into the world murdering all and sundry to take what they want - insisting their enemies are corrupt, while failing to tell the world that they're the one's responsible for destabilizing those very same enemies. The

Russian Mafia does. And thinks there's nothing wrong with their view point. They're off the rails. They have a problem.

The USSR and the Russian Mafia used our police as proxies to extra-judicially attack me, going so far as to have four separate cashiers in four separate stores over several months not scan all my food purchases, so store security could accuse me of shoplifting. No lie. I was just blessed to even notice it – I believe I was minutes away from being accused of shoplifting, but I quickly turned around on my mobility device, re-entered the store going against the flow of exiting customers, told the cashier the problem, and insisted on paying. Even after I'd returned, the cashier failed to scan every item he'd missed. I had to point out an item he'd missed to make sure I'd paid for it. The Russian Mafia was so determined to get me into the criminal justice system they told that cashier not to scan all my items. I had to ask him twice (after I'd already gone through the line) to please scan all my items.

Outside the store plain clothed operatives stared at me. I didn't know it was an entrapment setup until all the plains clothed security and plain clothed police stared at me, out of the corner of their eyes. Ethics saved me. God helped me notice that all my groceries hadn't been scanned. Ethics taught to me by my mother ensured I immediately returned to the cashier to pay what I owed.

The Russian Mafia has worked decades ensuring our police disproportionately kill people of color, mostly our men, *but in my case, the USSR attacked me in my family of color.*

They've worked decades to destabilize me, even, I believe, hooking my loved ones on drugs and/or stealth killing them.[125] A great many Americans, and tens of millions of people worldwide, sympathize greatly with Gaza Palestinians: including President Biden. He and his Administration are working as hard as possible to help build a Palestinian state that Palestinians control.[126] However, the Russian Mafia started the Israel-Hamas War and are stealth running the war, keeping Palestinian statehood out of reach so they can blame President Biden for the misery of two million innocent Palestinians, when the perpetrator is, as is often the case, the Russian Mafia.

The USSR made their bones in communities across America, and made some in the dominant culture who contracted them very happy. What the USSR didn't share with it's customers is that the Russian government is highly acquisitive, and has an agenda it doesn't discuss with it's allies. Making it's bones by serving the dominant culture while casting an envious eye across America, the USSR set into play it's plan to steal our country right out from under us.

Because the USSR's operatives lied so effortlessly and because the dominants they communicated with believed them harmless towards themselves, the USSR thought our people stupid. If you listen to or read Mr. Putin's opinions, he skirts

---

125 Maya King "Black Pastors Pressure Biden to Call for a Cease-Fire in Gaza," *New York Times* (1/28/2024).

126 President Biden and his Administration are working hard behind the scenes to deliver Gaza Palestinians their own nation. The Russian Mafia will, no doubt, tell Hamas this was the Russian government's intention all along. The Russian Mafia are world class, best in class liars. Their primary intention is to destroy Americans.

around saying the Russian government is the master race, but you can hear it in the language he chooses.[127]

Dominants, the Russian Mafia found, are usually concerned about, and resistant to, sharing power with less powerful groups. Most customers took the USSR and later the Russian Mafia operatives at face value. Operatives are courteous to dominant customers (as you'd expect an operation offering services would be) and espoused a similar racism as the customer. Because their skin tone is similar (and where it isn't the Russian Mafia entraps locals and use them as fronts), and because customers don't understand they've sub-contracted to human sharks, the KGB and later the Russian Mafia found it easy to con our people. One of the Russian Mafia's basic attack strategies is they never tell a victim they're being attacked,[128] even when the Russian Mafia is attacking-by-proxy aka attacking-by-camouflage. I've not witnessed them force an enslaved grandmother entrap her grandchild but that's in their bag of tricks. How can I possibly know that? Well, they forced both my parents to try to entrap me, deployed my 80+ year old mother the summer of 2023 to get intel from me so they could try yet another entrapment. They deployed elderly neighbors, women in their 80s and 90s, one as recently as December 2023 to harass me and had both those women smile in my face. No lie. Attacking someone's child by enslaving their mother and forcing that mother to try

---

127 Roger Cohen, "The Making of Vladimir Putin" *New York Times* (3/26/22).
128 General new war undeclared

to entrap her child, even though I was an adult in my 50s, is evil. I've seen entrapped parents forced to bring their children along on harassment attacks where I was the target. Young children wont' understand what's going on but I witnessed one adolescent with his plain clothed police officer dad as his father harassed me and that boy looked humiliated and embarrassed as he watched and trailed after his father as his father stalked an African American woman walking across a parking lot using a cane. This book is for that child, now an adult, since his dad harassed me over ten years ago. The young man has likely now been compromised by his own father, as my retired law enforcement father was forced to try to entrap me. When the Russian Mafia entraps our law enforcement, they don't tell them they'll have to entrap their own families. Hopefully we'll learn more after this book is published but the Russian Mafia can make most people noncommunicative. They already know about this book and are already getting their ducks in a row. That should distress me, that they own so many of our people they can get a head's up about anything and exploit or lie their way out of it but for some reason, I'm positive they won't take us. It's not only that I don't want them to, refuse to allow them to, but it's because they overshot. Went too evil. The world's governments understand the Russian Mafia ordered the Israel-Hamas War and there's no going back after you see a mafia murder/injure 30,000 children to unseat a political rival. Most West-allied government officials no longer believe there's anything but evil in the

Russian Mafia. Their success in installing pro-Kremlin governments, their naked ambition and their greed, their willingness to kill anybody's baby means West-allied nations must circle the wagons.

This strategy has been highly successful when they deploy it against American women and children. their force-dominants-to-entrap-subordinate's scam where, in my experience, I was having to evade an entrapment attack from my doctor as he offered to 'help' me.

The dominants who hired the USSR's operatives in the 1950s through the 1980s were led to believe that since the USSR operatives successfully eradicated and/or weakened power-requesting groups, spoke about those groups derogatorily, and smiled at and courted the dominants in the targeted country meant the USSR/Russian Mafia were trusted allies. That's not the case. The Russian Mafia believes no one is smarter than them and, being mafia, they tend not to respect other people or their laws. Dominants would be surprised to know the Russian Mafia and the USSR before them, think the dominant culture stupid. The Russian Mafia are criminals. Criminals believe anyone who trusts criminals are stupid. The USSR didn't tell their 20th American customers (nor the other customers the USSR acquired worldwide this century and last) that they were stupid to trust them. The Russian Mafia doesn't tell their customers they're criminals, nor do they call their customers stupid to their face.

The Russian Mafia's agenda for Americans: coup first,[129] then enslavement of us, then destruction of us. Our National Intelligence Council released a declassified Intelligence Community Assessment on March 15, 2021 entitled "Foreign Threats to the 2020 U.S. Federal Elections" dated 3/10/2021. In that report our security agency said, with a high degree of accuracy, that Mr. Putin ordered multiple agencies in Russia to stealth attack our democracy in support of former President Trump.[130]

Chapter 5.32 – The USSR was on the ground in America in 1971. Messing with human beings hard-wiring it really confuses us.[131] I understand why we've seen a spike in mass murders and drug overdoses. This is an advanced type of warfare and all of us must be warned

Post WWII, the USSR developed psychological warfare as a weapon and decided to deploy it to destroy and/or control their enemies, us, and anyone else. The USSR had operatives on the ground in American targeting children, women, and families as early as 1971. I was one of the children they attacked. This,

---

129 Julian E. Barne's "Russian Interference in 2020 Included Influencing Trump Associates, Report Says" *New York Times* (3/16/2021).

130 DNI Haines (declassified by) "Foreign Threats to the 2020 U.S. Federal Elections," *National Intelligence Council* Intelligence Community Assessment (3/10/2021).

131 Messing with victim's reality is a serious problem. It creates dissonance and confusion. Confused people are unhappy people and the Russian Mafia can do a lot with unhappy people. Currently, I believe they're manipulating Hamas to continue a war that's injuring and killing Palestinian children. Hamas is mostly Palestinians and were once Palestinian children. I'm positive they're unhappy with this war and I suspect they were manipulated and lied to, something the Russian Mafia excels at: lying to people to get them to cause harm.

despite the fact that I was 11, and that the USSR had promised not to attack civilians, women, and children under 15 when they signed the Geneva Conventions in 1948. They said they wouldn't attack civilians and children in war times, but here we were in peace in 1971 America, and the USSR was busy attacking me and mine. They entered my life when I was 11 and had made me homeless by the time I was around 12. Yet, in 1972 the USSR and our government were in detente, which was supposedly the easing of tensions between our nations, and in 1975 the Jackson-Vanik Amendment gave them, China and other nations normal trade relations with us yet all the while the USSR was on the ground attacking families in America, stabbing us in the back, confident they can enslave us. They're still confident they'll seize control of us this November.

The Russian Mafia, and their predecessor government, lie. They lie straight-faced and they believe anyone who believes criminals, which the Russian Mafia are, is stupid and deserves to be enslaved and destroyed. They won't tell us that to our face.

They are hypocrites, master hypocrites and, easily, the most dangerous mafia-government our species has ever produced. We are in their cross-hairs. And we will be destroyed unless we launch an effective defense.

If they stayed in their country with their beliefs that would be one thing, but they've attacked my people,[132] my

132 In this book 'my people' include all people living in America who are not direct operatives of the Russian Mafia or our other enemies. The millions of entrapped people unknowingly ambushed by Russia-controlled American operatives are my people.

country, destroyed my childhood, my family, much of my nation, attacked innocent Ukrainians and now, I believe, are responsible for the 2023 Israel-Hamas War attacking innocent Gaza Palestinians.

They don't believe that we'll consider waging nuclear war against them under any circumstances short of them attacking us. They want no women leaders to deal with in America because non-compromised women leaders think differently than non-compromised male leaders.

The Russian Mafia gives no warning. I've seen them force entrapped parents to use their newborns as beards (covers).[133]

Chapter 5.33 – The Russian Mafia has significantly negatively impacted my life

I am an ordinary person with no espionage training. I worked 20 years as legal secretary after the Russian Mafia harassed me into dropping out of engineering school. To write this book I had to learn how to write a readable, helpful book, learn how to assemble a laptop the Russian Mafia's hackers couldn't hack, and keep myself alive long enough to publish it despite the Russian Mafia's many, unrelenting stealth attacks meant to kill me, while they deployed one of my doctors and some of my immediate family to try to entrap me.

(1) The Russian Mafia prevented me from becoming an electrical engineering, (2) prevented me from voting in the

---

133 Which I've witnessed as they forced entrapped American parents to use their newborns as beards as the parent(s) harassed and/or tried to engage an unwilling me.

2008 presidential election (3) significantly slowed my ability to acquire programming competency by, I suspect, stealth eliminating basic to intermediate math courses at the city college I attend and has gone out of its way to attack me since I began taking programming classes, I think I began trying to learn C Programming in Spring 2022. Since I refused entrapment, they overworked me me, sickened and exhausted me so I couldn't continue my fallback career as a legal secretary, thus depriving me of a living wage and my government and my people of the thousands of dollars I paid in taxes annually. And I'm just one person. Extrapolate that out to hundreds of millions of people living in America. It means the Russian Mafia has deprived our nation of tens of thousands of professionals, especially people of color and women. It means that, nevertheless, I've been forced to endure them. There's a reason the Russian Mafia doesn't want Americans to know all their secret evil.

The Russian Mafia has stolen tens of billions from us in taxes our IRS needs to invest in infrastructure to assist impoverished nations better defend themselves against the climate catastrophe. By stealing our home from us when the KGB destabilized my mother, rendering her unable to work and the loss of her career, the generational wealth she would've accumulated by buying a modest house in Southern California in 1971 for $16,000 would now probably be worth $300,000, which my mother could use to pay for home health aides as she ages or bequest to her children if she so chose.

And we're just one family victimized and destabilized by the KGB and the Russian Mafia since the 1970s. I believe there are millions of us in Southern California alone.

Then there are the overdose deaths and/or the loved ones they hooked and/or stealth murdered. The Russian Mafia has done a job on me and my nation. All the billions from when they attacked innocent Ukrainians, huge chunks are owed to millions of Americans, Ukrainians, Syrian civilians, and I'm positive they ordered the 2023 Israel-Hamas War. If that can be proved Gaza Palestinians civilians are owned billions in restitution from the Russian Mafia.

The USSR (and the gang it controlled comprised of the KKK and local police), attacked and made me homeless when I was a child. After the demise of the USSR, the Russian Mafia continued and expanded the attacks, from my childhood in 1971 to the present, 2024.

Chapter 5.34 – Hard-wired & soft-wired blind spots Active Measures predominantly exploits

Most of us don't know we're under attack by the Russian Mafia and that they're exploiting what we don't know about the following: (1) our hard-wiring (instincts) controls most of our behavior but most people don't know we are a hard-wired species and how that impacts our behavior; (2) we are descendants of and so received our genes (DNA, hard-wiring) from ancestors who survived to reproduce because they listened to their dominants, and those dominants had a demonstrated ability to recognize non-overt threats, sound the alarm, and help their people effectively counter the threat. People who couldn't recognize non-overt threats and respond appropriately, or who didn't have access to dominants who could recognize and respond effectively to non-overt threats, did not survive to reproduce; (3) we are hard-wired and soft-wired to give elevated value to dominants from birth;[134] (4) we are hard-wired and soft-wired to listen when dominants sound the alarm about non-overt threats, and we ignore subordinate's warnings about non-overt threats (from birth nearly all people can recognize overt threats because of our pain receptors);[135] and (5) we are hard-wired to protect and defend the dominant's in our lives.[136] That means when the Russian Mafia entraps our dominants (parents, grandparents, doctors) and turn them on their families and communities, victims are hard-wired to not report them. Largely because the Russian Mafia forces enslaved dominants to label their attack as "help," and from

134 Dominants include parents, grandparents, extended family, doctors (teachers, landlord, police, other authority figures, etc.), who we depend upon for survival at birth.

135 People are hard-wired to recognize overt threats from birth but not hard-wired to recognize non-overt threats. Hard-wiring is difficult and expensive for an organism to accurately embed into genes so only do-or-die information is hard-wired in. Because the ability to recognize non-threats is just one step less crucial than the ability recognize overt threats, evolution piggybacked the recognition of non-overt threats to our hierarchical and socialization hard-wiring. We're already hard-wired to give added value to dominants so evolution did a tack on - we're hard-wired to listen only to dominants' warnings of non-overt threats. We ignore subordinate's warnings of non-overt threats.

136 From the first days of life babies know they need dominants, that they are essential to survival. Long after a baby is an adult, the residual of that baby-hood dependence is reflected in our hard-wired valuation of dominants who helped us live when we couldn't provide for ourselves. The Russian Mafia exploits that baby-love and need for their parents and uses it as a weapon against innocent people. This isn't the enslaved parent's fault – they are ambushed by a world class, best-in-class mafia.

birth we're hard-wired and soft-wired to perceive dominants as people who help us. By mislabeling the attack as not an attack nor as criminality/corruption[137] the Russian Mafia unfurls their attack camouflaged in the enslaved body, face, history, and trusted status of dominants who raised us. People are taught from birth to accommodate dominants, and who have no interest or intent to break the law, do not wish to appear disrespectful to a dominant. Thus, the Russian Mafia destroys us, our families, our communities and our nation while labeling us corrupt. When it is they're unfettered hatred and jealousy, their lies and secret war, they've aimed at our children, that is the epitome of evil.

The Russian Mafia targets families by entrapping authority figures (parents, grandparents, family doctor, priest, mentor, law enforcement) and deploying them to entrap their extended family, and community. Active Measures is genocide warfare because it deliberately destroys families by forcing parents to entrap their children then reveal the betrayal to the

---

137 My doctor offered in summer 2023 to help me get weight loss shots after I asked him if I qualified. He said no I didn't qualify but he offered to categorize me as diabetic to my insurer, telling me my insurer wouldn't question his categorization. Because I am not diabetic, I declined his offer. He didn't say he would be breaking the law or that I would be breaking the law if I accepted. He categorized his offer as 'help,' as in "I'll help you get …." He didn't say "If you accept my offer to lie and cheat your insurer you will be committing Medicare fraud and will be charged with defrauding Medicare and conspiracy to commit fraud." And those are only the charges I know about. The Russian Mafia, after me for over 53+ years, would've ensured they loaded anything else they could think of. I'm 100% positive I'm not the only victim they've offered this scam to which exploits our hard-wiring and our socialization. Which is causing our people and our nation to list, become un-moored.

children, thus breaking the family, the cornerstone of our species, our culture, our country and our government. Without our families, our culture and our government are destroyed.

The Russian Mafia is 100% committed to destroying us, not our government, us. Many Americans have been led by the Russian Mafia to believe (which mafia controls a significant portion of our media) that the conflict is between our democracy, our government, and the Russian Mafia, but governmental clashes are a tiny part of the problem from the Russian Mafia's point of view. The real problem is that the Russian Mafia has a need to dominant and enslave the world, to be top dog. They have a need, not a want. Any nation more successful than the Russian Mafia is a target of the Russian Mafia. It just so happens that because of our nation's sacrifices and hard work in WWII, and our alliances agreements – the benefits we negotiated should we help the world defend itself from the Nazis, that we exited WWII a manufacturing powerhouse. The USSR lost over 20 million soldiers, many of them murdered by the Nazis and brutal winter fighting conditions. After the end of the war, the USSR looked at what we'd negotiated for ourselves when we helped the world survive, and what they hadn't negotiated when they'd help the world survive, and they were enraged that they got so little from the war. They blamed our government for their government's failure to negotiate acceptable terms of engagement, although it's the government's responsibility to negotiate their terms, not another country's government's responsibility.

Their intense jealousy and rage from exiting WWII with what they felt was an insufficient international profile has built over the decades, as their government increasingly embraced their mafia and refocused on destroying successful nations and stealing their wealth (including destroying our children with fentanyl and other cheap poisons), rather than build up their own nation and people. The Russian government is a mafia who has unleashed KGB tactics against ordinary civilians, including against our children, and who has ordered Hamas to attack Israel, knowing innocent Israeli civilians and Gaza Palestinian civilians would be killed. This done to unseat our excellent President in November 2024, install one of their many puppets into our Presidency, and shove Israeli civilians, who have been resisting their right-leaning pro-Kremlin government, to embrace the Russian Mafia instead.

The Russian Mafia intends to destroy us, no matter what they say to our leaders. On the street where the rubber hits the road, where they've to my face shown me who and what they are, even in the lead-up to seizing power, they've demonstrated they are completely unworthy of control over any life form. That they'd proxy murder and/or injure tens of thousands of Palestinian children for blood pawns, to create pressure on a political enemy, to seize world dominance demonstrates their evil. These are grown men. Mr. Putin is the wealthiest man in the history of the world, but all he can see is our government is powerful and he needs to replace us with him. He and his Mafia are evil, entitled, spoiled. Baby killers. Unworthy to be called men.

In America, all I see is the devastation they're left in their wake. Attacks on our families of color, on our Caucasian families, on our native American and our Hispanic families, and all the other ethnicity's who make up America. We are in America's pre-Holocaust phase: if we allow the Russian Mafia to seize our presidency, they will destroy us and our nation. Mr. Putin intends to be the Closer of Active Measures this November. In this book I describe how I reached these conclusions.

We cannot survive or thrive under the stranglehold the Russian Mafia has on our people. Surrendering to the Russian Mafia, the very mafia proxy killing innocent Palestinian children, destroying our families, bombing innocent Ukrainians, and who uses fentanyl and other drugs to stealth murders tens of thousands of us annually. Surrendering to an enemy who is actively destroying our families is not an option. They will simply keep lying to us while they murder us.

And since it's been the Russian Mafia who has prevented our species from effectively fighting the climate catastrophe for decades, they will destroy earth's habitability too.

Chapter 5.35 – Spot checks in my 30s and age 40 as the Russian Mafia used proxies to attack me

The Russian Mafia finally succeeded in inserting two female operatives as office friends into my work life, after they'd harassed me from the word processing center which I'd loved before my supervisor and coworkers began harassing me. The problem was, even though the ladies were nice, their insistence on remaining in my life was off-putting. I was around 40, I think one was around her early 30s and the other was maybe late 40s. Legal secretarial work is hard and challenging. Certainly doable by a great many people, it requires concentration or it's very easy to leave something essential out of a filing or a service, or miss specific court rules and legal secretaries who do that are fired. Legal secretaries who can't pay attention to details can get a law firm sued by it's clients. Legal secretaries have no one backing them up. A competent legal secretary isn't chatting with their office friends across their desk, that's not this.[138] The Russian Mafia didn't understand or didn't care. They kept forcing me to move from

---

138 A legal secretary must focus, socializing at work is out. An attorney doesn't hear excuses – if a legal secretary offers one, expect to be fired. I was never fired or laid off because I took my job seriously. Attorneys can't micromanage a court filing and service because they don't know how to do it, just like I don't know how to write a compelling legal argument and which cases to site for the judge to take my argument seriously. If you don't mind focusing and attention to detail at work, and if you want a clerical position, I liked being a legal secretary. It's a busy job, you have to learn many different things, and in general attorneys didn't try to make me an administrative assistant. Revising and filing the documents trumps an attorneys need for an administrative assistant. The last thing an attorney wants is to miss a court deadline because a legal secretary was tasked with sending Christmas cards to clients. So a legal secretary is allowed to focus on doing her/his job well.

job to job, apparently hoping they'd get an opportunity to entrap me. They tried on my last three jobs for sure: (1) the last job they tried to get me to steal from the company using the business' credit card the first boss I had there assured me the credit card was in my sole care, that only I'd see and authorize the statement, that he wanted me to make purchases on his behalf, of my choice; (2) the job before that they had payroll give me an unnaturally large paycheck to see if I'd cash it (I didn't, I notified payroll); and (3) the job before that they positioned me to commit (a) tax fraud, (b) become a drug mule, (c) tried to get me to lie to the federal government and illegally accuse my supervisor of racism (she wasn't racist, they forced her to be cruel to me & had a woman of color "friend," invite me to join her federal lawsuit – I declined); and (d) immigration fraud.

The Russian Mafia's American operatives had different attacks at many of the jobs, especially my last three jobs, and they affixed two female office 'friends' to me at a West Los Angeles law firm but a legal secretary's job isn't a sociable job, it's where the rubber meets the road in law firms. Unless a law firm has employees who know how to timely and accurately prepare, file, and serve pleadings in various districts and courts, and are able to work unsupervised (attorneys and paralegals I reported to didn't know how to file the pleadings, they prepared and researched the legal arguments), that firm will be sued. Repeatedly.

I had no idea I was being pushed from job to job. I'd be hired at a company, like it and the people, and then, for reasons I never understood, my supervisor and co-workers began to harass and/or indicate I was unwelcome. I never did anything but my work to the best of my ability. I didn't know the KGB had been on me since I was 11 and was on me until their demise in 1991. To allow the Russian Mafia another bite at the apple (attack against me), I transferred to a secretarial desk after the Russian Mafia made the word processing center inhospitable to me at a West Los Angeles law firm. I remember one of the HR ladies sneered at me and I had no idea why. I was the quality of word processor and legal secretary I'd want in my employ if I owned a law firm. I did nothing but the best job I could, left my co-workers alone, concentrated on doing a good job and enjoyed doing a good job. The Russian Mafia's American operatives had two operatives ready and they affixed them to me when I transferred out of word processing to a legal secretary's desk (under duress because I was person non grata because of the Russian Mafia) but I left that desk to work at a downtown law firm, largely to get away from operatives who insisted on a friendship with me that made no sense. I didn't know the women were operatives at the time, I didn't know I was being hunted, their insistence on maintaining engagement with me was strange.

The compromised American law enforcement operatives who'd organized the whole let's-make-the-word-processing-center-inhospitable attack and get some 'friends' were so frustrated when I got a job elsewhere that they went loud, and began to overtly attack me nearly everywhere I went. The Russian Mafia allowed them this, certain no one would listen to me if I complained. It is horrific. I wouldn't wish it on anyone but the upside is, after 30 years, I was given the information that I had an enemy, where before I'd had all kinds of awful things done to me and happen to me despite my best efforts but didn't know why. Because of those 30 years of proxy attacks from 11 to about 41, I know how the Russian Mafia got Americans to attack the Capitol.

The USSR rewards decency by kicking the decent hard in the butt, using proxies, in preparation to trigger them as weapons to overthrow our government (after they've entrapped and/or influenced the victims into overthrowing our government). The Russian Mafia loves rewarding a decent, kind, hardworking, self-sacrificing person with extremely hard smack downs. Assuming the Capitol rioters were prepped like I was, the rioters treated everyone in their lives well, did the best job at work, showed love and respect to their family and friends, only to be lied, cheated and/or conned by supposedly reputable businesses, have contracts they'd paid for canceled, been cheated out of benefits they'd paid for. There are a thousand ways the Russian Mafia enrages ordinary people in order to embitter and disillusion them. I refused to be a raving maniac but I was pushed hard to become one.

It'll be good for Capitol attackers to understand they've been stealth attacked, probably for decades, by the Russian

Mafia who plans to use them to overthrow our government, to understand an enemy has destroyed their quality of life, that it was not their fault, nothing they did. It's extremely traumatizing for people who work their guts and heart out and get no praise. Our excellent President and his outstanding Administration and staff are constantly fed negative poll numbers to discourage them from continuing their service. I saved up money so my mother could buy another house and her handlers had her spend it and then used my siblings to drip suggestions of lies and cheating by my mom. No one in my family but me knows we're victims of a coup by the evil Russian Mafia. I hope my family reads my book so they'll finally know who's responsible for crashing and burning our once excellent family dynamic, so strong in our love and devotion as children that we survived that first attack as a family despite the KGB's illegal war against us and our nation. It'll be good for my loved ones to understand who destroyed their quality of life. We really are victims and tragically for our people and nation, we're far from the only ones.

The Russian Mafia are responsible for hooking some of my loved ones on drugs and causing them to overdose, and they're responsible for trying to hook me and other loved ones on drugs. I'm 64 and for decades I've directly experienced the Russian Mafia forcing entrapped parents, teachers, field trip minders, supervisors, landlord, bus drivers, store security, police, and many, many other people in America to covertly and/or overtly attack me. I've been hit by two vehicles by driven by Russian Mafia operatives and threatened by many more, who revved their engines when they saw me, and used their vehicles in many ways as weapons of intimidation. The Russian Mafia has attempted to blackmail with regards to bestiality and encouraged me to become interested in that lifestyle. They also had an operative encourage me to get a pet. They introduced bestiality based porn into my internet thread, left a disgusting message on my phone. I'm not into bestiality so it took me awhile to understand what they were going for. At the time of the blackmail I didn't know the Russian Mafia was my enemy so I was stumped as to what was going on. I knew I had an enemy, certainly by my late 50s, but why I had an enemy I didn't know until 2021 when my government allowed reporters to report books outing the Russian Mafia here. Finally, I was able to begin to pull all the attacks together.

Chapter 5.36 – The Holocaust is re-imagined in the 21st century – German Jews are now Americans, the Nazis are now the Russian Mafia, and we're pre-Holocaust

Our knowledge of history and our knowledge of the Russian Mafia, gives us an edge. We will save ourselves, and protect the innocent being evilly attacked by monsters who care nothing for Palestinian families.

We're in the pre-Holocaust phase. By covertly enslaving our dominants[139] and forcing them to entrap their

___
139 Dominants are people who have authority, power, or control over others under certain circumstances, or who provide an essential service in society. They include: parents in a parent-child relationship, teachers in a teacher-student relationship, landlords in the landlord-tenant

own children, the Russian Mafia has broken many of our families and destroyed the parent-child trust bond. Genocide – the deliberate destruction of our families as the Russian Mafia moves to destroy us. And because dominants are essential to our species' ability to launch an effective defense in response to covert attacks, we're fish in a barrel until we free our dominants or hire dominants.

The Russian Mafia hate us, which is why I say we're in the pre-Holocaust phase. If they insert their operatives into most of the top seats of our government, cabinet, security agencies and/or military, it's over for us. At my level in our society, a once working class African American woman who evades their entrapments, the Russian Mafia now feels no need to hide it's attacks on me. They don't believe anyone will pay attention to my warnings because their attack is keyed to our hard-wiring and our hard-wiring makes us deaf to subordinates who warn of non-overt attacks. I'm a subordinate. The Russian Mafia knows my people are hard-wired to ignore my warnings about their non-overt attack. They think they've got us. But this isn't 2 million years ago and our government and our allies' governments are working to adapt and more effectively counter an ascending Russian Mafia. Which means that, despite the Russian Mafia's slaves being thick on the ground here and elsewhere, leaders, especially those who's job it is to protect us, are willing to listen to any credible person raising

relationship, generals in a general-soldier relationship; doctors in a doctor-patient relationship, police in a police-citizen relationship, employers in a employer-employee relationship, bus driver in a bus driver-passenger relationship, etc.

the alarm about the Russian Mafia. So, unless the Russian Mafia shoots me, I'll get my story out. What will possibly happen is they'll ensure their operatives in our security agencies are my contacts. Sadly, one of the negatives of being hunted by the Russian Mafia for most of my life, and watching them destroy my family as it once was, and watching them destabilize my nation and attack our children, and bomb Ukrainians and attack innocent Israelis and Palestinians, is a difficulty to trust. I see the Russian Mafia. I can't unsee them. The are the most evil and ambitious mafia our species has ever produced.

I've endured, rough estimate, 4,000 attacks, as the Russian Mafia pushes me towards death from natural causes, since they can't entrap me and are doing what they can to murder me without appearing to do so. Because they're determined to be rid of me without actually shooting me, they stress me in ways I experience but which draw no attention to their operatives. Inside my apartment, they harass me with noise, and have tortured me for over a decade by deliberately interrupting my sleep cycle. As I type these words on 2/12/2024 at 12:01 pm, an operative is in the upstairs apartment standing over me in the above hallway. This is to get me to stop typing, to see if their sudden presence with their operatives obviously listening to me type, will stop me typing. It didn't. Nothing will stop me typing but death. I know the Russian Mafia intends to make that happen but my job is to warn my people, not worry about the endless threats from the

Russian Mafia in such a way that fear stops me from doing my job.

Many victims can't speak, they're enslaved or they committed suicide. These are our people. We won't abandon them.

I was a child when they first attacked me and I've witnessed a great many of their attacks on children living in America. Given their attacks on me, I believe more power in their hands will result in more brutality and horror against the innocent. Assuming I'm correct that the Israel-Hamas war is a proxy war, they know they're responsible for murdering by proxy an estimated 12,000 children but they don't care. They proxy kill Palestinian children as blood pawns, to unseat a political rival and close their coup against Americans. Since they hate us, and they're doing this evil to Palestinian's they don't even hate, we can expect far worse if we allow them to install one of their puppets.

I live in a predominantly Caucasian part of town. In what the Russian Mafia has planned for us, no ethnicity will be spared. They are pro-slavery and they hate women, from what I've seen, they hate all women but they appear to especially loathe women of color. People who are middle class, working class, upper middle class, none of it will matter, we'll all be slaves. I've worked with millionaire lawyers who are enslaved by the Russian Mafia. The Russian Mafia hate us and are sadists. If we let them install puppets here our suicide rates will jump up for all ethnicity's.

The USSR stealth harassed and threatened me using children as proxies when I was 11 years old living in America and the child forced to bully me was American. The majority of the children used as proxies when the USSR attacked me at age 11 and 12 were under 15. They've harassed me using operatives presenting as 'homeless' who didn't smell. Most homeless have very limited access to showers and bathrooms. Because of that limited access, homeless people smell. An operative fronting as homeless who doesn't stink isn't homeless.

Chapter 5.37 - I believe the Russian Mafia ordered the 2023 Israel Hamas War to unseat President Biden and speed up their destruction of us

The Russian Mafia is a mass murderer. They are, I believe, responsible for the Israel-Hamas War and, they are complicit in the murder and injury of over 30,000 innocent people in less than six months: Gaza Palestinians, Israeli's, minority workers in Israel when Hamas attacked. The Russian Mafia, if I'm correct, facilitated such evil to unseat their political rival, President Biden. Not only that, the Russian Mafia, if I'm right, has weaponized protesters to falsely accuse Americans and our President for mass death in Gaza when the Russian Mafia is the perpetrator, while Mr. Putin publicly calls for cease fire. This is the man and the Mafia/government who demand rule over our species. They don't think what they're doing is wrong because no one is outing them publicly but me. Our security agencies investigate before they publicly accuse a government of wrongdoing. I've been attacked by the USSR and Russian Mafia operatives for 53 of my 64 years of life. I recognize their signature: (1) the attack benefits the Russian Mafia's publicly stated goal of removing our government and replacing us with themselves in the world order; (2) massive numbers of innocent people are murdered or injured; (3) the Russian Mafia use it's proxies to accuse Americans or our government; (4) no one accuses the Russian Mafia of stealthily launching another in a long history of proxy wars against Americans.

Our security agencies investigate thoroughly before accusing anyone of anything. The Russian Mafia knows that but I'm surprised I'm the only person I've read about, in any of the newspapers, who is accusing the Russian Mafia of waging a proxy war in the 2023 Israel-Hamas War. That journalists are intimidated and newspaper owners don't want to accuse them is troubling.

Chapter 5.38 – The Russian Mafia's strategy behind attacking our dominants

Dominants are the glue that holds our species together. Without them there is no one to care for, raise and nurture our babies; lead us, help us develop policies to our benefit; help us navigate the climate catastrophe; help us defend ourselves against enemies; teach our children; police us; mentor and advise us (priests/preachers); help us be a more aware co-earthling and to share more equitably with the other life forms on earth. Dominants control their families and most of the assets on earth. Active Measures is a family-specific genocide, intended to destroy our families. Dominants control our families so they are a logical and efficient choice for the Russian Mafia to seize control of the largest number of us as fast as possible. The Russian Mafia doesn't control all 400 million of us but rough estimate I say they likely control around 100 million of us, enough to vote in one of their puppets. Former President Trump got over 74 million votes in 2020. While his numbers have probably gone down because of his convictions, the fact that he's still a viable candidate, despite the charges and convictions against him, demonstrates Active Measures in play, the deliberate weakening of our institutions, systems, and people. I'm not saying Mr. Trump is involved in Active Measures, that's not something I know. I'm saying he recently publicly said something along the lines that he'd invite the Russian government to attack our NATO allies for ...... When the presumptive Republican presidential candidate says that, it's a problem for people like me who support our NATO commitment. And when I read German history and the rise of the Nazi government I understand it is a bad idea to ignore what politicians say. German Jews who were paying attention to the Nazi party left the country when that government was elected, those who had the resources to leave, left. Most German Jews didn't have the resources to leave or they didn't have the connections. Antisemitism was an even bigger problem in the 1930s, and pre-Holocaust German Jews, the ones without the resources or connections to leave, hoped for the best. The best did not happen. Americans can't conceive of, just as German Jews pre-Holocaust could not conceive of, a government that tossed their babies into the air and shot them as if in target practice. But as Moishe the Beadle reported to his village, that is what they did. I can testify that, having read of many of the brutalities the Nazi's committed against innocent people, the Russian Mafia has them beat by a country mile. In my opinion, there is no mafia or government as evil. Admittedly, I am prejudiced because they've attacked me for 53 of my 64 years.

Working from the premise that my mother was destabilized when she was 29 (an operative was illegally placed in her bed). She was a young working mother of four in the KGB's sites because they were here hunting Americans, especially strong families of color, which is what our family was, largely because of our mother. The KGB destabilized her, our family's sole breadwinner, so they destabilized our family. Fortunately, our legitimate government gave us food stamps for food and a little money for housing so after the initial trauma we weren't starving on the street. My mother later bore her final child. As best I can tell, my mother and my four half-siblings are enslaved by the Russian Mafia, and have been told they've been entrapped by our legitimate government. I know that because, although my family isn't free to tell me they're entrapped, I know they are, and there's no way anyone in my family would remain a slave to a foreign enemy government if they knew they'd been deliberately entrapped by that government working genocide against us. All the victims the

Russian Mafia entrapped here will be very interested, and enraged, to learn this truth.

The Russian Mafia uses law enforcement's uniforms, employment and identifications as camouflage to convince my family that their handler's are controlled by our legitimate government, when their handlers are controlled by the Russian Mafia working a secret coup against our nation. When my family finds out, if they believe me, they're going to be stunned, and furious. If their handlers don't kill them. Suddenly killing us in a believable way would be a problem for the Russian Mafia but they're experts on cover stories. Of a six member family, my mother and her five children, only I remain unenslaved. And the Russian Mafia worked hard to acquire me but their main weapon was entrapment and I am an ethicist. They couldn't tell me they needed me to roll over because they're here illegally and our compromised law enforcement is acting extra judicially. Just being me, I hate lying so I don't. My mother was my first teacher and she hated lying, so I wasn't a liar. In a way, it's as simple as that.

Enslaving one dominant and four of her five children, presumably forcing her to entrap at least some of them, demonstrates why the KGB and the Russian Mafia target dominants. Destabilize one and that nets you five (in this example), by forcing the dominant to unwillingly destroy her family.

To exploit our blind-spot the Russian Mafia & before it the USSR spent the last 50+ years stealth attacking &

enslaving our dominants, in order to greatly reduce the number of dominants we have who can raise an alarm. So far, I'm the only person I've read or heard of who've called the 2023 Israel-Hamas War a proxy war. Our security agencies investigate and don't make allegations they can't back. The Russian Mafia knows that. I'm a dominant only in my relationships with my siblings because I'm first born in my family. The Russian Mafia has been on me for decades to entrap me. Because of their unrelenting entrapment attacks (which I didn't know for 30 years (11-41) until their operatives started harassing me to my face in public places, in grocery stores, on the sidewalk, on the bus, etc.), and despite my father's deployment to gas light me about the attacks (in my 40s and 50s), I was able to draw a connection between those sometimes very dramatic harassment and physical attacks from strangers, and my family's sudden and weird attempts to try to get me to do illegal acts in my 50s. In the beginning with my family I didn't use the word entrapment because the concept was foreign to me, despite me later understanding that I'd been unknowingly evading entrapment attacks since my 30s. To me, for the most part, those attacks had been people I knew and strangers trying to get me to behave in ways I normally wouldn't. The entrapment attempts were so strange, exotic, obviously criminal (Stacy, want to be a drug mule? (uh no)); (want to marry a foreigner for cash? (uh, no)); (how about joining a federal complaint against a cruel boss charging her with racism? (she hired me, I'm African American and about

40% of the word processing staff are people of color so, uh, no); (how about game show tax fraud? (the Russian Mafia forced my sick Aunt, supposedly in remission from cancer but clearly not, forced her out of her sick bed to get me to the taping of a game show. As we waiting outside of the building, she swayed on her feet in line she felt so unwell, and there was no seating. I couldn't convince her to leave so I stayed with her. I was stunned when my name was called at the taping I'd only gone for her because she said she wanted to go, but it was the Russian Mafia positioning me for income tax fraud (I was unsure if taxes were required on the $1,000 worth of prizes. I paid taxes on the prizes just in case, hoping the IRS would tell me if I overpaid. When they kept the extra money I assumed I'd had to pay taxes on the winnings. I unknowingly evaded entrapment by paying, just in case).) All those events were weird in my life. The KGB/FSB knew I did my own tax returns and knew I'd never won anything before. They hoped I'd not know taxes were required. I gave my mom the main prize, a cabinet she still uses today but which I wasn't fond of but was too surprised that I'd won to decline the prizes. I had no idea I was being hunted, no idea the game show was a set up. In my 30s, I'd been hunted and attacked by proxies for over 20 years, but the KGB was not in my face at the time. They used proxies to attack me. My theory is that once they got our security agencies and military focused on the Middle East after 9/11, they believed they could do what they wanted to whoever they wanted here. Today I was noise harassed by Russian

Mafia controlled operatives but I can't prove it. They were workmen working excessively loudly and the Russian Mafia always has a cover story. I have a book to write, so I'm writing it. It's more than possible the Russian Mafia already has a copy and are moving fast to counter step what I say. After this book is published and people hear of it, it's going to be hard for the Russian Mafia to come up with believable lies to people who's lives they've destroyed. While it's easy to think the Russian Mafia is smart, they inherited Active Measures, they didn't create it nor deploy it. They expanded it. They're going to have a Plan B but in the face of many tens of millions of victims in this country alone I don't know what they can say to stop Americans from feeling anti-Russian Mafia rage and refusing to vote for pro-Kremlin candidates. They'll say I'm lying, blah, blah, blah. But why would I lie? I'm not anti-Russian. I'm anti-Russian Mafia. I'm anti-men murdering babies and destroying families because they're sulking that they're not the big cheese and are willing to kill other people's babies to become the big cheese.

No one would possibly want them who knows what they are. Even their allies (slaves) China and India, who probably will never believe this book, wouldn't want them if they knew the evil the Russian Mafia is. This is just stupid.

It's stupid for men to behave as the Russian Mafia is behaving. They can't bear to live in a world they can't totally control. They won't accept our 'no' and leave the rest of us alone, but do crazy evil like start a proxy war that kills tens of thousands of innocent Palestinians, and kill/injure thousands of Israeli civilians, and act like it's not a big deal because no one is publicly calling them out on it. They think what they're doing is fine because they're murdering other people's babies, a sea of innocent blood to seize world dominance. They expect to get away with this evil. No.

The overt attacks beginning 2001 are horrific and terrifying but the upside is they gave me accurate intel: I learned I have an enemy.

Before then the Russian Mafia had used proxies stealthily so there was no way I could know to connect the dots. But after every grocery store and/or convenience store I went to I was harassed, I knew there was a problem, a big problem since I was apparently being followed everywhere, but why? Who? I knew the police were involved, they were some of the people who harassed me and I witnessed one forcing an ordinary citizen to harass me but the Russian Mafia doesn't explain it's attacks to anyone.

By attacking and enslaving our dominants, the Russian Mafia has greatly reduced the number of dominants we have available to us to raise the alarm. In a healthy society there'd have already been endless editorials accusing the Russian Mafia of orchestrating this war, facilitating the mass murder of innocent Palestinians to unseat our President. We're going to have to organize large groups of noncompromised dominants to get the word out about this type of attack and the message must not just be the evils the Russian Mafia throws out onto the innocent in the world, but to impact our audience, our presentations must be done by dominants in a way our species will activate itself.

Chapter 5.39 - Non-compromised dominants must sound the alarm but in ways our species will accept and recognize

We need to rethink our unwillingness to launch complaints against dominants trying to entrap us, like our parents, doctor, or any other previously trusted dominant in our lives. The Russian Mafia depends upon our silence about the dominants trying to get us to break the law. Dominants luring subordinates into corruption really screws with people's heads, especially our children. The Russian Mafia exploits every dominant/subordinate relationship and attacks people when they're in subordinate positions.

You can't automatically trust dominants at this time else the Russian Mafia will entrap and destroy you (the shoplifting entrapment set by the compromised cashier I barely evaded (had I not glanced at the receipt, you wouldn't be reading this book, I wouldn't be alive to write it).

That means it's a good defense strategy to assess your relationships in terms of dominants and subordinates, remembering that the Russian Mafia prefers to attack targets when they're in the subordinate position, based on my experiences being attacked by the Russian Mafia over decades. When I take the bus I'm subordinate to the bus driver. The Russian Mafia exploits my subordinate position to get the driver to ask me what stop I'm getting off. Based on that intel, which they couldn't themselves get from me, the Russian Mafia station harassing operatives all along the route to my destination and at my destination, their operatives ready to harass me. I live this, is how I know.

Chapter 5.40 - The Russian Mafia and before it the USSR, have worked, rough estimate 65+ years to destroy us. If you think they're going to stop attacking us, you're very wrong

Hoping for the best without preparing for the worse is a huge mistake, especially with the Russian Mafia who has dreamed of closing Active Measures for decades. The Russian Mafia is proxy killing Palestinians every day not because they hate them but because they're determined to exploit the horror the Russian Mafia is causing them to endure, merely for world dominance. If you think the Russian Mafia will slaughter Palestinian civilians, people they care nothing about but are they're forcing war on to use as blood pawns, and not murder us, a nation of people they've loathed for decades, you're mistaken. They loathe us not because we're loathe-worthy but because they have a huge inferiority complex and they're jealous of our government's financial power. They're bombing innocent Ukrainians because they covet Ukrainian's food production capacity and their fertilizer production capacity. The Russian Mafia are criminals: they see something they want, they steal it. Since they are part of the Russian government, they bomb and attack using the weapons of state available to them.

To get a sense of what our lives would be like with a pro-Kremlin puppet installed, think of the horror Haiti's are enduring, cross that with what innocent North Koreans are enduring, plus what innocent Ukrainians are enduring, plus what innocent Palestinian civilians and their children are enduring, plus what innocent people in African and South American nations are enduring, and you might, almost, approximate what the Russian Mafia has in store for us. Death is preferable to their dream for us. We fight or we die. Enslavement by them is not an option. Take my word for it, you do not want to go there.

To paraphrase Russian General Gerasimov, Chief of the General Staff of the Russian Federation, in his paper titled "The Value of Science in Prediction," published in Russian in 2013, in Russian military magazine *Military-Industrial Kurier*, footnoted below,[140] the General basically says that wars are different now. To me that means the Russian Mafia will do anything short of initiating nuclear war to install one of their

_____

140 Gen. Gerasimov, Chief of the General Staff of the Russian Federation,"The Value of Science in Prediction," *Military-Industrial Kurier* (2/27/2013). *"...wars are no longer declared...." "...The very "rules of war" have changed...." "...The role of nonmilitary means of achieving political and strategic goals has grown, and, in many cases, they have exceeded the power of force of weapons in their effectiveness...." " The focus of applied methods of conflict has altered in the direction of the broad use of political, economic, informational, humanitarian, and other nonmilitary measures – applied in coordination with the protest potential of the population...." "..Tactical and operational pauses that the enemy could exploit are_ disappearing..."*

puppets into the White House. It is my opinion with regard to the 2023 Israel-Hamas War that the only sides operating totally in good faith are us, and Palestinian civilians. We needn't treat other nations as if they're our enemy's but the USSR and the Russian government were once our trade partners, and now they're working to destroy us. Trade partners are trading partners. Just as we engage with them, so does the Russian Mafia. And the Russian Mafia has mastered blackmail and manipulating people via hard-wiring, have gotten it down to a science. Manipulating by hard-wiring is an entirely different type of warfare, and the Russian Mafia has deliberately hidden their abilities from our security agencies as they aim their warfare at American civilians.

Mr. Putin is on the clock and he's very highly motivated. He's lining up all his stealth attacks, the children he's destabilized here, so much evil to set into play. If they force hot war on us by not leaving our people alone, I prefer we go out fighting rather than be enslaved. The Russian Mafia has zero intention of comforting 400 million stunned Americans if they install one of their puppets. Just like they've no problem proxy killing innocent Palestinians to seize world power, they'll have no problem at all annihilating us. They will lie to us because that's just them. If we believe their lies, that would be a mistake. One we won't recover from.

Because the Russian Mafia attacks us in secrecy using proxies, stealth, and ambush, most people don't know our deterioration is because we're being attacked by an enemy foreign government. When they know, if they b1elieve, I think they'll have a negative opinion about the foreign enemy government attacking our children and the rest of us.

Chapter 6 – The Russian Mafia is dangling the Saudi deal as bait before the Biden Adm, encouraging them to put their energy into developing a Saudi alliance deal that leads to Palestinian statehood - but those negotiations will be stealthily extended by the Russian Mafia. We need to pressure Israel into an immediate cease fire because the Russian Mafia is running this war & is 50 steps ahead of us

While we're working to avoid the more obvious traps, we can save many Palestinian civilian lives by calling a cease-fire. The Russian Mafia intends for Israel to feel isolated & that's not something I want any people, but especially not our allies, to feel. <u>However</u>, I refuse to allow the Russian Mafia to install one of it's puppets here. Clues the Saudi deal is bait: (1) timing (the Saudi deal *helps* the Russian Mafia install a puppet in 2024 by focusing our attention on a Saudi deal rather than a cease fire, which implies the timing nor the deal are a coincidence), & (2) the Saudi deal has effectively stopped the Biden Administration from demanding an immediate cease fire. This is a psychological warfare, multi-operative-government attack the Russian Mafia intends to use to make their bones - & trumpet their world dominance. However, they're only dominant if we concede. How to survive? We must build a stronger, effective defense. Because General Austin was down last month, we have to assume the Russian

Mafia set something in play they'll hold back as a very nasty surprise.[141] I hope I'm wrong.

President Biden and his Administration must call for an immediate cease fire, sooner rather than later. If our alliance agreement with Israel gives us any wiggle room, if it allows us to pause in response to anything the Israeli government has done outside the terms of our alliance agreement, we use that wiggle room to call for an immediate, temporary cease fire. The Russian Mafia is using us not calling for an immediate cease fire to make their play here. I am soo sick of them I can't tell you. If we allow them to install a puppet here, the Russian Mafia will destroy us. So let's look at our alliance agreement with Israel and find where we can legally call for a temporary cease fire. Don't we have an emergency clause in there? I don't know is why I'm asking.

Calling the cease fire we explain why we hadn't earlier: that we'd been working as hard as possible to create a deal that

---

141 Speaking of nasty surprises the Russian Mafia has planned for us (but unrelated to General Austin or the dramatic take-down the Mafia has planned for us this 2024 election cycle (that dramatic take-down includes humiliating us and our outstanding President before the world and seating one of their puppets in the White House)), something is going on with the Russian Mafia and toilet paper. At least five times the last three years the Mafia had one of their operatives in my family order, bring, or deliver toilet paper to me, unrequested by me. A different family operative put something in my food to cause me to have explosive diarrhea, which forced me to use the toilet paper at her apartment, which I normally don't do. Colon cancer in America is on the rise. The Russian Mafia intends to kill me, and us, and there's a reason they had toilet paper delivered or given me or forced on me. I suspect colon cancer causing toilet paper is one of the weapons the Russian Mafia has developed. Also know the Russian Mafia pushed me to shop at a big box store (by a family operative years I knew she was an operative). The big box store is where many people buy their toilet paper here, and I suspect worldwide.

would lead to Palestinian statehood, which we'd hoped to push through quickly, but the deal is coming together far too slowly and we couldn't bear more innocents being killed while we work through slower-than-expected negotiations with the Saudi-Israel-American alliance. Given the excessive number of deaths of innocents, we decided to push for a cease fire.

President Biden strongly supports Palestinian civilians' right to their own country, & he's loved & respected the Israeli people for decades.   Yet he'd be the first to agree that American's survival come first with him. We are blessed to have him as our president.

After 65+ years of a stealth coup against Americans, the Russian Mafia is determined to successfully Close their coup this presidential cycle. We're on the ropes. If we allow the Russian Mafia to install even one of their puppets in the White House in 2024 or beyond, we are dead Americans.

Recently occurring backstory: the Russian Mafia stealth facilitated the 2023 Israel-Hamas War (our security agencies suspect it but suspicion isn't proof. The Russian Mafia didn't leave a paper trail), and is using Hamas (who are largely radicalized Palestinians), to facilitate the murder of tens of thousands of innocent Palestinians, to blame on Americans and to unseat our President. You better believe Hamas didn't sign on for the slaughter of thousands of Palestinian children, but our government doesn't talk directly to terrorists so the high value intel Hamas would be coaxed into telling our trained interrogators is unavailable to us. The Saudi deal coming into being months before the war is how the Russian Mafia had to play it to hide that the Saudi deal is part of their coup. If the deal came into view during the war it would raise more suspicion.

The Russian Mafia contains many Soviet Jews, some of whom had to sign off on Hamas' attack, murder and rape of innocent Israeli Jews. Bratva (the name of the Russian Mafia) brotherhood trumps religious and cultural considerations for at least some Russian Mafia members. Soviet Mafia Jews facilitating the murder of innocent Israeli Jews is a horror to me, but that's what happened. I noticed that one of the freed kidnapped victims was a Soviet Jew who thanked the negotiators who got her free. Those poor kidnap victims. If the Russian Mafia Jews threw Israeli Jews under the bus for their coup against us, you can just image what the Russian Mafia will do to hated Americans. As Russian General Gerasimov said in his 2013 paper "The Value of Science in Prediction," and I quote "...war is no longer declared..."[142] We can't assume any of the other players are acting in good faith, but we must act as if we believe they are. From what I can see, only our government and Palestinian civilians are believable.

_____

142 General Valery Gerasimov, Chief of the General Staff of the Russian Federation,"The Value of Science in Prediction," *Military-Industrial Kurier* (February 27, 2013). *"... wars are no longer declared…" "...The very "rules of war" have changed…." "...The role of nonmilitary means of achieving political and strategic goals has grown, and, in many cases, they have exceeded the power of force of weapons in their effectiveness…." "...*Among such actions are *the use of special-operations forces and internal opposition to create a permanently operating front through the entire territory of the enemy state…."*

The reason President Biden can't publicly castigate the right-leaning Israeli government when it behaves more like our enemy than our ally is because our alliance is a legal agreement. It outlines how we may and may not respond under a certain set of conditions. The Russian Mafia knows this and got the ball rolling by having Israel attacked. Israel is governed by pro-Kremlin, right leaning individuals, meaning the Russian Mafia knows what the Israeli government wants and what it will do under a certain set of conditions.

President Biden must give full-throated public support to a cease fire and denounce the Israeli government's atrocities against Palestinian civilians. ***Don't destroy or denounce the Saudi deal***, but rather respond to the Israeli government's attacks on Palestinian civilians as if the Saudi deal doesn't exist. If President Biden doesn't start publicly backing Palestinian civilians, he won't be re-elected and the Russian Mafia will slaughter us. Literally. Fortunately for us President Biden is pro-Palestinian and he and his Administration have been working very hard to get Palestinians their statehood.

The Russian Mafia is encouraging pro-Palestinian protesters world-wide to blame Americans and our President for the Israel-Hamas War, when it is actually the Russian Mafia who started the war. It's not in the Russian Mafia's interests to admit it, and it's in their interests to blame Americans. They know our security agencies will not publicly accuse them without proof. The Russian government lies, and uses proxies. That's who they are. General Gerasimov discusses the Russian government's strategy of using protesters as attack agents in his 2013 paper entitled "The Value of Science in Prediction," *Military-Industrial Kurier*, a Russian publication aim at the Russian military audience.[143]

If the Russian Mafia succeeds in installing any of it's puppets in the White House in 2024 or beyond, Americans will wish we were dead, the Russian Mafia will make our lives a nightmare, like they're making innocent civilian Palestinian's lives a nightmare. The Russian Mafia hates us and means us harm. They and their previous government, the USSR, have worked steadily for decades to destroy us and replace our system of governance with Russian style slavery. The Russian Mafia are slavers. They believe they are the master race and that all other races must bow down before them.[144] Mr. Putin, Don of the Russian Mafia and the richest man in the history of our species because he takes a cut of every drug deal, prostitution deal, gun running deal, money laundering deal,[145] etc., is determined in his lifetime to be the Closer of the coup he inherited from the USSR. He's extremely confident he can take us. I don't know how I know, I just know. He's only 71. With his great crime-wealth, unless we kill him, he'll fight death off long enough to enslave us. If we let him. It's essential we do not allow the Russian Mafia install a puppet as president here. Ever.

---

143 Ibid.
144 Roger Cohen, "The Making of Vladimir Putin" *New York Times* (3/26/22).
145 Karen Dawisha, *Putin's Kleptocracy*, Simon & Schuster (2015).

The Russian Mafia fears nothing, and is determined to destroy us. They prefer destroying by coup which is why you see increasing numbers of pro-Kremlin governments worldwide. The Russian Mafia will never stop attacking us, but it does not believe it can recover from a nuclear bomb dropped on the Russian government. Their post-WWII strategy has largely been nuclear bomb avoidance with lots of saber rattling. They'd bomb us in a heartbeat, but they're concerned for their government's continuance if even one nuclear bomb hits their government. The Russian Mafia knows we won't target ordinary Russians but our government knows a nuclear bomb targeting the Russian government would kill many innocent Russians. Our government isn't willing to risk killing innocent people to pay the Russian Mafia back for the many decades of horrors it's done to us. Last week Iran backed terrorists killed some our war heroes and injured many more. In retaliation, our government bombed appropriate sites. The overwhelming majority of civilians surrounding the bomb sites weren't harmed. In a nuclear war, retaliation doesn't work that way. Our government has decided, for now, not to retaliate against the Russian Mafia's stealth attacks against us using our nuclear arsenal.

The USSR named it's bespoke coup Active Measures, and it's been in play since approximately the mid-20th century. It's the reason our nation has deteriorated so precipitously. It's a secret war so there's no way ordinary Americans would know it existed.[146]

As a child stealth attacked by the USSR from 1971 on, I speak from personal experience when I tell you that you wouldn't enjoy the Russian Mafia having an even bigger presence in your life, although if you live in America, they already do negatively impact you, using proxies. This book is to inform people in America so we can better defend ourselves, to inform children worldwide about what's going on, and to inform the world how the Russian Mafia stealth attacks ordinary people so people can better defend themselves.

About a year after the USSR entered my life, they'd made me homeless by age 12. They'd stealth attacked my mother by inserting their operative in her bed using perfidy (an operative lying to a civilian as the operative does it's state's sponsored work),[147] which is illegal even in war times.[148] In times of peace, which we were with the USSR in 1971-72, what the USSR did to me was illegal under our laws and under international laws, as well as being immoral. It's sick for a foreign enemy government to stealth enter it's enemy's nation

---

146 https://en.m.wikipedia.org/wiki/Active_measures.
147 General Valery Gerasimov, Chief of the General Staff of the Russian Federation,"The Value of Science in Prediction," *Military-Industrial Kurier* (February 27, 2013). "*... wars are no longer declared...*" "*...The very "rules of war" have changed....*" "*...The role of nonmilitary means of achieving political and strategic goals has grown, and, in many cases, they have exceeded the power of force of weapons in their effectiveness....*" "*...*Among such actions are *the use of special-operations forces and internal opposition to create a permanently operating front through the entire territory of the enemy state....*"
148 Geneva Conventions of 1949, Additional Protocols. Specifically, Protocol 1 of the Geneva Conventions, Article 37. Prohibition against Perfidy. Article 37, paragraph (1)(c).

and target women, children and civilians, groups expressly protected by the Geneva conventions, to which the USSR was a signatory; that's illegal even in times of war. The USSR continued their stealth attacks against me until it's demise in 1991, after which the Russian government and it's Mafia continued the attacks, and expanded and intensified them. The Russian government continues "romance" attacks like it attacked me and my mother, as you'll see in the below footnoted 2023 *The Guardian* article.[149]

My mother was our family's sole breadwinner so when the USSR set her up, she lost her job, the only home mom owned, and I believe she would've suicided had her children not loved and needed her so much. She could barely explain what happened when I asked her, because the USSR uses proxies and she had no idea the USSR was in America targeting Americans.

This was around 1972, a time the USSR was supposedly adhering to the Geneva conventions they'd signed in 1948. But in 1971 the USSR was on the ground in America attacking children: I was one of their child victims. Neither they nor their subsequent government or Mafia have exited my life since I was 11; I am now 64. I'm pretty sure they've put a hit out on me. I stay in my apartment for the most part, but even there, they manage to harass me on a near daily basis.

_____

149 Shaun Walker, Pjotr Sauer and Tom Phillips "Panic and Emotional Pain as Alleged Deep-Cover Russian Spies Vanish," *The Guardian* (April 3, 2023).

Chapter 6.1 - Our alliance with Israel

Under the terms of our alliance with Israel, we agree to protect and assist them when they are attacked. To check that box, the Russian Mafia ordered Hamas to attack Israel, even though the Russian Mafia has quite a few Soviet Jews.

Although the Russian Mafia loves to have it's operatives report in our media how powerful we are (to raise our citizen's expectations, and then use other operatives to point out that our government really could do more if it wanted to (when the Russian Mafia pretty much knows how our government is likely to respond in any given stressor)), President Biden isn't a king. While he has power, he's also a working employee of the American people and as such he isn't free to explain his backing of the Israeli government because confidentiality clauses in our alliance agreement prevent government officials from discussing key components of that agreement – which the Russian Mafia well knows, and is why they planned and unfolded the war the way that they have.[150] So while Mr. Putin calls for a cease-fire,[151] knowing the type of alliance agreement the U.S and Israel have prevents us from doing that, unless certain confidential conditions are met, which the Russian Mafia has ensured neither Hamas nor the

---

150 Gen. Gerasimov, Chief of the General Staff of the Russian Federation,"The Value of Science in Prediction," *Military-Industrial Kurier* (2/27/13). "The focus of applied methods of conflict has altered in the direction of the broad use of political, economic, informational, humanitarian, and other nonmilitary measures – applied in coordination with the *protest potential of the population*." "..*Tactical and operational pauses that the enemy could exploit are disappearing...* "

151 Jason Neely and Andrew Cawhorne, Editors "Russia Calls on Israel and Palestinians to Cease Fire," *Reuter's* (10/7/2023).

Israel government will meet. So the Russian Mafia is just sitting back laughing it's head off and enjoying our country's dilemma, while it works to erode all of our President's support. That's one of the few things that makes sense. To pressure our government into a deal with the Russian Mafia. I pray our government makes no deal with those people.

President Biden worked as an attorney years ago. He respects the rule of law, which again, the Russian Mafia is using against him and against us in their determination to install one of their puppets.

The Russian Mafia strongly prefers confidentiality agreements, sealed records, secret plea deals because that gives them the opportunity to steal an advantage, usually by lying. I learned their preference years ago. Even before I knew the Russian Mafia is my enemy they attacked me using compromised law enforcement, and my landlord, showing me I have a powerful enemy able to weaponize my own law enforcement against me, and my landlord, among many other power players they've demonstrated control over as they attempted to entrap me. My landlord stealth attacked me by lying, and in our negotiations insisted I agree to seal the record in the Santa Monica Court. Stealth, blackmail, ambush, confidentiality agreements and plea deals are how I believe the Russian Mafia commandeered our entrapment mechanism to entrap, I suspect, tens of millions of innocent people living in America.

Seizing our law enforcement's entrapment mechanism, aiming entrapment at noncriminals, and coercing them to take confidential plea deals to stay out of prison and jail. Victims can't even say they've been entrapped or have a plea deal. I know at least six entrapped family members and none of them have told me they're entrapped and stayed out of prison/jail by taking a plea deal. I only know they're entrapped because they've tried to entrap me on multiple occasions. The Russian Mafia has tried for decades to entrap me using many strategies, which I describe in this book.

Using our Israeli alliance agreement that prevents President Biden from fully explaining his position to better defending himself, the Russian Mafia also built an entrapment mechanism that prevents noncriminals in America from reporting, complaining and/or demanding to know why our legitimate government (which the Russian Mafia is fronting as in entrapments here) is forcing them to entrap their own children and friends. Both my parents were forced to try to entrap me. Were it not for that plea agreement which I believe victims are forced to sign to stay out of jail, our legitimate government would know a lot more about how the Russian Mafia is running their coup here.

With it's proxy war murdering hundreds of innocents a day and the resultant pictures and video keeping President Biden's poll numbers low,[152] and by designing the war in such

---

152 Jonathan Weisman, Ruth Igirlnik, Alyce McFadden, "Poll Finds Wide Disapproval of Biden on Gaza and Little Room to Shift Gears," *New York Times* (December 19, 2023).

a way that President Biden is prevented from explaining himself, the Russian Mafia is having a really good March 2024. They couldn't care less that they're responsible for the murder of tens of thousands of Palestinian civilians, including babies. They believe all is good in their proxy war because no one knows what they're doing. Our security agencies know but must deliver proof. The Russian Mafia won't leave a paper trail.

Which Russian Mafia operative years ago in the American government pushed through that we not negotiate directly with terrorists? We need to throw that policy out. Not talking to Hamas directly is costing us, big time. We have professional interrogators. They'd be able to tease out that Hamas isn't happy about the high Palestinian death numbers. We could use that intel to help Palestinian civilians and their children. But the Russian Mafia doesn't want Palestinian civilians helped.[153] Not at all. In my opinion, the Russian Mafia is the most evil, cunning, cruel government in the history of our species. We better not let them take control of our government.

---

153 Andrew Osborn "Russia's Putin Tries to Use Gaza War to His Geopolitical Advantage," *Reuter's* (11/17/2023).

Chapter 7 - The pending U.S.-Saudi Arabia-Israel alliance - *timing is an important attack indicator in ambush, not to be ignored. An aggressor of the Russian Mafia's caliber can explain away everything but timing*[154][155][156]

We're trying to work an alliance with Saudi Arabia[157] to help Palestinians get their long overdue nation, and we want to demonstrate we're reliable partners to Saudi Arabia, putting

---

154 I am writing and publishing this book with enough time to, hopefully, help my people avoid voting in a pro-Kremlin candidate. No matter how the Russian Mafia's American operatives spin what I say, they can't explain away that I've produced a book in time to help my nation. The Russian Mafia, dangling a Saudi deal before the Biden Administration, will hide it's teeth behind proxies. The proxies will make a convincing argument but the timing of the deal, *neatly and conveniently timely*, is the tell. The Biden Administration, hating the suffering of innocent Palestinian civilians, reach for anything to help them. The dysfunctional House GOP won't approve the extra money President Biden requested, so our President is doing anything and everything to close the Saudi deal. Not only because he's up for re-election. President Biden is deeply empathetic. He is heartbroken by the deaths of Palestinian civilians and children. The Russian Mafia, working to push his negotiating team to exhaustion, will have the Saudi deal ebb and flow. A deal seemingly close, but then inexplicably far away. I recommend President Biden request the Israeli government do a temporary cease fire. The Israeli government won't like it but that will give President Biden's negotiators some breathing room. The Russian Mafia loves this time during their attacks, where they feel they have all the cards. They are sadists. They love harming our President and his Team. Causing innocent people pain is the closest some of the Russian Mafia's members ever get to true happiness. That Palestinian children are dying everyday mean not a thing to them. And because we aren't negotiating directly with the terrorists, we're not leveraging all our negotiating skills to extract all the information we can, if we interacted directly with them.

155 Ivan Nechepurenko, Anton Troianovski, Vivian Nereim, "Putin to Visit Saudi Arabia and U.A.E. on Wednesday," *New York Times* (12/5/2023).

156 Stanley Reed, "Saudi Aramco Abruptly Drops Plans to Expand Oil Production," *New York Times* (1/30/2024).

157 Mark Mazzetti, Ronen Bergman, Edward Wong, Vivian Nereim "Biden Administration Engages in Long-Shot Attempt for Saudi-Israel Deal," *New York Times* (6/17/2023).

another layer of stay-in-your-lane pressure on President Biden and his Administration to work precisely within the U.S.-Israel alliance agreement. The confidential parts of our alliance agreement with Israel prevents President Biden and his Administration from explaining more fully to voters their position opposing a cease-fire, or pressuring the Israeli government even more than they already are behind the scenes, to stop the bombing of Palestinian civilian domiciles.

Isn't it interesting that this deal began to come together a few months before the Israel-Hamas War? There are no accidents with the Russian Mafia. *Here, right now, is the culmination of approximately 65 years of a carefully planned stealth war against Americans.* The Russian Mafia thinks 50 steps ahead. They aren't perfect, they make mistakes, but they're not making mistakes right now. For instance, if we weren't in delicate negotiations with the Saudi's, we'd be pushing the Israeli government a lot harder publicly but something in our alliance agreement is stopping us. It's multiple specifics at play the Russian Mafia knows and is exploiting to our detriment, which our government may not tell us. It's not just the confidentiality of our alliance agreement with Israel the Russian Mafia is weaponizing against us, it's that the Russian Mafia created a situation, using Saudi Arabia as bait, which motivates us to put our best alliance-face on display. *And President Biden is not free to discuss these confidential matters which, again, their confidentiality just happens to benefit the Russian Mafia, just like the plea deals entrapped Americans are coerced into benefit the Russian Mafia.* Happy coincidences that help the Russian Mafia install one of their puppets in the White House aren't coincidences. The Russian Mafia creates it's opportunities.[158]

Chapter 8 - The Russian Mafia deploys pro-Palestinian protesters to harass us and our leaders

The Russian Mafia routinely use nontraditional warfare against their enemies. They urge and/or facilitate pro-Palestinian protests[159] worldwide and have some of their operatives blame Americans and our government for the extreme loss of innocent life caused by Israel's bombs. In reality the 2023 Israel-Hamas War is yet another proxy war by the Russian Mafia, this one intended to unseat our President. Russian General Gerasimov discussed how the Russian government uses protesters to weaken enemies in his 2013 paper "The Value of Science in Prediction," published in Russian in the Russian military magazine *Military-Industrial Kurier,* target audience the Russian military.[160]

---

158 Steven Erlanger, David Pierson, Lynsey Chutel "Iran, Saudi Arabia, and Egypt Invited to Join Emerging Nations Group," *New York Times* (8/24/2023).

159 General Valery Gerasimov, Chief of the General Staff of the Russian Federation,"The Value of Science in Prediction," *Military-Industrial Kurier* (February 27, 2013). General Valery Gerasimov states: *"… wars are no longer declared…" "…The very "rules of war" have changed…." "…The role of nonmilitary means of achieving political and strategic goals has grown, and, in many cases, they have exceeded the power of force of weapons in their effectiveness…." "…applied in coordination with the **protest potential of the population**…."*

160 Ibid.

Chapter 9 – "It's not personal, we just want world dominance" - apparently the Russian Mafia doesn't think mass murder of children to unseat a political arrival so they can murder even more children disqualifies them from world domination

A man aims his gun at you, orders you to hand over your wallet. After you do so, he shoots you in the head. As you fall dead to the ground he turns away and says "It's not personal, it's just business." What's 'just business to him is your death to you.' This is an accurate analogy of how the Russian Mafia behaves in the world.

The USSR found itself in a quandary after stealing American's intellectual property to build it's version of a nuclear weapon:[161] that using the weapon would mean their demise. America has more than one bomb and the USSR, and later the Russian Mafia, could expect the American people would retaliate if the USSR bombed them – and that we'd likely target the USSR government. It understood it could not recover if hit with a nuclear bomb. So how to seize control of the world as the USSR/Russian Mafia is determined to do?

To evade being bombed by Americans in retaliation for the USSR bombing America, the USSR understood it had to (1) control the American president and the people(s) who authorize a nuclear bomb deployment. They decided installing puppets in key positions like the presidency is essential. People they installed would never bomb them. That's why you're

---

161 Calder Walton "How the Soviets Stole Nuclear Secrets and Targeted Oppenheimer, the 'father of the atomic bomb.' *The Conversation* (July 24, 2023).

seeing so many pro-Kremlin governments being embedded: installing puppets is the new coup. When puppets aren't necessary it's because the Russian Mafia helped the current leader in the targeted nation ascend to power and has removed his competitors.

The Russian Mafia is determined to install one of it's puppets in the White House in 2024, and beyond. The Russian Mafia has also decided it needs to flip the current world order to a system it prefers. The Russian Mafia is a slaver. It believes, because it has ambushed and lied and cheated to innocent ordinary people, that means it's destined to rule our species, even as it wages mass murder against Palestinian children, it thinks, because there's no hard proof yet, that it's gotten away, again, with murdering people of color, children, and that the Russian Mafia won't be held accountable. They actually believe that we'll allow them to enslave us. No. We won't.

The Russian Mafia intends to flip the world to slavery where they and their slave-allies rule the world and everyone else is slaves. Flipping the world order means the Russian Mafia wouldn't have to worry about being bombed and it would be world master, like the Nazi's believed they were the only ones worthy of world domination because they murdered so many innocents. The entitlement of some people is beyond comprehension.

Chapter 9.1 - Nuclear avoidance runs through the USSR & Russian Mafia's strategy from post-WWII to the present – that may change when they fail to secure the U.S. Presidency this November but they're not big on sacrificing themselves so China & India can reap the rewards. To the Russian Mafia, China & India are slaves. They're not sacrificing their lives for their slaves' benefit, even to destroy Americans. They inherited an up & running genocide. Mr. Putin didn't build it or deploy it. Active Measures is their big play. When they fail, they'll do stealth terrorism attacks here but they've got no big fall-back play

So, how to place puppets in other countries and how to enslave the world? By lying to everyone about what your intentions are. By illegally entering and stealth seizing control of a country, making war against innocent, ordinary people, including their children, with no warning, using psychological warfare and other nontraditional warfare the old KGB passed on to the Russian Mafia.

Stealth entering a country and studying every aspect of it. Blackmailing more power players in that country and forcing them to pass laws the USSR and later the Russian Mafia exploits, like Jackson-Vanik.

Corrupt the targeted country and it's people, flip the medical system into a for-profit system instead of one that helps sick people, unless that sickness is a revenue stream for the medical center; stealth attack ordinary people to entrap them and destabilize and criminalize them wearing our legitimate government's uniform until people are so destabilized they don't know whether they're coming or going.

Ensure pro-destabililzing and anti-defense laws are passed, make sure the country fails to understand it's being attacked and make sure there's insufficient affordable housing so poor elderly people sleep in their vehicles.

Flood countries with cheap deadly drug poisons targeting their young. Entrap parents and force them to entrap their own children. Coerce/influence vulnerable people you've destabilized using entrapment so that they never have their own private life back, but must always obey a slaver, fronting as a law enforcement officer but who's actually owned by the Russian Mafia.

And tell Americans in every way possible that we're corrupt, or stupid, or whatever. And convince your homies, your slave-allies, that you've found a way to destroy Americans and put that in play facilitating the murders of thousands of Palestinian children, and it all comes down to what you're willing to do to ascend to power. And apparently, there is nothing you won't do. And apparently your allies think it's perfectly fine for you to kill innocent children because, after all, it's not their children. You won't tell them that their children will be your slaves too. You're not big on disclosure.

The Russian Mafia has commandeered part of policing mechanism and aim it at noncriminals. Innocents are ambushed and coerced into taking a deal the Russian Mafia has no authority to offer but using compromised American law enforcement, fronts (lies) that they are our legitimate government.

The Russian Mafia controls how our police police in America, control some of our jails and prisons, have tried to get me placed in them for decades, and retaliated and began overtly harassing me in the street when I proved to be difficult to corrupt and entrap.

To flip our nation from an imperfect but working on it democracy to slavery, the Russian Mafia needs to frustrate and stealth attack millions of us. It's far easier and faster to make that happen by entrapping dominants like parents, teachers, supervisors, etc. and siccing them on subordinates, forcing the dominants to stealth harass and entrap those subordinate to them. So that's what the Russian Mafia is doing here. Probably in much of the world, they're extremely covetous

To seed your spies you got Jackson-Vanik passed in 1975 and less than fifty years later you're working to Close your coup, determined to install one of your puppets in 2024.

Currently, you're evilly forming African Americans, Evangelicals, anyone and everyone into a weapon to destroy our democracy to flip it into slavery. You won't be allowed to do that but you're insisting on having your try: evil means nothing to you, destroying the lives of our children and our families like you're attacking Palestinian and Ukrainian civilians. Why you want to rule a species you obviously, demonstrably hate, makes no sense. You make no sense. Your aspirations make no sense. You are insane, Russian Mafia.

Our security agencies can't prove you started the 2023 Israel-Hamas War but for sure they and our allies know you did. I know you think you've got everyone on lockdown, all the power players but no one except you thinks the world will stand by and let you enslave us. Why do you think we would do such a stupid thing. Just because you've ambushed your ascendancy doesn't mean we'll let you enslave us.

Chapter 10 - The Russian Mafia works the shadows to destroy us – the Israel-Hamas War serves as an example

The Russian Mafia exploits and manipulates secrets, confidentiality agreements, plea deals, sealed records, anything and everything it can to destroy us and terrorists affiliated with the Russian Mafia may have high value intel. The Russian Mafia navigates the space between what people know and what they don't know, rather like a how a middle-person exploits the lack of knowledge between how much it costs to produce something, and how much a buyer will pay for it. Middle-people get rich that way. The Russian Mafia is seizing control of our species that way. The Russian Mafia is a pro-slavery entity. If we allow it to place a puppet in the White House in 2024 or beyond, if we allow it to seize control of our species, we will wish we were dead. A current example of what the Russian Mafia does after it has seized control of multiple related parties like Hamas, the Israeli government, and Iran is to start the 2023 Israel-Hamas War. This a proxy war structured by the Russian Mafia to unseat President Biden and install one of it's puppets into the White House. That's the endgame, no matter how much dust the Russian Mafia kicks up to cover their tracks.

Currently, the Russian Mafia is successfully exploiting ordinary American's ignorance about our legal alliance with the Israeli government, how it limits our response. It matters especially now, as we're trying to work a closer alliance with Saudi Arabia. Interesting, isn't it, that we're especially careful because we're trying to secure an alliance deal with the Saudis, prove to them that we're a nation of our word. But what if the disagreement and distance reported to be between the Russian government and the Saudi kingdom is a set up. There are no accidents with the Russian Mafia.

We know what the Russian Mafia wants, world dominance. The Russian Mafia hates us and our government, has been working steadily, purposefully, deliberately to destroy us since 1970. The Biden Administration is the final administration the Russian Mafia intends to have no control over.

Why do we have a policy of not negotiating directly with terrorists. Understand, the Russian Mafia is deeply embedded in our country, ion our government, in our families, in our businesses. The Russian Mafia has directed policy and law here, as well as ordinary people. There's a reason Mr. Putin always wears a smirk when he discusses us – he believes he'll be overtly managing America in his life time, he expects to install one of his puppets into the presidency in 2024. Currently he's working to get African Americans to not vote for President Biden to protest Mr. Biden's position on the Israel-Hamas War. The Russian Mafia doesn't tell us that it is in fact the perpetrator/instigator of that war.

The Russian Mafia works the shadows, it succeeds by hiding relevant intel from people who need the intel. Just like the Russian Mafia (1) exploits the confidentiality clauses in our alliance with Israel to get American voters to turn away from President Biden so the Russian Mafia can install one of it's puppets in the White House, is similar to how it (2) prevents ordinary Americans from telling our legitimate government they are entrapped and being forced to entrap their own children, after being coerced into taking a confidential plea deal from law enforcement fronting as American, (3) just like the Russian Mafia manipulated Hamas (which is largely made up of radicalized Palestinians), into starting a war that's resulted in the murders and injuries of tens of thousands of Palestinian civilians, including children, just like the Russian Mafia.

The Russian Mafia exploits and creates one side not having critical intel so the Russian Mafia can do it's dirt.

That's why I say to my government, it's much better that we deal directly with terrorists. We have trained interrogators and terrorists if directly questioned would provide us useful information.

Chapter 11 - I demand justice, financial restitution & a formal apology for myself, my loved ones, and for all the people stealth attacked by the Russian government

1. <u>For myself</u>. I demand justice, full financial restitution and a formal apology from the USSR for it's deliberate destruction and destabilization of my childhood home and my childhood and my childhood family in America, which home and family were fully operational and effective in nurturing me and protecting me before the USSR's operatives stealth attacked me and my family, using Perfidy[162] and other illegal means of war against me in my childhood, when I was under the age of 15, in direct contravention of the Geneva conventions, which the USSR agreed to abide by,[163] and against the genocide conventions the USSR agreed to abide by as signatory.[164]

2. <u>For my half siblings raised with me or who I lived with during my childhood &/or during their childhood</u>. I demand the same justice, restitution, and formal apology for my siblings I was raised with, all three of whom alive at that time were under the age of 15. I demand justice, restitution, and a formal apology for my mother, who was approximately 28-29 at the time she was stealth attacked by the USSR, and an operative placed illegally in her bed using Perfidy, which is an illegal war crime during times of war and times of peace.[165]

---

162 Geneva Conventions of 1949, Additional Protocols. Specifically, Protocol 1 of the Geneva Conventions, Article 37. Prohibition against Perfidy. Article 37, paragraph (1)(c).

163 Ibid.

164 Convention of the Prevention and Punishment of the Crime of Genocide art. 2, 78 (UN.T.56 277, 9 December 1948).

165 Geneva Conventions of 1949, Additional Protocols. Specifically, Protocol 1 of the Geneva Conventions, Article 37. Prohibition against

And I demand justice, restitution and a formal apology for my youngest half sibling born in 1976, who's quality of life has been destroyed by the Russian Mafia, Russian government and the USSR, her health attacked and the quality of her childhood largely destroyed by psychological, emotional and sexual attacks by children and adult half siblings controlled by the USSR, Russian government and/or the Russian Mafia.[166] I charge that my brother in law, my sister in law, and my extended family were subjected to illegal entrapment by the Russian government and I demand justice, restitution and a formal apology from the Russian government.

3. My mother, as a female, had the right to expect no enemy government would attack her and/or her children and/or their home. I demand justice and restitution for my civilian mother who was set up and lied to by USSR operatives including the operative placed in her bed as her lover,[167][168] the operative "palm reader," and the operative "California friend," and/or any other additional individual whose full name I don't know at the time of typing this document (1/16/2024), using Perfidy, an illegal war crime, encouraged my mother, who is

---

Perfidy. Article 37, paragraph (1)(c).

166 Convention of the Prevention and Punishment of the Crime of Genocide art. 2, 78 (UN.T.56 277, 9 December 1948).

167 General Valery Gerasimov, Chief of the General Staff of the Russian Federation,"The Value of Science in Prediction," *Military-Industrial Kurier* (February 27, 2013). ***"The very "rules of war" have changed. The role of nonmilitary means of achieving political and strategic goals has grown, and, in many cases, they have exceeded the power of force of weapons*** in their effectiveness.

168 Shaun Walker, Pjotr Sauer and Tom Phillips "Panic and Emotional Pain as Alleged Deep-Cover Russian Spies Vanish," *The Guardian* (April 3, 2023)

untrained in espionage, and who, as an ordinary citizen, full time surgical technician employed at a hospital in a city in Southern California in 1970, and a mother with full custody of four young children she had financial responsibility for, the USSR and it's gang composed of KKK operatives and/or it's extended family of the KKK and/or of the USSR's spouses and/or children, and law enforcement and/or their children affiliated with the KKK and USSR gang who controlled the block we lived on in a Southern California city in 1970.

4. <u>My father</u> forced to try to entrap me, and/or who was forced to try to influence me over decades, I demand justice, financial restitution, and a formal apology from the Russian government.

5. <u>My aunt</u> forced to try to entrap me, and/or who was deliberately hooked on drugs and forced to try to influence me over years, I demand justice, financial restitution, and a formal apology from the Russian government.

6. <u>My cousin</u> who was deliberately hooked on drugs, and <u>for her six or seven children</u>, deprived of their mother, I demand justice, financial restitution, and a formal apology from the Russian government.

7. <u>My niece and nephews</u> stealth attacked as children and deliberately made homeless, I demand justice, financial restitution, and a formal apology from the Russian government.

8. <u>My nephew</u> attacked as a child and deliberately hooked on drugs, I demand justice, financial restitution, and a formal apology from the Russian government.

9. <u>My beloved grandfather</u> murdered by the Russian government's operatives because the Russian government knew I would've gone to live with him after they exhausted me, I demand justice, financial restitution, and a formal apology from the Russian government.

10. <u>My mentor</u> murdered to emotionally and psychologically destabilize me during Finals, I demand justice, financial restitution, and a formal apology from the Russian government.

11. <u>My cousins</u> deprived of their mother who was stealth murdered by the Russian Mafia, and for <u>my cousin's children</u>, deprived of their grandmother by the Russian government, I demand justice, financial restitution, and a formal apology from the Russian government.

12. <u>My aunt and uncles</u>, deprived of their sister who was murdered by the Russian Mafia when she was unable to persuade me to become a drug mule, and <u>for my aunt who's child was deliberately hooked on drugs</u>, I demand justice, financial restitution, and a formal apology from the Russian government.

13. <u>My uncle and aunt</u> who were deprived of their father who was murdered by the Russian government, I demand justice, financial restitution, and a formal apology from the Russian government.

14. <u>For my cousins and half sibling</u> who was deprived of their grandfather who was murdered by the Russian

government, I demand justice, financial restitution, and a formal apology from the Russian government.

15. <u>For my nieces and nephews</u> who were deprived of their grandmother because her handler(s) controlled who she could socialize with, I demand justice, financial restitution, and a formal apology from the Russian government.

16. For my <u>stepmother</u> who I witnessed was forced to get me from point A to point B during my visits, I demand justice, financial restitution, and a formal apology from the Russian government.

17. For my <u>mother's ex-husband</u> forced to try to engage me, I demand justice, financial restitution, and a formal apology from the Russian government.

Chapter 12 - Entrapment attempts

Below are some of the entrapment attempts the Russian government has used against me the last approximately 15 years. The most recent one was by my general practitioner within the last seven months, during the summer of 2023. The Person Used column is the person the Russian government used to try to coerce and/or lure me into committing the crime(s). These attempts were made on different dates unless otherwise indicated.

Entrapment Attempt	Person Used
(1) told to cheat Social Security by exaggerating my medical problems	Parent, retired law enforcement
(2) told to cheat Social Security by exaggerating my medical problems	Parent

Entrapment Attempt	Person Used
(3) told to cheat Medicare by exaggerating my medical problems	Parent, retired law enforcement
(4) told to cheat Medicare by exaggerating my medical problems	Parent
(5) doctor offered to lie to Medicare on my behalf, the lie unsolicited by me, to get me something he said I wouldn't ordinarily qualify for	Doctor, current
(6) doctor offered to lie to Medicare on my behalf, the lie unsolicited by me, to get me something he said I wouldn't ordinarily qualify for (interestingly, I had to go to the ER after a different operative frightened me so much with a noise attack my heart beat at 200bpm and I couldn't breathe. I requested to go to the ER closest to my apartment, but the paramedic said it was closed and took me to the next closest ER. There an ER nurse was super friendly. We got to discussing weight loss after the medication they gave me helped slow my heart rate & I could talk. I told her about trying to lose weight but that I didn't quality for the weight loss drugs because I'm insulin resistant, not diabetic and she encouraged me to ask my doctor again. When next I saw my doctor I asked him and he offered, without my asking, to lie for me so that I could get the weight loss drug, say that I am diabetic, even though I'm not. Why this is interesting is it demonstrates the Russian gov can direct me to an ER where they control more of the employees and then have someone they own there suggest I behave a certain way and when I asked my doctor for a	Doctor, current

weight loss med, he was ready to offer it illegally. Because most Americans don't know how deeply embedded the Russian government is here, they'd never believe that the Russian gov controls at least one paramedic, at least one ER nurse, and at least one doctor; that the Russian government controls those three individuals who work for three different companies but can manipulate them all in order to try to entrap me. That experience is what you need to know. The Russian gov double teams, quadruple teams. I've been quadruple teamed before by the Russian government. The more people in the target's circle the Russian government controls, the more likely the Russian gov is to triple and quadruple team the target and innocent people need to know this is a thing the Russian government does.

Entrapment Attempt	Person Used
(7) unsolicited recommendation that I apply for Disability and Unemployment simultaneously	Sibling
(8) unsolicited recommendation that I apply for Social Security and Unemployment simultaneously	Sibling
(9) told to cheat Medicare by applying for it while simultaneously lying to an insurance company	Sibling
(10) cheat Social Security by applying for it while simultaneously lying to an insurance company	Sibling
(11) told to cheat Medicare by applying for it while simultaneously lying to an insurance company	Parent[169]
(12) told to cheat Social Security by applying for it while simultaneously lying to an insurance company	Parent[170]
(13) told to cheat federal program that provides low cost internet/phone service to income qualified customers. When I said I didn't qualify my sibling told me to let the program decide if I qualified or not	Sibling
(14) told to cheat federal program that provides low cost internet/phone service to income qualified customers	Parent
(15) invited to join a federal complaint charging racism after Russia-controlled American operatives had my boss treat	Supervisor, co-worker

169 Called parent to complain that sibling was with me asking me to lie. My parent told me to lie to help my sibling. I refused. This entrapment attempt was a double team by using my beloved parent and brother on the same day.
170 Ibid.

Entrapment Attempt	Person Used
me cruelly for no reason, out of the blue, for many months. They then had a woman of color invite me to join her lawsuit where she charged racism	
(16) asked to commit immigration fraud when a plain clothes African American policeman approached me after work and asked me if I could help his friend who was being oppressed in another country and who needed to come to America and get a green card. A brief unconsummated marriage where I'd be paid	Plain clothes police officer I'd never seen before
(17) income tax fraud – Russia-controlled American ops had my very ill cancer survivor aunt (who they later killed with a drug overdose when she was unable to interest in me into becoming a drug mule), get me to a game show and had me win prizes totaling about $1,000. They knew I did my own taxes and expected I'd leave the value of the prizes off my income tax;	Aunt
(18) Tried to set me up for shoplifting by having cashier at a big box store fail to ring up my maybe 10 food items in my shopping cart;	Cashier at big box store & store security
(19) Had a high end store not ring up my groceries when I handed them to him;	High end grocery store known the world over
(20) Had a grocery store cashier give me twice as many stamps as I paid for;	Grocery store 1 cashier
(21) Had a grocery store cashier at another store refuse to ring up a package of tortillas. After I busted the shoplifting entrapment attempt Russia-controlled	Grocery store 2 cashier

Entrapment Attempt	Person Used
police kept it going for several stores, letting me know they knew where I was any time day or night.	
(22) I applied for a covid rental assistance I qualified for. Russia-controlled American operatives kicked my completed second application out of the system while having their people tell me there was no problem. When Russia-controlled American operative finally let me know there was a problem, they told me all I had to do was to press submit. And now, they told me, the program had been expanded, I could qualify for six months instead of three. This time I went to the library to re-submit the application since apparently there'd been a problem when I submitted it on my phone. There'd been no problem. Russia-controlled American operates slow walked my application so instead of	Russia-controlled American operatives, hackers & covid program operatives

getting three months rental aid, which I qualified for, I only got two months and was told to reapply. I reapplied. Nothing happened for weeks and although I continued calling the program, I was told everything was fine. When I was finally told I needed to resubmit the application again, and that I could apply for six months free rent, instead of the original three, and more assistance with my electricity bill, and gas bill, I quickly resubmitted it. What the people answering the phone weren't allowed to tell me was that Russia-controlled American operatives had changed the wording on the application, so that I no longer qualified. After I realized the form was different from the original form, I called the organization. Russia-controlled operatives blocked my phone from getting through. Americans must know this: Russia-controlled operatives can prevent you making calls on your phone (many times I placed calls Russia-controlled operatives wouldn't allow to get through). Russia-controlled operatives can turn off your electricity and then surge it, to destroy your household products, and their hackers can unsubmit forms you've completed and submitted.

(23) given the opportunity to embezzle from the law firm I worked at when they gave me a credit card to make purchases of my choice for my boss, and I'd be the sole person reviewing and authorizing the statement;	Attorney I worked for, & law firm
(24) free movie app w/illegal access to the latest movies put on my phone without my permission 1st time	Sibling
(25) free movie app w/illegal access to the latest movies put on my phone without my permission 2nd time	Sibling
(26) offered illegal use of a fire stick (Amazon product) - sibling;	Sibling

Chapter 13 - I believe the 2023 Israel-Hamas War was authorized by the Russian Mafia to unseat President Biden

I believe, but at this time can't prove, that the Russian government authorized Hamas[171] to attack innocent Israeli's to

---

171 Hamas is controlled by Iran. Iran is owned by the Russian government. Any terrorist organization controlled by the Iranian government will do a favor for the Russian government. I doubt, but cannot prove, that Hamas attacked Israel without running it by the Russian government. Had the Russian government said no, Hamas wouldn't have acted. Also, I published the first draft of this book on October 5, 2023; Hamas attacked on October 7, 2023. It's quite possible my messy first draft

launch the Israel-Hamas War, in order to unseat President Biden.

If I'm right:

(1) the Russian government is responsible for the murder of 15,000+ Palestinian children, blood pawns to destroy our government, and unseat the Russian government's principle political enemy;

(2) a mass murderer of babies ascending to world domination is a big mistake for our species; and

(3) Palestinian babies won't be the only babies dying at the hands of the Russian government.

The Russian government, in charge of the world, would mean a bloodbath for our species and if you think it's bad while the Russian government ascends to power, you won't want to see what they'll do to secure world power. Tens of millions of Americans and our allies would be enslaved (even more than already are), and massacred.

The Russian government are masters of psychological warfare. They lie for fun and profit. I'm not exaggerating. America's international standing and self-worth have been

_____

had enough in it to alarm the Russian government. Someone, I believe under orders by the Russian government, attacked the electrical system where I live on September 29, 2023, with the intention, I believe, of destroying the chip in the computer where I usually charge my battery. The Russian government is deeply embedded here and knows what's on deck from traditional publishers, outing them. They didn't know this book was coming and I believe their operatives attacked the grid where I live, and sped up Hamas' attack against innocent Israelis, the Russian government adjusting to a book they hadn't known was coming. They've got very fast response times but if I'm right, they're tied into pre-existing business relationships here. When new people come online the Russian government has to scramble.

pummeled because the Russian government has played us for over 50 years. If you think we're stupid, you're wrong. The problem isn't our stupidity, it's our enemy's brilliance, hatred of us, insanity, and willingness and skill at lying straight-faced, while laughing at us for believing them, and thinking us weak for wanting peace. They don't want peace but instead demand dominance and enslavement. They're sadists who hate us. That's a bad combination for anyone subordinate to the Russian government. How do I know? I've been targeted and attacked by them for 53+ years. I've survived to warn you. Please be warned. Whatever happens, never surrender your ability to defend yourself. Anyone asking you to surrender your weapons as a sign of good faith, intends to destroy you.[172] To survive we need to understand these guys, and we need to understand them yesterday.

To survive the threat they are, we must provide our people with a quality education. That is the intention of this book. I'm learning how to write a book to tell you my story. I've lived through decades of the Russian government's attacks but writing about it requires a whole other skill set, so please cut me some slack while I find my way as an author.

Although the USSR signed the Geneva conventions and anti-genocide treaty before 1950, by 1970-1971 they were on the ground in America attacking me and my family.[173]

---

172 The Russian government loves to brutally murder and terrorize people.
173 Those treaties promised the signatories wouldn't target civilians, women or children under 15, in times of war. Implied, but not stated in the Geneva conventions was that signatories wouldn't target civilians in times of peace. Despite tense times between the U.S. and the USSR in

The USSR used their embedded operatives to form alliances with the KKK (an organization they considered pro-Kremlin like they consider racists, sexists and homophobes pro-Kremlin), and used their operatives to attack me and my family, a family of five with four children under 15, and a single, divorced mother aged 28-29, our family is African American, and my extended family includes children of Puerto Rican descent who the Russian government attacked when they were children, to make them homeless.

Not only did the USSR make me homeless during my childhood and my half siblings I was raised with, the subsequent Russian government continued attacks against me and my extended family. They deliberately entrapped my youngest brother's children, to make them homeless, to string at least one of them out on drugs, alcohol, and which lead to the death of at least one of my youngest brother's grandchildren, all the consequence of the Russian government specific attacks and destabilizing strategies against me and my extended family. Considering the thousands of other Americans I've witnessed controlled by the Russian government, multiple thousands of people who've been forced

---

the 20th and 21st centuries, we've not been at war, so the USSR has no international law to stand on. It doesn't care. In addition to international laws, Americans are protected under our domestic laws which neither the USSR or the Russian government respect. The Russian government is Mafia. What that means is it doesn't care about our laws and if it can get away with destroying us under the radar, it will. And that's exactly what it's doing. Currently, the Russian government is avoiding doing anything that can make it appear anti-people of color in America but a more racist, sexist, homophobic government I've never heard of.

by their handlers to harass and/or attack me, my family is not the only one under attack.

We're not going to make it out alive unless we, at minimum, educate our people, and preferably defend ourselves. Demanding justice, demanding restitution and forcing the Russian government to stop their stealth attacks are also demands I make against them.

After the initial attacks against me and my family the USSR, and later the Russian government continued to expand it's attacks and expanded attacks onto my extended family, loved ones, and/or beloved friends. In the case of one of my beloved friends, the Russian government killed him, I believe, to emotionally destabilize me during Finals week of my programming class in Spring 2023. I believe that's the case because they made sure to tell me the morning I left the ER, from the attack they'd made against me the day before. The Russian government has worked unrelenting since it entered my life to destabilize my housing and my education. The only reason the USSR had for destroying us was because we were a family and happened to be people of color, and at that time the USSR was embedded here destroying anyone with strength prioritizing a person of color, likely because we had fewer resources, but I suspect the USSR was told to destroy any women and/or girls they found to be strong and who had no backup. For example, you'll notice Native American women have a relatively high disappearance rate and police don't usually discover what's become of them. My theory is the

Russian government may have been trafficking Native American women as early as the 1970s, 1980s, etc. Because I defended my younger brother when he was attacked, and because I defended myself when I was threatened by a child operative controlled by the USSR, the USSR destabilized me and my family by setting my mother up, using an operative they illegally placed in her bed, using Perfidy, which is a War Crime, to make us homeless, interrupt our schooling and remove us from schools and friends we loved. For example, I was taking German and learning how to play the saxophone. My schooling interrupted, I never returned to learning German or learning to play the saxophone. The saxophone I'd been using to practice was borrowed from the school I'd been attending before I was threatened by a child controlled by the Russian government, KKK, and/or local police, and before we were made homeless by the USSR. The Russian government targets children's housing and schooling and as I've found from direct personal experience, when children's housing is interrupted, that child rarely springs fully back. For example, in my case I never picked German back up, even though before we lost our housing I had earning a grade of A in German.[174]

The USSR formed alliances with police affiliated with the KKK, and had a field day attacking and destabilizing American children which I know from personal experience. I

---

174 Geneva Conventions of 1949, Additional Protocols. Specifically, Protocol 1 of the Geneva Conventions, Article 37. Prohibition against Perfidy. Article 37, paragraph (1)(c).

know because I lived it. I'm one of the children in a family the USSR illegally attacked.

The USSR *was* here illegally, the Russian government *is* here illegally, the KKK lynched people illegally, and the relatively few police who were klans men back in the day acted *extra judicially*. So, even though I'm routinely stealth harassed and/or attacked as recently as the present (January 2024), Russia-controlled American operatives and embedded USSR operatives have been attacking me for over 53+ of my 64 years of life, illegally.

Chapter 14 - The Russian government is working a coup here

The Russian Mafia/government is working a coup here. For the coup to work victims must (1) *believe* without a shadow of a doubt they were entrapped (2) *randomly* by (3) *American* law enforcement and that the arrested person (4) *is corrupt*.[175]

The Russian government uses American law enforcement officers they've entrapped and force them to entrap noncriminals, allowing targets to assume the officers arresting them are employed and directed by law enforcement that our legitimate government must have authorized the entrapment. Our government does authorize entrapment but not the types of entrapment the Russian government is running.

---

175 Janis Berzins "The New Generation of Russian Warfare" "...the Russian view of modern warfare is based on the idea that the main battle space is in the mind." Aspen Institute | Prague, *Aspen Review* (March 2014). New Generation Warfare: https://en.m.wikipedia.org/wiki/new-generation-warfare.

For example, I know without my government having to tell me, that they'd never accept as a legitimate entrapment either of my parents, much less both, were used against me as weapons to try to entrap me, with one of my parents a retired police officer. And yet the Russian government had both my parents tell me to lie and/or exaggerate my health problems to Medicare. The Russian government also had my current doctor, twice, encourage me to lie to Medicare to obtain a benefit he told me I didn't qualify for but which he offered to lie for me to get: the lie unasked for by me.

Apparently our government doesn't now require an explanation on how the entrapment occurred so the Russian government has entrapped law enforcement, perhaps some of government officials, and our citizens to convince our non-compromised government that America is loaded with liars and thieves, when the truth is the Russian government is working a their coup against us, to destroy us, yet again, and to make us as miserable as they possibly can, including attacking our children.

As Mr. Berzins says in New Generation Warfare, the battlefield the Russian government exploits is the mind; both the minds of our government officials and the minds of citizens untrained in espionage.[176]

I believe a general overview of the successful entrapments are submitted to our government, emphasizing the

176 Janis Berzins "The New Generation of Russian Warfare" "...the Russian view of modern warfare is based on the idea that the main battle space is in the mind." Aspen Institute | Prague, *Aspen Review* (March 2014). https://en.m.wikipedia.org/wiki/new-generation-warfare.

victim's intended theft of government funds, but it's doubtful the compromised law enforcement are asked to describe how the entrapment was secured.

The Russian government has deep knowledge of and mastery our systems, having lied, conned and inserted themselves here illegally last century.[177] What we now have is a blood bath where innocent Americans hate our government, because they've been lead to believe by Russia-compromised American law enforcement, that our legitimate government is forcing Americans to entrap their own children.

A signature Russian government move is to harm innocent people and blame our government for it. You can see echoes of it in the Israel-Hamas War which, I believe, is a proxy war the Russian government facilitated to unseat President Biden, insert a puppet in the White House, and destroy us. No headline or protests I've seen blame the Russian government for the 2023 Israel-Hamas War because the Russian government and it's Mafia do their dirt in secret and suggest to gullible protesters that we are to blame, and our President.[178]

---

177 Craig Unger, "*American Kompromat*" (Penguin Random House LLC 2021).

178 Gen. Gerasimov, Chief of the General Staff of the Russian Federation,"The Value of Science in Prediction," *Military-Industrial Kurier* (2/27/13). "The focus of applied methods of conflict has altered in the direction of the broad use of political, economic, informational, humanitarian, and other nonmilitary measures – applied in coordination with the *protest potential of the population.*" "*..Tactical and operational pauses that the enemy could exploit are disappearing...*"

Chapter 15 - The Russian Mafia exploits hierarchy & chain of command to attack us – we must break that cycle

On the ground in America, once the Russian government seizes control of the authority figure, whether it's a parent, doctor, employer, lover, teacher, etc., nearly everybody influenced by that person becomes compromised. We've got to break that cycle or we're toast. They exploit it every chance they get because it's cheap, ongoing, unrelenting harassment that pushes me ever closer to death. The Russian government embraces murdering their enemies.

Because the USSR *was* breaking the law, and because the Russian government *is* breaking the law, the war, coup, and genocide are carried out in secret: *no paper trail.* If the Russian government went loud, like they're doing against Ukrainians, we'd bomb them. The Russian government doesn't want their government bombed. They don't believe they can recover from being bombed.

Chapter 16 - The Russian Mafia's current strategy

(1) The Russian government is committing stealth genocide against families in America by deliberately destroying my family, and other people's families in America, using force, lies, coercion, illegal drugs, etc., and other fraudulent means, to force family members to participate in entrapment scams whereby the Russian government targets noncriminal family members and friends, the intention being to later have influencers influence the entrapped person into flipping into

389

the GOP Base. I've seen this done to two of my family members, my mother and my sister so it's a triple and/or quadruple team effort to seize illegal control of our voting system.

The Russian government hid itself behind systems/institutions/individuals our law enforcement trusted and forced our law enforcement to entrap family members, but did not limit entrapment to families.

The Russian government forced law enforcement to entrap supervisors, teachers, car mechanics, auto insurance agent, individual employees and/or individuals with access to a targeted individual, doctors, law enforcement, security officers (plain clothed and uniformed), lawyers, Human Resource employees, grocery store employees, authority figures, etc.

The Russian government forced entrapped parents to groom and shape their own children into illegal weapons and forced them to attack other citizens and/or children, forced entrapped parents to use their children under the age of 15 in a military capacity to harass and/or threaten adults and/or children.

Chapter 16.1 - The Russian government forced entrapped parents to entrap their adult children[179]

The Russian government and/or it's allies forced children with entrapped parents to attack, engage, befriend and/or threaten other children and/or adults.

The Russian government forced an entrapped parent and/or family member to beat and/or harass a child the Russian government wanted to harass, enrage, frighten and entrap.

Not only has the Russian government destroyed my family as it once existed, but I've witnessed it's ongoing destruction of, rough estimate, three thousand other Americans and/or migrants it controls, including, rough estimate, 1,000 children in varying stages of grooming and/or militarization by parents and/or operatives the Russian government controls.

The Russian government forced people inside and outside of my family to try to lure and groom me into breaking the law, and tried to addict me to drugs and/or alcohol.

To get me into the criminal justice system, the Russian government had people it controls throughout our society attempt to entrap me and use other means to criminalize me, going so far as to have cashiers at four stores within the last several years omit to scan products I'd placed in my shopping cart, omit to scan products handed to me by the cashier, and omit to scan products I'd placed on the conveyor belt to be scanned by the cashier, in the Russian government's attempt to

_____

179 I've been directly, personally victimized by this scam, both my parents, one a retired law enforcement officer, were used to try to entrap me, after the Russian government's American operatives deliberately exhausted and sickened me with overwork and abuse.

have me leave the store whereby they would have security guards charge me with shoplifting.

The Russian government creates it's opportunities. For people who won't steal, the Russian government has compromised cashiers and store security work together to get the victim arrested. I'm writing from direct, first-hand experience. I've experienced these attacks personally.

The Russian government has worked unceasingly for decades to put me into the American criminal justice system, presumably because they control individuals in that system.

The Russian government forced my entrapped parents to tell me to break the law and/or do other self-harming behaviors. Fortunately I was 50 when they told me to do these behaviors, so I had a mind of my own. But I've witnessed newborns with entrapped parents[180] and, assessing how the Russian government has destroyed my family, children born into families with entrapped authority figures have no chance of a normal life and childhood, which necessitated the writing this book of warning.

When I refused to be entrapped into breaking the law, the Russian government used my parents and other people within my nation as weapons to attack me, including uniformed American law enforcement, plain clothed law enforcement, supervisors and co-workers from at least six jobs, my landlord, operatives the Russian government controlled who I thought were my friends, and other individuals I

---

180 I knew the parents were entrapped because they stealth harassed me and/or attempted to engage me in a way I now recognize as suspicious.

interacted with including doctors, teachers, and strangers the Russian controlled throughout my country. The Russian government has used thousands of people as weapons against me to psychologically, physically, and/or sexually attack me. It's used operatives to terrorize and threaten me. It has tortured me by attacking my sleep cycle.[181]

American families form the foundation American culture, society, and government. By attacking families, the Russian government seeks to destroy my people, our culture and our government;

(2) Deliberate destruction of my career and quality of work life; (3) Destruction of my natural interaction with my society; (4) Physical attacks by Russia-controlled American operatives; (5) Thousands of psychological attacks and threats spanning decades; (6) Stealth harassment by American law enforcement and plain clothed security spanning decades; (7) Threats of and feigning of physical assault, including groups and/or individuals swarming me to deliberately frighten me; (8) Loud noise and startling attacks where Russia-controlled American operatives appeared suddenly to frighten me, and which such an attack resulted in my experiencing an elevated heart rate of 200 bpm necessitating a trip to the ER; (9) Sexual assaults by Russia-controlled American operatives; (10) Torture (via deliberate sleep deprivation spanning a dozen years); (11) Russia-controlled operatives in the apartment

---

181 Convention on the Prevention and Punishment of the Crime of Genocide, adopted 12/9/1948 – General Assembly Resolution 260A(III).

above mine, and operatives outside my apartment had eyes on me inside my apartment whereby operatives followed me room to room, honked their vehicle horn when I opened my front door (even at 2am), and operatives honking their horns as I entered and/or exited a room; (12) Medical-related attacks; (13) Vehicle-related attacks; (14) Blackmail attempts; (15) School-related attacks; and (16) Etc.

When I began writing presidents in 2016, or other agencies begging for help, American law enforcement increased their harassment against me. Because it seemed to me most people with access to me were compromised, the legal assistance service in my area I believed was compromised, and no president and/or agency I wrote to was willing to help me, I began a publishing business intending to write this a book, describing the attacks against me. Because Russia-controlled American operatives were aware of me long before I knew they were my enemy, they were able to prevent my getting the technological help I needed until 2023, which is why you're just getting this book now.

The Russian government is determined to destroy us, and determined to avoid being bombed by us. To that end, the USSR and the Russian government chose a slow-moving coup believing it gave them the best chance of destroying us while given them the best possible chance of evading retaliatory bombs we might try to drop on them in self-defense.

In their coup, the Russian government intends to install and maintain a pro-Kremlin president in the White House, along with pro-Kremlin advisors. That president and/or advisors would be the "inside-job" weapons used to dismantle the parts of our government (security agencies, miliary, etc.) the Russian government considers hostile to their world domination aspirations.

I am an ordinary person untrained in espionage. This book describes how the Russian government uses psychological warfare, hybrid warfare, and stealth warfare to destroy ordinary people, while the Russian government camouflages itself within the skin of victims it has entrapped who have access to us. Our loved ones, authority figures like our teachers, doctors, supervisors, landlord, grocery store employees, friends, etc. The Russian government uses the people who have access to us, and who we trust as weapons to lure, groom, entrap, and/or psychologically attack us.[182] I don't believe our family, friends, teachers, doctors, landlord, and work colleagues know an enemy foreign government is their handler, but instead I think they may believe that people with access to them, American law enforcement, and by extension our legitimate government, is calling the shots.

By telling my story I hope to help ordinary people worldwide (including children) better protect themselves

---

182 Imagine my surprise when my beloved retired law enforcement parent told me, unprompted, to exaggerate my medical problems to Social Security Disability; imagine my shock when my other beloved parent who raised me to be an ethicist told me, unprompted, to lie to Social Security Disability. The Russian government entrapped my parents and forced them to try to entrap me. The Russian government has commandeered our entrapment system and aims it at noncriminals.

against predatory, proslavery governments like the Russian government.

Chapter 17 - The Russian Mafia commits genocide & other crimes against the American people

This book includes a partial list of how Russian operatives and Russia-controlled American operatives covertly (and in my case covertly and overtly) attack ordinary Americans, from 1971 to the present. I list many specific entrapment attacks I've endured by the Russian government and it's operatives. In this book ordinary Americans ("ordinary Americans") includes: (1) American citizens, including incarcerated citizens; (2) Native Americans; (3) immigrants living in America, either legally or illegally; and (4) the children of groups (1) through (3). In this book ordinary Americans under attack by the Russian government live and/or lived in America between 1971, to the present: I chose this time frame because it's the time frame the Russian government began, and continues, it's attacks against me, and it's within this time frame that I've witnessed the Russian government attacking other ordinary Americans, including children I've witnessed being groomed and/or used as weapons by the Russian government in it's approximately 1,000 attacks against me where entrapped parents used their children as beards and/or as co-attackers, under orders of the Russian government, against me, over multiple decades. In this book the Russian government ("the Russian government") includes: (1) the Russian government in existence in 2023; (2) the former government of Russia, known as the USSR, which formally ended in 1991; (3) Russian government operatives in America; (4) Russia-controlled American operatives; and (5) Russia-controlled non-American operatives. In this book the Russian government's attacks against Americans between approximately 1971 to the present are my primary focus because those are the decades I've been attacked by the Russian government, but I also briefly discuss USSR attack strategies against America before 1971.

I charge the Russian government with unrelenting stealth warfare against ordinary Americans beginning in approximately 1971, starting when they forced their operatives into my life when I was approximately 11, to the present: spanning approximately 53 years. For the majority of my life I've been attacked by the Russian government.

The deliberate targeting of civilians in war time is expressly prohibited by the Geneva Conventions. We're not at war with the Russian government but it is at war with us.[183]

The Russian government's (1) stealth attacks in peace time against ordinary Americans who are civilians, including stealth attacks against our law enforcement officers who are also civilians; (2) entrapping ordinary American civilians, (3) forcing entrapped parents to entrap their own children; (4)

---

183 General Valery Gerasimov, Chief of the General Staff of the Russian Federation,"The Value of Science in Prediction," *Military-Industrial Kurier* (February 27, 2013). General Valery Gerasimov states:
*"In the 21st century we have seen a tendency toward blurring the lines between the states of war and peace. **Wars are no longer declared** and, having begun, proceed according to an unfamiliar template." "**The very "rules of war" have changed....**"*

flooding our nation with highly addicting and deadly fentanyl,[184][185][186] and not just our nation.[187][188] Ecuador's children are being gunned down, forced into a gang when they're 7th graders. When they refuse to join the children are told their families will be murdered. No child can deal with such evil. Mexico's two most powerful cartels, Sinaloa and

---

184 The Russian government and it's Mafia are one of the most successful illegal drug dealer systems in the world; targeting anyone, including children in West aligned nations is part of the Russian government's & it's allies' strategy to destroy us. According to Craig Unger in his book *House of of Trump House of Putin*, "...Putin and his cronies [...] were allowed to make billions through drug trafficking, extortion, elaborate financial schemes, the sex trade, arms deals, and the like...." Mr. Unger also states "...made Vladimir Putin the richest man in the world, with wealth, according to Hermitage Capital Management CEO Bill Browder, approaching $200 billion." Craig Unger, "*House of Trump House of Putin*," (Penguin Random House UK 2018).

185 David Pierson, Edward Wong, Olivia Wang "U.S. Raises Pressure on China to Combat Global Fentanyl Crisis," *New York Times* (7/7/2023). The Chinese government sees fentanyl as leverage to use against us, I believe essentially this is the deal they're offering to us: you leave the South China Sea and Taiwan to us, and we'll stop the flow of fentanyl precursors to the cartels.

186 Julie Bosman "Fentanyl Cuts a Bitter Swath Through Milwaukee," *New York Times* (12/12/22). Glenda O. Hampton, Executive Director of Gateway to Change, a drug treatment center in Milwaukee said "I've seen a lot of terrible drugs. This is the worst."

187 "There is no question there is interconnectivity between Chinese organized crime and the Chinese state," said Montoya. "The party operates in organized crime-type fashion. There are parallels to Russia, where organized crime has been co-opted by the Russian government and Putin's security services." Sebastian Rotella and Kirstein Berg "How a Chinese American Gangster Transformed Money Laundering for Drug Cartels," *ProPublica* (10/11/2022).

188 "At no time in the history of organized crime is there an example where a revenue stream has been taken over like this," said retired DEA agent Thomas Cindric, a veteran of the elite Special Operations Division. "This has enriched Mexican cartels beyond their wildest dreams." About six years ago federal antidrug agents saw early signs of what would become a tectonic change as it's investigations tracked drug criminals. Former Senior FBI official Frank Montoya Jr, who served as a top counterintelligence official at the Office of the Director of National Intelligence said "we suspect a Chinese ideology and strategic motivation behind the drug and money activity." He also said, "to fan

Jalisco Nuera Generacion, and the Albanian Mafia are just some of the cartels operating in Ecuador, another example of how allies of the Russian Mafia destabilize South, Central and North American countries to fulfill Mr. Putin's agenda to destroy us. Prisons are gang headquarters and recruitment centers. "You fill these prisons with people that need to survive there," and unknown source said. "Many are rejected by families due to drug use and the only way to survive was to become part of a gang."[189]; (5) forcing entrapped parents to groom (train) their children to harass ordinary Americans the Russian government wants attacked; (6) forcing entrapped parents to use their children, some as young as days old infants, as shields while the entrapped parent harasses an ordinary American;[190] (7) forcing entrapped parents to use their children under the age of 15 years as a co-attacker and co-conspirator against an ordinary American; (8) building a child army in America and forcing that army to attack ordinary Americans, including other children, the Russian government

_____

the flames of hate and division, the Chinese have seen the advantages of the drug trade. If fentanyl helps them and hurts this country, why not?" Sebastian Rotella and Kirstein Berg "How a Chinese American Gangster Transformed Money Laundering for Drug Cartels," *ProPublica* (10/11/2022).

189 Julie Turkewitz "How a Peaceful Country Became a Gold Rush State for Drug Cartels," *New York Times* (7/12/2023). "...huge increase in violence including children gunned down outside their homes or schools." "It's so painful," said a community leader. The leader's neighborhood has been transformed in recent years, with children as young as 13 forcibly recruited to criminal gangs. They are threatened. "You don't want to join? We will kill your family," the leader describes how 7th graders are forced to join gangs. "We're in a terrible crisis, both in the prison and outside in the streets," said Ms. Morales, member of Committee of Relatives for Prison Justice.

190 This was one of my personal experiences.

wants destroyed, and (9) forcing entrapped parents to lure their own children to commit crimes against the State are War Crimes, Crimes Against Humanity, and Genocide against the American people. I charge Genocide because the Russian government deliberately destroys our families when it forces entrapped parents to entrap their own children; and when the Russian government forces entrapped parents to use their children under 15 years as co-attackers when the entrapped parent is forced to harass and/or attack the ordinary American the Russian government wants harassed and/or attacked. Genocide is the deliberate destruction of a people made even more heinous when done by ambush, with no war declared, against civilians and against children with no means to protect themselves.[191] The cornerstone of our species is our families. By attacking our families, the Russian government strikes at the very heart of us, who we are as a nation and as a people, with the clear intention of destroying us.

United Nations Human Rights – Convention on the Prevention and Punishment of the Crime of <u>Genocide</u>, adopted 12/9/1948 – General Assembly Resolution 260A(III)

Article I – "...genocide, whether committed in time of peace or in a time of war, is a crime under international law which they (Contracting Parties) undertake to prevent and to punish."
Article II – "genocide means any of the following acts committed with intent to destroy, in whole or in part, a national, ethnical, racial or religious group, such as:
(a) Killing members of a group (this includes torture);
(b) Causing serious bodily or mental harm to members of a

---

191 Gen. Gerasimov, Chief of the General Staff of the Russian Federation,"The Value of Science in Prediction," *Military-Industrial Kurier* (2/27/2013).

United Nations Human Rights – Convention on the Prevention and Punishment of the Crime of <u>Genocide</u>, adopted 12/9/1948 – General Assembly Resolution 260A(III)

group; (c) Deliberately inflicting on the group conditions of life calculated to bring about it's physical destruction in whole or in part; (d) Imposing measures intended to prevent births within the group;

(e) forcibly transferring children of a group to another group.

Article III – The following acts shall be punishable;

(a) genocide;

(b) conspiring to commit genocide;

(c) direct and public incitement to commit genocide;

(d) attempt to commit genocide;

(e) complicity in genocide.

Article IV – Persons committing genocide or any of the acts enumerated in article III shall be punished, whether they are constitutionally responsible rulers, public officials or private individuals. Importantly, the victims of genocide are deliberately targeted, not randomly…."

<u>Crimes Against Humanity</u> – Rome Statute of the International Criminal Court – Article 7 – International Criminal Court – Rome Statute of the International Criminal Court, (7/1/2002)

For the purpose of this statute "crimes against humanity" shall mean any of the following acts when committed as part of a widespread or systematic attack directed against any civilian population, with knowledge of the attack:

a) murder; b) extermination; c) enslavement; d) deportation or forcible transfer of population; e) imprisonment or other severe deprivation of physical liberty in violation of fundamental rules of international law; f) torture; g) rape… and any other form of sexual violence; h) persecution against any identifiable group or collectivity on …. national, …. cultural, ….. or other grounds…."; i) enforced disappearance of persons; j) the crime of apartheid; k) other inhumane acts of similar character intentionally causing great suffering or serious injury to body or to mental or physical health.

Crimes Against Humanity – Rome Statute of the International Criminal Court – Article 7 – International Criminal Court – Rome Statute of the International Criminal Court, (7/1/2002)

"Attack directed against any civilian population means a course of conduct involving the multiple commission of acts… against any civilian population, pursuant to or in furtherance of a state or organization policy to commit such attack."

War Crimes mean grave breaches of the Geneva Conventions of August 12, 1949 including:

(a)(1) willful killing; (2) torture or inhuman treatment including biological experiments; (3) willfully causing great suffering, or serious injury to body or health; (4) extensive destruction and appropriation of property, not justified by military necessity and carried out unlawfully and wantonly; (5) compelling a prisoner of war or other protected person to serve in the forces of a hostile Power; (6) willfully depriving a prisoner of war or other protected person of the rights of a fair and regular trial; (7) unlawful deportation or transfer or unlawful confinement; (8) taking of hostages.

(b)(1) Intentionally directing attacks against the civilian population as such or against individual civilians not taking direct part in hostilities;

(2) intentionally directing attacks against civilian objects … objects which are not military objectives….;

(3) intentionally launching an attack in the knowledge that such attack will cause incidental loss of life or injury to civilians or damage to civilian objects or wide-spread, long-term and severe damage to the natural environment which would be clearly excessive in relation to the concrete and direct overall military advantage anticipated;

(4) attacking or bombarding by whatever means, villages, dwellings, or building, which are not military objectives…..

….

(11) killing or wounding treacherously individuals belonging to the hostile nation or army;

(12) destroying or seizing the enemy's property…."

(13) declaring abolished, suspended or inadmissible in a court of law the rights and actions of the nationals of the hostile

War Crimes mean grave breaches of the Geneva Conventions of August 12, 1949 including:

party;

(14) compelling the national of the hostile party to take part in the operations of war directed against their own country,…";

(16) committing outrages upon personal dignity including humiliating and degrading treatment;

(17) utilizing the presence of a civilian or other protected person to render certain points, areas or military forces …. from military operations;

(18) conscripting or enlisting children under the age of 15 years into the national armed forces or using them to participate activity in hostilities…….

(e)(1) intentionally directing attacks against the civilian population as such or against individual civilians not taking direct part in hostilities.

War Crimes are grave breaches of the 1949 Geneva Conventions, and include crimes directed against civilians. War Crimes against (a) persons requiring particular protections, including women, civilians, and children under 15 years of age; (b) …; (c) war crimes against property and other rights; (d) prohibited methods of warfare; (e) prohibited means of warfare. Examples against prohibited acts include cruel treatment and torture; intentionally directing attacks against the civilian population; conscripting or enlisting children under the age of 15 years into armed forces or groups or using them to participate actively in hostilities.[192]

---

192 How I'm stealth attacked in my apartment and tortured by being deliberately deprived of sleep. Russia-controlled American operatives order my neighbor(s) to make excessive noise, especially when I'm studying or trying to sleep. Deliberate sleep deprivation is one of Russia-controlled American operatives favorite attacks because it's a very harmful attack that only negatively impacts me. Deliberate sleep deprivation is labeled as torture so you need to understand the Russian government tortures innocent people it wants sickened or killed but when they don't want a murder to be an obvious hit to a medical examiner. Everyone needs to sleep. The ones I have aren't that great but they're better than nothing.

Civilians are protected under the Geneva Conventions of 1949 and Additional Protocols. Specifically, Protocol 1 of the Geneva Conventions, Article 37. <u>Prohibition against Perfidy</u>.
Article 37, paragraph (1)(c)

"The feigning of civilian, non-combatant status," by an enemy agent constitutes Perfidy when an agent of a foreign enemy government uses civilian, non-combatant status as cover to attack and destroy civilians in a targeted enemy nation.[193] Perfidy is a form of deception in which one side promises to act in good faith with the intention of breaking that promise...." Perfidy constitutes a breach of the laws of war (including distinction and proportionality), and so is a war crime as it degrades the protections and mutual restraints developed in the interests of all parties, combatants and civilians.

Perfidy – the use of unlawful deception. Acts of Perfidy are deceptions to invite the confidence of the enemy to lead them to believe... they are entitled to or is obliged to accord, preferred status under the law of armed conflict, with the intent to betray that confidence."

Chapter 18 - The Russian Mafia weaponizes 'romance'

Whitney Houston sang How Will I Know so beautifully it's one of my favorite songs by her. When she sings 'how will I know,' backup responded 'girl trust your feelings.' But I argue in a world where the Russian Mafia is clawing it's way to ascendancy, are world masters of psychological warfare, no

---

193 The Russian government and it's allies in Southern California, the KKK and local police in approximately 1970-71 placed a Russian-controlled operative in my mother's bed, using as her boyfriend/lover, with the intent to destroy her, her livelihood and her children, which family KKK families in the neighborhood where they lived wanted the family out of the neighbor, and the Russian government used the destruction of the family, including children, to ingratiate and/or embed themselves into the gang.

one can rely just on their feelings. If you do, the Russian Mafia will have you for lunch.

The KGB used the "mentor romance" attack against me nearly 35 years ago, and still deploys this attack.[194] My supervisor was nice, handsome, excellent at his job, and happily married. He didn't say so I just got that vibe. He'd interviewed and hired me. He and I worked alone in the word processing department during the day shift and he had the radio on, turned very low, playing beautiful romantic music while we worked. I remember Mariah Carey's "Vision of Love" was out around that time, she is an amazing artist, that was her breakthrough hit. Don Henley's "The End of the Innocence" played, and Heatwave's "Always and Forever."

I had no idea it was a set up. I was around 30. My supervisor trained me on how that law firm preferred their documents to look. One day, not long after I'd started, an attorney stormed in and insulted me about a document I'd

194 Shaun Walker, Pjotr Sauer and Tom Phillips "Panic and Emotional Pain as Alleged Deep-Cover Russian Spies Vanish," *The Guardian* (April 3, 2023). "Campos Wittich was a Russian spy living under an assumed name in Brazil. At least six such suspected spies have been unmasked… in various locations over the past year, suggesting there could be….more. He was a deep cover spy working for an elite intelligence programme…." "She (the victimized girlfriend) suspected nothing at all," said the friend of Campos Wittich's partner." "She is just a lovely woman looking to create a family and get married to a man she thought was the man of her life." "She is really scared of this situation and hurt by all the pain of having an abrupt cut-off in a relationship that was perfect in her eyes," said one of friends. The victimized girlfriend of the spy is a veterinarian. She had no suspicion she was involved with a Russian spy. The romantic partners of both spies have been left devastated, said their friends and acquaintances in Athens and Rio."…Campos Wittich may have been in an earlier 'embedding' stage of their missions, which could have gone on for decades if not detected," reporters state.

completed for him. I always did excellent work so he may have complained about how long it had taken me to complete the project. New, I was still learning the systems there, and, presumably, had the attorney told my supervisor the document was urgently needed, my boss would've done it. My supervisor told the attorney to never speak to his staff discourteously again.

Understand that in my 20s, the KGB had had a supervisor physically assault me. After I transferred out of her department, my next supervisor sexually assaulted me, and co-workers treated me rudely for no reason. A promotion was dangled before me but after I pulled away from my supervisor's uninvited kiss on the mouth in his office, I didn't get the promotion. That was at the studio. Then I worked part time at a production company, nights to accommodate engineering school in the day and where the teaching assistant in my math class accused me of cheating during a math test and threw me out of the test. I protested but he insisted. This setup was before the mentor-romance setup. In class another operative was 'apparently' looking at my test as I took it and I had no idea. Forced to leave the class, I dropped out of school, having no idea how to convince a teaching assistant that I hadn't cheated, when he admitted that he'd seen another student cheating off my exam yet insisted I leave the exam. I was an ordinary person, bothered no one, was trying to re-acclimate to school after having worked full time for several years. I had a part time job at the production company, did my job, left.

I had no idea I'd drawn a world class evil security agency. The mentor-romance attack went like this: my supervisor was very kind, treated me courteously, took up for me against an abusive attorney (that attorney was such a setup – had he not arrived to tear into me, I'd have no way to understand I'd endured a 'white knight' mentor-romance attack. In the law firms where I worked, attorneys didn't 'storm into' word processing centers accusing word processors; when they had a problem they took it up with the word processing manager. With me, the KGB's deployed the 'white knight,' someone who takes up for you, a psychological hit if the attacker is a higher status person, such as an attorney when I was a legal word processor.)

Since I was 11 the KGB has attacked me emotionally, tried everything it could think of to emotionally destabilize me, little kid versus the KGB. I know it sounds crazy, but it's the truth. They used proxies. Proxies never tell you they're attacking you because their handlers, the KGB and now the Russian Mafia, are here illegally and can't come and kick down your door.

In psychological warfare, it's necessary for the 'mentor' to be kind, generous with praise, and supportive, and will stand up for you when you're unfairly treated. Of course I fell in love with him. After the abusive supervisors I had before and after him, he seemed so wonderful. What saved me was his marriage and my ethics. Hurting someone's wife by having an affair with her husband didn't work for me, ethically. Had he been single, that would've been a problem for me. So I left the job, wrote and told him I'd fallen in love with him and that's why I left. Not knowing it was a romance attack by the KGB I loved him for decades. Now I don't love him because he was used by my enemy to attack me.

The Russian Mafia deployed my compromised mother and sister to encourage me to date an operative who I'd known in high school. I was uninterested but they tried off and on for a few years. I didn't know he was an operative, that my mother was enslaved and that my sister was enslaved. I lived all these awful experiences. And worse.

My mother and I have been victims of the Russian government's "romance/mentor" entrapment attacks, and I think other family members may have been victimized in this way as well. Now I know the Russian Mafia is well known for these kinds of stealth assaults. The attacks I know about target women but likely men are targeted by female operatives as well. In the case of my mother, the USSR's romance attack cost us our modest home and my mother her career. We became homeless as a direct result of the USSR's "romance" attack on my mother – I was around 12 and I remember the operative the KGB put in her bed. I remember at least one instance where he used disinformation against me. That was my first time but far from my last being stealth attacked by the KGB. As you might expect from my experiences, I see these kinds of attacks from the point of view of the victim because I was victimized by them.

It's not only that my supervisor was handsome, knowledgeable, smart, and defended me against attack, the KGB used hierarchy as a weapon. Most people assign an additional psychological value to dominants. It's leftover behavior from infancy the KGB noticed and developed into a weapon. Supervisors are considered dominants in a supervisor-worker relationship. When you have a supervisor who is knowledgeable and kind, that rings all kinds of emotional bells, especially if they've been deliberately targeted for most of their life, as the KGB attacked me. Most people, no matter how powerful, retain some part of their subordinate selves, which is how the Russian Mafia enslaves millionaires and billionaires. Those people, now dominants, didn't begin life that way.

Chapter 18.1 – A favorite way the Russian Mafia enters a target's life

All people begin life as subordinates and the Russian Mafia uses that intel to destroy it's enemies and control it's slaves. To attack a target they study who the target interacts with, who is in a dominant position in a relationship with the target. Then the Russian Mafia (and before it the KGB) entraps the dominant and uses the dominant as camouflage from which the Russian Mafia destroys the target. That's their modis operandi. The Russian Mafia is into slavery. It uses as a weapon against innocent people, the added value that we assign to dominants in our lives.

The Russian Mafia, using proxies, attacks targets using unrelenting waves of difficulties (they've been doing that to the Biden Administration since day 1), until the target goes under.

The Russian Mafia always has a believable explanation at the ready for a casual glance. I've been attacked by the Russian Mafia for decades. I have a sense of how they work. But I know, positively, that they exploit people too. And I strongly doubt most of the slaves they control here know they're controlled by an enemy foreign government.

Chapter 19 - My family backstory: My mother[195]

Mom has had a difficult life, but she was scrappy because she had a good first 12 years. Abandoned at birth by her mother, who'd endured a long, excruciating labor with no pain meds (mom was a large baby and my grandmother's first child), grandmother experienced sudden, severe postpartum depression and pushed just born mom nearly out of the bed or (or something similarly alarming). Fortunately my great grandmother, who I never met, had helped my grandmother through the birth and agreed to take care of her grandchild, her first. Great grandmother fell in love with mom who was adorable.

Great grandmother was a firm but loving and kind parent, very organized. She worked as a maid at a hotel downtown. She was illiterate, was married but mom never saw

---

195 Before the USSR attacked me in 1970-71.

her husband. Great grandmother had two daughters, my grandmother and an aunt I never met who died fairly young.

Great grandmother remarried and the couple were happy but her husband was laid off when automatic elevators were installed. He'd been a bellboy and where they lived, Jim Crow was rife. He tried to find work but no one would hire him. He'd been a bell hop and now that work was gone due to technology. Great grandmother worked at the hotel as a maid and earned enough for the family but her husband grew increasingly depressed because no one would hire him. In those days, life was extremely difficult for men of color. He drank and unfortunately when he drank he became violent. He and my great grandmother fought, there was a knife and her husband accidentally stabbed her. He hadn't meant to kill her but he was a violent drunk. He went to prison for ten years and mom's quality of life was destroyed.

Mom was forced to live with her mom, who herself had been abandoned as a child, as her mom had left her with neighbors as great grandmother left the area looking for work, I want to say maybe in the Great Depression but I don't know for sure. Abusive grandmother was very young, a toddler, and she believed she'd been abandoned. Her mother was there, one day, then she wasn't, and apparently someone in the extended family abused my grandmother, turning her mean.

She died young of a heart attack. I miss her. She was a kind grandmother to me but a horrific mother to my mom.

Mom's grandmother was unexpectedly murdered when mom was 12. Forced to live with her mother, my grandmother beat her unrelentingly, and grandmother's common law husband attempted sexual molestation. Despite the horrific beatings and the inappropriate touching of a 12 year by a grown man, mom complained, loudly, to the bullies' faces and told anyone who would listen that she was abused. The neighbors paid attention and witnessed what grandmother's boyfriend got up to when she wasn't there. Mom was released and allowed to go live with her great aunt.

My mother had a great 12 year childhood, then a nightmare when she was 12 to 16. Boys discovered her while she lived with her great aunt, mom became pregnant and in those days in her family, unplanned pregnancy meant she had to get married. Teenagers forced to marry with no idea how to be a couple, and more children than adults, their marriage quickly fell apart. I was their only child together.

Mom was beautiful and knew little about reproduction and she quickly had 4 children and three ex husbands: everyone was young and trying adulthood for the first time. Getting married because of unplanned pregnancies didn't help a marriage sustain itself.

Mom worked to earn money. Marriages didn't last, she knew nothing about spousal or child support. Besides her children, her great love was singing and she had a beautiful voice. Offered a gig as a backup singer mom could find no one she trusted to care for her four young children so she packed away her dreams and looked around to see how she could raise her children. A surgical technician training course was on offer and she tried it, and loved it, earning excellent grades and became well liked by her teachers. Because of her hard work and aptitude she was offered a job in Southern California upon completing the course. She accepted the job and left the Midwest, cautiously optimistic her life would get better.

She was an outstanding young mother, adhering to how her grandmother had raised her, and eschewing the beatings she'd gotten from her mother. She was organized, young, confident strong. She loved her job in California, saved part of her salary working as a surgical technician and put a down payment on a small house. That was maybe a year and a half after we'd arrived in California, so I was maybe 10-11.

Mom was a very hands on parent, loving, firm, busy but supportive. We relied on her exclusively. Mom hadn't been told before she bought the house but the neighborhood was a KKK neighborhood. Our neighborhood was mostly Caucasian with a few, maybe 10% African American families. We kids and mom had no problem with the change. We were kids and flexible and mom wasn't racist so ethnicity didn't matter to her. She loved her job, her kids were happy in school, she was buying her first house, she was happy. We were happy.

In addition to the KKK, another even more ominous entity, unknown to us controlled the neighborhood, the operatives of the United Society of Soviet Republics, USSR. They were in Southern California working a coup against

413

Americans, stealth because a foreign enemy government working a coup on American soil against Americans is illegal. Decades later I read of foreign nations here trying to do dirty deeds but when I was 9, 10, 11, 12, I knew nothing about the USSR, though mom had likely heard of them.

She didn't know they were accessing us to decide what to do with us, they ran the neighborhood. To decide, they had one of their child operatives we thought was our friend hit my brother while we were playing. Surprised, no one had ever hit us that I could recall, I hit the child back. I was 11, he was maybe 12 or 13. We'd been playing in his front yard, which was right across the street from our house. We left, went home and told our mom who told us not to play with him again.

Seeing that I'd hit the boy back in retaliation when he hit my 10 year old brother right in front of me, the USSR had another child operative threaten me on the school bus. They wanted to see my response but I was 11 and didn't know this was a test. The child, older and bigger than me, threatened me on the way home after school on the school bus. When we executed the back door of the bus, him right behind me, I looked around our bus stop for a weapon to defend myself. Spotting a brick I picked it up and turned to face him and waited for him to hit me. The eldest in my family I was taught not to start fights and not to hit other children, but that I was allowed to defend myself if attacked. So I waited there, intending to retaliated. I'd never seen this boy, I probably thought he was a child in the KKK family.

My mother's boyfriend, a man I believe was an operative planted in mom's bed to destabilize her, suddenly honked his horn and waived me to him. I dropped the brick and walked to his car. He told me to get in so I did. He asked me what had happened and I told him a boy said he was going to beat me up so I picked up the brick to hit him with it after he hit me.

The operative took me home and informed my family, saying I was tough, but it hadn't been as he represented. I was 11 and he was an adult. I didn't know how to counter his version of the story.

By the time I was, I think 12, we were homeless. It had happened when my mother sent us four children to visit our grandparents and fathers back east over the summer. Mom had gone on a car trip with the operative and apparently he and his friend had robbed a gas station, something like that. Although my mother said she had no idea he was robbing anything, and had stayed in the car with the operative's girlfriend while her boyfriend entered the gas station store to buy whatever, something happened and my mother was charged. Perhaps as an accessory? I still don't know.

By the time I returned to California, my 29 year old mother was bedridden, devastated, couldn't explain what had happened, and was in pain. She no longer had her hospital job but when I asked her why she became upset so I stopped asking. Our water and I think utilities began to be turned off. I asked our neighbors could I fill a bucket from their water hose and they gave me permission. My brother filled buckets as well. Because we hadn't been there when whatever happened had happened, and because the USSR, when they have people take plea deals, which apparently mom couldn't discuss, I still don't know.

But now I'm 64, not 12, and I've endured so many attacks by not just the USSR, but also the Russian Mafia. The reason mom couldn't say what had happened was, she may have been coerced into taking a deal or lose her children, and she couldn't bear that. Mom worked. I think she years later told me the operative was a heroin addict but I barely knew what that was.

Now that I've endured thousands of attacks mostly from people I don't know, as well as entrapment attempts by my family, I understand why mom was confused: the Russian government is running stealth attacks here with no warning whatsoever to ordinary people who've no idea there's an enemy foreign government on the ground. This book is intended to inform Americans and people worldwide what they can expect if they are foolish enough to allow the Russian Mafia control of any part of their country. Be forewarned: the Russian Mafia is highly acquisitive. As they behave in Ukraine, that's who they are. Additionally, they began the 2023 Israel-Hamas War as a proxy war to unseat President Biden and install one of their puppets in the White House. If we let that happen, our lives won't be worth living. I hope this book helps ordinary people understand what we're up against. I urge us to fight, with everything in us. I've actually had plenty of experience attacked by the Russian Mafia. I'd have been dead long ago but they were worried that if they flat out killed me, someone in my family might be tempted to ask a non-compromised officer to investigate. I'm the only on in my family not compromised. At this time in their coup, the Russian Mafia wants no one to know anything about who they are and how they are to overthrowing our government and destroying our people. All that government dysfunction, the housing shortages, the deterioration of our nation before our very eyes, the fentanyl drug flows, the overwhelming majority has been caused by the USSR and later the Russian Mafia, working their secret war here. The only thing we've got the Russian Mafia doesn't want dropped on them are nuclear bombs and they're in the process, as best I can tell, of trying to install as many pro-Kremlin cabinet members and/or ensure future presidents are owned.

Let me be clear. Death is preferable to anything the Russian Mafia has to offer. My people will be tempted to work with them. Go head. They'll just lie, smile at you and kill you. I'm warning you in the hopes you save your life and your children's life. Things are bad. They're going to get worse. Americans don't understand how seriously evil the Russian Mafia is. The Don of the Russian Mafia is Mr. Putin. He'd throw anyone under the bus to seize world power. Quite a few of the Russian Mafia's members are Jewish, yet at least some of them had to go along with the Russian Mafia ordering

Hamas to attack Israel, knowing Israel Jews would be killed. Bratva trumps culture, religion. When Jews attack Jews, that's a messed up Mafia. Additionally, Hamas is mostly radicalized Gaza Palestinians. They are not happy with over 10,000 Palestinian children killed, yet if anyone's breaking rank I haven't heard about it. It doesn't help that we don't deal directly with terrorists, which sounds like some BS a Russian Mafia operative got passed through.

We are in trouble my people and although we've got trusted allies, we have a really, really, evil, hungry to kill us enemy government. Please read my book and decide for yourself. Despite the many, many attacks I list here, this is less than half. My father was used to influence me and he's compromised. I didn't know. So I had to remember these attacks and I couldn't remember each and every one of them. We're in trouble. I hope this book helps you survive, and I hope people around the world read it and I hope it helps them survive. If they've got anything the Russian Mafia covets, they're in trouble.

Post Script: my mom would've made it. She'd survived her horrific childhood, had born four children in 7 years largely because birth control wasn't available to her and she knew so little about men and how to navigate sex with them. She was practical. When she couldn't get safe minders for her children so she could go on the road, she chose plan B, and she was right. Had the USSR left her and her children alone, she'd be in her 80s, retired, a home owner, not being forced to stealth entrap the only one of her children the Russian Mafia has not been able to compromise.

My mother is a heroine. She rebuilt her life after she'd been dealt really nasty hands by having her beloved grandmother murdered and her mother abuse her, a man groping her when she was 12. She got out of marriages that didn't work and went to work to support her children. But the USSR was here and they destroyed her quality of life like she was nothing, forced her to take the only dog she'd had to the pound, made her cry, humiliated and bewildered her like the Russian Mafia does to this day.[196] The Russian Mafia is evil. What they did to my mother, my extended family, what they doing to innocent Gaza Palestinians, including babies, they will do to you unless we effectively defend ourselves.

Chapter 19.1 - Mom, age 27, American Migrant, Parent of Four (1969)

My mother was 27 in 1969, with four young children she had sole custody for; a divorcee who had no idea her life was about to be negatively impacted by the USSR when she accepted a job offer from a hospital located in Southern California.

She relocated from the Midwest after she'd completed and excelled in a surgical technician training course back east, had driven across country with her children and another adult to split the cost of gas and the driving. The two adults had the front seat and us kids slept sitting up in the back seat. I think

_____

196 Shaun Walker, Pjotr Sauer and Tom Phillips "Panic and Emotional Pain as Alleged Deep-Cover Russian Spies Vanish," *The Guardian* (April 3, 2023).

we drove straight across the country with no motel stops until we reached Los Angeles.

We children slowly adapted to the varying plant pollens our bodies had never encountered before we headed West. The pollens caused our sinuses, eyes, and mucus membranes to swell and leak fluid that sealed our eyes shut in our sleep. When we tried opening our eyes in the morning we found we couldn't: the dried, gluey, sticky pollen barely budged as we viewed the world through swollen slits of puffy eyes that watered and hurt and made us cry. Our mother began to gently bathe our eyes open every morning, and reassured us that everything would be okay. I was 9, my siblings were 8, 6, and 2: two boys and two girls.

Chapter 19.2 - Mom age 17-27 (1958-1969) unplanned pregnancies, marriages, divorces: an artist's heart & her careers

Mom was young to have four children under 10. In the late 1950s and early 1960s when she was in her teens and early to mid-20s, young women in her family were forced to marry if they had unplanned pregnancies. My parents had dated as teenagers and had an unplanned pregnancy so were forced to marry.

The liked each other but were too young for marriage. Mom said they were such children they'd argue over which breakfast cereal to buy. In the early days of their marriage she said they misread each other's intentions, and hurt each other's feelings without meaning to. There was no adult they both trusted to help them navigate misunderstandings and hurt feelings. Mom said she tried to get her mother in law to referee but my paternal grandmother didn't want to take sides. My father was her first child and my grandmother adored him.

My mother was very beautiful and multiple young men expressed interest. After a misunderstanding by the time one of my parents wanted to make up, the other had already moved on to a new relationship.

Still married to my dad, but separated because they were unable to talk honestly to each other, or hear what the other person was trying to say, Mom's next pregnancy by her new boyfriend was unplanned. Her new boyfriend had been eyeing her while she was with my dad, had helped her get a job, and find a new place to live. Still a teenager, she was too inexperienced to ask her new boyfriend to wear a condom.

Mom's ex-husbands were young like her, I think her oldest ex-husband was 30 when she was 27 or so. She had the least sexual experience between her and them but she was either their first or second wife. From what she told me when I asked her when I became an adult, people in their late teens and early 20s dated, had unprotected sex, became pregnant (1950s/1960s), got married, discovered who their spouse was during the marriage, and divorced when they discovered their personalities and/or lifestyles didn't mesh.

Mom's three marriages and divorces occurred when she was 18 to 27. The marriages seemed to have occurred because of unplanned pregnancy. I'm assuming that sometime in her mid to late 20s she discovered the pill.

I asked if she received child support or spousal support from her former husbands, and she said no. She birthed and raised five children, four of the children she was married to the baby's father, but she received no spousal or child support. That's an extremely difficult accomplishment considering she was stealth attacked by an enemy foreign government from the age of 28 to her present age, and considering the Russian government hates Americans, hates women, and hates people of color.

Mom told me that for the first few months of my life my father paid the rent on the room mom rented in a boarding house. I'll confirm the weekly rate but I think she said it was $8 per week, so my dad contributed under $40 per month for maybe the first three months of my life. Their marriage was, sadly for me, over during my infancy.

To survive financially, mom worked low paying jobs. When she was pregnant she didn't always work, maybe through part of the pregnancy, and she may have worked part time while married. Her dad served our country in WWII and was honorably discharged, but I don't believe she received any benefits from his service.

Mom's first two husbands served in the military but my dad served after he and mom separated. I believe I was a toddler during his service, but I don't know for sure. I gained no financial benefit as a child of a veteran, with an impoverished teenage mother. Even though my parents separated during my infancy they didn't divorce for awhile. I'm not sure when mom's second husband's military service ended but I believe he was honorably discharged as well.

Mom told me she worked minimum wage jobs at a dry cleaners, at a burger joint, etc., before and after the birth of her children. She couldn't afford to hire an attorney to advocate for her to get spousal and/or child support. Of her three former husbands, one occasionally sent Mom money for his son, to buy him clothes, or he'd send his baby toys for Christmas when we first moved to California. But occasional clothes and Christmas and birthday presents for your toddler aren't child support. Child support is providing month in, month out for your child's needs. As my brother got bigger, I believe his father sent him money, occasionally, as a gift. Occasional gifts are very nice but they're not child support. Had my mother received spousal or just child support from her three ex husbands, the USSR wouldn't have been able to make us homeless.

Chapter 19.3 - Mom's childhood from birth to 12 (1941-1953)

My mother was raised by her maternal grandmother for the first 12 years of her life. My mother's mother, who I believe was 19, experienced post-postpartum depression and rejected my mother, her first child, at birth. My mom was her grandmother's first grandchild, was born in her apartment, and had been helped into the world by her grandmother. The baby became the center of her grandmother's world. Her grandmother worked as a maid at a hotel and was highly ethical and illiterate (she signed her name with an X). Mom told me that she'd read the newspaper out loud to her grandmother as soon as she'd learned to read. That's may be where she got her love of reading, because her grandmother needed her to read the news to her. Mom passed her love of reading on to me.

My mother's father had served in WWII and was honorably discharged. My mother was his only child; he was, I think Mom said, 30 when my grandmother became pregnant at 18. He was unmarried but tried to get custody of his baby from my great grandmother, but my great grandmother loved the child so, raised her from birth, she refused to surrender the baby. Her father saw that she was very happy living with her grandmother and with her maternal aunt, so my maternal grandfather resigned himself to visiting her regularly when he was in the area.

Chapter 19.4 - Mom's childhood ends at age 12 - my great grandmother is murdered (1953)

After my great grandmother was murdered my maternal grandfather again sought custody of his 12 year old child but my maternal grandmother wouldn't agree. At that time Mom said, a form of welfare existed (this was around 1953) and my grandmother knew she could apply for it if mom lived with her, so she refused my maternal grandfather's repeated custody attempts and moved regularly to evade him. My maternal grandfather developed an illness, he died about 10 years later when mom was around 22 or 23. I'm sad I have no memories or photos of him but Mom said I met him when I was a toddler. It's highly likely, in the last decade of his life when Mom was 12, he was beginning to feel unwell and couldn't afford, nor had the strength, to keep fighting my maternal grandmother for shared custody. He died in his early 50s. Mom said he was very handsome, was good to her, and had fought to get custody of her multiple times.

Chapter 19.5 - Mom, age 12-16 (1953-1957) is physically & psychologically abused by her mother & threatened w/rape by her mother's boyfriend

My mother was a beaten slave in her own mother's house,[197] of value only because: (1) welfare paid my grandmother a small stipend to have her there, and (2) my 12 year old mother was forced to work as an unpaid au pair for my grandmother, caring for her 3, then 4 half siblings. My mother told me her baby brother initially thought she was his mother because she'd been his primary caregiver when he was an infant.

---

197 I don't know why my grandmother beat my mother. She died unexpectedly in her early 60s, years before I knew much of this story so I had no opportunity to question her. She was a wonderful grandmother to me and a monster mother to her first born child. The only clues I have for the abuse were my mother told me that my grandmother had been abandoned by my great grandmother when my grandmother was a toddler. My great grandmother had left my toddler grandmother with neighbors in order to go find work, I'm going to guess this was sometime in the 1920s. She left with no explanation to my toddler grandmother or no explanation my grandmother could understand or remember. My great grandmother appeared years later to reclaim my grandmother, but by then my grandmother had possibly been abused by any number of the people she'd been left with. Toddlers must be potty trained, they get sick, need care. I'm guessing my great grandmother sent the family money but that doesn't mean they treated my child grandmother well. That's one clue, that she was an abandoned toddler, bewildered and abused by someone in the fostering family and/or their acquaintances. While my great grandmother was away working, she had another child with her husband, who was my grandmother's father I guess, but I don't know his situation. When my great grandmother reclaimed my grandmother she took both her daughters with her to the city where she'd found work. Another clue why my grandmother beat my mother may have been out of jealousy. My great grandmother loved my mother, helped birth her. My grandmother could see it. My mother was loved by her grandmother from birth, while my grandmother had been abandoned by the mother she loved. Few impoverished children receive therapy. My grandmother probably never knew why she beat her child. For sure, my mom never knew why she was so viciously beaten. She'd done nothing to be socked in the stomach by her mother.

In addition, there were the frightening sexual assaults, the inappropriate touching of her body no child should have to endure. My mother told her mother about the inappropriate touching, outed the perpetrator when he touched my mother right in front of her my grandmother, but my grandmother did nothing. It wasn't until the neighbors warned my grandmother about her boyfriend, after they saw him lock my mother in the house so she couldn't escape when he was there alone with her, when my grandmother wasn't home. After the neighbors spoke up, and because my mother reported every inappropriate touch to her mother, or to anyone who would listen, my grandmother finally took a stand against him and the groping stopped. However the beatings from my grandmother continued. My mother longed to escape.

Because my mother rightfully continued to protest the unrelenting physical abuse she received from her mother, and the continual groping from my grandmother's boyfriend, I suspect my grandmother grew concerned social services might get wind of it. Grandmother wanted to continue receiving that small stipend.

My mother complained to her perpetrator's faces: to her mother's face, to her mother's boyfriend's face, and the neighbors would, of course, know that he was inappropriately touching a 12 to 16 year old.

Chapter 19.6 - Mother escapes at 16, goes to live w/her great aunt (1957)

At 16 my teenage mom was allowed to go live with her great aunt where she could finally breathe. Initially when her aunt approached her, she'd duck or jerk away, fearful of being punched. But her great aunt was not abusive so my mother eventually relaxed.

Chapter 19.7 - Mom age 17 (1958 -- the gift & beauty of boys, but no therapy after years of abuse)

Allowed to live unharassed after four years of abuse, she blossomed. Her great aunt was nice to her, she enjoyed school and made new friends. She loved singing and joined the school choir, dreaming of becoming a professional singer. Her singing voice reminded me of Roberta Flack in the 1970s: I loved her singing voice.

Boys began paying more attention to her. She noticed they liked her and she liked them back. She knew almost nothing about sex. Initially, she liked them because they were nice to her. She hadn't known nice for the last four years and she appreciated kindness even more than most people appreciate kindness. Her great aunt gave her permission to date. She enjoyed going out with a boy on a Friday or Saturday night. This was 1957, 1958, 1959. She was 16, 17, 18. Her date might take her to a movie, buy her a cheeseburger and a milkshake, and flirt, a little shyly, with her. She'd flirt, shyly, back. Dates began to press her for sex. She was uninterested in sex. She knew very little about it but she liked boys, liked being with them, liked their kindness and handsomeness, liked that they said she was pretty. They were often on their best behavior because they were trying to woo her into dating them exclusively.

Because she'd endured so much trauma by 16, had been physically abused, sexually threatened from 12 on, had abruptly lost her beloved grandmother, and the only stable home she'd ever known, and had no therapy to help her navigate those traumas, she especially appreciated the kindness her dates showed her.

She knew little about the male teenage sex drive, she genuinely appreciated and highly valued the kindness her first dates showed her. Initially, she didn't understand their kindness was agenda-based, that they were nice to her because they wanted sex. She thought they were nice to her because she was a nice person, that they liked the person she was. As the pressure on her for sex began to build, to avoid sex but to keep boys in her life because she genuinely liked them for their companionship, and for the fun they brought to her life (she loved to dance), she developed a strategy of dating multiple boys without having sex.

She'd noticed the more she dated a particular boy, the more that boy pressured her for sex, but she found that when she dated multiple boys, that very same boy reset back to his sweet, kind, undemanding self, hoping she'd eventually choose to date him exclusively. My mother was quick to notice and come up with a workaround. While he was on his best behavior, hoping for exclusivity (which in those days meant he might expect kissing, heavy petting, sex, depending on his girlfriend), she'd get a fun, flirty date with him instead of a wrestling match or pouting.

Her great aunt, seeing the number of beaus, grew alarmed. Mom tried to explain her strategy but the meaning got lost in inter-generational translation. Her great aunt was maybe her mother's aunt, maybe even her deceased grandmother's aunt, I'm unsure). Her great aunt put her foot down: my mother had to choose one boy and date only him. My mother liked my father and that's how I got here. Mom said that had she been allowed to continue her successful dating strategy of going out with a wide variety of boys to evade sex pressure, she'd not have become pregnant. I believe her.

Chapter 19.8 - Mom age 28 (acclimating to Southern California in 1969-1970)

My mother had no family in California to live with temporarily that she knew of, but her second ex-husband's brother had divorced, and his ex-wife and their children lived in Southern California. The ex-wife and children agreed to let us live with them temporarily, to find our feet. There were a couple kids near my age in that family, and some older teens. I interacted more with the kids nearer to my age and thought they were nice.

After enduring a scary skid row motel experience where someone tried to break into our rented room while we were inside, we headed out of Los Angeles to the city where we'd settle, not far from the hospital where my mother was to work. We lived with the host family for, I want to say a year, while mom acclimated to her new job, enrolled us kids in school, and saved up money to buy her first, and only, modest house. Being my mother, she'd, of course, offered to pay the host family for sharing their home with us, and paid rent while we were there.

Our host family, I think the mom was an LVN in a hospital, socialized with soldiers stationed at March Air Force Base which has since been closed down. She took Mom with her to help her meet people. Mom made friends, of course, being young, beautiful, and friendly. At least one of the soldiers wanted to marry her but he was mean to one of my little brothers (who reported the incident to Mom), so Mom declined his proposal.

Chapter 19.9 - Mom was independent because she'd had to be

In her teens and 20s mom didn't view men as income sources, she'd worked a variety of low paying jobs from her teenage years on. I'm positive I got my work ethic and pride-in-craft from her, as well as an inability to see men as income sources. Decades later, the Russian Mafia dangled a doctor and a doctor's income in front of me, a man they'd entrapped and intended to use to try to entrap me.[198]

_____

198 One of the Russian Mafia's go-to attacks is to install an operative (someone they control) into the bed of a target (someone they want to control). I've never seen anyone fail to be entrapped after a compromised bed partner was installed in their life. I think the Mafia's success rate is close to 100% because sex/love/mentor attacks are so

My entrapped mother and sister were used by their handlers to push me towards the slave whose name they brought up several times over several years and encouraged me to date him, my sister referenced his income multiple times. That strategy didn't work because I'd been raised by my young mother to believe I could earn my own income, and to view men as individuals and gifts in and of themselves, not as income sources. I knew how I felt: before I realized the man was an operative, I knew I didn't view him romantically. There was no reason to marry a man I didn't' love. I told my family I wasn't interested, but they'd occasionally suggest I pursue him.

Chapter 19.10 - Mom's dad

To my knowledge, the only man who'd tried to take care of Mom had been her beloved father but he'd been rebuffed by both my great grandmother and grandmother because they selfishly refused to share custody, so mom had largely missed out on a deep relationship with her father, although he did try.

---

successful. Never, ever, lean into a sex relationship if you know you're doing any type of work negatively impacting the Russian Mafia. People are so romantic, both sexes, but especially women of color, who tend to highly value male mentors. In women of color cultures I'm familiar with, men are deeply revered. Any man of any ethnicity, gay, straight or trans, who wants to feel how much women need and adore men, attend an African American church and present as male. My mother has always strongly preferred men, maybe because her father was largely missing from her life, and she attended church pretty much weekly the first 12 years of her life. After her grandmother was murdered, she sought out a church for emotional comfort, even when she lived with her abusive mother, because her grandmother had loved church so. My mother in general adores men, and that's not a put down. One thing you see in many African American churches is female's deep need and emotional pleasure in men.

Mom had been so abused by her mother, who worked as a maid and cook later in life, and who was not educated, so mom knew almost nothing about laws and protections for ex-spouses and children. Because pretty much any money she got she had to earn through jobs, she taught me, by example, that a woman works for her income, and that working and pride in your work, is the norm.

Chapter 19.11 - How my mother's work ethic positively influenced my approach to work & made it harder for the Russian Mafia to get me fired

Mom *loved* being a surgical technician. She wore scrubs, she was extremely busy, but when I asked her questions about her job as she drove home after work in an old subcompact Datsun she loved (gifted to her by the husband of one of her kind supervisors at her new job), mom, having stopped to offer 11 year old me a ride as I walked home from the school bus stop up the long hill towards our house, she'd answer my questions. What I got from her answers and her deep satisfaction was her love of her job, and the pride she took in doing it well. This was before the USSR attacked us, so this pride-in-work was another gift my mother gave me, one of many, before the USSR destabilized us.

Children learn in this way from their parents, asking a question here and there over years, and the parent answering honestly, and the child see's the parent's honesty. Through my mother's example I learned that a female worker works both inside and outside the home, and takes pride in whatever work she does, always doing her best. That's what my mother showed me, and that's what I emulated and that's why the Russian Mafia had to entrap and/or push my supervisors and co-workers to harass me out of jobs.

Chapter 19.12 - My mother feels cautiously optimistic about her fresh start in California. No one deserves what the USSR and later the Russian Mafia did to her

When my mother moved to Southern California with her children in tow, she knew next to nothing about the Russian government. A hardworking, busy young woman with four children under 10, she was recently divorced and a newly minted surgical technician. She'd worked hard and excelled in the training to such an extent that she'd been offered a job upon completing the program. She loved her job. Her children loved their schools and were adjusting to the move. She allowed herself to feel cautiously optimistic.

Chapter 19.13 - Mom buys a modest 3 bedroom house at 28 & we settle in (1970)

After working and saving a bit, mom used her wages earned as a surgical technician to put a down payment on a modest home not far from work. It was the first and only home she's owned and the USSR conspired to steal it from her by setting her up.[199] She didn't know when she bought the house that the neighborhood was controlled by a gang consisting of USSR operatives, the local KKK, and the local police.

_____

199 Unless she becomes a homeowner in her 80s.

Settling into our new neighborhood, playing with the dozen or so neighborhood kids in either their front yard, or our front yard; playing on a neighbor kid's slip & slide (where I got a concussion after accidentally hitting my head when I slipped instead of slid), we had so much fun. Our mother signed me and my siblings up for Saturday afternoon matinees which we loved, where we saw all the kid movies with a theater full of other children for a low, parent-friendly price. I think Saturday matinees were to help parents get a leg up on gathering family necessities without hauling children around. Mom had been allowed by her grandmother to attend Saturday matinees in the 1940s and she'd loved them, so she made sure to search for kid-friendly treats she thought her children would enjoy.

When mom lived with her mother from 12-16, she'd been repeatedly beaten, forced to feed, care for and clothe her siblings, while evading her mother's boyfriend's groping hand's while both of them used foul, insulting, racist language when referring to her. There'd been no Saturday matinees for her during those horrific years. That nightmare behind her, Mom loved seeing her children's joy in the movies. At that point, when Mom was 28, before the USSR installed it's operative into her life to destroy her, was I believe the happiest she'd been since her grandmother died. She had her beloved children with her. We were thriving, loved our friends, our schools and teachers, our house, and of course our mother. She was thriving. She was buying her first house, she loved her job.

A husband of one of her supervisors gave her a little subcompact to run errands in, her Datsun. She loved that car.

Chapter 19.14 - My mother's approach to socializing & parenting

Mom had a social life but she always kept it separate from her children, probably because she'd been threatened with rape from puberty to 16, which she didn't tell me until decades later. She didn't drink or do drugs but she smoked. I hate that smoking hurts so many people, and I especially hate that it hurts my mother. There's now a new therapy that helps people stop smoking I hope my mother will try it.

Mom had an active social life she kept separate from us. We saw her friends as we'd enter and/or exit the house to go play with or return from playing with other neighborhood children on the weekend. We knew she had friends (women and men, parents of neighborhood children, soldiers from the Base), but they weren't in our house for the most part, in terms of sleep overs. I want to say I saw the end of one card party once, but I don't have a strong memory of it. The only time I heard what I later learned were sex sounds were when the operative the USSR embedded in my mother's life spent the night, so maybe I was 11 or 12 and he'd come after Mom had put us kids to bed. During my childhood I almost never saw Mom drink alcohol, maybe once at a family Thanksgiving gathering. Mom was serious about keeping her social life separate from us. Because she'd always kept her social life separate her behavior was normal to us. As young children we had no idea about sexual predation while we were in our mother's care. That's because she'd been a child victim of repeated sexual groping attacks and threats of rape.

Chapter 19.15 - A weekend in the life of a working mom

While we were at the movies, Mom was working. She'd clean the house, do laundry, buy groceries, get meals started for Saturday/Sunday night dinner (if she wasn't taking us kids to a diner which she did occasionally). She'd plan our meals for the week, buy school supplies our teachers asked us to bring, and she'd run errands. She was very organized as she worked full time and was the only parent on the ground for four young children who depended entirely upon her. She combed our hair, mine and my sister's and divided it into two or three braids. I think she did that at night and we'd wear a hair net to bed. She'd comb her own hair. She'd brush our brothers' hair until she taught them to do it for themselves. She'd wash, iron and lay out our school clothes for us every night on the living room couch so she and we knew what we were to wear to school that day: four children, four sets of clothes, five days a week. She woke first and woke us up and we'd get ourselves up, dress in the clothes she laid out for us, having taken our baths the night before. We'd brush our teeth, wash our faces and help any younger sibling needing help with those tasks. We'd eat, usually cereal with milk. She'd buy us the cereal we liked and/or wanted to try. Usually there were at least two types of cereal and we kids liked them because she bought them for us. My favorite was cocoa puffs although I also liked cocoa pebbles. One of my siblings liked Trix. I think Mom bought Cheerios too. Mom gave us lunch money to buy school lunch. I think she usually left for work before we left for school. Then we'd leave. We knew to lock the door behind us because she'd taught us how to take care of ourselves for years, as she often worked, or attended school or ran errands. I think she probably dropped my baby brother off at kindergarten because he was 4, and I think we three oldest children, 11, 10 and 8, got ourselves to our bus stops. My two siblings went to elementary school near our house, and I went to junior high when I was 11 in 6th grade. I wanted to attend their school with them but I was thrilled to be in junior high and in 6th grade.

Chapter 19.16 - How Mom bought school clothes, school shoes and school supplies for four children on a single income in 1969-1971

At the beginning of every school year Mom bought us two packages of underwear and socks, and a few new outfits and separates, using a store's lay-away plan or money she'd saved for our school clothes to ensure her children had clothes that fit them for the year. (Usually she took us to Sears; I think occasionally to Kmart and another department and discount store -- they were affordably priced department and discount stores that stocked children's clothing, shoes, and school supplies in Southern California.)

School clothes shopping was pretty much the only time Mom took us clothes shopping. She took all of us at once so she wouldn't have to guess after last year's growth spurt. Both feet would be measured, avoiding the necessity of a return trip if one shoe pinched a child's slightly larger foot. Apparently that had happened because she insisted both our feet be measured (by a retail assistant, or if clerks were busy helping other customers, Mom measured our feet). She insisted we try on both shoes in the store. She'd press on the big toe of both shoes to ensure they could accommodate a years growth spurt.

Back-to-school shopping was exciting and fun for us kids because we got to choose a couple new outfits, the only time that happened over the year, and we also got to weigh in on our sibling's choices, talk them into an outfit, separate or shoes one of us especially liked on them. I remember Mom took me school shoe shopping and I chose big brown platforms (thick platforms were the style then). I loved them, she loathed them. Still she let me get them. I wore them to school but they were lost with most of our belongings when our house was repossessed.

Taking four children back-to-school shopping was exhausting so Mom let us beg her into the rarely permitted McDonald's drive thru. When we went school clothes shopping we knew we'd get new clothes, shoes, underwear, and a hamburger too. School clothes shopping was something Mom always did, and something we kids always looked forward to, like when she'd take us Christmas present shopping and have the kids being bought for go off with a sibling so she and the present buyer siblings could go through checkout and bag their purchases.

Physical education and home economics classes were taught then and Mom bought whatever our new teachers told her we needed. She received a list, I guess through the mail, because she always knew what was required.

Mom let us choose a new lunch box every year: sometimes she gave us lunch money to buy school lunch, sometimes we brought a sandwich from home, usually it was kid's preference.

Mom supported our family on her surgical technician salary. I think I was around 11 or 12 and one of my teachers told our class we needed specific supplies for an upcoming assignment. I told mom, who was apparently financially tapped out at the time. She told me to call and ask my father. The supplies cost $20 or less. By that time my father had remarried and had two children. I believe he was out of the military, so he was around 31-32. When I asked him for the money for school he said he didn't have it. That was the first and last time I asked my father to provide school supplies. Throughout my schooling, from kindergarten through 12th grade, my dad provided no financial support to me or for me. My mother didn't tell me this. I only learned that was the case after my dad refused to send me money for school. Around that time I began to understand I needed to start thinking about how to earn money. I think I began selling Christmas cards, door to door, to some of our neighbors. A few families bought the cards from me. Besides the few paid babysitting jobs I got (I occasionally babysat for two families in our neighborhood, one African American family, one Caucasian American family), and the Christmas cards I sold a year or two, that comprised my sole earnings.

Much of what my mother had nurtured and created was destroyed by the USSR after it attacked her and destroyed our quality of life. Over fifty years ago it began it's Active Measures coup against innocent American families and, very likely, innocent immigrants living here as well.

Looking at my nation now, and at the many teetering democracies worldwide, the Russian Mafia has sped up and worsened the destruction the USSR's attacks perpetuated against ordinary people. Bequeathed to the Russian Mafia, the USSR sought in it's ego, hatred, and jealousy of Americans, to destroy the decent quality of life our government strove to supply to us, which the Russian government is incapable of creating for it's own people. Since the Russian Mafia and it's government can't create, they destroy. The Russian Mafia has our species on course to destroy earth habitability. I'm not exaggerating. Entrapping and enslaving the leaders of billions of people worthwhile, the Russian Mafia refuses to allow it's allies to work with us to effectively control the climate catastrophe. No habitable planet means no us. The Russian Mafia and the Russian government deny self-extincting their own species and life on our planet.

Chapter 19.17 - Trick or Treating in a kid-friendly neighborhood

We went trick or treating and loved it, our Mom bought us whatever costumes were reasonably priced or on sale at Zody's or Thrifty's (discount stores in Southern California in the 1970s), but we used old pillow cases in which to carry our loot so there'd be no chance of bag breakage. Mom bought four costumes, bought groceries and prepared our meals, paid her mortgage, paid our utilities and water bills, and bought our school supplies. I was 9, 10, 11 and 12 and had little concept of money at that time. My younger siblings knew less even than me. We relied completely on our brilliant, hardworking, organized mother.

Because our neighborhood and several of the surrounding neighborhoods were packed with kids, nearly every house on our block offered kid-friendly candy. Miniature candy bars were commonly given and kids loved them, but a favorite and far rarer but occasionally given treat were full size candy bars offered by adults who'd forgotten to buy Halloween candy, then panic bought full size bars, loving to see children's excitement when they received a full size candy bar.

Though candy was at every grocery and drug store, our mother didn't allow us much of it and we rarely went grocery shopping without her. I don't remember her once offering McDonald's unless we asked. She'd take us occasionally but she believed home cooking was best for her children and she insisted, and cooked, so we ate most of our meals at home.

Now I understand expense played a role in her food preparation – then as now it's much more affordable to buy groceries and prepare meals at home, especially when feeding a family of five on a single income.

Mom occasionally baked home made, slightly sweetened bread from scratch which was delicious, twisted it somehow, added nuts and raisins. We always begged for more.

For trick or treating usually our mother walked with us, or one of the other parents in the neighborhood did, either an African American or Caucasian American parent walked with a group of maybe 10 or so kids. Mom insisted on checking our candy to ensure it was safe to eat, that was nonnegotiable. Even when another parent walked with us and Mom stayed home to hand out candy, we weren't allowed to eat our candy or anything given to us while trick or treating. It never occurred to us to ask another parent for permission to eat candy as we went from house to house. We knew our mother wouldn't like it. There's a lot to be said for hands-on quality parenting from birth. Much misbehavior can be eliminated if you have at least one parent who knows what they're doing, or knows what not to do, are loving, firm, and consistent. Perfection isn't required of parents but it helps if a parent understands something about human development, are kind, consistent, and understand the value of building self-respect and self-confidence in children. Mom had strong views on raising her children because she'd experienced such a nightmare loss pre-puberty and if possible, her experiences during puberty were worse.

When we got home, Mom allowed us several pieces of our choice after she checked our selection hadn't been tampered with. Working in a hospital probably made her even more careful about checking Halloween candy, but she rarely told us of bad things done to children by predators using Halloween candy as bait. She laid down the law and we followed it. She said "no eating until I check it" so that's what we did. We obeyed her because she was our mom. She'd done much of the heavy lifting of socializing us when we were infants, toddlers, and 5, 6, and 7. When we moved to Southern California, only one of her children was a toddler and he, of course, obeyed like the rest of us did.

Mom let us go through our haul. We'd spread it out on our beds, separate out our favorites, compare what we'd gotten with what our siblings had received, and trade among ourselves for favorites. We'd ask our mom what she wanted and she'd occasionally select something.

Seeing what we'd received was one of our favorite things about Halloween and there were always surprises. You'd think since we'd watched adults and kids put candy into our pillow case that we'd know what we got, but there were always pleasant favorites we hadn't noticed being dropped in. Our mother taught us to politely thank each benefactor, no matter what we thought of the candy.

Then she'd take the four pillow cases from us, get us in our baths and through teeth brushing, braid our hair for school (put a hair net over her girl's hair), put us to bed, then go through the candy after we were sleep. She threw out anything suspicious; there was always partially opened candy she never let us eat. And she wouldn't let us eat the fruit (around that time predators sometimes hid needles and razor blades in it). She'd parcel out a piece or two of our choice until we forgot about it. Then she probably threw the rest away.

Chapter 19.18 - Warning from an African American neighbor family that KKK families were our neighbors

A family member of one of our new neighborhood playmates offered to pierce my sister's and my ears so we could wear earrings. We asked our mom and she gave us permission.

After we'd gotten our ears pierced, one of our African American neighbors told my mother there were KKK families in the neighborhood and that one of those families had pierced our ears. After the neighbor's warning, my mother told us kids that there were KKK families in the neighborhood.

Born and raised in the Midwest, we kids had never heard of the KKK. We were already friends with the neighborhood kids, including those in the KKK families. Mom didn't raise us to be racists so we didn't understand racism.

When we were naughty in the Midwest city where we'd been raised and an adult reported it to her, she'd always tell us who reported our misbehavior. I remember one time a Jewish neighbor reported to Mom when one of her children misbehaved, that child was questioned and upon admitting the misbehavior, he was punished. He never repeated that misbehavior. The adult reporting the behavior was treated respectfully by our mother, treated courteously as Mom treated all our neighbors.

Chapter 19.19 - How my 28 year old mother explained the KKK to her children in 1970-71

Mom had to tell us what racism was.[200] She explained that some people don't like other people because their skin is browner. As a 10-11 year old it was just one more fact in a world full of facts. We liked all the neighborhood kids

_____

200 When I was a young child, 4 to 6 or so, I loved Elvis Presley. His movies were played on television in the Midwest. I think I saw Viva Las Vegas but maybe that was later. I loved him, my first experience being a fan but because I was a child, I didn't love him as a fan loves, I loved him as children love, with a deep child love. He was beautiful and he could sing and dance. The movies he was in always showed him singing and dancing and doing both exceptionally well. I didn't know the word fan. Someone, maybe an uncle, witnessing my adoration as I watched a movie, told me an old story about how Mr. Presley allegedly viewed African American people. I didn't know the word racism or anything about race, and my uncle didn't use that word. I only knew that I loved Elvis Presley and my uncle told me he wouldn't like me because of my skin color. I didn't learn until years later that Mr. Presley denied that story, so I believed my uncle. The pain felt so bad. Someone I loved didn't love me for a reason I didn't understand. I said to my broken heart, okay, he dislikes me so I don't love him anymore. I stopped watching his movies and forgot about him. When he died young, I again heard that rumor and that he'd always denied it. I'd forgotten him so completely that I hadn't known he was still alive until his death. Racism hurts children, and adults and everyone in between. The only entities I know who love racism are the Russian Mafia and it's government. They're using it to destroy the world order to remake the world in their image. Racism is an effective weapon for them so they love it. Plus, they've drunk the kool-aide, the Russian Mafia embraces racism, sexism and homophobia. They're also pro-slavery and believe they're the master race. They also believe everybody not them is stupid. This arrogance is because they've lied to the world, to our governments, and to ordinary people the world over so successfully, the lies they told us straight-faced, and/or blackmailed our leaders into backing, they believe us believing them means they're the master race, that everyone else in the world are meant to be their slave, and their mastery of psychological warfare has elevated them. But it hasn't. They are just another cruel, evil variant of our species, ascending at the worse possible time for our species and earth habitability. I wish this wasn't true, but I'm telling you the truth. Not for you to hate them, but for you to see them. They're not big on transparency. They're big on world dominance. That's their nirvana. They hate everyone not them. Their mastery of psychological warfare

including the kids in the KKK families. Everyone was nice as far as we were concerned. No one was rude to our face. Yet. We didn't know it was coming. We hadn't experienced being attacked so we didn't understand.

The African American neighbor who'd warned us didn't warn us about the USSR, so we had zero idea the USSR was an aggressor on the ground in our new neighborhood. The USSR-KKK local police hybrid gang showed us their presence but they used child proxies. We kids had never heard of the USSR or children in stealth armies or child proxies.

The Southern California KKK parents and their children didn't introduce themselves as affiliated with the KKK. The USSR didn't introduce itself at all except through attacks – it was here illegally, as the Russian Mafia is here illegally and in other countries, illegally.[201] Had it not been for an African American neighbor telling my mother, we'd have never known the KKK controlled our neighborhood. They were nice to our face.

Chapter 19.20 - The USSR inserts their operative into my mother's bed & into her children's lives: perfidy, a war crime

The USSR and the Russian government sign agreements, then ignore them – in 1970-1972 the USSR commit the war crime Perfidy against my mother and against her children. Placing

_____

gave them their edge and their willingness to lie about everything, straight-faced.

201 Shaun Walker, Pjotr Sauer and Tom Phillips "Panic and Emotional Pain as Alleged Deep-Cover Russian Spies Vanish," *The Guardian* (April 3, 2023).

their operative in my civilian mother's bed and into her children's lives under false pretenses was raping her and stealth attacking children under 15, a war crime the USSR had agreed not to commit when it signed the Geneva conventions in 1948.

The USSR's operative entered Mom's life treating her and her children kindly and generously. I don't' remember what the operative gave me for Christmas, but I remember he gave my little sister a coat, I think as a Christmas present. The USSR's operatives, with their secret police the KGB, would've studied mom and had no problem designing an operative who appealed to her. The reason we're on the ropes in 2024 is because the Russian government and their Mafia are masters of psychological warfare.[202] My mother was 28, an ordinary woman, smart, but with no espionage training. She'd completed high school and surgical technician training and excelled in surgical technician school. She had no experience being attacked by a foreign enemy government or any knowledge that Americans were being attacked. The USSR's operative placed in my mother's bed was African American. Most people even today wouldn't expect an African American male to be controlled by the USSR. However the Russian government has been working these types of attacks for decades. They're still ongoing because they work. One of my hopes is this book helps people better understand, and better

_____

202 The USSR/Russian Mafia master psychological warfare and don't believe in warning victims they're being attacked.

449

protect themselves against the Russian Mafia.[203] It's not about being smart, it's about understanding your enemy. Remember an FBI official reopened the investigation into Mrs. Clinton's email before the 2016 election? That official had espionage training, yet I believe the Russian Mafia played him. Another official with espionage training was played, who was a professional with espionage training was compromised. Espionage training helps, probably a lot, but it's also important for all of us to be on the same page, to know the Russian Mafia is on the ground in America, hunting and destroying innocent people. We have a best-in-class enemy who intends to destroy us.

I respect my government's position to avoid nuclear war. Because the Russian Mafia has compromised so many of our systems and people, they're confident they will destroy us and so they won't stop stealth attacking us.

I believe the KGB's perfidy sucker punch nearly caused my mother to suicide. I believe only her love for her children and our love and need of her held her to earth. Her youngest child was 5. Again, she refused to abandon us even though she wanted to crawl into a hole and face no one. I stopped asking her questions about what happened because she struggled to find answers. After her childhood of trauma, to be attacked by the KGB, and later the Russian Mafia, I believe only her deep love for her children gave her purpose. I'm so grateful she

_____

203 Ibid.

stayed alive but what the KGB did to her, and to us, was beyond evil.

Chapter 19.21 - Mom had got a puppy

I think the puppy mom got at our new-to-us house in Southern California in 1971 was the first one our family ever had. When the USSR attacked our family mom couldn't work for awhile, she couldn't pay the mortgage, utilities or water bill, or feed us or herself. She couldn't feed our dog. We became homeless, all our furniture lost when our home was repossessed and our mom was too sick and too poor to rent a U-haul and storage. She had to take her beloved to the pound. She cried. The USSR made my amazing, hardworking, beloved mother cry and they made her bedridden for awhile when she was 29 years old.

Chapter 20 - The 2023 Israel-Hamas War correlates to the Russian Mafia's pattern of using civilians, including children, as blood pawns

The Russian Mafia's deep disdain for life and disrespect of civilians, women, children, and people of color is again on display in the 2023 Israel-Hamas War, which is a proxy war to unseat our President and install one of the Russian Mafia's preferred candidates in the White House in 2024. As per usual with the Russian Mafia, it uses lies, lack of transparency, and disinformation to ensure the world blames Americans and our President for a war the Russian Mafia facilitated in order to seize world domination, and which it is content to have Americans blamed for.

That's what you'll see with victims of the KGB or the Russian Mafia. They attack in ways that people without espionage training cannot hope to piece together. Not because victims aren't intelligent but because the Russian Mafia has seized control of individuals and organizations and operate they manipulate in secrecy. Entrapped people aren't allowed to warn the innocent. Just as our government is legally constrained by confidentiality clauses in our alliance with the Israeli government, which constraints the Russian Mafia knows of and exploited, our President, because Israel was attacked, is legally confined to what he can publicly do and say. The Russian Mafia made sure Hamas attacked Israel, and then created yet another weapon aimed at us as the Russian Mafia facilitated worldwide pro-Palestinian protests, everything designed to work their coup here. If we let the Russian Mafia install one of their puppets, we'll wish we were dead.

The summer of 2023 my mother told me her meet story with the USSR operative who she to this day doesn't know was an operative unless she reads this book, and believes me. My mother is a slave of the Russian Mafia, who believes she's entrapped by our legitimate government. The Russian Mafia lies best in the world, the most believably and are masters of psychological warfare. There is no telling what her handlers have told her over the decades.

Chapter 21 - Me

I am my parent's only child, my mother was 18 when I was born, my father 19. They both went on to remarry, divorce and have more children. Mom was 19 when she bore her second child. She had four children by the time she was 25, and had ended her third and final marriage by the time she was 27.

Mom worked inside and outside the home, prepared our meals, cleaned the house, washed and ironed clothes for herself and her children, combed our hair, and worked outside the home in a variety of low paying jobs. And she pursued her passion, singing. She sang in night clubs, sang well enough to be offered a touring gig as a backup singer for a renowned singer who'd had a hit, but she had no one she trusted to watch over her children.

She and my grandmother were on better terms by my mother's mid-20s but her mother wouldn't watch four grandchildren. My grandmother's common law husband no longer lived with her but he paid the mortgage on her modest home as they shared a couple of children together. My grandmother worked as a maid and most of her children were still at home. Her oldest son may have begun to be in trouble with the law by that time, though I'm not positive. I was 7. Adults in my family didn't tell little kids that kind of information in the 1960s.

Mom's marriages didn't last as they were mostly the result of unplanned pregnancies with young men she liked but who weren't ready for monogamy, which at least one of her young husbands failed to tell until after the marriage; upon learning of his cheating, mom divorced him. While married her husbands helped to financially support her and their children but once the marriage was over so was the financial help. She received no spousal or child support so she worked to feed her children. That's not to say her ex-husband's didn't give her a little money for food if she asked them but they didn't regularly, routinely give her money for their children's food, clothing and housing, nor did they provide her with spousal support.

Mom loved singing but she was unwilling to abandon her children to whoever was willing to take us in order to pursue her dream. That was the quality of parent we had before the USSR destroyed that parent, a parent who would never consider harming her children or putting herself first. My paternal grandparents and my maternal grandmother would've taken me; my brother's and sister's father would've taken them but my mother didn't want to split us up; and my mother's last ex-husband would've taken his son, but again, mom's children were very close emotionally and she didn't want to separate us. She'd raised us to be a family together. She'd lost her grandmother when she was 12 and she didn't want her children separated. As young children, we had no idea we were half siblings. So Mom moved on to Plan B.

Unable to pursue her dream of a singing career mom made the practical decision to train for a different career. Surgical technician training was on offer in the town where she lived, and she was told jobs were plentiful. She enrolled, loved it, and excelled in it. She completed the training by the time she was 27 and was offered her first job in Southern California.

Chapter 21.1 - I was a child protector & a child lieutenant

Mom needed babysitting help and I was her oldest child. She had to run errands, go pay a bill, buy groceries, do a million tasks that a single working mom or a single mom in school has to do. She couldn't bundle four small children in winters with many feet of snow into and out of a car, leave us in the car while she took care of business in extreme temperatures. It was easier to leave us at home. So she built in me an increased sense of responsibility from my infancy.

Mom was super busy from 18 to 27, and needed help that wasn't available to her. At 18, 19, 20, 21, and 22, she had three babies, her first two marriages had ended, her beloved father got sick and died. She'd had no therapy after enduring four years of severe child abuse, nor was she given comfort after her grandmother's sudden death, but instead she entered a home where she was beaten for nothing that she'd done. She completed high school thanks to her determination, intelligence, and her great aunt allowing her to come live with her, but a high school diploma doesn't prepare a parent for three babies needing them 24/7, without spousal or child support.

My brother was born before I was a year old. Mom always told me she'd wanted a girl as her first baby. She taught me to touch and handle my smaller, more fragile siblings very gently, very carefully, as most parents the world over teach their older children to handle their infant siblings. She taught me to love them.

When she stepped away she taught me that as their big sister, I was their protector, that that was my job when she wasn't there. Within the first year of my life, I was socialized by my mother to *be* a protector and a defender. My mother didn't call me protector, and an infant and toddler, I didn't know those words. I was a protector because that was my job when my mother wasn't around and my siblings were in my care. Humans are air-breathing mammals. Before we take a breathe we don't think "I need to take a breathe now." We just breath. When you're trained to do a job, whatever that job is, it becomes normal to you.[204]

---

204 The Russian Mafia, the KGB, the FSB, entities who prioritize the mastery of psychological warfare favor children because children are a blank slate for evil entities who want to build human weapons. Evil is a point of view. When children are trained to reach for behavior and not socialized to view that behavior as bad but good, that's how I believe the Russian Mafia begins to build human child weapons. The evil don't confine themselves to children, which is why I believe the Russian Mafia ensure our interactions in our prisons, jails, and/or with our police officers and prison guards are horrific experiences, to traumatize and destabilize our people as hard as possible. For decades the Russian Mafia has tricked, schemed, lied, cheated and run entrapment scams against me to put me into the criminal justice system, or even into the personal bankruptcy system, which apparently they also control. I believe Mr Putin and the Russian Mafia kidnap Ukrainian children to make them weapons. I also believe the American child armies I've been attacked with for decades, despite the USSR promising not to harm children under 15 in the Geneva Conventions, is yet one more indication that the Russian Mafia view children as disposable pawns. If you start training children young enough to do evil, evil is normal for them. Evil is a value system. There's a reason we've seen a significant uptick in mass murders committed by late teens, 20 something year old's over the last 20-30 years. You're mistaken if you think we've seen the last of that. ***The Russian Mafia destroys over decades and destroying Americans is their favorite thing***. I'm positive the Russian Mafia is busily training and abusing children in America as I type these

As mom socialized me she didn't teach me that I was small and vulnerable, <u>so I didn't know that I was</u>. Rather she taught me to protect the smaller and the more vulnerable around me. That was my first identity.

I don't romanticize power because I've always had it, it was gifted to me by my birth order and by my young mother's need. My family, especially my siblings, taught me when I was a small child that power is work, responsibility, and hard.

I was babysitting my siblings by the age of 6 or 7, and I tended not to repeat mistakes. For instance, I was a toddler and was with my mother as she ran errands. She smoked and used the car's cigarette lighter. She frequently pushed it in and out it'd pop. I'd seen her do it many times.

She left me in the car by myself as she ran a quick errand, I was maybe 3. I pushed in the cigarette lighter. Out it popped. I pulled it from it's holder as she always did. I looked down the barrel and saw for the first time a pretty and interesting orange glow at the base. I stuck my finger in it. The pain was the worse I'd experienced up to that age and I never forgot it. Unsurprisingly, I never repeated the experience, but I also never discussed the cigarette lighter in my sibling's presence, not wanting to draw their interest to it when I saw for myself the pain it caused. When I made mistakes I tended not to repeat them, even from the youngest age, *and most importantly for a babysitter,* I tended not to allow my siblings to make the mistakes I made. So when mom left her children in

_____
words. My mother trained me for good. The Russian Mafia doesn't train children to do good.

my care, they were usually returned to her undamaged. This a desirable trait in a babysitter. Mom didn't tell me I was helpful and responsible, she treated me like I was helpful to her, and responsible. I behaved protectively with my siblings, whether mom was there or not. I've never felt sibling rivalry towards my siblings, and only pride in their achievements.

Chapter 21.2 - My favorite adults during childhood

I love my mom but as her oldest child, she put me to work. Out of necessity I was her lieutenant. That forced me to mature faster than I normally would have. She set high standards of behavior for me because she needed the help. I struggled to meet her standards since I was little more than a toddler, but fortunately she never changed the guide posts and I was able, just, to be the quality helper she needed.

Basically, she needed me to help watch my siblings so they didn't get hurt, while she ran an errand, washed the dishes, did a load of laundry, cut up vegetables for dinner, fried chicken, vacuumed the floor, or did the zillion other tasks a working and/or school enrolled mother of four must do. Oh, and she worked outside the home. She needed me to treat my siblings gently and kindly, and to not harm them. I was in her rotation of babysitters by the time I was 6 or 7. I was not her only babysitter, but I was one of them.

Chapter 21.3 - My young childhood wasn't all about service

Mom made sure I learned how to swim by signing me up at the community pool, and when a teacher or a new stepmother treated me cruelly, she had my back. Her love for us was reflected in the high quality of her care and raising of us, which I took for granted because she'd existed when I was born. I assumed, if I thought about it at all, that all children received such quality care. She is entrapped by the KGB/Russian Mafia, and, having known her from my birth, and now knowing the Russian Mafia is working a coup here and has been for decades, I see the negative ways slavery has impacted her parenting, and definitely her quality of life. *In brief, as a parent and as a person, her first loyalty is to the Russian Mafia.* That is why I charge the Russian Mafia and the Russian government with genocide against people living in America. Unlike other coups across our species' history, generally, coups didn't demand parents to abandon their children, to destroy their families. In this coup, the Russian Mafia is destroying families in America. That's what makes their attack genocide. Families are the key formation of our species. Destroying our family, you destroy our species. The Russian Mafia doesn't care.

My mother was an excellent parent; loving, firm, consistent. My world revolved around her and I had unquestioned confidence in her because, not only did she house, cook, feed, clothe me and braid my hair, and do the same for herself and my siblings, she made sure we got our vaccinations timely (an experience we didn't appreciate at the time – I still remember my 6 year old brother running screaming down the clinic hallway, fleeing yet another vaccine).

Mom was in her early 20s doing the best she could; I didn't know that then, I know it now. Because of the sequence of events that had befallen her, my mother was, at times, a very stressed drill sergeant in terms of needing my help. She said I was her only child who could tell when she was stressed. I don't know how I could tell: she didn't yell at us and she wasn't violent towards us. I probably heard something in her voice. She later said I was more sensitive to my environment than my siblings. For sure I'm not psychic[205] but I can generally tell when someone is upset even when they're not displaying it overtly. That's why I knew when the American operative lost his temper and went loud on me. Female operatives I thought were my friends had been tasked with getting me to a restaurant and this operative, he looked like a plain clothed police officer, was listening in on our conversation. I said something risque (it was girls night out, I didn't know anyone but my "friends" was listening. I didn't want to be the only woman saying nothing about sex. The "friends" were fishing about my sex life but I didn't know that. They both brought up their sex lives but I didn't know I was

---

205 I was hunted by the USSR and later the Russian Mafia for 53+ years
    and didn't know I was being hunted until the Russian Mafia's
    American operatives began to overtly attack me to my face when I was
    around 40-41, sometime after 9/11.

being double teamed. (I've been quadruple teamed so beware that's something the Russian Mafia does here.) I'm not trained in espionage, the attacks I'm discussing in this book are from my experiences, unless I state otherwise.

At that time, the Russian Mafia had failed multiple times to entrap me, spent over ten years setting up four entrapments that I'd evaded unknowingly. On top of that, the USSR had spent twenty years working to destabilize me. I didn't know how I was evading them. I knew many bad things had happened to me but I had zero idea I was being hunted. I didn't know I was evading them.

American operatives controlled by the Russian Mafia were increasingly frustrated, and when the Russian Mafia and/or the American law enforcement they own are frustrated with a victim, they let the victim know. The Russian Mafia and the American law enforcement they own are very entitled, treat me like a criminal because they want me to be one.

After hunting me for years, they'd gotten two operatives to befriend me and I was leaving the firm where those operatives worked. Previously the Russian Mafia and the USSR had deployed coworkers and supervisors to harass me, and the Russian Mafia certainly continued that tradition on my next, and last two jobs.

I'd given notice at the law firm I worked at in part to get away from unnaturally sticky "friends," but I didn't know they were operatives, I just didn't understand their determined attachment to me.

My "friends" and I laughed at my joke, and glancing up I saw a man I didn't know staring at me in disgust. I looked away, realizing that he'd been eavesdropping. Later that evening I stopped at the grocery store as I walked home after the meal. At the grocery store I was overtly harassed several times for the first time directly[206] by uniformed and plain clothed security and/or law enforcement. I believe the angry eavesdropping operative ordered a change in attack strategy in retaliation because I dared to unknowingly resist entrapment by an enemy foreign government. Before they went loud, I had no idea there was a "they." The next nearly 25 years was horrific but the good thing to being overtly attacked is that you know you have an enemy. Because of their overt attacks I'm able to write this book of warning to the world. Because of their overt attacks I was able to understand that my family was being used to entrap me after the Russian Mafia sickened me when I was 50. Had the Russian Mafia not gone loud, I doubt I would've been able to piece together that my mother and siblings were entrapped. By the time they were deployed to entrap me, to try to get me to lie and cheat for gain, I'd already been overtly attacked by many people, out of the blue, and had no idea why. But I knew I had an enemy. Knowing you have an enemy when you have one is high value intel. Which is why the

---

206 Until I was 40-41, the USSR and the Russian Mafia had abused me using proxies like supervisors or coworkers; cheated me using the mechanic and a law firm who stole from me; or having a teachers assistant and a "student" double team to accuse me of cheating on a test. That night was the first time the Russian Mafia deployed security and/or plain clothed police I'd never seen before to harass me. It's been extremely difficult, as you can imagine.

Russian government no longer tells it's victims it's attacking them.[207]

Chapter 21.4 - My childhood – mom's stress load

Because of her stress load and the unhappiness it caused her, mom wasn't my favorite adult when I was a child. She was one of my favorite people but my favorite adults during my childhood were a maternal aunt, and my grandparents, because they loved me, hugged me, smiled at me a lot, never criticized me, gave me anything I wanted to eat during my visits; offered me candy, lemonade and ice cream, all of which I much preferred to babysitting my brothers and sister, even though I loved my half-siblings.

When I was at home, which was most of the time, I assisted mom because she needed the help. When I was with my aunt or grandparents, I was on vacation. I didn't know that word when I was 6 and 7, but I knew which experience I preferred. Mom let me visit my favorite relatives regularly but after a day or so she wanted me home, I think to help watch my siblings while she cooked, cleaned, ironed, ran errands, worked outside the home, and ran the household.

Cinderella was my favorite story during my childhood because I could relate to her as a worker family member. While I wanted more grandparent time, overall I had a good childhood living with my mother, who I trusted. And I loved

---

207 General Valery Gerasimov, Chief of the General Staff of the Russian Federation,"The Value of Science in Prediction," *Military-Industrial Kurier* (February 27, 2013). *"….wars are no longer declared…"* *"...The very "rules of war" have changed…."*

my siblings. I spent a little time with my father but he was in the military during part of my childhood. I spent more time with his parents and extended family than I did with him. I adored my paternal grandparents.

Chapter 21.5 - Babysitting my fearless young siblings – learning to negotiate with them when I was 6, 7, 8, 9; my mother and my siblings taught me that power is responsibility, and hard work when done right

As a child, I had to come up with a simple strategy to keep my siblings safe while our mother ran errands. They were wild young ones, fun, fearless, and thought themselves invincible. In my memories of them as very young children, they wanted fun, quests, games, stories: they were used to being held in safety and wanted something different when our mom wasn't around. They were very strong-willed, all of them dominants. When our mother left, they wanted adventure. So I learned to negotiate: as long as they didn't hurt themselves, or anyone else, or the furniture, I wouldn't boss them around (they hated being bossed around and I felt zero need to boss them. My only need, built by my mother, was to keep them and myself safe until she returned). Most of the time my mother returned from errands to find undamaged kids and furniture, and reasonably happy children because they hadn't been micromanaged. Years later one of my sisters told me how much she'd appreciated during her childhood my not bullying her.

At 6, 7, etc., I didn't know how to articulate my strategy but basically in my mind there were 3 teams: my mother, me, and my siblings. (1) My mother needed her children kept safe when she wasn't around; (2) not all my siblings were interested in being kept safe when our mother wasn't around but they did prefer to not be in pain, and they hated being bossed; and (3) my job was to return undamaged children and undamaged furniture to my mother, and preferably, happy children because they hadn't been bossed around. I don't remember my siblings ever telling my mother I'd abused them while I watched over them. It never occurred to me to harm them, physically or psychologically.

Because of the needs of my mother and siblings I learned how to work successfully with people when I was 6, 7, 8, 9. The key is respecting others needs, being responsive to their needs. Listening. Respect. Valuing. It helps if you love the people you're caring for, and it helps if you care about their comfort and happiness.

Chapter 21.6 - My maternal grandmother

To her children, our mother was the center of our world, and she was the law. I wanted to stay with my maternal grandmother back east but my mother had refused. She had very good reasons for not allowing me to stay but I was 9 and she was 27 when I asked. She didn't know how to tell me that my beloved maternal grandmother had beaten her for no reason for years, had punched her in the stomach for no reason when she was between the ages of 12 and 16, leaving her gasping for breathe and about to pass out from the pain. My grandmother's common law husband witnessed the attack and said he'd report my grandmother to the authorities if she ever hit my mother like that again. He "working up to being a child rapist" thought my grandmother's assault against my mother so severe that he threatened to report her. Because of his threat, my grandmother never again gut punched my mother, but she still beat her.

Mom didn't know how to tell me when I was 9 that the man who occasionally visited my maternal grandmother, who always smiled at me because he saw how much I loved him, and who sometimes gave me a whole quarter for the ice cream truck, had tried to rape mom when she was a child.

When I asked to stay with grandma, mom told me no, and explained why decades later. I had no choice but to accept her no. I wasn't the type of child to sulk. It took me awhile to adjust my view of my grandmother and her common law husband after mom told me how horrifically they'd abused her. Of course I believed her. She was very specific, her stories consistent. After she told me I was angry with her for letting me love them so deeply when they'd been monsters to her. But I had to remember she was a child herself when she endured the traumas. She'd had no therapy. How does a survivor describe that level of trauma to a 9 year old. Mom didn't know how to discuss it.

Mom worked, went to school, married and divorced her husband, had a few girlfriends she played cards with when she had a few hours to play cards, and she tried to make a go of it as a singer (I loved her voice). I got no overt spoiling from her but I knew she loved me because of the way she treated me.

The Russian Mafia, when they use her as a weapon against me to extract information and/or to fish to see if they can entrap me, use her as camouflage. The way they exploit my mother's and my relationship, and the way I've witnessed the Russian Mafia's exploitation of other American children operatives is how I know they're committing genocide against people living in America. The Russian Mafia intends and is working to destroy us.

Chapter 21.7 - From the Midwest to Southern California: new teachers, new friends & tetherball

Back east my friends had been African American. In Southern California most of my new friends were Caucasian as they were the largest ethnic majority in our new city and comprised most of my classmates. I had no problem with the change.

My mother raised me and my siblings to respect all people irrespective of ethnicity. Where we'd lived back east our neighbors had mostly been African American, with a sprinkling of Polish American and Jewish American families. In Southern California our neighbors were mostly Caucasian American, with a sprinkling of African American families -- we were one of the sprinkles.

Back east my teachers had been African American, in Southern California my first teachers were Caucasian. I learned from and obeyed my teachers because they were good at their job and helped me understand math, which I loved, and reading, which my mother had taught me to love.

Chapter 21.8 - 5th grade in Southern California (age 10, 1970)

I completed 4th grade back east and began and completed 5th grade with the host family in California, where I was taught tetherball and played against other ten year old's who appeared to have mastered the game by the time they were five. I was repeatedly trounced; even little kids age 8 ran the board, knowing exactly how hard and at what angle to hit the tethered ball so it arced well over my head, even as I jumped trying to intercept it. I watched it, out of reach, as it returned to the little Californian playing against me, and they'd hit it hard to send it powering around the pole. I don't remember winning even one game, but I enjoyed trying. Just like my playmates in the Midwest my California friends liked hopscotch, jump rope, and playing with their yo-yo. I'd had to leave my old, beloved, well broken-in bike back east my Mom had bought me a few years back as we couldn't haul it in the trunk of the car, but Mom promised me a brand new bike which she delivered after she put a down payment on and we moved into our new-to-us, modest house.

In California I played with a wider array of ethnically diverse kids. I liked the kids near my age in the family we stayed with temporarily. I liked my 5th grade teacher. My siblings liked their school. Mom loved her job. We were happy.

Chapter 21.9 - We moved out of our host's house and across town into our own house after I'd finished 5th grade

We made friends with all the children in our new kid-packed neighborhood, most of the kids were Caucasian, which was the norm in the Southern California city where we lived.

Chapter 21.10 - 11 year old in junior high school - 6th grader

My 6th grade teacher read to our class out loud and she chose good stories. I loved her reading period. I'd lay my head on my desk and listen to her tell me a story.

Walking down the hall in 6th grade an African American teacher approached me and asked me to join band. I'd never thought of learning to play an instrument, no one in my immediate family did, and I had no instrument. He said he had plenty so I went to band practice and asked my mother's permission, which she granted. That was my introduction to music and I learned how to read music and play instruments starting with him.

There was a very nice janitor who worked at the school, a near retirement Caucasian gentleman who was very nice to me.

The standout adults I remember at that school were my very proactive band teacher who facilitated my learning four instruments starting with the first one he supplied, because he needed more kids in band; the kind janitor; and my teacher who read stories out loud to our class.

Chapter 21.11 - Paid babysitting – who knew? It wasn't until I was 11 & hired by one of our new neighbors that I received remuneration for babysitting. I'd been babysitting my siblings for years and not received a penny. I guess Mom felt room & board & an occasional cuddle were plenty compensation

One of our African American neighbors asked me to babysit her and her husband's two young children. I asked my mother for permission as she, I, and our neighbor walked up the street to our new neighbor's house. Mom gave me permission and I was paid, I think, $.50 or $1.00 per hour, I forget which. The oldest in my family at 11, I was an old hand at babysitting but had never been paid. I had no idea people were paid to babysit. Getting paid by our new neighbor made a very favorable impression on me. Her little ones were adorable and well behaved.

When I watched my younger siblings for my mother as she ran errands and/or was on her way from work home, they were headstrong and adventurous and I had to protect them without seeming to until our mother returned to relieve me of the responsibility. I never felt I had power over my siblings. Rather I felt a responsibility to keep them safe while they were in my care. I didn't articulate my thoughts, I didn't know the words then but even as a child I interpreted power as responsibility, not as dominance over others.

Chapter 22 - How the Russian Mafia uses racism as an "in" to embed themselves in the dominant group of a nation, only to later destroy that nation. They've worked this attack against Americans (using the KKK). They've worked it against Italians, French, Israelis, Chinese & Indians, to name a few

When she was a young mother, Mom taught us by her actions that people of all ethnicity's are to be respected. We didn't know that wasn't the norm at that time. Now, the Russian Mafia has mom spout the racist dog whistles her influencers teach slaves. My beloved mother is a slave and, especially when you hear racism from someone who you knew was not a racist, that's a red flag the Russian Mafia has their hooks in them.

Racism is a very high value weapon the Russian Mafia deploys in many nations to gain a foot hold in the nation. In America, they exploited racism in the KKK to gain a foot hold. General Gerasimov alludes obliquely to it in his paper titled "The Value of Science in Prediction."[208] He says "the use of special-operations forces and internal opposition …. to create a permanently operating front through the entire territory of the enemy state…."

---

208 General Valery Gerasimov, Chief of the General Staff of the Russian Federation,"The Value of Science in Prediction," *Military-Industrial Kurier* (February 27, 2013). *"….wars are no longer declared…" "...The very "rules of war" have changed…." "...The role of nonmilitary means of achieving political and strategic goals has grown, and, in many cases, they have exceeded the power of force of weapons in their effectiveness…."* "...Among such actions are *the use of special-operations forces and internal opposition to create a permanently operating front through the entire territory of the enemy state…."*

Racism has been very good to the Russian Mafia. The Russian Mafia deploys racism as an "in." They offer the dominant, most powerful religion and/or ethnicity the service of eradicating (who are being asked to share power with a minority) or destabilizing the minority. The Russian Mafia does so, using the blood of minorities, one of it's favorite things to do as it believes it is the only master race in our species. Then, having made their bones with the dominant group, they, over time, steal the country out from under the nose of the dominant group.

Chapter 22.1 - The USSR's 1st illegal attack against my civilian mother's children[209]

One day while playing with one of our new friends who was maybe 12 or so and who had a deformed leg (one shorter than the other so he wore a shoe with a much thicker heel but still limped), when out of the blue and without provocation he hit my brother right in front of me. We'd been playing and laughing: me, our friend, and my three siblings, then "kow-pow."

---

209 All my mother's children were under 15 in 1971-72, and my mother was and is a civilian, so the USSR's attacks were in direct contravention of the Geneva conventions, which the USSR had signed decades earlier, & against American law. It is illegal in America to attack innocent, nonviolent children, out of the blue, whether the attacker is child or adult. The USSR used children to attack us because an adult physically attacking children would've stood out more. Child on child violence to the casual eye the USSR would be interpreted as a dispute among children, even though when we were attacked, there had been no dispute. The USSR placed an operative in my mother's bed and used that adult as a weapon because, again, to the casual eye, no one would perceive an African American males as owned by the USSR. The USSR knew, and exploited, that no one see's children as weapons but when I was a child, the USSR used children as weapons against me.

We weren't hitters and weren't used to kids hitting us. That was the first time I remember a kid hitting us. Our mother was a very hands on parent and our only parent on the ground in California. She didn't tolerate fights. Even her two oldest children, me and my brother, 11 and 10, didn't fight. We wrestled and played with each other and our siblings. After a second of stunned silence, I hit the boy back, once, on his back/shoulder, then we left his front yard where we'd been playing, crossed the street to our house, opened our front door, and went to tell our mother what had happened. Being hit by another kid was a big deal. We didn't describe the attack in ethnic terms because we didn't think in ethnic terms. The child was Caucasian which our mother already knew. Rough estimate, 90% of our neighbors were Caucasian; Caucasian Americans were the largest ethnicity in the Southern California city we'd moved to, and their large numbers were normal to us. That had been the reality on the ground when we'd arrived a couple years ago, just as in the part of the Midwest city where we'd once lived African Americans were the largest ethnicity. Differences in skin color were no big deal to us. The sky is blue, part of the country is mountainous, part of the country is flat, part of the country has many lakes, in different parts of the country various ethnicity's have larger numbers of individuals than in other parts of the country.

We didn't correlate ethnicity with violence. We were children who'd largely been protected by our mother from violence, but we were so young we didn't know our species can be very violent. For sure we didn't know the USSR was in our country working a coup. We didn't understand ethnic-based hierarchy, the USSR's hatred of Americans, and that coward enemy governments like the USSR attack civilians, women, and children, even after promising they won't.[210] All we children knew was one of our friends had hit one of us for no reason.

We were an African American family of well-mannered children, two boys, two girls. We had a hardworking, amazing, devoted mother. Because we kids were born in the Midwest, not the South, and had never heard of the KKK until 1971-72 we didn't understand ethnic-based protocol. For instance, our mother raised us to treat all adults with respect, and to be courteous and kind to other children. If you're a parent, you know how much work it is socializing children. My mother put in that work starting when we were very young. *While we were taught to give respect and courtesy, we were raised to expect courtesy and respect in return.* That became a problem when the USSR and it's KKK allies expected us to take a beating without defending ourselves. We saw ourselves as the same as other children, only with browner skin.

Some African Americans raised in the South at that time had it drilled into them that Caucasian people were socially superior and they were taught to never confront them. My mother's children didn't get the memo. We were

---

210 Geneva conventions

respectful, courteous, friendly children who adapted easily to a more mixed ethnic environment. Our neighbors in the Midwest had been mostly African American but had also included Polish people and Jewish people. We were taught to respect adults, irrespective of ethnicity and we were taught to respect others, in general. We assumed that we too were worthy of respect and courtesy.

The out of the blue attack surprised my brother and all of us, we'd been playing with our friend, not arguing. Our once-friend was older than my brother and I believe me. My brother was 10, the hitter was maybe 12 or 13. At age 11, my mother had raised me to care for my siblings when she wasn't around, so I hit the boy, once. Then we left. I didn't kick him, punch him, push him to unbalance him, or call him names. It didn't occur to me to do any of those things. I hit him with my open palm on his shoulder/back as he turned away after hitting my brother.

We kids would've been stunned to learn that was only the beginning of the USSR's stealth destabilizing attacks on our family. They've attacked at least three generations of us, focusing on destroying innocent women and children, one of the coward Russian Mafia's favorite things to do. And these are the men determined to rule the world, cowards that knowingly start a war where most of the people killed are civilians, women and children – the 2023 Israel-Hamas War.

The child was the first of a series of children the USSR used as proxies to attack me or my family. The USSR, and later the Russian Mafia, lean into ambush, it's their favorite strategy, even when attacking innocent children. The day of that attack, I all of us were younger than the attacker. I was 11, my brother 10, my sister 8, my brother 4.

Children under 15, civilian adults, and civilian women are protected by the Geneva conventions (which the USSR agreed to abide by decades ago) *in war time.* The USSR cowardly attacked us in times of peace and destroyed, an unknown tens of millions of us since they began Active Measures against us in the 20th century.

The Russian Mafia and it's government operate here illegally. The last thing they can do at this time, before they've installed and can maintain a puppet, is to publicly report their attacks. If they did so, our government and our people would be given a heads up, and the Russian Mafia refuses to warn anybody.

The Russian government published their ambush strategy in 2013, in a paper by General Gerasimov's aimed at the Russian military entitled The Value of Science in Prediction," *Military-Industrial Kurier.* In that paper the General admits Russia no longer declares it's wars. Additionally, the General discusses nontraditional warfare and the Russian government's embrace of it. Although his paper implies the Russian government made these changes in the 21st century, he lies. The USSR used many of these attack strategies against Americans children at least as early as 1971.

I know because I'm one of the children who was attacked using these strategies.[211]

USSR-KKK gang members didn't come to our door demanding an explanation about why I'd hit their child operative. Instead, the USSR installed an operative in my mother's bed, attacks the Russian Mafia still work to this day as you can read in the footnoted *The Guardian* 2023 article.[212] Not long after the first child attack, when I was 12, the USSR had removed us from the neighborhood, using perfidy, an illegal war crime of installing an operative controlled by the USSR into my mother's bed. That's illegal. Again, neither the USSR nor the Russian Mafia, care anything about our laws.

---

211 General Valery Gerasimov, Chief of the General Staff of the Russian Federation,"The Value of Science in Prediction," *Military-Industrial Kurier* (February 27, 2013). "*...wars are no longer declared...*" "*...The very "rules of war" have changed....*" "*...The role of nonmilitary means of achieving political and strategic goals has grown, and, in many cases, they have exceeded the power of force of weapons in their effectiveness....*" "*...Among such actions are the use of special-operations forces and internal opposition to create a permanently operating front through the entire territory of the enemy state....*"

212 Shaun Walker, Pjotr Sauer and Tom Phillips "Panic and Emotional Pain as Alleged Deep-Cover Russian Spies Vanish," *The Guardian* (April 3, 2023). "Campos Wittich was a Russian spy living under an assumed name in Brazil. At least six such suspected spies have been unmasked… in various locations over the past year, suggesting there could be….more. He was a deep cover spy working for an elite intelligence programme…." "She (the victimized girlfriend) suspected nothing at all," said the friend of Campos Wittich's partner." "She is just a lovely woman looking to create a family and get married to a man she thought was the man of her life." "She is really scared of this situation and hurt by all the pain of having an abrupt cut-off in a relationship that was perfect in her eyes," said one of friends. The victimized girlfriend of the spy is a veterinarian. She had no suspicion she was involved with a Russian spy. The romantic partners of both spies have been left devastated, said their friends and acquaintances in Athens and Rio."...Campos Wittich may have been in an earlier 'embedding' stage of their missions, which could have gone on for decades if not detected," reporters state.

Chapter 22.2 - The USSR's 2nd attack on my mother's children, this time the USSR specifically targeted me

After school one day I was on the school bus headed home, when a bigger, older boy I didn't know who may have been 12-14, threatened to beat me up after we got off the bus. I was in 6th grade. I think this threat occurred after the neighborhood boy hit my brother.

He sat behind me looking serious and I had no doubt he meant what he said. I don't remember what I had with me, probably my book bag, a school book from one of my classes, maybe a pencil and paper, nothing to defend myself. I'd never been threatened to be beaten up before and no one had ever beaten me up. I'd never seen my mother beaten up, nor anyone beaten up.

My mother had explained about the KKK so I probably thought he was in a KKK family. We were sitting towards the back of the bus. I was a skinny brown girl, my hair braided into pigtails by my mother. To an 11 year old, most experiences are new. I was scared by the threat, but not terrified. New things are normal for kids and unpleasant new things aren't rare in a kids' life. Getting vaccinated comes to mind: shots hurt but kids are ordered by their parent(s) to accept the pain so we do. Kids are built to survive the newness of life if they have a stable, supportive family.

My mother's children had her and we had every confidence in her because she'd had our backs from birth. She explained clearly what she expected from us. Starting a fight was unacceptable. If we had a problem with a teacher or another child, we were to tell her and she'd sort it out. But if we were attacked, she expected us to defend ourselves.

Accepting that I'd been threatened, I decided I needed a weapon. The threat was a test by the USSR but I was 11 and I'd never heard of the USSR. The USSR and the Russian Mafia ambush and attack children and adults but at 11 I'd heard of neither entity. The boy exited right behind me after I exited the bus from the back door, our bus stop was a field. I looked around for a weapon and spied a brick a few feet away. I walked to it and picked it up. It was a bit heavy but I thought I could manage.

I turned to face the threat, and waited for him to hit me. Because I was the oldest, I'd been raised to not hurt other children. Even bigger, older children were not to be attacked. My mother raised me to view protecting children as an honor and a duty. At 11, I didn't think those words but they shaped my behavior.

I was quiet and introverted. It never occurred to me to pick a fight with another child, from preschool through the 12th grade. I knew I had the right to defend myself if someone picked a fight with me at school but that had never happened. After he hit me, I intended to hit him with the brick. I didn't consider crying, begging him not to hurt me, or running away. I didn't consider his ethnicity or that he was bigger than me, and a boy. I think when children are loved and cared for, they understand they have worth. I automatically assumed, based on the love I'd received my entire life from my mother, my grandparents, and my extended family, that I was of value and, of course, I would defend myself if attacked by a bully.

Before the fight started, someone honked their horn. My mother's boyfriend, the man I now believe was an operative -- for the first and only time was at my bus stop. He waved me over to his car. I dropped the brick and walked to his car. He told me to get in so I did. He asked what had been happening and I told him a boy said he was going to beat me up. My mother's boyfriend drove me home, told my family what had happened, and exaggerated.

That was my first experience, though not my last, of hearing an operative's disinformation. He made me sound super tough, like I was a tough kid. He gave me a rep I neither valued nor wanted. To me, the almost fight was a matter of letting someone beat me up or defending myself. I'd chosen to defend myself, many children would've chosen that. I'd been scared but not terrified. The brick would've really hurt him. I was happy not to hurt a child.

Defending my family & myself caught the USSR's eye but they were here, on the hunt, running their Active Measures coup, destroying anyone they could. They're cowards. Most governments can ambush civilians, women and children: to do so is flat out evil and proves nothing but cowards, and especially horrific after signing the Geneva conventions.

No telling how many tens of millions of wonderful people in America the USSR and the Russian Mafia destroyed and imprisoned here out of jealousy and hatred. Mom needed many of the traits she developed in me, years before the USSR sickened and destabilized her.

Chapter 23 - The USSR and the KKK in Southern California in 1971-72 – Mom's children don't know a world class monster is stalking them, in their country illegally

My mother hadn't yet taught her children (oldest child 11, youngest 4) the unspoken rules of ethnic hierarchy in 1970-72. She was super busy working a full time job and raising four children on her own, 24/7, when the USSR gut punched and ambushed her. She hadn't been raised in the South, her kids didn't normally get into fights because this was the first time the USSR, the KKK and local police attacked us. They used child proxies they controlled to assault my brother and threaten me. Taken together, especially since the USSR worked here illegally, in stealth mode working their Active Measures coup, there was no way my 28-29 year old Midwest African American mother with no training in espionage could've anticipated the USSR's destabilization of our family and known it was part of the USSR's and later the Russian Mafia's coup. Even now, well over fifty years later, most Americans have zero idea the Russian Mafia is working a coup against us called Active Measures, that it's built around subversion, and that they've been stealth destroying us for decades, which is why our nation is on the ropes (affordable housing shortages, police brutality, children not even attending school, classes set up to ensure students fail (I have recent experience of that in my city college courses), overdose deaths off the wall, families destabilizing, etc.). Because the Russian Mafia is here illegally working these attacks, and because the victims who know about the attacks are nearly 100% entrapped and prevented from discussing their entrapment, like the Biden Administration is prevented from discussing the confidential terms of our alliance with the Israeli government, it's really, really hard for ordinary people to understand what's happening. Basically, we're screwed unless we start warning people. We're on the very edge of the Russian Mafia seizing control of us, our nation, our government, and the world. As a person who's experienced over 50 years of abuse from the USSR and the Russian Mafia, death is better than being their slave.

Having lived among KKK families in my childhood, and having been attacked by the USSR and the Russian Mafia in Southern California for over 53 years, I've yet to hear nor read about other victims, even though my extended family are victims. That's because the Russian Mafia, if nothing else, has mastered blackmailing, entrapping and coercing ordinary people, using whatever is on the ground they can use as a weapon to prevent victims from talking.

My mother, like most people even in the 21st century, simply didn't know there's an enemy foreign government on the ground in American targeting ordinary people, civilians, even children. She definitely didn't know the USSR had operatives embedded in our country before 1970. Even now, most Americans have no idea how deeply and widely the Russian Mafia is embedded here, have never heard of the Jackson-Vanik Amendment and what those over half a million Soviets immigrants did here. Most people here see our nation's deterioration, but because the Russian Mafia has seized parts of our media, and using dark money, has exploited and weaponized many of our systems, Russian Mafia owned operatives say we're corrupt, and Americans have no idea the Russian Mafia is feeding them lies.

I hope this book helps all people worldwide understand the Russian Mafia is making a play to control our species, that ordinary people are under attack, who's attacking us, why we're under attack and the way the Russian Mafia is attacking us. The Russian government is so evil, it won't even bother warning children it's stealth attacking them.[213]

Over fifty years later after their first attacks against me, in August 2023 my mother and I were talking on the phone, and she told me how she'd met her then-boyfriend, the operative. She said she'd gone to a palm reader who'd told her to be on the lookout for him. Not long after that, one of my mother's new California friends asked if she could introduce the man to my mother. After experiencing decades of more entrapments than I can remember, the palm reader and my mother's "friend," sound like Russia-controlled operatives to

---

213 General Valery Gerasimov, Chief of the General Staff of the Russian Federation,"The Value of Science in Prediction," *Military-Industrial Kurier* (February 27, 2013). General Valery Gerasimov states: *"Wars are no longer declared.." "..The very "rules of war" have changed."*

me, setting her up with the man the Russia-controlled gang used as the weapon to destroy my mother's quality of life. Whatever the USSR did to her, she never fully recovered, probably because they got her to doubt her own strength.[214]

Chapter 24 – Entrapment - how I believe the Russian Mafia stops people fighting entrapment

Once the USSR convinces the entrapped target that they were entrapped because they're corrupt, and because they have a criminal heart, and are unworthy, the innocent target, who truly is innocent, but who has no experience with psychological warfare, no one they can go to for advice,[215] and has been set up, are convinced they're corrupt, and with the help of their compromised handlers, they're never allowed to be free again.

Chapter 25 - In America in 2024, the Russian Mafia has, I believe, millions of American slaves, people they've entrapped

Once the Russian Mafia has entrapped victims, they can never escape the Russian Mafia, fronting as American law enforcement who themselves are slaves of the Russian Mafia. The success of this scam is why the Russian Mafia believes they're geniuses and the rest of us are stupid. We're not, but anyone in a police uniform coming at us puts most of us in a

---

214 Janis Berzins "The New Generation of Russian Warfare" "...the Russian view of modern warfare is based on the idea that the main battle space is in the mind." Aspen Institute | Prague, *Aspen Review* (March 2014). https://en.m.wikipedia.org/wiki/new-generation-warfare.

215 After my mother was destabilized, she got entirely new friends/influencers. I suspect something similar is done to some Evangelicals, to rural Caucasian people and to some African Americans.

subordinate position. The Russian Mafia loves working targets in that position – it's their money position

This scam works because the Russian government is a liar and a thief and uses psychological warfare against civilians untrained in espionage. An analogy is a 30 year old man fighting an 8 year old boy. Of course the boy loses in a fist fight, not because he's weak but because the contest is unequal. Of course, the USSR is destabilizing us given that they're ambushing us without warning.[216] *I believe my mother would've committed suicide except that her young children were there, loving her, needing her.* So, she got slowly to her feet after the USSR's sucker punch, and carried on as best she could. At some point the Russian government entrapped her. I don't know when.

Had the Russian government not sicced their operative on her, mom would've continued working in the career she loved, raising children she loves in the only house she ever took a mortgage on, and she'd be enjoying her retirement instead of being forced by her handlers to entrap her oldest non-compromised adult child: me.

The USSR made me homeless when I was around 12, and my mother and my siblings as well. The USSR has remained the aggressor in my life ever since and I've seen it seize more and more of my nation and my people. This book is written in order to reverse that trend and take back our nation

---

216 General Valery Gerasimov, Chief of the General Staff of the Russian Federation,"The Value of Science in Prediction," *Military-Industrial Kurier* (February 27, 2013). General Valery Gerasimov states: "...*Wars are no longer declared....*" "*The very "rules of war" have changed..*"

from the Russian Mafia. That won't be easy. I advocate we bomb the Russian Mafia.

Tragically for my family and our nation, the Russian Mafia has spread their influence and harm to, I believe, tens of millions of people worldwide. I've personally witnessed thousands of entrapped people, most of them Caucasian as I live and worked in cities with big Caucasian numbers.

Chapter 26 - Mom goes down

I was around 12 that summer. My mom had sent her children back East to get some grandparent and father love. By the time we returned to start school, after the USSR had finished with her, my 29 year old mother was bedridden, had been arrested and released from jail, had lost her job, and was deeply traumatized. When I asked her what had happened, she couldn't explain. Now that I've endured 53+ years of covert and overt attacks by the Russian government, and know the Russian government to be evil masters of psychological warfare who feel no need to explain their attacks to their victims, I understand why my mother couldn't explain what had happened -- she probably didn't fully know what had been done to her. The Russian government isn't big on explaining to targets, they're big on destroying families. It's one of their favorite things, their happy place -- especially destroying American families, especially families of color, but in all honesty, the Russian government hates all Americans. If my extended family is allowed by their handlers to read this book, I doubt if anyone will believe it. The Russian government will have spun so many lies, victims can't conceive their lives have been destroyed as part of a coup by a foreign enemy government. The Russian government is illegally embedded here, what they're doing is illegal under our laws and under the rules of war. The KKK illegally attacks people of color so they won't talk; and the local police were involved extra judicially, meaning my mother was never going to be told the complete truth.

Her attackers lied to her, used stealth and proxies[217] and as a 29 year old surgical technician and mother of four children, with no support system, she had no experience with psychological warfare and she went down.[218] She won't know until she reads this book, if she reads it and believes it, that the Russian government destroyed her quality of life as part of their coup to destroy Americans.

Setting her up and entrapping her was illegal but the Russian government is Mafia and cares only about seizing world domination, and making their bones in front of the world. All they care about is destroying Americans and flipping the world order. Their allied governments are enslaved by them, is what I believe. Look at how the Russian

---

217 Her then-boyfriend was African American. No one would think that an ordinary, seemingly nice man was controlled by a Russian operative/KKK hybrid gang. This book is probably the only one describing that such a gang existed in Southern California in the 1970-1971. The reason being is the Russian government is working a decades long coup here and entrapped civilians are lied to and not told they've been entrapped because the Russian government is working a coup.

218 Shaun Walker, Pjotr Sauer and Tom Phillips "Panic and Emotional Pain as Alleged Deep-Cover Russian Spies Vanish," *The Guardian* (April 3, 2023).

government has murdered ordinary Syrians, Ukrainians, Palestinians (using Hamas as a proxy), Soviet Jews, Israeli's (using Hamas as a proxy), and countless other individuals in Russia-bordering nations powerless to stop the Russia government.

In a world of bullies, the Russian government is the apex bully, the most destructive government our species has ever produced. It uses media and lies to try to pound us into dirt but our government is learning and working to minimize harm to civilians by putting more helpful articles in our newspapers and better controlling the Russian government's access to minimize the Russian government's exploitation of civilian's lack of knowledge. I believe the Russian government stealth attacks civilians worldwide, including children, to overthrow or weaken a targeted government and/or to support and/or install puppets to control that government and it's people. They believe the end justify the means, no matter how many babies they bomb, they think as long as they can lay the blame on Americans, they can do what they want. Anyone who believes the Russian government has anything to offer but death, destruction, slavery and cruelty hasn't been paying attention.

My mother was our family's sole breadwinner so after the Russian government's sucker punch we quickly became homeless: she was too unwell to work. I suspect the Russia-controlled gang made sure she couldn't resume her surgical technician career. When I asked why she didn't find another surgical technician job, she couldn't explain.

Chapter 27 - Our water is turned off

Unknown to me, our house was being foreclosed. I guess mom couldn't bear to explain it, especially since she was ill. The USSR would've made sure she felt only shame & responsibility. She has no idea she was set up by the USSR. No telling what lies her handlers told her.[219] If she's allowed to read this book, it'll be interesting to hear, if her handlers allow her to discuss it, her response.

When the USSR/KKK/police gang stealth attacked my mother and sickened her (which she'll hear for the first time if she reads this book, in her 80s), she was unable to work or to function well for awhile, rough estimate, several months. She lost everything but her children, was too unwell to pack our belongings beyond a few sets of clothing for her and her children, and a few pots and pans, etc.

Mom was impoverished to rent a U-haul or a storage facility, so we lost most of our belongings, our bikes, our beds, most of our treasured family photos, our living room furniture, our goldfish tank, our refrigerator, stove, kitchen table, etc. But we were together and that's what mattered most to us. I don't

---

219 Palestinians endless hunt for potable water. The USSR forced a similar fate on me. By sickening my mother, our family's sole bread winner when I was 12, she was unable to pay the mortgage, buy food, pay the water bill. Our water was turned off while we lived in our repossessed house, our mother devastated by an enemy foreign government she'd no idea had deliberately destroyed her life. She was nothing to them, and they intend to do the same to hundreds of millions of other Americans and migrants in America.

remember her explaining much to me or her children. That was the first and only house mom bought so she probably didn't know what house repossession was like.

I think she was too traumatized to explain and she was in physical pain, I won't detail what was going on with her health but she was in physical pain as well. I love her unconditionally and was highly sensitive to her pain and mental state. It was obvious she was in distress. I raised no subject that would upset her more. It was obvious she didn't feel well, and later it was obvious she was in physical pain. I focused on supporting her and my siblings as best I could, and doing whatever she asked me to do, like going to our neighbors to ask them if I could fill a bucket with water from their yard hose so we could flush our toilet.

I saw none of our neighbors reach out to us and I wondered why, but my mother had told me our neighbors included the KKK. She didn't know the USSR was also our enemy. I likely put it down the lack of help to the KKK. Knowing what I know about how the Russian Mafia stealth attacks, I'm positive my mother was made to feel deep shame for whatever "crime" she was set up for.[220]

---

220 Our ages when we were stealth attacked by the USSR using the war crime Perfidy were: mom (28), me (11), brother (10), sister (8), baby brother (4). All of us were civilians, all of us children under the age of 15 and so were protected by age on the Geneva conventions. Our mother is female civilian so was out of bounds according to the Geneva conventions and anti-genocide agreements, to which the USSR was a signatory. Even during war times, much less during peace, the USSR attacked my family illegally. In 1970-1971, the U.S. and the USSR were at peace. The Russian government is here illegally, so it will never admit it's attacking civilians here. I got final confirmation when the operative upstairs (who apparently died December 2023, my theory is

490

Chapter 28 - How the Russian government weaponizes shame and humiliation

The Russian Mafia weaponizes guilt and shame to destroy innocent people. To build and use shame and guilt as a weapon, the Russian Mafia must convince victims 100% that the entrapped person is corrupt, otherwise victims will continually look to be rescued and vindicated.

I know multiple people who are entrapped, including my parents and half siblings, and none of them give off the vibe that they've been illegally entrapped and intend to be vindicated. Any yet, I know 100% they were entrapped based on the Russian Mafia's multiple decades, multiple entrapment attacks against me.

The Russian Mafia, using our victimized police as camouflage, put all their money on convincing entrapped victims that they haven't been victimized. Far more important than plea deals and confidentiality agreements, which are just tapping mechanisms, the real attack is convincing innocent people that they're guilty. Even my entrapped father gives off no vibe that he was set up, and he's been trained in policing.

It's essential to the Russian Mafia that our legitimate government or ordinary people here know the Russian Mafia has seized control of our entrapment mechanism. I don't know

---

she died at the hands of the Russian government because they certainly wouldn't have allowed her to be questioned by legitimate American law enforcement). She'd came and stand above me in her apartment, day or night, when I typed comments critical of the Russian government in the *New York Times* as recently as 2023. She was American. I wondered what her handlers said to her when she was ordered to harass me when I typed critical comments about the Russian government in the *New York Times* comment section.

what they've told police who arrest noncriminals and offer them plea deals that force them into slavery. Hopefully after this book drops compromised police will talk. But they may not.

Mom was made ill by USSR operatives who set her up, unknown to her, in her late 20s. Poor health and the stress and trauma of the ambush attack and losing her home and her career meant Mom couldn't work for awhile, but by 33, 34 or so she'd gotten a full time factory job and worked there until she hurt her back, I want to say 15 years or so later.

Chapter 29 – The USSR worked hard to emotionally destabilize, enrage & humiliate me between 1970/71-72-1991

Nearly everything was left behind when our house was repossessed. She was around 29 and I was 12 when the USSR made me and my family homeless.

Chapter 30 - The next few years after the USSR's stealth attack

A few years after that my mother beat me for no reason when I was around 15, something highly unusual for her. Up until then, I believe my mother spanked me once when I was a child, after which I embraced ethics which meant that I tell my mother the truth, including that I planned to do something naughty and get out of being punishment by admitting the transgression. I've never been a schemer so it didn't occur to me to be bad and later admit it to get out of being punished. I consider her beating me out of the blue for no reason, around the time the Russian government was using three girls in high school to threaten me, but who I avoided because I worked part time to help my family survive financially, and I studied the rest of the time.

I consider her beating me as proof she was compromised by then. Nowadays I witness Russia-controlled American operatives forcing parents to immediately begin grooming their children, but my mother and our family were so traumatized by the USSR attack, it took my mother several years before she was able to take direction from her handlers, which is what I consider her beating me to be: directed by the USSR because it was determined to destroy American families.

Nowadays, the Russian Mafia begins grooming children as soon as their parents take a plea deal. Parents aren't told when they take the deal that the Russian will enslave their children as well but from what I've observed, that's what happens. I've witnessed it in my family in that both my parents and I believe all the siblings I was raised with are now entrapped. Most of my nieces and nephews have been attacked and destabilized, I believe under order of the Russian Mafia, and as bad, I've seen for myself at least a thousand children I don't know, most of them Caucasian, have been used in the stealth children's army the Russian Mafia is running here. In addition to the approximately 1,000 children used to stealth attack and/or harass me either alone, when they were on school outings, or when they assisted their parents as their parents harassed me and/or were used as beards (cover) by their parents as their parents harassed and/or attacked me, over the two plus decades the Russian Mafia went loud in my face, I, rough estimate, have had thousands of mostly adults (3,000), and children (1,000) harass and/or attack me including being hit by two vehicles driven by operatives, being kicked, called the n word, being stomped on, being sexually assaulted, many, many hundreds of covert physical assaults as operatives deliberately brushed against me on the bus, a man tried to kick me in the face on the bus, noise harassment over decades, plains clothed police feigning jogging into me, plain clothed police walking into me and forcing me off the side walk so I had to step down into the street, plain clothed police walking directly into me so I had to step around them, and on and on. Then there were the uniformed police, waiting for me in their vehicle as I arrived at the bus stop, sitting and staring at me, daring me to stare back at them, harassment increasing when I dared to write to former presidents begging for help to stop the unrelenting harassment, etc.

I am 64. The USSR entered my life when I was 11. I've done my best to be a good, decent person despite the fact that the Russian Mafia made sure I was lied to, cheated by supposedly reputable businesses, conned, beaten, stomped on, kicked, called the n word, hit by two vehicles & threatened to be hit by 5 to 10 additional operatives using their vehicles to terrorize me; even supposedly reputable businesses cheated me; that most of my supervisors and many of my coworkers from my full time jobs inexplicably turned against me for nothing that I'd done. I can't tell you how many uniformed police harassed me and I've been harassed and/or attacked by, rough estimate, 4,000 people, perhaps 1,000 of them children.

I know I'm not the only person attacked by the Russian Mafia in America in this way. This book is to tell people, including children, that the Russian Mafia is here, running scams, trying to get as many of us hooked on drugs as possible. If you have friends steering you into drugs, even if it's someone you feel loyalty towards, someone who defended you from school bullies, work bullies, someone you trust because they entered your life being kind to you, that person is very possible sincere but it's also likely they're owned by the Russian Mafia. I want the world's people to at minimum have the information that the Russian Mafia starts wars now without declaring them. If nothing else, ordinary people the world over need to know that information.

Although Ukrainians aren't blessed to be bombed, at least they know they have an enemy. In America and I believe in other countries, the Russian Mafia destroys by proxy. When the Russian Mafia operates in a country other than Russia, or in it's allies' countries, the Russian Mafia is a criminal entity illegally operating in a victim country. For other fifty years the USSR and later the Russian Mafia operated in the United States. I know this because they attacked me at 11 and have continued to harass and/or attack me overtly. The Russian Mafia intends to control the world. It attacks ordinary people illegally. What the Russian Mafia is doing is illegal, so they'll never tell you. They control many parts of the media so you'll hear a lot of anti-American propaganda. It's like a mass murderer killing you and telling you you really want it. The Russian Mafia is a criminal organization seizing world domination. Ordinary people need to be warned.

Chapter 31 - How I evaded entrapment – an ethics shield built by my young mother, re-embraced in my adulthood

I unknowingly evaded the Russian Mafia's many entrapment attacks in adulthood by embracing the ethics my mom taught me in my childhood. It was not easy, and most of the time I didn't know I was being attacked until long after the attack (the Russian Mafia uses proxies. It's illegal for the Russian Mafia or the Russian government to attack Americans in America. Since they refuse to respect our laws, they run stealth attacks using proxies). I'm able to write this book outing the Russian Mafia because I'm unentrapped. I believe there are tens of millions of people worldwide entrapped by the Russian Mafia. You don't read their accounts because the Russian Mafia doesn't allow innocent victimizes they own to describe the Russian Mafia's methods. I'm sure the Russian Mafia has compromised our entrapment system and I believe that most compromised law enforcement running entrapment attacks believe our legitimate government authorize the entrapments. Legitimate entrapments are run by our legitimate government. From what I can see, scam Russian Mafia controlled entrapments force entrapped victims, after convincing them they're corrupt, to entrap their loved ones/friends, including their children.

The first 12 years of my mother's life with her grandmother and aunt were happy, where she was loved, nurtured, cared for, and taught a high degree of ethics. My great grandmother, an illiterate maid who worked at a hotel and who attended church regularly, was an ethicist. She insisted my mother attend church with her, and taught her to value honesty. That was how my mother was raised and where she'd been happy, so when she had her children she emulated her grandmother's parenting strategy rather than her abusive parent's beatings.

My young mother taught ethics to me and my half siblings using the carrot method: if we told the truth when she questioned us we got no spanking. She was consistent, was never enraged. Even when she spanked us she never lost her temper. When we misbehaved, or were accused of misbehavior by an adult who told our mother, our mother always talked to us, even when we were barely out of toddler-hood (5, 6, 7).

She was truly a magnificent parent, best in class. That is the parent the Russian government stole from me.

She told us the allegation, told us who'd told her, and asked us if it was true. At that time in my childhood, when I was under 9, no adult in the Midwest ever reported misbehavior that hadn't occurred.

We were children so we didn't know our mother had asked her neighbors to please report any of her children's misbehavior to her. My mother had learned the value of using neighbors' eyes when her mother's neighbors had observed her mother's common law husband preparing to rape my mother, had reported his behavior to my grandmother, and had saved my mother from being raped when she was a child and teenager. Mom knew the value of neighbors' eyes. As her young children, we knew nothing of the horrors she'd endured as a child.

When our mother questioned us with regards to the allegations against us, we had the choice of telling her the truth or lying. She told me if I told her the absolute truth, I'd receive no spanking. She didn't lie: she listened to my confession and cautioned me to not repeat the misbehavior, and I received no spanking. I much preferred a quiet talking to over a spanking so ethics were a hit with me. She didn't call her parenting strategy ethics, she didn't call it anything. It was just the way she'd been raised by the grandmother she adored, and the approach she preferred in comparison to the beatings she'd received from her mother.

Decades later, when the Russian government launched it's many entrapment attacks against me (which unbeknownst to me, were based on getting me to lie and cheat my government for gain), they ran smack into the shield of ethics bequeathed to me from my young mother and my great grandmother, an illiterate ethicist I never met. That shield is what's protected me during this phase of the Russian government's coup against us, and allowed me to mostly unknowingly evade entrapments attempts so I can warn the world.

Chapter 32 - Despite being raised to be honest, I'm not a perfect person -- mistakes I've made, lies I've told

The two lies I've told in my life I told as a young adult -- the guilt I felt easily overpowered the value of what I'd lied to get; I hated the guilt – something I'd not had to live with in my mother's house. I've not lied since.

1ˢᵗ lie

I lied when I was approximately 19 or 20 when I applied to engineering school and failed to tell them I was already in college at another school. I applied to engineering school using my high school transcript. I lied because some of my freshman college grades were poor. I was the first person to attend college in my immediate family, I had no mentor or guidance. When I'd sought help from my high school guidance counselor, she was so racist she discouraged me from applying for scholarships and grants even though my grades were good, and my mother's wages qualified me for financial aid. She stopped me coming to ask her for help by accusing me of stealing $20 from her purse. I was devastated. I'd stolen nothing. I'd gone to her office to ask for help, maybe 5 other students were there. I believe I was the only African American student asking for help that afternoon. I guess she stepped away from her office without closing her door and $20 was stolen, or perhaps she lied. Before her accusation I'd stop by her office to ask for help and guidance, which she grudgingly and largely refused to give, even though her job title was guidance counselor.

She told me to stop applying for so many scholarships and grants. For perspective, to give you an idea of who and what she was, I received one major grant for college, which I used to pay for tuition, school fees, textbooks, food and housing. I also received one scholarship of I think $300. I don't remember receiving any other financial aid, though I admit it's been decades. I worked while attending college, for spending money. I told my high school counselor my mother was a factory worker and our family's breadwinner. I told her I worked part time to financially assist my family. She didn't care.

My mother rented a decrepit house for us because that's all she could afford. She just managed to feed and cloth us, and keep the lights, heat and water on; she couldn't afford to send me to college. I worked part time because my family needed me to work and contribute, and I needed to work to buy necessities, like a bed.

After the accusation from the guidance counselor, I didn't know who to turn to for career guidance. My parents graduated high school but didn't attend college. They both got specialized training, my mother became a surgical technician, and my father, was a security guard in the miliary in his 20s. After being honorably discharged, he worked as a security guard, later was a jail guard (which he hated), and worked his way into retiring as a police officer.

My parents are intelligent people but they didn't attend college, so couldn't help me with career choices. A college student friend who lived near me gave me a solar energy book for my birthday when I was 19 or 20. I loved that book and discovered an interest in solar power and electrical engineering. I applied for engineering school using my high school grades and was accepted. I felt guilt for not reporting my two or so years as a college student. I hated feeling guilty. I'd done nothing previously to feel guilt, so the guilt from lying was a big feeling. That guilt, and not ignoring it, helped me save my life and is one of the reasons I'm able to write this book.

2nd lie

The second and final lie I've told was to gain financial to attend engineering school. I told the lie when I was approximately 23-24 years old. I worked full time but earned a modest salary. I lived paycheck to paycheck, earned enough to pay rent, and cover basic living expenses. I helped my family financially by paying for my young sister's camp when she visited me for the summer, and I kept my old VW bug running.[221] I wanted to attend school but hadn't enough to pay

---

221 Unknown to me at the time, the USSR was stealth attacking me and causing me to have bills I couldn't afford. This was part of their multi-decades effort to enrage me. Had the Russian government not been a player in my life, I would have saved the thousands I spent being cheated by a business I worked for, by the operative I bought the car from, by the insurance company who convinced me to buy hit and run

for school, rent, etc. I learned of a financially stable family who sometimes helped students attend college. I wrote to them and told them I'd been accepted to engineering school but that I couldn't afford to go. I sent them proof of my acceptance. I lied by failing to tell them I'd graduated from college and was trying for a second degree. They agreed to finance my schooling, I think they sent me $3,000 to pay for school. I used the money they sent me to pay for school tuition, fees, and textbooks. But I hadn't told them the whole truth, that this was my second degree, so I felt guilt. After Russian-controlled American ops hounded me out of engineering school, I got a full time job, saved up, and repaid the family the money they'd sent me, telling them I'd not completed engineering school. They thanked me, and returned the money to me, telling me to spend it as I wanted. I gave it to my mother, along with other savings, to help her put a down payment on a house for her. She didn't buy another house and used the money for other things she needed or wanted.

The guilt I felt from those lies taught me lying for gain made me very unhappy and wasn't worth the guilt. I learned these important lessons by my mid20s, and had no idea my decision to stop lying would impact anyone but me. I had such a strong ethical line because my young mother, and her grandmother before her, valued ethics highly. Were it not for

---

insurance only to refuse to pay for repairs when my car was hit and run. I spent $3,000 on those repairs alone. I accept responsibility for my lie but had I not be deliberately cheated by USSR operatives running their attack against me, I would not have had to ask for financial aid from a kind family.

those two women, I'd be entrapped just like millions of others by the Russian gov. I'd be prevented from writing this book.

The Russian government is working to prevent this book, but I've no handler blackmailing and threatening me with prison for writing it. I had no way of knowing the Russian government was working a coup in my country and entrapping ordinary people by conning, grooming, and luring them. I didn't know when I made my choices in my mid-20s that Russian-controlled American operative's main weapon against ordinary people here is entrapping them and getting them to lie to the government for gain.

Chapter 33 - The Russian Mafia attempts to blackmail me

At the modest three bedroom house my mother bought when we relocated to Southern California, her first and only home -- my mother got a puppy. She loved that dog and hated when we lost our home. She cried when she had to take the dog to the pound. We were soon to be homeless, couldn't feed ourselves, had nowhere to go, and couldn't afford a dog, so my mother took our dog to the pound.

I don't remember an earlier dog in the Midwest, so it was our family's first puppy. I think I was 10½. One day I was playing with the puppy as I used the bathroom, the puppy was on the floor (this was decades ago so I don't remember exactly), the puppy licked my private parts. I was surprised by the sensation and told my mother what happened because I told my mother everything. I don't remember what my mother said but she didn't make a big deal out of it. My subsequent years were chaotic as my mother was set up by an operative sicced on her, causing her to lose her job and became ill. We were homeless, allowed to stay with a Christian lady acquaintance of my moms, then we left that shared space. My mother found and rented for us a series of decrepit houses, whatever welfare would pay for. Welfare didn't contribute much for housing but it did pay for our housing. We were grateful to have a roof over our heads, and we were grateful to be together. After my mother felt better she got a full time job in a factory.

When I was in my late teens or early 20s I lived for a while with a family who had a dog. My responsibility including caring for the dog when the family was away. While caring for the dog I allowed the dog to lick my vaginal area. I'd remembered the pleasant sensation I'd felt as a child. But I found as an adult I didn't feel those sensations. I understood I wasn't interested in having a dog lick my vagina. In my childhood I hadn't known I have nerve endings in my vagina that are sensitive to stimulation. As an adult I know my body has sensitive places. As a child I associated the pleasant sensation with the puppy. As an adult, even with a similar experience, I understood the pleasant sensation was from nerves being stimulated, not because the stimulator was a puppy. Because I'd been homeless as a child, and as a teen I'd focused on helping my family survive, I hadn't made time for a boyfriend, so the puppy experience in my childhood was a loud experience in my childhood, where I learned I could feel

504

pleasure on my body. Before that experience, I hadn't known I could feel pleasure, I hadn't been aware of the concept of physical pleasure. As an adult I saw for myself by comparing my adult experience with the dog, to my childhood memory with the puppy, and my experience with my fingers on my body as a teen, that it wasn't the dog, it was that my body could experience pleasure from being stimulated.

An experience with a dog in my late teens or early 20s isn't something I'd ever mention to anyone, not even my parents, who I told everything. I'm only telling it now because Russian-controlled American ops either videoed me during the experience, or questioned me when I was sedated post surgery, and asked me about it. I hadn't told my parents about my adult experience because I saw for myself that I didn't have a problem. I'd been curious from my childhood experience, but I saw from my adult experience that I wasn't interested in bestiality. Had I seen that I was interested in a bestial relationship I would've told both my parents and asked for help. They would've been upset but I had every confidence in them that they'd help me if I had that problem. I saw I didn't have that problem so I didn't mention the experience to them.

I had no idea anyone knew but me. I certainly didn't knowingly or willingly tell anyone, but Russian-controlled American ops definitely know because they've been trying to blackmail me with the information.

I've lived in this apartment over 35 years. I came to it as a roommate to a man I'd previously worked with. I'd left the company. He lived here with his roommate and after she left, he called and asked if I wanted to be his roommate. I declined. I'd had a roommate experience and hadn't liked it. I was renting a bachelor one room apartment. It was small but had no roommates. After about six months I reconsidered and became his roommate. We were roommates for five years when he left to get his own apartment. I asked the management company if I could rent it in my name and they said yes. That's how I came to be in a rent controlled apartment in Santa Monica, California. But you should know that I was lured here and that there are cameras I can't see, that I'm followed room from room and harassed daily. The harassment didn't begin until after the Russian government made me sick but they made me sick when I was 50 and I'm 64 now.

The operative ("op") who lives upstairs, and other ops nearby, have tortured me for well over a decade by interrupting my sleep so I now have sleep disorders. When I was offered this apartment as a roommate I was about 29 and the Russian government had been after me since I was 11, so for nearly 20 years. They wanted me here so they could keep eyes on me and get a better idea as to why they were failing to entrap me. I had no idea. I know I've said that in this book, but I didn't know I've had world class KGB-affiliated enemies since I was 11 years old. Because our government invested in powerful weapons, the Russian government was forced to work a multi-decades-long, covert coup. Rather than bomb us outright as the Russian government is bombing Ukrainians. Ukraine cannot be allowed to be stolen from the Ukrainian people. For many reasons they are very important but one of the high value services they provide is their farmer grow much more food than Ukrainians eat and sell the grains and fertilizer to impoverished nations, without price gouging the poor. Ukrainians must be protected.

When my roommate left I took a breathe and relaxed, enjoying living alone after having been a roommate for five years. The Russian government hacked my laptop and sent porn to it. I had no idea. I looked at the site suddenly appearing on my computer not even knowing I'd been hacked and that my laptop was getting porn sites because my computer was hacked. I ordered some of it, 95% of it traditional male/female sex. The site also offered dog and woman sex, which I'd never heard of. Curious, I ordered several of those.

I told my parents and my siblings I'd ordered porn online, and asked my father if it was illegal to order. He laughed, and said no. I also bought some at video stores in my area. I was in my early 30s, worked a lot, had no boyfriend. After watching it for a few years something about it began to bother me, I can't tell you what. For some reason, I got the feeling that some of the women in the videos were forced, that people were being made to shoot these videos. No one told me, I don't remember reading anything about it, I just got a feeling that people were being forced to do this. After that I didn't watch them again and threw nearly all of them away. I know I have a few porn dvds around here somewhere but I haven't watched them or looked for them in decades, and can't tell you where they are.

From the porn I ordered Russian-controlled American ops got a sense of what I was curious about but I stopped ordering it and stopped shopping for it. They tried for several years through a family member op to get me to get a dog, fishing. But I always said no, not because I knew anyone was fishing but because I worked a lot and wasn't home enough to properly care for a dog. I wasn't here to walk it and I live in apartment with no yard. I couldn't afford a dog either. Russian-controlled American ops were hoping I'd want one for sexual gratification but I'd learned from my experiences with the puppy and from my adult experience that I was uninterested. There was no one to discuss it with because I'd not told my parents, as I'd not had a problem. However, Russian-controlled American ops wanted me to have a problem.

The Russian government began harassing me about bestiality when I was in my 50s, my late 50s, in response to me telling my father I intended to write a book about being harassed, attacked, tortured, etc. They've on at least two occasions had ops in a group of two or three, all with dogs, swarm and crowd around me while I was on my mobility device. They've had Russian-controlled ops call and/or email me leaving disgusting emails or messages. They hacked my phone (or it may have been a laptop) and sent a advertisement for a man interested in having sex with a horse. These stealth attacks occurred in my late 50s, maybe early 60s. By that time

I knew I had enemies since my early 40s, but it took me awhile to understand I was being threatened with being outed as being into bestiality.

My response is, I'm not into bestiality. I never was, but since the Russian government wants to go there, let's go. If I were into bestiality, I'd admit it. I'd willingly stand before the world, knowing I'd be loathed, and tell the world the Russian government is destroying innocent children in my country, I'd describe how they're doing it, and how I know, because they made me homeless when I was 11 years old. I'd tell what they've done to me, to my family, to my nation, to my government, and to the tens of millions of other innocent people they cowardly stealth attack worldwide, all for power that will mean nothing on an uninhabitable planet. Nothing but my death will stop me telling the world about the attacks the Russian government is doing here, to ordinary, unsuspecting families and to their children.

It's impossible to blackmail me. If I did something wrong I'd admit it, and still out the Russian government for it's Crimes Against Humanity.

Chapter 34 - The Russian Mafia goes loud, having strangers stealth harass and some physically attack me nearly everywhere I go

With the demise of the USSR in 1991, the new Russian government, instead of just trying to make me as miserable as possible, began a series of entrapment attempts which they've continued even as recently as a few months ago, when my general practitioner tried to entrap me into defrauding Medicare. I'm 64, so I've been victimized by the Russian government for about 53 years of my life. They show no sign of stopping their attacks against me, or my country.

Not only have they targeted and harassed me for decades, but I believe they've murdered some of my loved ones, set up my niece and nephews to be made homeless, and ordered destabilization and harm to my extended family, and unrelentingly attack my country. The Russian government intends to rule the world and we, ordinary Americans, are in their sights.

This book tells the story of how the USSR began a coup against the American people in the 20th century to illegally target civilians, women, and children in direct contravention of the Geneva Conventions of 1949, to which the USSR was a signatory. Using my life as an example of the Russian government's destruction of my quality of life, the USSR began it's attacks against me when I was about 11, and continued until I was around 31, when that government ended in 1991. I was born in the Midwest and have lived in Southern California since I was 9. The Russian government, then known as the USSR, became attacking me when I was 11. I am now 64. After the USSR ended, the current Russian government picked up where it had left off, but instead of focusing just on enraging me and attacking my finances, the Russian government also began precision entrapment attempts against me as well as physical attacks.

Although the USSR entered my life threatening me with physical violence, and, I believe, was the reason my beloved mother inexplicably beat me, out of the blue, for no reason, when I was around 15[222], and my younger brother, when I was around 16, twice picked fights with me, out of the blue, for no reason. Other than those assaults, and the assault by my first supervisor at the studio where I worked in my early to mid 20s, and the multiple sexual attacks the Russian government had various operatives use against me from my teens to the last one was in my early 60s, for the most part the USSR focused on enraging me, and stealing money from me by having their operatives destroy my property and not compensate me for it.

Although the USSR had it's ops assault me, the Russian government kicked it up to a whole other level. They had one of their female operatives kick me, hard; had another operative stomp on my foot, hard; had an operative try to kick me in the face on the bus; had two operatives on two different occasions hit me with their vehicles, when I had the walk sign and rode my mobility device in the cross walk, during the day. When I looked at the op as he edged his vehicle into me, he smirked at me as his car pinned my mobility device under the bumper of

_____

222 This was the last time my mother hit me. Until that time I'd maybe received one childhood spanking, so I was shocked when my mother whipped me when I was 15. That whipping was why I left home as soon as I could at 17. When I left, the Russian government destroyed my youngest sister's childhood. Had I known then that my family was under attack by a foreign enemy government I would've never left my loved ones. I didn't know. Entrapped people aren't free to discuss the terms of their entrapment.

his car. The first assault was by a pickup truck and a police officer was already on the scene, his car and camera pointed at me, so I later understood, after the trauma of being hit by the truck, that it was a setup. To harass me many cars rev their engines when near me, or speed up when they see me jaywalking on a quiet street, or refuse to stop when I had the right of way in a crosswalk, or as I move my mobility device to cross in the cross walk in front of them they rev their engines while looking the other way as if they don't see me and are about to do a peel out.

Around 2001, when I was 40-41, thinking of those attacks, there was a major increase, and they went loud. The Russian government wanted me to know I was being terrorized so they had strangers on the street, on the bus, in grocery stores, plains clothed and uniformed security, harass me; enter my space. I was an office worker, I wore conservative office clothes, office appropriate shoes and clothing but you would've thought I pushed a shopping cart into the stores, instead of just my purse. Some where I'd spent thousands buying presents for my extended family over the years, I was suddenly stealth harassed.

Before then I hadn't known. Two office friends who I now know were operatives where inexplicably "sticky" which was why I left that job, to get away from them. At the new job, Russia-controlled American operatives tried to get me attached to another worker. I knew nothing about operatives then but I'd left my previous job because two ladies seemed to have inexplicably formed an attachment to me for no reason. So I simply ignored the operative at my knew job, even though I didn't know she was an operative. She behaved "sticky," too often where I was, too interested for no reason.

Recalling the difference in the Russian government's attack strategies against me: from 11 to 17, the USSR was into threatening, harassing and beating me. At that time only my mother beat me[223] but the USSR had four other students threaten to beat me up, a bigger boy when I was 11, and three girls altogether in high school. They had my high school counselor accuse me of theft to get me to stop seeking her advice to get financial aid for college. And my younger brother inexplicably picked fights with me twice.[224]

Russia-controlled American operatives are so furious at me for not being entrapped that they attack me on a near daily basis, even though I mainly stay inside my apartment. They're enraged, as if I belong in prison and they're not doing their job because I've managed to remain unentrapped. I cannot tell you how weird it is being an innocent person and being hunted in your own country by a foreign enemy government who've compromised law enforcement who are supposed to be

---

223 I was around 15 when my mother beat me so she was around 35. She beat me for no reason so I'm going to say that, looking back, that was the first time I saw evidence that she was entrapped. I'm going to assume her handlers forced her to beat me after I evaded the three high school bullied they were using to threaten to beat me up.

224 My brother was maybe 15 when he picked two fights with me. I was minding my own business preparing a meal and he picked fights and wouldn't leave me alone until I picked up a weapon and gave chase. When he ran, I stopped chasing him and returned to meal prep. He's entrapped so maybe he's been entrapped since he was 15 and perhaps his handlers forced him to harass me.

protecting you. My own police don't see me, they see what their handlers tell them I am. And if you know anything about the Russian government, they are world class liars. I've refused their scams and they retaliate to attack me, like the Mafia they are.

They forced my parents, siblings, doctor, boss, plains clothed police, and strangers unknown to me to work to entrap me. Using my loved ones as camouflage and weapons to destroy me is evil, as is destroying my ability to trust my family. The Russian government uses my own society against me, use it as a weapon to stealth mistreat, abuse, harass and torture me in every way they think they can get away with.[225]

The Russian government forced my father, a retired police officer, to tell me to exaggerate my health problems to Social Security Disability after Russia-controlled American operatives deliberately made me sick and prevented the first several doctors I visited to diagnose me. While he worked as a

_____

225 The Russian government's American operatives let me know they monitor my every move in this apartment. To protect my thoughts I write on notebooks, but the width of the page so I can write underneath one of the pages and retain privacy. One of the doctors they control, and they control many, I was describing my symptoms and reading my notes and she asked me why I had written my notes the length of the paper. It was not something an non-compromised doctor would notice, but the Russian government wanted me to know that they're a participant in all my doctor's office visits. The reason is the Russian government doesn't want me to seek medical care, so at nearly every medical visit, they'll have a medical provider do something sketchy, offer to lie for me, behave in a way that I'll know they're compromised. The Russian government intends to kill me using "natural causes." We need to destroy that government. It is so deeply embedded here they know we'll never get them out. The only way we can is by destroying that government. It's either them or us and they got a head start last century and Mr. Putin and his government have simply built on that.

police officer, my dad was used to gas light, misdirect, and influence me because the Russian government had found me resistant to corruption. My father told me to not write this book, sneered at me and implied people would think me crazy for writing it because, at the time I decided to write it in my late 50s, I didn't know who the perpetrator was beyond the fact that police were extra judicially harassing me and ordering people I didn't know to harass me. Had I exaggerated my health problems to Social Security Disability as my father told me to do, and had I broken the laws my mother told me to break, I, like them, would be enslaved by the Russian government. The Russian government has led entrapped people to believe our government has entrapped them and is forcing them to entrap their own loved ones. I don't think that's true.

I believe our legitimate government is notified in some way when people take a plea deal to stay out of prison, but I don't believe: (1) they know our entrapment system has been commandeered by the Russian government; (2) that parents are being entrapped for the specific purpose of entrapping their own children; (3) that families are being targeted because parents more easily take a deal when their children are threatened with social services, (4) children of entrapped parents are groomed by the parent's handler's to target[226] and

---

226 When they were children, the Russian government sicced "friends" on my niece and nephews, had their school friends tell them they stole laptops from school and asked my niece, nephews, and their mother if they could hold the laptops for them, and if they would, their school friends would give my niece and nephew a laptop for the family. One of the children in my family said no, but another one said yes. It was a setup. The children in my family were, I want to say 9, 13, 15,

harass children and adults in families the Russian government is working to destroy; (5) that the Russian government has formed underage children into child armies which they use to harass, engage and/or befriend targets; (6) that people are being illegally entrapped because the Russian government uses authority figures to tell them to break the law; and (7) entrapped people and their children are being coerced and/or influenced to vote in Russian government puppets.

I believe tens of millions of innocent people, including migrants living in America, and millions of innocent people worldwide, are entrapped by the Russian government, forced to vote and/or harm others under orders of the Russian

---

somewhere around there, and would've never stolen computers from anyone. The Russian government got them charged as accessories. They were poor, they lost their public housing, never returned to school. One got his GED the other two got no more school. The Russian government destroyed their lives because they knew I loved these children, I spent money on them and helped them as best I could. But I didn't know my family was being hunted and the Russian government hunted children in my family, including me. They'd made me homeless at 12. Those three victims are now adults. They're underemployed, have substance abuse problems and none can afford to marry because they can't afford a family or to pay rent. The Russian government attacks our children and our children have no idea why their lives are destroyed. My mother was attacked when she was 28, minding her own business, raising her children. The Russian government destroyed her quality of life and turned her to religion, one of the Russian government's favorite scams. Not because religion is a scam, the way the Russian government exploits it is a scam: victims are trained to blindly believe and the Russian government need only control the head guy, tell him what to say and do. Right when our nation was about to take off and soar after the 1960s, the Russian government has done everything it possibly can to destroy our potential. It's horrific what they've done to my people and my country and it's worse that people don't understand the Russian government is destroying so many lives because it's jealousy and ego driven, is a slaver government, and intends to rule the world. The Russian government has no business ruling anything. It is unworthy to rule.

government. The Russian government views itself as one of the better examples species.[227]

In addition to the Russian government destroying my family as it once was, and, I believe, murdered three of my loved ones, I've witnessed the Russian government's attack and/or militarize several thousand Americans and/or immigrants civilians, including I estimate approximately a thousand American children and/or immigrant children under 15, using them to provide cover for their entrapped parent's stealth harassing attacks against me. I've witnessed children under 15 being active partners assisting their entrapped parent and/or care providers to harass me. Entrapped parents, teachers, individuals, and school field trip minders, and the children they control, are forced by Russia-controlled American operatives to harass, attack, torture and/or hostilely engage me on the street, in grocery stores, on the bus, in my apartment, when I'm with my family, in my doctor's office, at bus stops, at the post office, nearly everywhere I am, including when I go to the bathroom inside my own apartment, or use a bathroom while out running errands, Russia-controlled American operatives have people harassing me nearly 24/7. I'm not exaggerating. Russia-controlled American operatives treat me as if I run a major cartel, they go out of their way to have operatives sneak up on me when I throw out trash or am reaching for food items at the grocery store to startle me.[228]

---

227 Roger Cohen, "The Making of Vladimir Putin" *New York Times* (3/26/22).
228 Russia-controlled American operatives have gone out of their way to startle me for decades. The first time I was aware of it was when I went

Their intent is to stress murder me, causing my death by "natural causes." They don't quite control our country enough to shoot me in cold blood, which, I'm convinced they do in many other countries. Based on the thousands of attacks the Russian government's American operatives have made against me, they have no fear our legitimate government is a threat to them. Probably because when they attack me they know no legitimate law enforcement is around. Maybe it's because they consider themselves legitimate law enforcement.

I've been harassed and/or stealth attacked by, rough estimate, several thousand strangers, I also witnessed a young Caucasian man in his 20s, when I was approximately 48, forced by a Caucasian uniformed police officer in a marked police car to harass me. The forcing was done right in front of me and the compromised police officer didn't care that I saw what he did. After that experience, and after I observed during a very short police strike in California where I wasn't harassed by anyone for a day, I understood that some American police are compromised, and were the managers of the people attacking me. Uniformed police have harassed me for years. When I drove they paced along side me, when I walked they drove police car after police car in a procession at least three

_____

to an event with my sister and her husband, it was a Halloween event maybe 20 years ago and operatives sprung out at me pretty unrelentingly. Even recently, I was reaching for water at the grocery store and an operative suddenly appeared out of nowhere to startle me. Russia-controlled American operatives have sent operatives to my apartment to startle me, they are actively trying to frighten me to death. Since they couldn't get me to break the law, they're working stealthily to murder me. It will be a hit and I'm not exaggerating.

times, when they saw me on my mobility device, they turned on their siren, when they knew I was approaching a bank, they'd have a car's alarm go off. When I arrived at a bus stop or waited for a bus, police in police cars would come and idle their vehicle and look at me, harassing me in all kinds of stealth ways they meant for me to notice but not the casual onlooker. When I worked downtown police officers, both uniformed and in plains clothed would be at my bus stop, some appeared to dare me to stare back at them. I didn't because, they were clearly hostile and wanted a reason to attack me as I sat at the bus stop.

It wasn't until 2021 when I read *The Guardian* article[229] that I learned the Russian government is the puppet master controlling some of our law enforcement and is working a coup against us. I'm writing this book to warn you and to inform you about specific types of entrapment attempts the Russian government use against me.

---

229 David Smith, "The Perfect Target: Russia Cultivated Trump as Asset for Forty years – ex-KGB Spy," *The Guardian* (January 29, 2021).

Chapter 35 – For decades, the USSR and the Russian Mafia attack me via proxies at work, beginning in my first job after college, ending with my last job in 2010. Nearly all my supervisors, even temporary supervisors (who were operatives placed just to pinpoint harass me since I tended to come to work, and sit and do my work, the Russian Mafia had to put harassers right next to me. I wasn't the type of employee who wandered the halls. I was usually working unless I needed to go the bathroom

I was the worker the supervisor wanted to keep, the co-worker who started no drama with other co-workers, the worker who produced work where attorneys didn't complain about work quality, so no one was itching to get rid of me until the USSR and the Russian Mafia pushed them. There are no perfect workers but I quietly did my work, the best I could, proofing it, because delivering the highest quality work possible is what my mother taught me from my childhood. She showed me with her quality parenting, and she alluded to it when she described her work.

My supervisors never, on their own, tried to push me to leave any company. The Russian Mafia, using compromised American operatives and law enforcement, spread disinformation, had to, otherwise my supervisors and co-workers wouldn't have harassed me out of the company and/or department. I produced quality work and always treated my supervisors and co-workers professionally and with courtesy. At work, I was there to work, and that's what I did.

The Russian Mafia forced my supervisors and co-workers to stealth harass me. I'm positive those same types of attacks continue to be used and are the reason behind the huge employee turnover during and at the end of covid.

Chapter 36 – The USSR & Russian Mafia harassed me to leave jobs I loved in order to created a new opportunity for them to entrap me: they didn't care that I loved my job and liked my co-workers. Apparently American operatives were paid per entrapment attempt

The USSR and the Russian Mafia pushed me to leave companies after they saw no way to entrap me there. I didn't want to leave until I began to be abused. The USSR and the Russian Mafia pushed my supervisors and co-workers to create a hostile work environment. At a Westside law firm where I'd worked maybe 5 or 6 years, first in word processing and later as a legal secretary, the Russian Mafia pushed my supervisor and co-workers in word processing to such an extent that my supervisors began accusing me of shoddy work. After the first accusation I began keeping a before and after copy of my work: making a xerox copy of the document before I began it, and kept a copy of the document after I'd finished working on it (temporarily). The next time I was accused of shoddy work, I pulled out my proof. I have never seen such rage on a supervisor's face but they stopped accusing me of shoddy work, understanding that I had no problem proving my work quality.

Chapter 37 - The Russian Mafia sickened & exhausted me & sicced my loved ones on me

Based on personal experience, entrapment attacks the Russian government has attacked me with: trying to get me to cheat on my taxes (they had me win something valued at a little over $1,000 on a game show to see if I'd cheat on my taxes. I did my own taxes at that time, had never won anything so they hoped I wouldn't know I needed to file taxes on a prize valued at over $1,000. I claimed the prize on my taxes just in case the IRS wanted to know, I had no idea it was a setup). They forced my beloved aunt, in remission from cancer and barely able to stand to get me to the game show, and who unknown to me they'd hooked on opioids to make her more amenable to entrapping me, as that was the only way they could think of to get me to a game show because I worked a lot. They later killed my aunt with an opioid overdose after she failed to facilitate my becoming a drug mule, which is the job the Russian government wanted for me.

The Russian government tried to get me to file for Disability, then tried to get me to file for Unemployment, simultaneously. The Russian government have their operatives harass me unrelentingly on the street, bus, in grocery stores, etc. often as operatives fronting as homeless; when I go the doctor to find our why I feel so sick, they had doctors they control who seemed to have never attended medical school, as I now know I presented classic symptoms for insulin resistance, obesity, migraine, frequent urination, numb and

522

tingling toes/feet (now my toes don't tingle. Had the Russia controlled doctors diagnosed me when I initially reported the symptom, I would've been better able to halt the peripheral neuropathy I subsequently was diagnosed with[230]). At my job they had the employers they'd entrapped work me to exhaustion. I'd never been sick before beyond reoccurring sinus infections, and colds/flu, and knew almost nothing about Social Security Disability, they had an attorney they control at

---

230 Russia-controlled American operatives did their best to not have me diagnosed. I collapsed before I began to be diagnosed with any of my health problems. When I was tested for sleep apnea , Russia-controlled American operatives rigged the test. Not even lying. I went to an internationally renown medical center in Los Angeles. The sleep tech hooked me up to the leads and while the person did so I began to feel extremely sleepy. At that time I 99.99% of the time didn't feel like that, even when I was trying to sleep. I think I was tested around my mid50s, sometime in my 50s, and I'd been harassed on the street since I was around 40-41, uniformed police harassed and/or directed their operatives to harass me (I'd witnessed it), and multiple family members had tried to entrap me so I knew I had an enemy but my enemy hid their identity from me. Because I'd been overtly attacked for approximately 15 years when I began to feel sleepy, I was very alarmed, given that sleepiness wasn't normal for me. I put two and two together and deduced that whoever hated me didn't want me to be diagnosed with sleep apnea, which I knew I had because I woke myself up snoring. To rig the test my enemy, who I knew controlled police, had arranged to have a sleep inducing gas pumped into the room. I can't describe the terror. Because I was so frightened due to being stealth attacked, despite the sleep inducing gas, which I couldn't smell or see, I was unable to sleep. I stayed there that night but had to return for another session because sleep apnea can't be diagnosed unless the patient sleeps. On the second test, no sleep inducing gas was pumped into the room and I slept as best I could. That test diagnosed me as having sleep apnea. The doctor who diagnosed me encouraged me to get on psychotropic medications when I told her police were harassing me. She also wrote an incorrect doctor's report. When I told her it was incorrect she refused to fix it. I complained to the medical center and higher. The medical center said that because any changes in her notes wouldn't change my diagnosis of sleep apnea, the doctor wouldn't be required to correct her notes. I believe I filed a report to a California agency reporting the incident, but nothing was done.

a free legal service nonprofit in Santa Monica tell me to apply for Disability; after I did, I was denied.

While trying to figure out how to live out of my car, except I had no car, the Russian government had a cousin of my mother's (who has since died but who apparently had worked in social services back east) advise me to appeal, so I do. Before that I thought I'd have to live in a shelter, or maybe rent an old camper and live in the desert. I expected to be homeless and although I knew I was under attack, I hadn't yet figured out that the exhaustion attack was orchestrated by my enemy. I went to a shelter in Santa Monica to ask for a bed but was told none were available but they could probably find me a bed at a shelter within a thirty mile radius but they couldn't guarantee my safety there. For years I've given money to homeless people at bus stops, outside stores, etc. I'd sat at a bus stop years ago when I worked, sat next to a homeless gentleman, gave him some money and asked why he wasn't in a shelter. He said he didn't feel safe there. His response was the first I heard that some shelters aren't safe. Before he told me, I hadn't known.

After I appealed, the Russian government had one of my loved ones recommend a law firm that helps people appeal their Disability denials. That law firm agreed to take my case, but shortly after Disability reversed it's decision I was the victim of identity theft. I suspect someone affiliated with that law firm was the perpetrator but I don't know for sure. The only reason I learned my identity was stolen was because I continued to file taxes. One year after I filed the IRS told me I'd already filed for that year. Since I knew I hadn't I told the IRS and they informed me a fraudulent tax return had been filed using my social security information. They refused to tell me the name of the perpetrator when I asked.

After the law firm agreed to take my Disability denial case but before my identity was stolen, I thanked my loved one for helping me find a law firm to represent me in my appeal. I had over twenty years legal secretarial experience but for the most part I worked for firms who practiced general business law, so I knew almost nothing about other types of law. For instance, when I successfully evaded entrapment attempts by Russia-controlled American operatives it was solely because of ethics, not because I saw the trap. Some of the compromised police seemed to think I saw them. I didn't. Not until I was 40-41 and even then, my father successfully gas lighted me, telling me that stores were experiencing a big increase in shoplifting so to ignore the attacks. So I did. I ignored those store attacks for nearly all of my 40s because my father told me to and he gave me a reasonable explanation for the store-based harassment. An important feature of the Russian government's stealth war is they always offer a reasonable sounding explanation, as well as have an influencer on hand to get their targets to turn their gaze away.[231] I cannot tell you how big an

---

231 When I fell into poverty after I as deliberately sickened by Russia-controlled American operatives, I qualified for food stamps. If you've ever qualified for food stamps you know it means you're destitute and the government will provide you with enough food to keep you alive if you shop for cheap foods. There was a period of time after my

asset it is to the Russian government that they in my experience have a compromised influencer ready to feed their target lies. I was overtly attacked on the streets and nearly everywhere but inside my apartment (I began to be attacked in my apartment in my early 50s). My father was used to gas light

---

temporary disability ended and my Disability appeal was still under review where I was destitute. My appearance reflected this. When I left my apartment to buy food or pay bills, my clothes were often holey and I looked poor. During this time, someone in law enforcement, I'll probably never know who, was, I believe listening in on my Whatsapps, reading my emails and/or listening to my phone calls with entrapped family members, the officer decided to investigate what I was saying. I sorry, I don't remember how I figured this out, but I think it was related to the fact that operatives the Russian government routinely use to harass me occasionally pull back from hard core harassment and torture, and when they do I know it was because someone was looking over their shoulder, spot checking them, and they were warned. An non-compromised law enforcement officer discussed me with someone at the law firm where I last worked and was told I was let go due to appearance. That was a lie but it's an important lie. I wore office appropriate clothes when I was employed, but had no money to routinely buy clothes when I was made ill. The non-compromised officer who briefly looked into my situation was told what he knew to be true during his assessment of me, that my clothes had holes in them. The Russian government knew my clothes had holes in them that he could see so they told the officer something the officer could see. But the officer hadn't known me when I was a legal secretary, and so didn't know what my clothes looked like then. By point is, the Russian government pitched their lie to what the non-compromised officer knew to be true, and what that person would accept as the truth. This is critically important. It's why spies seeking asylum here are sometimes granted asylum: they tell our security agencies what the Russian government knows our security agencies know, and a little more, making it worth while to accept the person seeking asylum but not enough to halt the Russian government's core attack on us. The Russian government's core attack on us is entrapping ordinary people, forcing them to entrap their family, and forcing the entrapped people to vote for whoever the Russian government wants seated in a power position. I think had Russian operatives seeking asylum here told our security agencies the Russian government's core attack strategy, our security agencies would know about it. Our security agencies know a lot, far more than I know about the Russian government, but I know how the Russian government works on the

me, to shift my gaze, to feed me believable lies.[232] It wasn't until my late 40s when the overt attacks were so obviously unrelated to shoplifting that I understood I wasn't being harassed because of a store.

I thanked my sibling for recommending the law firm who accepted my Disability case, and the Russian government

---

ground here against ordinary people, and I don't think our security agencies knew that before I started telling them. We're in trouble. The Russian government knows how our security agencies think, and how police think, and the Russian government intends to destroy us. I'm not a hawk. I don't say we need to destroy the Russian government lightly. I'm loyal to my people and my country so of course I don't want us defeated by our enemy but just because our enemy defeats us isn't a reason, for me, to push for their destruction. I don't want us defeated but I wouldn't push for the destruction of an enemy just because they defeat us in strategy. I see what they are willing to do to get their way and their behavior is evil. Their evil strategies have benefited them so they won't be phasing those out. We live on a planet we need to function a certain way. We have only one planet habitat. The Russian government, in it's blind ambition, is forcing us and it's competitors to more closely embrace oil and other fuels that we now know is leading to our extinction, and the extinction of many other life forms. We're in charge here. No other species can destroy the Russian government. I know it will cost us, probably tens of millions of innocent American lives. I don't want that for my people. But. If we don't destroy the Russian government our species won't survive. Not only that, if the Russian government seizes control of our species, it enforces slavery for everyone not them. An enslaved America is horrific. An enslave any other country is horrific, but slavery is the Russian government's sweet spot. We're an inexperienced species with much to learn, but it's clear the Russian government is a problem for life on our planet, and for our species.

232 The Russian government inserted my father into my life with the express intention of using him to influence me. My parents dated as teens, became pregnant, and their marriage fell apart. They both went on to remarry and divorce. I love my father but he was busy serving in the military, working, remarrying and having additional children. He wasn't into me, I was into him. When I was 11 and asked him for money for school supplies, which was probably $10 or $20, he refused. He had a job and he didn't provide my mother with child support. As far as I can tell after my first few months of life, when he did initially pay my mother's boarding room rent, he wiped his hands of me after my mother began dating someone else and became pregnant. As far as I know, the only child support my father supplied in my 64 years of life

promptly weaponized my gratitude to use against me by having her, unprompted, advise me to file for Unemployment. Had I applied for Unemployment while my Disability appeal was pending, I'd be saying to Disability that I'm too sick to work, and saying to Unemployment I'm available for work. That's a contradiction suggesting fraud. At that time that sibling hadn't been used by the Russian government to lure me, I think that was the first time. I remember thinking after her suggestion she must not have understood what she said. I quietly ignored her advise. Had I followed her advice, the Russian government would've entrapped me.

The Russian government used her, with no prompting from me -- in other words I didn't ask for her opinion or advice but simply thanked her for helping me find a law firm who'd take my Disability case. Because she has access to me, and because the Russian government used her to help me find a law

_____

was the first few months after my birth when he paid for my mother's room at the boarding house. I reached out to him as I grew up but he was young and busy and working and trying to find his way. By 11 I lived in California and he lived across the country from me. In my 30s, Russia-controlled American operatives had been stealth attacking me for over 20 yeas with no entrapments landing. I believe they forced him to respond to me in order to use him to influence me from that point on. We talked politics occasionally on the phone, which I loved. In my 40s the Russian government used my father to gas light me when I told him I was being attacked in stores. When I sought his advice as I considered career options, considering the military and policing, he told me not to do either. Since he was a veteran, a police officer and my father, I listened to him. I believe my father is a good man. I believe my mother is a good woman, and I believe my siblings are as well. The problem is, when a parent is entrapped by the Russian government, the Russian government uses them as a weapon to influence the target. That's been my experience. I love my parents and my siblings. I adore them. But they are slaves of my enemy and are exploited by my enemy to harm me. The Russian government has made very clear to me that it intends to destroy me. I'm writing this book to warn the world.

firm to handle my case, in other words, something they knew was helpful to me, they moved immediately to exploit their prior attack. I was disabled because the Russian government worked me to exhaustion. When I went to doctors seeking help, the Russian government wouldn't allow those doctors to diagnose me. The Russian government created my health problems then used those problems to try to entrap me.

Chapter 38 - The Russian Mafia creates it's opportunities by using people you love against you

They'll break something of yours & have their operative offer to help you; they'll bully you & have their operative rescue you; then use that operative to destroy you.

The Russian Mafia uses gift giving or assistance as a weapon to destroy you. They create a problem for their victim, then insert their operatives into that person's life to destroy them. Their kindness hides a knife. At no time will anyone tell you anything until you're entrapped and forced to take a plea deal, where the Russian mafia uses it's compromised American law enforcement operatives to convince you that you're corrupt and weren't set up, which of course, you were.

(1) Brother operative – the Russian government manipulated my gratitude to my brother use as a weapon to destroy me, and manipulated my mother's position as an authority figure to try to entrap me. My brother drove to visit me when I was unwell for the first time in my life. I was unable to work, and had little disposable income. He drove an hour to spent time with me maybe every six weeks for 6

months. He gave me a little money, and bought me a few groceries. I felt so grateful. The Russian government then used him to try to entrap with insurance fraud, Disability fraud, and conspiracy to commit both. He asked me to lie on the insurance form, to say I had no health problems, telling me he'd get the best rate if I said I had no health problems. I said no.

I understood my brother asking me to let him take a life insurance policy out on my life, after he explained his new job included selling life insurance policies, and he wanted to make a good impression on his new boss. But then he insisted I lie to the insurer, tell them I had no health problems when I knew I did.

I'd never been sick before, I knew almost nothing about Disability or life insurance, but it seemed to me that Disability would confirm whether I was sick beyond the many medical reports I submitted to my attorneys who handled my appeal. I didn't understand then that my brother was trying to entrap me, or set me up for entrapment. What I thought then was Disability would, of course, investigate my claim of Disability. I sure would if I was in charge of Disability. Had I told an insurance company I'm fine and Disability I'm not, that's a lie.

Although I didn't know my brother was an operative then, and although then in my early 50s, maybe 51-53, I had no idea I'd evaded multiple entrapment attacks.[233] When I told my

_____

233 By the time of my brother's entrapment attempt, I'd been given an opportunity to embezzle from the previous law firm I'd worked at but, of course, I didn't do it. That entrapment was so in-my-face weird, I couldn't help but notice it. I didn't think of it as entrapment because it was the first time something like that had been shown to me so

brother no, he acted like I was being unreasonable. Now I understand he behaved as his handler directed, as the Russian government directed, but then, I didn't know. I called my mother to complain while he was still visiting me. She told me to do as he asked. I said no again. We three didn't speak for awhile.

It's important for readers to understand the sequence of events: the Russian government took I estimate 6 months before having my brother ask me to do anything. During that time, they had him visit me, talk to me, made me think he visited me out of concern. The gift of his time, his concern, which I interpreted as love, groceries and a little money, were things I appreciated. Then, they lowered the boom on me, tried to exploit my love and gratitude to him, and my mother's position of authority, all she's done for me in my life, the Russian government stole when they donned her skin, to try to get me to self-harm. The Russian government are liars, thieves, slavers and cheats.

(2) Sister operative – the Russian government manipulated my gratitude to my sister to use as a weapon to destroy me. Another relative did a similar entrapment, but different. After I was denied I appealed because a cousin of my mother told me to. My sibling recommend a law firm she'd seen ads for who helped people appeal disability denials. I called them, met with them and they agreed to represent me. I

---

obviously. I didn't know what to think except it was strange. I didn't know when my brother and mother told me to commit insurance and Disability fraud that I'd been successfully evading entrapment attempts for over 20 years.

thanked my sibling. Not long after that we were talking on the phone and she recommended I apply for Unemployment. I listened to her. I'd never been on Unemployment, which Russia-controlled American operatives knew and counted on my not knowing the rules. To qualify I thought I'd need to be able to work. I was appealing my denial from Disability, who I had told I was unable to work. Had I followed my sibling's advice, I would've been charged with Disability and Unemployment fraud, and conspiracy to do both, which the Russian government was herding me towards. Then I didn't know that, it took me a while to figure this all out. The Russian government isn't big on transparency, is operating here illegally and immorally; the entrapped police operatives it controls are acting extra judicially, so none of the people running the attacks are explaining to victims. And they're all doing this out of sight of our legitimate government.

My siblings are compromised. The Russian government doesn't tell you your loved ones and friends are compromised, the Russian government forces your loved ones and friends to try to entrap you and/or to lure you to a specific location.

When your friends, extended family, co-workers, teachers, police, grocery store, employers, pharmacists, parents, landlord, doctors, loved ones, bus drivers are entrapped they belong to whoever owns them. An enemy foreign government, the Russian government, owns our people and is determined to destroy us.

So many of us are dying daily. The Russian government isn't listed in the newspaper as the culprit because when people die they can't explain what was done to them and how they came to make the choices they made. They're not there to tell you about the Russian government-controlled influencer who teased and lured them into trying a drug for the first time and they died. Or their parent told them to lie and cheat a government program which fraud resulted in a prison sentence where they were raped and brutalized by Russia-controlled American prison guards, after which they turned to drugs just to cope. The Russian government leaves dead bodies in it's wake. Don't fall for the drug scam, don't be curious. And if someone is nice to you and then asks you to lie for them, or if someone is nice to you and recommends you sign up for something you don't believe you qualify for, don't do it. If you're talking to them, don't agree to do it.

Agreeing to defraud the government is a crime and it's a bad idea to lie to anyone, I speak from experience. To be convicted of conspiracy you don't need to do the crime, just agreeing to do the crime is a crime. I was a legal secretary but I didn't learn about conspiracy until sometime around my mid-50s. Russian-controlled American ops are running a secret war against ordinary people.[234] The Russian people aren't the

---

234 General Valery Gerasimov, Chief of the General Staff of the Russian Federation,"The Value of Science in Prediction," *Military-Industrial Kurier* (February 27, 2013). "In the 21st century we have seen a tendency toward blurring the lines between the states of war and peace. *Wars are no longer declared* and, having begun, proceed according to an unfamiliar template."

problem, the Russian Mafia is the problem. It's here and working to destroy as many of us, or get as many of us as possible into the criminal justice system where they can have entrapped guards and/or inmates beat and/or rape you to death. Kalief Browder, a 16 year old teenager held at Rikers Island for three years without trial, insisted he didn't steal the book bag he was accused of stealing. He refused to accept a plea deal because he was innocent. I believe him. I've experienced four different shoplifting entrapment set ups at four different stores by the Russian Mafia in California, within the last four years (I'm typing this book in 2024), and I've had similar experiences to his. He said law enforcement in unmarked cars came and parked near his mother's house and drew their finger across their throat to indicate to him that they would kill him as he sat on his porch; I've had many officers, in police cars, stare at me as I sat on my mobility device, or stood at a bus stop years before I owned a mobility device, daring me to stare back at them, which I didn't do. For years I've had Russia-controlled operatives honk their car horn as I enter the bathroom in my apartment, which can't be seen from the parking lot. The Russian Mafia wants me to know they have eyes on me even inside my apartment, to make feel always afraid. The Russian Mafia greatly enjoys making people feel afraid. One of their operatives honked when I opened my front door at around 2 am, to let me know they had eyes on me. They've harassed me more times than I can remember as I type this book in March 2024.

I believe Mr. Browder when he said he was harassed after his release and I believe he didn't steal that book bag.[235] The Russian Mafia is here and they're working to kill us, all Americans, all migrants seeking safety here. Please be careful. Our police are being used as a weapon by a foreign enemy government to steal their, and our, nation from us. Our police must be made to understand how they're being exploited and they we must overhaul law enforcement from the ground up. That would be challenging at any time but it's especially hard, but necessary, given the numerous operatives the Russian Mafia has embedded here and the numbers of entrapped innocents they control. At any time day or night the Russian Mafia can have any number of people commit mass murder here, they've already tested their system: the mass murders we've experienced. The Russian Mafia has messed up so many of our innocent people; years of psychological abuse from their families or other people with access to them. I was blessed the USSR didn't get a hold of me until I was 11.

I am blessed to have had the quality of mother I had during my childhood. I've witnessed quite a few American children, most of them Caucasian as I live in a Caucasian dominated area, and the Russian Mafia uses people who they control in the area to harass me – those children already have a serious set back because their parents are entrapped. Unless we get the Russian Mafia out of our country, those children will

235 Michael Schwitz and Michael Winerip, "Kalief Browder, held at Rikers Island for three years without trial, commits suicide," *New York Times* (6/8/2015).

never know the quality of parenting they deserve to have because their parents, like mine, are slaves of the Russian Mafia.

(3) "Love" Russian government style – the Russian government manipulated my gratitude to my supervisor to use as a weapon. I worked in a word processing center and my boss was very nice, kind, handsome. To persuade me I had romantic feelings for him, despite him being happily married to a lovely woman, the Russian government had an operative attorney enter word processing yelling at me, accusing me of bad work on a job. My supervisor came to my defense, stood up to the the attorney in front of me telling him to never treat to one of his employees so discourteously again.

I didn't know it then but this type of psychological warfare is called a White Knight attack and has been used for decades to harm targets, specially women. I did fall in love with him, he was kind, handsome and a great worker. He was also married. So I left that job and told him why afterwards. Again, ethics saved me. I refused to break a woman's heart by having sex with her husband. This White Knight setup was made against me when I was around 30, so over 30 years ago, but the Russian government still runs these scams to the present day[236]

_____

236 Shaun Walker, Pjotr Sauer and Tom Phillips "Panic and Emotional Pain as Alleged Deep-Cover Russian Spies Vanish," *The Guardian* (April 3, 2023).

Chapter 39 - Office supplies theft grooming attempt. The USSR attempted to groom me into becoming a thief using office supplies to prime the pump when I was in engineering school

Having no idea I had been stealth attacked by proxy since I was 11, when I decided to go to engineering school, I gave notice at the studio where I worked, having no idea the Russian Mafia controlled the ground there, was the reason two previous supervisors had abused me, was the reason some co-workers were abusive. I was blessed because around that time, maybe 1986, the Russian Mafia hadn't been given it's head yet against me, I think the USSR and their KGB strategists controlled operatives on the ground here, but hadn't yet sent our security agencies on a wild goose chase in the Middle East, so the USSR were relatively cautious. They were still highly aggressive, but there was a marked difference and increase of attacks against me when I was under 30 and over 30.

Under 30, the USSR worked to encourage me into criminality by having my supervisor and coworker telling me I could take the office supplies, including postage, that I wanted, and then had me work alone in an office open to me, where many of the executives left values in their office. The USSR then waited for me to steal but since I'm not a thief, I wouldn't. That enraged them, so they had operatives at my school attack me and I dropped out of school

Indeed, the reason the studio hired me was to destroy me. So many workers have no idea the Russian Mafia controls the ground here. I'd learned about the job by going to a career center board at a school affiliated with the college I'd graduated from.

Aged 22 or so, I had zero idea I was being hunted, just like most targets of the Russian Mafia don't tell the targets. Even as I type these words laying on my bed around 6:30 p.m. on 1/29/2024, under the cover of preparing the apartment upstairs for the next tenant, an operative is making as much noise as he can. Operatives awoke me around 11:20 a.m., as you can imagine when under attack for decades, my natural sleep cycle has been destroyed by the Russian Mafia.)

I was lured by the Russian Mafia with the offer of a job to a production company so they could entrap me. I theorize Carrie Fisher was lured by the Russian Mafia similarly, where foes who were really her foes swarmed her with drugs, knowing she was a recovering addict. My supervisor and a coworker say I could help myself to office supplies, including postage, so I did. I now believe the Russian government was trying to groom me into a thief. I worked alone at the company at night and nearly all the offices were opened, with expensive items laying about. It felt odd. I didn't know I was being set up but if felt weird to me. No business had ever offered me free office supplies and postage before. After I left that company I didn't take office supplies from any other firm because no other firm offered that perk to me. I think, next time, if a supervisor and/or coworker tell me to help myself, I won't do it unless the business owner gives me permission.

538

When my beloved grandmother died in my mid-20s I was greatly saddened. I'd bought her a birthday present for her upcoming birthday I never got to give to her. She died suddenly in her sleep in her early 60s, unexpectedly. She left no will. After her death, I took one of her rings, she had many; I took a few pieces of inexpensive costume jewelry and an old mirror she'd always kept on her dresser since I was a child. She had an old 22 pistol I took but I haven't seen it for many decades. I tried to learn how to use it and clean it but I never got around to it. I've no idea where it is. Because employees of the landlord have a history of entering my apartment without a 24 hour notice or warning, it's possible one of the companies employers entered the apartment years ago and stole it. In addition to knowing employees at the management company enter without notice (one man terrified me by walking into my bedroom out of the blue decades ago), I've returned to the apartment after running errands and found things moved, or doors closed or items that weren't in the apartment when I left (like medications an operative stole from me at the ER when I had a medical emergency). That's why I tend to take important things with me on my mobility device, out of fear they'll be stolen by my Russia-controlled American operatives.

Chapter 39.1 – Attacked by Russian Mafia-controlled plain clothed security at a big box store

I was lured to a big box store by an entrapped family member long before I knew I was under attack. The big box store had a return policy that initially required no receipt and customers could return items with no deadline. The big box store then changed their return policy and didn't warn customers. When I worked, I spent thousands of dollars there for extended family. When I couldn't find my receipt to return an expensive bike, they wouldn't allow me to return it and I was too busy working to find the receipt. After I was made disabled, I found the receipt and returned the bike because I had no money to pay rent. I also returned other items not knowing the store had changed their return policy, again, because I had no way to pay rent and understood the store had a return policy that allowed customers to do so.

Long before the product returns in my early 50s after I was disabled, store security harassed me and got even worse after I became disabled. Standard issue is the Russian Mafia positioning operatives in front of foods I usually buy there to block my access to the product, and just stand there blocking my access, as if they don't see me waiting for them to select a product and move out of the way. When I was spending many thousands of dollars over the course of years, plain clothed security harassed me, tried to initiate confrontations with me, picked on me, tried to initiate conversations with me as if they were a customer. One walked up behind my shopping cart as I pulled it on my mobility device and began pushing it and walking with me, like he was shopping with me, all kinds of inappropriate behavior from plain clothed law enforcement officers I'd never seen before. To harass me the Russian Mafia placed an operative in front of me in line at the pharmacy to start an altercation with me, had the operative began an animated conversation with one of the pharmacy clerks. I said nothing and the operative eventually walked away. The Russian Mafia give their operatives a heads-up after it got bus drivers to find out where I was headed.

I estimate I returned maybe $3,000 worth of items, all with receipts, because I couldn't pay rent after I was made disabled in my early 50s. That store has compromised staff. I shop there because I can't grow my own food in an apartment. Workers there have deliberately sought to either entrap me, tried to groom me to lie and expect more money on returns than I was due (one manager became annoyed when I insisted on no more than what I was due: that was the day of the shoplifting entrapment attack). Operatives there have harassed me for decades.

Chapter 40 - The Russian Mafia Illegally weaponized my immediate family to entrap me

When the Russian government had my sister recommend I apply for Unemployment even though she knew I'd already applied for Disability, had I followed her advice, I would've faced a long prison sentence. They didn't tell me they'd entrapped my sister and were using her as an "in" to get me to self-harm. The Russian government's strategy is to not declare war and to let civilians figure out they're being attacked.[237] Another entrapment attack was to get me to lie to Disability, to

---

237 General Valery Gerasimov, Chief of the General Staff of the Russian Federation,"The Value of Science in Prediction," *Military-Industrial Kurier* (2/27.2013). *"...The very "rules of war" have changed...."*

exaggerate my health problems. My retired police officer father told me over the phone, when I called to tell him the attorney at the legal assistance nonprofit advised me to file for Disability. He told me to exaggerate my symptoms, which hadn't occurred to me, and which I declined to do.[238] He didn't offer me a place to live in his home.

---

238 The Russian government deliberately entraps authority figures of victims they want entrapped, and force those authority figures to influence the target. My father who I love and have known my entire life, was a law enforcement officer, he retired as a law enforcement officer. I was 50 when he told me to lie, so he probably was retired by then. The one-two punch of having a parent tell me to lie, a parent who is a beloved authority figure in my life, combined with that parent having worked as a law enforcement officer is an example of how evilly effective the Russian government is in exploiting and weaponizing our entrapment system, as well as human hierarchy, their intention to enslave and destroy us. From childhood we're forced to obey our parents. Fortunately I was middle age when the Russian government worked this particular attack on me, but imagine this attack against a child. I've witnessed at least 1,000 children being exploited by the Russian government as their entrapped parents were forced by their handlers to use them as beards as the parent stealth harassed me, and/or the children worked with their parent to harass me. Recall the USSR began stealth harassing me when I was 11, had my mother beat me when I was around 15, and had my younger brother pick fights with me when I was 16. And my niece and nephews when they were children were set up by the Russian government's Americans operatives to become homeless and have never recovered from that attack, have substance abuse issues, are underemployed and undereducated, with only one of them earning a GED, the other two, now adults, without even a high school diploma. I believe the only reason my nephew earned his GED was because when he was a child I paid to have him tutored, and he loved the tutoring. That little bit of assistance pushed him to get his GED after he his family lost their public housing. The Russian government is attacking our children. Imagine a child's confusion when they're told by their entrapped parent to do a bad thing. A child would obey, we're taught from infancy to trust and obey our parents. When the Russian government entraps a parent it hijacks the children as well, because children are emotionally and financially dependent on their parents for survival. The Russian government's intention is to destroy the trust between parent and child, and between citizen and our government. Poll after poll shows young adults have lost faith in our government. One of the reasons is the Russian government deliberately and stealthily destroys essential

If I'd done what my father advised, I'd have committed fraud and conspiracy, which would've meant a long prison sentence, something my retired law enforcement officer father would've known before he suggested it.

Entrapping an ordinary American with conspiracy to commit a crime is a favorite of the Russian government. It means a long prison sentence; you don't even have to commit the crime, you just have to agree to the crime. That's what the Russian government was going for in their attacks against me, pushing me hard to agree with either of my parents, both who were forced to tell me to break the law, my father over the phone, my mother in her car.[239] Our legitimate government

relationships in a target's personal life, in ways our legitimate government can't see. And because civilians know nothing about this secret war, what is visible is the colliding failures as our children increasingly fail in school and increasing overdose. This is how the Russian government is committing genocide against us, by deliberately destroying the parent-child relationship, the cornerstone of our culture and our species.

239 I suspect operative's cars are bugged. I've had more than one operative suggest and/or do something shady when I was a passenger in their car. Also, something's up with the Russian government and toilet paper. The Russian government, who has worked to stealth murder me for decades but hesitated because I'm the only person in my family not entrapped, my immediate family knows and the Russian government is concerned that if I turn up dead, and my family know they've been forced to try to entrap me for decades, my entrapped family may slip in their grief and talk with non-compromised police. So the Russian government is working to kill me under the radar. One thing that stands out is toilet paper. An operative in my family has been used at least three times to bring and/or order me toilet paper. Colorectal cancer rates have been climbing and scientists don't know why. One of my relatives put something in my food that caused me to have diarrhea, the person laughed. I had to use the toilet paper there. My suspicion is the Russian government has found a way to put cancer causing pathogens on toilet paper. I don't know if that's the case and I can't prove it but what I do know is Russia-controlled American operatives hate my guts, for nothing I've done, and have worked to kill me under the radar for decades, startling me, interrupting my sleep. They are anti-Stacy

doesn't get these details from compromised American police. The Russian government leaves a lot out, because it's working to destroy us.

Another entrapment attack was the Russian government tried to get me to lie to Disability and to an insurance company, simultaneously. The Russian government had my brother closest in age to me visit, maybe once very six weeks or so for about six months. He bought me a few groceries, maybe $20-$30 worth, gave me a little money, maybe $20-$30 worth, and spent time with me. I thought he was so sweet to drive an hour to see me, buy me a little food, give me a little money, and spend time with me. I wasn't working because of health problems and spent my days alone, so I appreciated his visits even more.

Then the Russian government sprung their trap: he'd told me months ago that he'd started a new job and he asked if I'd let him take out a life insurance policy on me. That's a little strange but he explained it's his job now. He's visited me and

----

Hackney. So when an operative brings me toilet paper, drives a distance to bring it when they're sick, has toilet paper delivered to me and brings me toilet paper, all without me mentioning it, given the Russian government violence against me, I figure they've found a way to poison toilet paper and want to try it out on me. Quite a few well-loved celebrities have died young of Colorectal cancer and one of the Russian government's favorite things is to stealth murder one of our younger popular celebrities/musicians and watch our nation grieve. I believe they've stealth murdered many ordinary people, musicians, celebrities and politicians here. And I'm suspicious of the many celebrity deaths in South Korea. Any ally of ours the Russian government punishes.

been so nice, for the most part[240], so I said yes. He would be paying the premium as I was jobless.

He said I needed to tell the insurance company I was in good health so he'd get the lowest monthly rate. I'd told him months ago I was sick. I repeated it. He sneered at me and said I wasn't. It was super-weird: my brother has known me since he was born. I've never lied to him. He's never known me to lie but suddenly he's someone I don't recognize: he's the Russian government. By that I mean he's owned but I didn't know it until that moment. At that time I didn't think in terms of operatives, entrapment and slaves and being owned. Now I do because I've had a decade more of attacks but then I didn't because it was all new to me, having my family attack me.

_____

240 I'm insulin resistant which I control with diet. I told my brother. My blood sugar when I was diagnosed was 6.3. 6.5 is diabetic. After that diagnosis I gradually learned how to eat fewer carbs. I told my brother and he told me during his visits I wasn't pre-diabetic. This is something you'll notice from Russia-controlled American operatives: they gas light you. It's psychological warfare, to make you doubt yourself. That's one of the reasons they push victims to religion, they need to make people doubt their ability to steer their own lives. My brother gas lighted me and both my parents did as well. My parents are authority figures but my brother isn't. He's my younger brother, but the Russian government uses whoever and whatever can to make you doubt reality, so heads up on that. Gas lighting examples include, my brother telling me I'm not insulin resistant when I am. My father told me plains clothed security wasn't harassing me like I told him they were but said stores had experienced a big jump shoplifting and that was why I was harassed in stores. When I told him I'd spent thousands of dollars in a particular big box store over years, shopping for my extended family, so obviously I wasn't a shoplifter. My father got a bit testy with me when I said that. And my mother told me I hadn't been sexually assaulted after I told her I had been. The Russian Government is all about psychological warfare so anyone you trust enough to say things to, they'll use to gas light you, especially if that person is an authority figure in your life or the Russian government thinks they're an authority figure to you.

Also, this entrapment attempt was in my early 50s, sometime between 2010 and 2012, rough estimate. I knew I had enemies because police had been attacking me for years, but who and why I had enemies, I had no idea. I didn't learn who my enemy was until 2021, ten years after this entrapment attempt I learned who owned my brother. I knew something was wrong but the Russian government does not explain. It hits hard and mean and the target is left emotionally gasping and trying to understand what happened. My brother looked like my brother but he wasn't my brother.

He actually expected me to lie to an insurer so he could benefit from my death, while he knew I had appealed Social Security Disability and was hoping they'd believe me. I was surprised. I was surprised a lot by the Russian government in my 50s made much worse because I didn't know who was attacking me. I knew there was this force attacking me since I was around 40-41 when strangers began harassing me in stores, grocery stores, on the bus, etc., and I knew the police were part of it because they attacked me – in my late 40s I witnessed a uniformed police officer force a young man to harass me, and the officer didn't care that I witnessed what he did, but I didn't know who the puppet master was. I thought it had to be some entity with deep pockets because hundreds of people were harassing and attacking me, including the police, but I was stumped. Someone with deep pockets was the government, one of the law firms I'd worked for? The problem was I did nothing illegal for my government to abuse me and I did

546

nothing to law firms to make someone want to stealth destroy me).

I reminded my brother about my Disability appeal, he ignored that. I knew almost nothing about Disability, but it seemed obvious to me that Disability would check if people were lying. Telling an insurer I was well, while telling Disability I was disabled, seemed a very bad idea to me. My brother refused to understand.

I called my mother during his visit to complain about his demand and my mother shocked me by telling me to "do as your brother asks." I said no, and the three of us didn't speak for awhile.

Had I done what they told me (and please don't do it), I would've gone to prison for Disability fraud, insurance fraud, and conspiracy to commit both frauds. The Russian government might've offered me a plea deal and I'd have been their slave, like my some of family members are. Not all plea deals are run by the Russian government. Just like the Russian government inserts stories it wants run in our media but doesn't stop media from running other stories, many plea deals are on the up and up.

If you're offered a plea deal that doesn't insist you help law enforcement in future cases that are unrelated to yours, and if your arrest is known to your family, your neighborhood, and/or your boss, then I theorize the plea deal is from our legitimate government,[241] but a plea deal where you're asked to

_____

241 I've never been entrapped so I can't tell you step by step the process
    after you're arrested. If you have the option, seek legal advice before

help the police in any way on any case moving forward, is not a plea deal to accept. The Russian government ambushes people and targets the working poor and/or people especially vulnerable, like parents who have children and would do anything to avoid putting them in foster care, which, I theorize, is why the Russian government targets families. Entrap one parent, the entire family is entrapped, something the Russian government won't admit to anyone taking one of their scam plea deals.

Chapter 41 - The Russian Mafia attacks an American child: Amber -- the making of a weapon

When the child was a toddler of around two, her primary day-to-day caregiver, an older sister, graduated high school and left home for college. The toddler (I'll call her Amber) was devastated and clung ever tighter to her beloved mother, who worked full time at a factory. Amber was an adorable baby, beautiful, sweet natured, obedient. Although her oldest sister had moved to college housing, the college she attended was in

accepting a plea deal. I suspect many people don't have that option. I know people are guaranteed counsel but I think that's in certain crimes. I don't know for sure, but it seems likely the Russian Mafia will have noncriminals charged with crimes that don't guarantee them counsel. Russia controlled entrapment setups aren't, based on what they've so far tried to entrap me using, violent crimes. The overwhelming majority of entrapments the Russian government attacked me with were theft-related crimes with our government's various social safety net programs the victim, or bank or insurance fraud, white collar crimes. They also tried drug related entrapments to set me up to be a drug mule, to buy drugs and/or hook me on drugs by having operatives leave a drug in my apartment after they left, hoping I'd be curious enough to try it. Those types of crimes, had I bit, would've included conspiracy charges because the Russian government had their operatives suggest many of those crimes to me. I'm noncriminal, which is the group the Russian government is targeting in America.

the same city, so Amber and her sister maintained a close relationship and enjoyed many outings. Still, Amber's oldest sister no longer lived at home.

The Russian government exploited Amber's lack of protection, her mother's need to work full time, and the family's ignorance that they, along with the majority of Americans, were under stealth attack by the Russian government. Russian-controlled American op's ordered that Amber be psychologically and sexually attacked, using Amber's entrapped half siblings, an entrapped cousin, a neighborhood child, and school bullies to destroy Amber's childhood from the age of about seven on. The hard core attacks began I believe after Amber's sister graduated from college and moved to Los Angeles to find work. Amber was about six years old when her sister moved to LA. Her sister lived again, briefly, with Amber before she moved to LA. Amber mentioned no sexual assaults or bullying to her sister when she was around 6. While in college her sister volunteered at Amber's kindergarten to spend more time with her. She seemed well adjusted, happy, not bullied. That all changed.

Amber is the youngest of five children on her mother's side, was conceived by her divorced mother in her early 30s. Amber's mother and her boyfriend, Amber's father, decided not to marry. Amber's father is a widower who had three underage children at home when he began a relationship with Amber's mother. On her father's side, Amber has two half brothers and a half sister she occasionally visited at her father's home.

At some point during her childhood when she was under the age of 10, Amber was sexually assaulted over the course of several years by her maternal half brother, a paternal half brother, a visiting maternal male cousin, and a female neighborhood school friend. Amber told her mother, who was her confidante. Her mother stopped the attacks. The brother who sexually attacked her is entrapped. Most of Amber's immediate family is entrapped.

Amber was bullied in elementary school. She was bullied in high school by several girls. She was bullied by two maternal half siblings, a sister and a brother, who are entrapped. Amber's mother is entrapped. When Amber was a teen she worked part time at a fast food restaurant where her brother was her supervisor. He psychologically abused her so horrifically, her menstrual cycle became irregular, and she'd shake in fear of him. He, I believe was deliberately sicced on her by his handlers. Her maternal female half sibling is a beautician. That sibling abused Amber so horrifically, she was traumatized for well over a decade.

Amber developed health problems in part from the intense psychological abuse, and also because of physical ailments. Diabetes runs strong on her father's side of the family and unknown to her for years, she was diabetic, although not diagnosed until her 20s. Diabetes attacks her teeth, and she has a serious eye disease. Her mother had a difficult menopause and physically assaulted Amber, complained to Amber's siblings about Amber, which made her brother and sister behave even more cruelly towards her.

Yet despite all that, Amber loves her mother, she loves her siblings, and avoids the cruel ones as best she can. Her older half sibling in LA wasn't Amber's confidante, Amber's mother was her confidante. What Amber didn't know was that much of her family is entrapped by the Russian government, and those attacking her were told to do so. Why target children? Well, the Russian government manipulates innocent children very easily. Most children want love, a secure, comfortable home, food, school, and to not be bullied. When you strip those benefits and protections from young children, they are traumatized, looking for a savior, a rescuer, and cling to that person. The Russian government deliberately traumatizes children in preparation to use them as weapons. The spike in our mass shootings are, I believe, are a reflection of the Russian government's stealth attack on our children. I say I believe it because I can't prove it but I strongly believe it is so. Entrapped parents, I've seen for myself as I was attacked by strangers in the street, are forced to groom their children, to see them as tools, to use them as weapons. I know because Russian-controlled American ops have forced entrapped parents to bring their children along with them, to use as beards and worse. The parent(s) are forced to harass me in view of their children, as I sit on my mobility device. Using their parents in this way, the Russian government grooms some of our nation's children to believe that cruelty to a disabled woman is normal. I've seen for myself by the many attacks I've endured by Russian-controlled American ops, that parents aren't teaching their children ethics. I believe the Russian government has inserted itself into the parenting of some of our children[242], and that entrapped American parents are unable to protest because they are now slaves. I suspect that's why the Russian government has stolen tens of thousands of Ukraine's children since the Russian government bombed Ukraine. I suspect that's why our nation is experiencing a spike in mass murders done by young people the last multiple decades. The Russian government targets children, makes them weapons the Russian government controls then unleashes them.

As a child, I was attacked by the Russian-controlled American ops affiliated with the KKK in the city where I lived. Amber too, was deliberately attacked in her childhood.

The way Russian-controlled American ops work here is they pound on targets, ordinary people, women, men, children, just pound away, and never say why. Amber to this day has no idea why some of her siblings were and are cruel to her. Russian-controlled American ops apparently aren't allowed to explain why they're attacking their victims, and all the Russian government cares about is that our people are pounded into dust.

_____

242 My mother beat me when I was around 15 out of the blue for no reason. I now suspect that she was ordered to do so by her handlers and if I'm correct, my mother has been entrapped since she was 33 or earlier.

Chapter 41.1 - The Russian Mafia hates women & deliberately destroys girls

The Russian government makes attack dogs out of people. They make people, especially girls, feel as threatened and frightened as possible, then hook them up with a "white knight," a so-called protector, who is also a victim of and owned by the Russian government. People are so beaten down, and have no idea why or by who, they don't know whether they're coming or going. Nothing in their life is easy or works as promised. When they behave in positive ways, the Russian government makes sure they receive negative results. And ordinary people have no idea their quality of life is being deliberately destroyed. Because the Russian government does all this stealth,[243][244] and they're effective at it. I believe that's how the Russian government got so many ordinary people to participate in the attack on the Capitol.

The Russian government works to keep people emotionally and psychologically off balance and miserable because miserable people can get rid of their government and have no idea they've been used as a weapon by an enemy government. The Russian government loathes Americans and migrants who live here, and most Americans have no idea the

243 General Valery Gerasimov, Chief of the General Staff of the Russian Federation,"The Value of Science in Prediction," *Military-Industrial Kurier* (February 27, 2013). General Valery Gerasimov states: "…. Wars are no longer declared …" "Among such actions are the use of special-operations forces and internal opposition to create a permanently operating front through the entire territory of the enemy state…."

244 Shaun Walker, Pjotr Sauer and Tom Phillips "Panic and Emotional Pain as Alleged Deep-Cover Russian Spies Vanish," *The Guardian* (April 3, 2023).

Russian government knows our favorite foods and controls what some of our grocery stores stock[245] – or control some of the stores in Southern California. I can't speak to all the stores in the country but I know the Russian government controls grocery stores stocks here so I assume California isn't the only state the Russian government controls to that level.

Keeping Americans and migrants as bewildered as possible, as overworked and as stressed as possible, Russian-controlled American ops can form those victims into a weapon of rage. You saw that rage aimed at the Capitol on January 6th. I believe there are millions of innocent people the Russian government is deliberately enraging, with the intention of destroying our people and our government.

I know about the Russian government's rage building strategy because they used it against me, then tried to exploit the rage they thought they'd built to get me to aim it at a boss they'd forced to be cruel to me. All to entrap me. To destroy me. The Russian government is a monster. And we're in their sights. Our allies are in their sights, and our neighbor countries as well, who the Russian government and it's Mafia are deliberately destabilizing. The Russian government and it's Mafia spend so much time destabilizing the world it's no wonder their own people are neglected and oppressed.[246]

---

245 The Russian Mafia lured to a specific store, one easier for their operatives to harass me.
246 Anton Troianovski, Yuliya Parshina-Kottas, Oleg Matsnev, Alina Lobzina, Valerie Hopkins & Aaron Krolik "How the Russian Government Silences Wartime Dissent," *New York Times* (12/29/23).

Chapter 41.2 – Post-attack the USSR/Russian Mafia push victims to religion

Mom turned to religion for comfort,[247] led there by the USSR's operatives working stealth. A woman showed up and befriended her and assisted her into religion, where my mother has remained ever since. My mother doesn't know some of those people were operatives, she thought they were her friends because the Russian Mafia lies to good people.

To this day my mother doesn't know she was steered to religion. After the Russian government sucker punched her they pushed her towards religion because they can more easily control people in specific types of religions. My mother lost her beloved grandmother, was beaten unrelentingly and threatened with rape by her mother's boyfriend, yet in true American heroine fashion she pulled herself up by her bootstraps and learned a skill, a trade, only to be entrapped and enslaved by the Russian Mafia,who forces her to try to entrap her last non-compromised child. That same Mafia is facilitating the deaths of thousands of innocent Palestinians and is the very same Mafia intending to seat one of it's puppets in the White House in November 2024. Over my dead body. Even if they murder me, I'll have published this book of warning so my people, my species can save themselves.

To acquire billions of potential soldiers, the Russian government need only control the Chinese Communist Party and Mr. Modi's party. Controlling the leader means the

---

247 Ruth Graham, Charles Homans "Trump is Connecting with a Different Type of Evangelical Voter," *New York Times* (1/8/2024).

Russian government controls all who that leader controls. That's why you see China is now a surveillance state: the Russian government wants to control every one of it's assets. It views people as things it owns. It isn't selling cheap oil, etc. to China and India out of the kindness of it's heart. The Russian government will take Chinese and Indian people and use them as hackers, soldiers, anything, everything. And the Chinese Communist Party and Mr. Modi's party will let them, because the leaders are owned.

People accustomed to being flock are much easier to control than people who've not been oppressed. If you read about ordinary people in modern Russia in the newspaper, they're so oppressed most dare say nothing.[248] The Russian government doesn't even bother reporting an accurate number of Russian soldiers who die in Ukraine. If the soldiers aren't reported deceased the Russian government doesn't pay out benefits to the dead soldier's loved ones. That level of oppression doesn't happen by accident. It happens because the Russian government is a very hands-on slaver.

The Russian Mafia is a racist, sexist, homophobic slaver government who sees itself as the world ruler and everyone else as subordinate. It is truly evil. Is in the process of making it's dream of world domination a reality. My mother raised me to accept people, to not be racist, sexist, homophobic, etc. but decades after she taught me these basic truths, she's now entrapped by an evil Mafia government who

---

248 Ibid.

feeds her lies and partial truths and that's the world she's trying to navigate. My mother is magnificent and the Russian Mafia has made her a slave. They are evil.

As you see by Amber's story, the Russian government isn't hands on just in Russia. They're very hands on in America. Most Americans have no idea the Russian government and it's allies control the electricity into our apartments and homes; control who we can reach on our cell phone, control what products our grocery stores stock, control the quality of teaching we receive in school, control the quality of medical care we receive, control whether we're accepted for affordable housing, and control our freeway traffic. I suspect the Russian government also controls one of the banks the State of California uses.

Amber experienced unstable housing for most of her life. Housing in Southern California is expensive and, although Amber's mother is compromised, people the Russian government entraps are still harassed, traumatized and agitated, Russian-controlled American ops keep them off center, anxious, frightened. The Russian government loves to frighten innocent people. It's one of their favorite things. Frightening Americans and migrants in America, they love that, to have Americans under their boot.

My theory is Amber and her mother received the benefit of affordable housing when they agreed to try to entrap me. Amber graduated from university and graduate school despite a serious eye disease that made it very difficult for her to see consistently and clearly. Yet she fought her way through to graduate. Her husband also earned a masters degree.

Amber has multiple health problems and the Russian government controls some of our medical care providers. They attempted to hook her on opioids to make her more amendable to entrapping me (another thing the Russian government loves to do is hook people, especially Americans, on drugs). Russian-controlled American ops had Amber's surgeons fail to satisfactorily perform 5 to 7 knee and other surgeries, over about four years. Amber lived in excruciating pain. Even now she has pain. She was forced to rely on her husband for basic care. He also has health conditions.

After the failed surgeries and years of pain, Amber now protests less, is my theory, when she's tasked with getting information from me for her handlers or attempt to entrap me. Recently she was used by Russian-controlled American ops to get the names and dosage of my newest prescriptions. She did that by telling me she was in pain. Her handlers knew I'd recently been to the ER and to the doctor's office and they wanted the names of the new medications I'm taking to, I suspect, swap in a few fentanyl tainted ones. I've experienced tampering of a prescription pre-pickup from a pharmacy before. With my new prescriptions, neither the medical office nurse, nor pharmacy cashier I spoke with told me my prescriptions had been sitting at the pharmacy for weeks before I picked them up. This particular pharmacy is closed on weekends. It'd be nothing for Russian-controlled American ops to get into a pharmacy and slip in a few similar looking fentanyl tainted pills into my new prescriptions. Many people don't even look at their medications when they take them.

Russian-controlled American ops wanted to have the right size and strength of my medications so I'd notice no difference. I'd just not wake up the next day, making the Russian government very, very happy.

Chapter 41.3 - Amber is enslaved by the Russian Mafia

The Russian Mafia are woman-hating pimps, blackmailing their way to world domination. Nearly worldwide women's rights are in decline as a direct result of the Russian Mafia's ascendancy.

An example of an attack: Amber let me know she'd been trying to call me. Because I'm harassed so unrelentingly by Russian-controlled American ops when my phone is on, I usually have it off, which the Russian government exploits by having those it doesn't want me talking to try to reach me via phone when they know my phone has been switched off.

People I know email me and I call them. When I called Amber she asked me to hang up so she could call me on her house phone. She said her cell phone drops calls. I acquiesced. When my phone rang Amber's name didn't register but another woman's name did so I didn't answer. After going back and forth, Amber said that must be the name of the person who'd had the number before her and that she'd go into her computer to insert her own name. I said I thought this was her home phone. Her home phone goes through her computer, so it's hacked. Amber doesn't have to tell her handlers the medications I'm taking, they've hacked her computer, and had her tell me to hang up and let her call me back. Her handlers had her explanation all ready when she said calls are dropped on her phone. Russian-controlled American ops have their cover stories ready, and those cover stories make sense.

After living years in excruciating pain, Amber does what her handlers tell her to do. How do I know? Amber and/or her husband have been used to try to entrap me. Multiple times. I've been hunted so long by the Russian gov, I now recognize entrapment attacks. I've been alive nearly 64 years. For 53 of those years I've been targeted, threatened and attacked by the Russian gov. If there's an expert on stealth harassment and entrapment attacks by the Russian government against ordinary Americans, I'm probably that expert.

I'm not dead because when Russian-controlled American ops offered me free drugs of choice I politely declined; when surgeons they own pushed me to have an unnecessary hysterectomy, I refused. When their ops offer me sugar and carb laden foods in the ER, I politely refuse. When their ops try to lure me out of my apartment to be attacked, by throwing an impromptu rave, twice, north of Wilshire in a well off city, in a part of the city not known for raves, and police don't show to break it up, I remain inside. And when I'm concerned my medications or water or anything the Russian government can access in America has been tampered with, I don't take those meds and I drink bottled water. There are medications I need but the Russian government has demonstrated in a very clear way to me, that they control the ground in Southern California. I'm determined to write this book. I must warn you. America and the world is full of baby Amber's and I can't do nothing while children's lives are destroyed.

Amber's path from innocent babe to a weapon deployed by the Russian government is ongoing. She has no idea that anyone has been a puppet master in her life. I suspect most victims here and around the world have no idea the Russian government is responsible for destroying the life they should have had. Amber told me she's considered suicide. Many victims of the Russian government have suicided, unable to live the unbearable life of being used as a weapon to harm their friends and families. So many South Koreans have suicided I'm suspicious the Russian government or it's allies are involved. I'm grateful Amber is alive and with me. Even though she's compromised, I'm grateful she's alive, no thanks to the Russian government who would kill her except they might need to use her in some way. None of this was her idea. None of this was our police's idea. None of this was the idea of compromised American and migrant parents. This evil is the Russian government's idea.

I know some of Amber's life, and I know some of my mother's life, and I know how the Russian government has attacked me for decades. Russian-controlled American police, though some are on-the-ground managers of the Russian government coup here, are also entrapped. This coup won't stop until we begin reporting it so everyone knows the evil the Russian government does.

Chapter 41.4 - The Russian Mafia deploys Amber as it uses her to attack me

Amber was used to push me to accept an op as a roommate. I refused. She and her husband were used to get me to a theme park's Halloween event where I was harassed so unrelentingly, I was traumatized for days. Amber was used to try to entrap me into unemployment fraud, suggesting I apply for it after I'd already applied for Social Security Disability, which she knew. Had I done as she suggested, I would've been charged with both Unemployment and Disability fraud. When you submit a claim for disability you're telling Social Security Disability you're unable to work. If you subsequently tell Unemployment you're available to work, you're lying to one of them, they may assume you're lying to both of them.

Amber recommended I apply for Unemployment after she'd recommended a law firm I hired to help me appeal my disability denial. I had no idea how to appeal my denial, so I accepted her recommendation. That firm agreed to represent me and Russian-controlled American ops used my gratitude to Amber, that I now had legal help thanks to her recommendation, to use her to set me up for Unemployment and Disability fraud, where she advised me, out of the blue and unprompted, to apply for Unemployment.

Chapter 42 - I charge the Russian government, the Russian Mafia, & the USSR with genocide against the American people

I, Stacy A. Hackney, African American citizen of the United States of America, on March 6, 2024 charge the Russian government with facilitating genocide against my people, who are the people of all ethnicity's in the United States, including Native Americans, immigrants living in America, legally or illegally, between 1971 to the present, as well as people incarcerated in America between 1971 to the present. The basis for my charges include:

1. The Russian government and it's Mafia and their allies, and drug cartels, deliberately target and destroy individual Americans via the drug trade, specifically by facilitating the fentanyl drug flow into America, causing the deaths of, on average, hundreds of Americans and/or migrants per day, most who are in the prime of their lives, which the Russian government does with the intention of: (1) emotionally devastating families, communities, and our nation; (2) significantly reducing the number of military personnel available to serve our nation in current and future wars; (3) weakening our nation's financial strength and making it harder for us to fight the climate catastrophe; (4) making it harder for us to fulfill our pledge of financial assistance to impoverished nations overwhelmed by the climate catastrophe; and (5) weakening our species' ability to survive the climate catastrophe;

2. The Russian government targets and destroys families in America by entrapping parents and forcing entrapped parents to entrap their own children. The Russian government's intention is not only to destroy our families and our nation but to destroy our culture, because the cornerstone of our culture is our families;

3. The Russian government compromises, corrupts and entraps some of our law enforcement with the express intention of forcing them to run entrapment attacks against ordinary citizens and/or migrants, facilitating the Russian government's destruction of my people. After ambushing and entrapping noncriminals Russia-controlled American operatives then force, blackmail, influence and/or coerce some of the victims to vote for candidates the Russian government wants installed into positions of power in furtherance of their coup against us, as well as force some of the victims to further assist compromised police to entrap more noncriminals;

4. The Russian government forces police to ambush and entrap parents with children and/or the family's primary bread winner and/or care provider. After the entrapped parent accepts a deal (in order to avoid having their children's home life be disrupted, and to avoid having their children sent to a possibly abusive foster care environment, all the negotiating and offers to the entrapped parent made by uniformed law enforcement and/or persons with ID representing that they are authentic law enforcement, when in fact, the Russian government controls these specific entrapped American

officers), victims are groomed and blackmailed. After the negotiating is done and the deal made, entrapped officers, entrapped parents, entrapped bread winners, and/or entrapped care providers, groom the children (and/or adult vulnerable, dependent family member) into a stealth child army (and/or adult weapon), forcing children, teens, or vulnerable adult to engage, harass, bully, influence, befriend, obtain private intel from, etc., Russia-targeted children, family, and/or adults. The compromised child is taken along with the entrapped parent and used as a beard as the parent stealth harasses a target, and/or the child assists the parent in harassing the target; and/or the vulnerable adult financially and psychologically dependent on the bread winner and/or care provider is groomed into a weapon by Russia-controlled American operatives to target whoever the Russian government wants destroyed. I've been attacked in these ways, which is how I learned about them. The knowledge was painfully learned. I hope to spare you from the pain of betrayal and just gift you with the knowledge. I'm theorizing about the negotiating because I've never been entrapped but rather have decades of experience being attacked. I pieced together the parts I haven't experienced using the parts of the experience I endured.

5. The Russian government facilitates homelessness in America in a combination of ways, including (1) using Americans they control to resist affordable housing being built in specific neighborhoods, (2) flooding our nation with fentanyl, (3) using drug dealers to offer free drugs of choice to

get people hooked, (4) using teen and/or young adult influencers to encourage friends/acquaintances to experiment with drugs, (5) having drug dealers offer drugs to poor clients on credit to get the client hooked, (6) using supervisors and bosses they control to stealth harass workers to make them miserable, and so more vulnerable to operative influencers encouraging overworked, stressed employees to reach for peace and calm they hope to find in drugs, (7) facilitating the lack of affordable housing by encouraging investors to buy out starter home housing stock, (8) encourage landlords they control into greed and cruelty, using them to push out low income renters the Russian government wants destabilized, and (9) the Russian government hides it's presence here making it extremely difficult for ordinary people untrained in psychological warfare to connect the dots and understand the Russian government is working a coup against Americans.

6. Psychologically destabilizing groomed children of entrapped people even further and turning them into mass murderers targeting ordinary people in grocery stores, jogging, etc.;

7. The Russian government entraps business owners, supervisors, and managers and force them to stealth harass employees;

8. At grocery stores the Russian government controls, it directs security to harass customer(s) the Russian government is targeting;

9. The Russian government corrupts and entraps our medical professionals who are directed to prioritize earnings over patient health, encourage procedures and surgeries the medical provider knows isn't in the patient's best interest, use compromised doctors to encourage patients into self-harming choices, including encouraging a patient to lie to Medicare for gain.[249] I've had family members undergo repeated unsuccessful medical procedures, I believe, to get them hooked on pain medication to make them easier for Russia-controlled American operatives to control, in my case, to get loved ones to try to entrap me. I've experienced luring, grooming and/or entrapment attempts by more than one doctor.

10. I believe the Russian government deliberately hooked one of my loved ones on opioids for the sole reason that she was my favorite aunt and they'd failed for over two decades to entrap me and tried to use her to do it. They tasked her with getting me to be a drug mule, which is impossible for anyone to do. After she failed, they overdosed her on opioids. She was a cancer survivor and they used her on two entrapments and killed her to emotionally destabilize me. The

---

249 Russia-controlled American operatives well know after decades of their failing to entrap me that I'm uninterested in cheating anyone, and the Russian government's operatives have hacked my phone where I discussed this issue with my mother on Whatsapp, so now they have my GP offer to lie to Medicare for me in order to stealth harass me and make me avoid seeking our needed medical care. They did another version of this type of attack with my heart specialist doctor. They'll have the medical care provider behave in an off-putting way. After decades of their attacks against me using my medical care providers, the Russian government now have them deliberately make me uncomfortable so I'll avoid going to the doctors, to speed up their murder of by "natural causes."

Russian government I believe killed two more of my loved ones, my paternal grandfather who I was close to and who they killed after they heard him tell me he'd taken out a life insurance policy and left me the beneficiary. By that time, in my late 40s, the Russia-controlled American operatives had decided to work me to exhaustion and they didn't want me to go live with my grandfather, which is what I would've done. So they murdered him, devastating me. The last murder of a loved one was a long ago mentor they murdered in May 2023. I know they killed him because of how they let me know. They'd used the manager of this apartment to noise harass me and startle and frighten me so badly my heart rate elevated to a high rate necessitating a trip to the ER. I had two ER trips and the next morning went the pharmacy to pick up my prescriptions and operatives made sure I learned about the death that morning, to emotionally destabilize me so I couldn't complete the course. I know this sounds crazy but Russia-controlled American operatives go into a frenzy of attack when I take programming classes. Decades ago when I was in engineering school they attacked me, and I'm learning beginning programming now and they've done clusters of attacks to get me to fail two finals, a midterm (by turning off my hotspot internet, despite me being paid up for over 6 months into the future) and a quiz fail. Something about me learning engineering, math and/or programming puts them into attack mode. In Spring 2023 they had me attacked many times while I was winding up the reading and class assignments.

During Finals week I was attacked, rough estimate, at least 10 times, often with noise attacks. I'd start a quiz and my neighbor would suddenly start playing his movie very loudly. The day I had to go to the ER, the manager of the building where I lived walked right up to my front door to suddenly loudly noise startle me. Something about startling me, Russia-controlled American operatives have been attacking me that way for decades. When I go out, they'll do setups where the bank I'm approaching has a car alarm going off as I arrive (the attacks are ubiquitous, normal seeming, everyday), police will see me and suddenly start their sirens, firemen on their trucks suddenly turn on their alarms, it's not something I can prove, it's stealth noise harassment, and it's really obvious to me so I always carry ear plugs when I go out because Russia-controlled American operatives invariable attack me with noise. To emotionally destabilize me during finals week Spring 2023, the Russian government let me know they killed my long ago mentor and made sure I knew Thursday morning. My Final program was due the next day, I'd spent hours in the ER unable to sleep. I had a quiz plus I needed to complete some reading. Russia-controlled American operatives have tortured me for over a decade by deliberately interrupting my sleep cycle, I can't sleep normally and because of the attack by the manager, I hadn't slept for maybe 36 hours. At the ER I'd asked the doctor for a sleep aid, he said he would and then he didn't, meaning Thursday morning in June, I think the 1st, I got out of the ER, was told someone I loved had died, I needed to sleep and I needed to complete my Final program, complete a final quiz and complete the final reading. Without sleep I couldn't do any of those things. The prescriptions the ER doctor wrote did nothing to help me get sleep. It just so happened that months ago I'd dropped an ambien pill on the floor and hadn't picked it up. That day I picked it up, wiped it off, took it and that was how I completed my Final, my quiz and my reading, despite the operative neighbors next door suddenly playing their movie loudly just as I began my timed quiz. These attacks work against me because the Russian government has eyes on me inside my apartment. I have no idea how they do it but I know that they do. The upstairs operative within the last several months would come stand in their apartment, over my bed, as I typed a negative comment about the Russian government. Sometimes they'd drop something heavy after I typed something about the Russian government. Any time, day or night, they'd drop something, make noise, drag their furniture when I typed a negative comment about the Russian government. It was obviously retaliatory.

11. Something about my having scholastic success threatens Russia-controlled American operatives because they go out of their way to noise attack me at critical times of study or testing. Since these people are attacking me extra judicially, no one attacking me has ever bothered to explain why. The Russian government is embedded here illegally. There's nothing they can say to justify attacks on civilians in America.

And the American operatives who I believe are law enforcement, I've no idea how they can believe I'm a criminal. It's obvious I'm not.

12. Deliberately destabilizing and demoralizing people to make them turn to religion, where the Russian government can more easily control them by controlling their religious leader;

13. Deliberately inserting an Russia-controlled operative into a divorced mother's sex life, which mother had four young children, for the express purpose of destabilizing and destroying that family's ability to remain housed in that neighborhood: the Russian government hybrid gang consisting of the KKK and local police wanted me and my family removed because when they had Caucasian children they controlled hit my younger sibling and/or threaten to beat me up, I stood my ground. I was 11 when the KKK/Russian government hybrid gang[250] had a maybe 12 year old slightly bigger, older boy attack my 10 year old brother, and they had a 12-14 year old bigger, older boy threaten to beat me up when I was 11 on the school bus after school. With the bully behind me as I exited the bus, when I picked up a brick to defend myself, to them that meant war, even though they started it, and were embedded here illegally, they've attacked me my whole life, as part of their coup, doing their best to put me into the criminal justice system, having a cashier at a big box store

---

250 Janis Berzins "The New Generation of Russian Warfare" "...the Russian view of modern warfare is based on the idea that the main battle space is in the mind." Aspen Institute | Prague, *Aspen Review* (March 2014). https://en.m.wikipedia.org/wiki/new-generation-warfare.

failing to ring up the food in my grocery cart so the store security could accuse me of shoplifting. The cashier was stunned when I returned to tell him he hadn't rung up all my items. He asked me if the exit employee told me and I said no, I saw for myself and returned to tell him. The Russian government is here, is mean, and it intends to destroy us. I survived to warn you.

14. The Russian government used the opportunity to destabilize my family to convince the KKK/police alliance that the Russian government's operative was highly effective in removing a family the KKK wanted removed from the neighborhood; thus embedding itself deeper into a gang in Southern California comprised of Russian government operative(s), local police, and a local branch of the KKK;

15. Using Perfidy (lies/misrepresentations), which is prohibited under the rules of war, and thus is a war crime, a Russia-controlled American operative represented himself as a civilian, noncombatant to a civilian mother of four young children, with the express purpose of removing her from a neighborhood the KKK, the local police and a Russia-government operative wanted removed from the neighborhood in Southern California in 1971. Not only was 1971 a time of peace in America and so the targeted woman had no way to know she was a target of a foreign enemy government of the United States, but even if she had known she was under attack by a foreign enemy government, civilian woman are given special protections under the 1949 Geneva Conventions. So the

Russian government's operative contravened those 1949 Geneva Conventions to lie to and so rape (sex under false pretenses means the civilian did not agree to have sex with a Russia-controlled operative). The Russia-controlled operative was infiltrated into the woman's life by a palm reader and a recent "friend," she'd made in Southern California. The 28 year old civilian woman was attacked for the express purpose of destroying her career and removing her from a neighborhood where she'd been buying a modest home and raising her four children. The attack/removal was, I believe, by request of the local KKK/USSR hybrid gang families in the neighborhood who wanted her removed. These types of stealth attacks under the guise of romance by the Russian government's operatives continue to this day.[251] Those families

---

251 Shaun Walker, Pjotr Sauer and Tom Phillips "Panic and Emotional Pain as Alleged Deep-Cover Russian Spies Vanish," *The Guardian* (April 3, 2023). The USSR and the Russian government have been running these types of stealth attacks for many decades. My mother and her children were the victims of such an attack around 1971. Campos Wittich was a Russian spy living under an assumed name in Brazil. At least six such suspected spies have been unmasked… in various locations over the past year, suggesting there could be….more. He was a deep cover spy working for an elite intelligence programme. "She (the victimized girlfriend) suspected nothing at all," said the friend of Campos Wittich's partner." "She is just a lovely woman looking to create a family and get married to a man she thought was the man of her life." "She is really scared of this situation and hurt by all the pain of having an abrupt cut-off in a relationship that was perfect in her eyes," said one of friends. The victimized girlfriend of the spy is a veterinarian. She had no suspicion she was involved with a Russian spy. The romantic partners of both spies have been left devastated, said their friends and acquaintances in Athens and Rio."...Campos Wittich may have been in an earlier 'embedding' stage of their missions, which could have gone on for decades if not detected," reporters state."...are now in detention, including a Russian with a Brazilin cover identity, jailed in Brazil after a decade-long mission in which he obtained a Master's in the U.S….."

disapproved of two incidents involving the woman's children: (1) when a Caucasian boy of approximately 12 or 13 hit one of the woman's younger children, a boy of 10, his sister, a child of 11, hit the attacker back. The children had been playing in the attacker's front yard, had had no previous hitting episodes, but after the attack promptly left and informed their mother, who directed them to no longer play with the child who started the fight; and (2) a Caucasian boy believed to be in one of the KKK's families, who was 12-14 years of age, a boy bigger and older than the woman's oldest child, who was a skinny 11 year old, me, threatened to beat me up on the school bus as we were heading home after school.

16. I was surprised by the threat but composed: her mother had explained to her that the KKK meant some people hated others because their skin was darker. The child believed the threatening boy must be one of the people who hated her because her skin was darker than his. At their bus stop, a field, the girl exited the bus with her book bag and school book, looked around the field for a weapon to defend herself from the threatened beating, spotted a brick, walked to it, picked it up and turned to face the child who threatened her. At that moment the Russia-controlled operative, her mother's boyfriend, honked his horn and waived her to his car. He'd never meet her at her bus stop before and he never did again. She dropped the brick and walked to his car. He asked her what had been happening and the girl said a boy on the bus said he was going to beat her up, so she picked up a brick to

defend herself. When the operative took the child home he informed her family about what had happened. The child (me) remembered all these years later because the operative made me sound like I was some hellion. I wasn't. I was a quiet, introverted child. The reason he labeled me as such was he knew he was going to destroy my family, so he lied about me and created an image about my 11 year old self for the KKK/Russia government hybrid gang that wasn't true, but which served their purposes. Many decades later, after I've been lied on by other operatives, I now know how the Russian government misrepresents the truth to further their agenda. He was told to be a my bus stop by his handler, to prevent the boy threatening me from being hurt, so he was there. The operative then misrepresented to the USSR, the KKK, the local police and my family what had happened. I was 11. I had no idea what he was doing or why, nor did I have a way to counter him. He was a Russian government operative. I was a child. From the threat from the bigger boy, and the lies the Russia controlled operative told my family, to the present, the Russian government has been stealth attacking me. My government, I don't believe, has any idea of the danger I'm in nor why. It's because the USSR embedded operatives here beginning last century and set then into ingratiating themselves into systems like the KKK,[252] and destroying families with the intention of destroying Americans. The Russian government intends to kill

252 Janis Berzins "The New Generation of Russian Warfare" "...the Russian view of modern warfare is based on the idea that the main battle space is in the mind." Aspen Institute | Prague, *Aspen Review* (March 2014). https://en.m.wikipedia.org/wiki/new-generation-warfare.

me. I'll do my best to stay alive but I can't expect I'll succeed against a trained operative(s).

Chapter 43 – One of our species most powerful mafias

As Ms. Dawisha implies in her book *Putin's Kleptocracy*,[253] and Mr. Unger implies in his book *House of of Trump House of Putin*,[254] the Russian Mafia is increasingly a major player in the Russian government, and it's my opinion Mafia's don't behave a citizens expect would prefer. Mr. Putin is called President but he's actually the Don of the most powerful Mafia in the history of our species. He's the richest man in history, reportedly[255] receiving a portion of every deal, including revenue from drug flows, weapons flows, and human slavery, every deal the Russian Mafia makes.

Unlike in the movies where Mafia's are shown as out-powered by the State, the Russian Mafia, through it's Don, controls thousands of nuclear weapons. Americans don't need to be afraid of what the Russian Mafia controls in Russia but our people must understand what's at stake for us.

To the Russian Mafia, all governments are competitors and competitors are to be destroyed and/or enslaved. That's a problem for us because we're uninterested in becoming slaves,

---

253 According to Karen Dawisha in her book *Putin's Kleptocracy* (Simon & Schuster, 2015) "the FSB (Federal Security Services) has 'absorbed' organized crime …" She also states "….Russia under Putin had become a virtual 'mafia state' in which state structures operate hand in glove with criminal structures to their mutual benefit, …"

254 According to Craig Unger in his book *House of of Trump House of Putin* (Penguin Random House UK 2018) As Oleg Kalugin, former head of counterintelligence for the KGB told Mr. Unger…. "the Mafia is one of the branches of the Russian government today."

255 Ibid.

yet that's exactly what the Russian Mafia has planned for us, enslavement and/or death.

I know this because since I was an American child of 11 and 12, in 1970-71, the USSR entered my life to destroy me and my family, even though I've lived in America my entire life and even though the USSR signed the Geneva conventions, anti-genocide laws and other rules of war promising they wouldn't attack children under 15, civilians and/or women during times of war. During times of peace the USSR entered my nation and attacked me, against the domestic laws of my nation, and against the international laws to which it was a signatory.

This was before the Russian government, then known as the USSR,[256] formed an alliance with the Russian Mafia. If you know anything about Mafia, you know it's a criminal organization, and that it ignores laws put into place to aid ordinary people survive. The Mafia is beyond a gang, is far more powerful and violent than most gangs. Mr. Putin, President of the Russian government, and Don of the Russian Mafia, the most powerful Mafia in the history of our species, supports Mr. Trump's candidacy for United States president.

In Ecuador, ordinary people are having their quality of life destroyed by gangs.[257] Given my own personal, first hand experience of the Russian government's lies, I have zero confidence in any promises it, or it's Mafia, make, and given

---

256 In 1970-1971.
257 Annie Correal "Ecuador's Attorney General Took on Drug Gangs. Then Chaos Broke Out," *New York Times* (January 13, 2024).

that the USSR itself showed me who and what it was beginning during my childhood, it makes absolute sense to me why the Russian Mafia's preferred candidate is leading in the polls.

This is bad news, but there's worse: the USSR bequeathed to the present Russian government a coup and genocide attack against Americans already in progress, when the Russian government came to power in the 1990s. This is why you find part of our government listing, and why so many seemingly sane people are backing a pro-Kremlin candidate.

Rough estimate, the 600,000+ Soviet immigrants, spies, and/or slaves, the USSR embedded here beginning in the 1970s via the Jackson-Vanik Amendment[258] have not been sitting on their laurels. Some of them have been busily destabilizing our nation. If we don't get a clue real soon, the Russian Mafia will steal our nation right out from under us and they'll be pushing slavery at us real hard.

Lest you have a romantic view of what America will be like managed by the Russian Mafia, the Israel-Hamas War offers plenty of clues.

Despite the fact that the Russian Mafia has a good mix of Soviets Jews, they didn't stop Hamas' slaughter of Israeli Jews on 10/7/2023. And despite the fact that Hamas is owned by the Russian Mafia and despite the fact that most of Hamas

_____

258 Craig Unger, "*House of Trump House of Putin,*" (Penguin Random House UK 2018) regarding the USSR's exploitation of the Jackson-Vanik Amendment passed in1975 to seed our nation with 600,000 Soviet immigrants, spies, slaves, an unknown number of whom were sworn to aid the USSR.

are radicalized Palestinians, so were Palestinian children with Palestinian parents, grandparents and extended family, the Israeli military have reportedly killed 23,000+ Palestinian civilians.

This intel is critically important. In the Russian Mafia, Bratva membership trumps everything, with Mr. Putin the Don. This means that cultural ties, religious ties, everything falls away from the Russian Mafia except for Bratva membership.

Hamas is allied with the Russian Mafia but were not allowed to reach a cease fire despite overwhelmingly large numbers of victims are Palestinians.

If we allow the Russian Mafia to take us, these men don't care even about other Jews. Their god is the Bratva. And Hamas, manipulated into starting a war, must stand by watching thousands of it's own people slaughtered. There is no future for us if we allow ourselves to be enslaved by the Russian Mafia. Anything they say will be lies. I'm living who and what they are. They are liars. We believe them, we will die.

The Russian Mafia is working a coup here. This is the first coup in the mid-to-late 20th century & 21st century we've experienced and it is frightening.

Even though a bombing war is horrific, a stealth coup in my opinion has the horrible feature that innocent citizens don't know they're under attack. For example, the 50+ Iowans would've voted differently had their caucus been framed as a stealth attack by the Russian Mafia attempting to steal our nation from us, which is what is actually happening.

Chapter 44 - Framing can bring accuracy and clarity

Although many of the Russian Mafia are Jewish, they didn't stop Hamas from attacking Israeli Jews. Most of the Israeli victims were Jewish. So, no matter what your culture, religion, or anything else, the Russian Mafia prioritizes Bratva membership over anything else. Mr. Putin is Don of the Bratva.

Mafia's never see concede dominance to competitors and to the Russian Mafia, we, and all other nations not the Russian Mafia, is a competitor.

Chapter 45 - My near-term ancestors from the last several hundred years say to me "death is preferable to slavery"

It is my understanding the ethnicity of my near term primary ancestors are African, Caucasian, and Native American.

My African ancestors say death is preferable to slavery. My Caucasian ancestors were, I believe, indentured, the experience a misery to them. My Caucasian ancestors say death is preferable to slavery. My Native American ancestors say death is preferable to slavery. All of my near term ancestors say no to slavery.

I know what some of the KKK did. They let in the KGB, the Russian Mafia. I forgive them. They were lied to and manipulated, just like the Russian Mafia continues to manipulate it's slaves worldwide.

Caucasian American males are among my ancestors. It doesn't matter that all Caucasian American males don't know it. I know it. They are my people. I refuse to abandon my people to the evil Russian Mafia who has fed some of them lies. The Russian Mafia will say anything to anyone to seize world dominance.

The Russian Mafia feels a deep seated need to return to slavery, have their heart set on enslaving Americans. Whatever deal they offer, we will regret if we accept, for the Russian Mafia is incapable of honoring any deal, and they cannot help but lie. If I am ever voted leader of my nation, the Russian Mafia will never take us. Also, I will spend my life warning my people of the evil the Russian Mafia is perpetuating against us.

Chapter 46 – We've been attacked by the KGB running Active Measures here since last century

Who's destroying Americans?

By the late 1960s, early 1970s, America was poised to soar but we've been prevented from soaring by our enemy, the Russian government, the Russian Mafia, and before that, the USSR, which is the name of the previous Russian government which ended in 1991.

The USSR stole our nuclear weapon intellectual property and created their own nuclear weapons decades before the Chinese government was accused of intellectual property theft.

By lying, blackmailing and cheating, the Russian government, then known as the USSR began their genocide and coup against us in the 20ᵗʰ century. Indeed, the USSR was in America in 1970-71 attacking me and my family and making us homeless when I was 11-12, decades *after* it had signed the Geneva conventions, anti-genocide laws and other rules of war related laws the Russian government now says it no longer honors as of the 21ˢᵗ century. In fact, the USSR was breaking the Geneva conventions, anti-genocide agreements and other rules of war protections in my life as early as 1970-1971, while publicly admitting to breaking those laws in 2013.[259] Because of my own direct personal childhood experience I submit that the Russian Mafia, who runs the Russian government,[260] is an unreliable trade partner.

The Russian government stole access to our computer technology[261] via the 1958 Lacy-Zarubin Agreement (some of our leaders may have been blackmailed into such high value sharing with the USSR). In the 1970s, the coup de grace benefiting the USSR was the Jackson-Vanik Amendment, whereby Russia seeded upwards of half a million spies, operatives, and/or Soviet immigrants here.[262] That law was

259 General Valery Gerasimov, Chief of the General Staff of the Russian Federation, "The Value of Science in Prediction," *Military-Industrial Kurier* (February 27, 2013).
260 Karen Dawisha, *Putin's Kleptocracy*, Simon & Schuster (2015).
261 Via Lacy-Zarubin (1958), the USSR was also granted access to our education, science, medicine, film, dance, music, tourism, and scholarly exchange, and other systems. To me, and I admit I could be wrong, but this access only makes sense if someone was blackmailed into sharing such high value, critically important information during the Cold War.
262 Craig Unger, "*American Kompromat*" (Penguin Random House LLC 2021).

passed sometime around 1975 and you'll have noticed that our nation has been crashing and burning ever since.

If you're American you'll notice the decline of our country the last 50s years, the spike in homelessness and drug use, fentanyl, the dysfunction in our government (largely based in the GOP,[263] but not exclusive to the GOP), all that, and far worse, because the worse: the majority of the Russian government's and Russian Mafia's attacks are hidden.

Why we've failed to achieve our potential since the 1960s? Because the Russian government and their Mafia are running a secret coup against us and committing genocide against us. The Russian government's coup is aimed at destroying our government, while their genocide attacks are aimed at destroying us as a people. The Russian government calls this warfare <u>active measures</u> and it's cornerstone is <u>subversion</u>.[264] Subversion is when an enemy works in secret by

---

263 Operatives are not all compromised law enforcement. Most of the operatives who attack me aren't law enforcement but rather are people who've been compromised, presumably by compromised law enforcement, but I didn't witness how they were entrapped so I'm assuming they were entrapped by compromised law enforcement. The majority of our law enforcement aren't entrapped. An analogy is our military: while a small percentage of our military are compromised, most aren't. Another example: something is up with our House GOP. But if you look at that group's voting choices, about half appear to be compromised and the other half not. If you look at House Democrats, many fewer appear to be compromised but you'd be mistaken to believe none are.

264 https://en.m.wikipedia.org/wiki/Active_measures.

proxy[265] to destroy a nation and it's people. Using proxies,[266] the Russian government prevents whatever improvements we and our government are trying to achieve. For example, Americans naturally invest in students, in our working class, in women, in people of color. Those investments, when carried out, strengthen us, which is the opposite of what the Russian government and it's Mafia want so they weaponize Russia-controlled operatives here (who are Americans living in America but who are entrapped by Russian government's operatives but who they may not even know they're controlled by the Russian Mafia) to slow walk &/or degrade those

---

265 Proxy is when the Russian government uses someone it controls to do something it wants done. For example, when I was a teenager the Russian government tried to get four young adults to beat me up, people I didn't know. For a variety of reasons, none of those people were successful, but my mother had been entrapped and so the USSR forced my mother when I was 15 to beat me, out of the blue, for no reason. That beating caused me to leave home when I was 17. Had my mother usually beat me I wouldn't have been so shocked by her attack around 1975, but up until that time I may have received a spanking when I was around 4 years old. The beating (it wasn't severe but it did occur), was weird even at the time because my mother couldn't articulate why she was hitting me. Now I believe she was ordered by her handler to attack me. The USSR wanted me beaten and they couldn't get other young people to do it so they used my mother. Of course I wasn't going to retaliate against my mother so I took the beating, but I never forgot it. That is an example of the USSR using someone it controlled, my mother, as a proxy. The USSR wanted to beat me but had no legal authority to be in our country. It controlled my mother so it used her to beat me. Another proxy was my younger brother, who, for unknown reasons, tried to pick fights with me when I was around 16. My beloved mother and beloved brother are entrapped so, given that their behavior was so strange then and out of line with how they usually engaged with me, I'm going with the theory that my mother was entrapped since I was 15, and my brother was entrapped since I was 16. They were likely entrapped sometime earlier but the actual entrapment would not have been told to me.

266 Which are people and/or institutions it's compromised who may not even know they're controlled by the Russian Mafia.

investments &/or the people the improvements were meant to aid, stopping the investments from working as intended, to the point where a significant number of students in America under-achieve academically; the positive marker's of home ownership & relative health etc. of people of color are significantly lower than those of Caucasian Americans;[267] many working class &/or working poor are barely financially surviving, to the point they carry exorbitant credit card debt; [268] & women no longer control their reproductive rights.[269][270] All these setbacks and many many more,[271] were/are deliberate attacks against us by the the Russian government/Mafia. No one is marching here in protest against the Russian Mafia because it uses proxies[272] and most ordinary people have little knowledge of proxies &/or psychological warfare.

How may we survive? Build and launch an effective defense prioritizing sharing accurate intel, timely.

Am I insane or psychologically unstable? No – I've lived 53 of my 64 years stealth attacked by our nation's enemy, the USSR, and later, the Russian government, while living inside America, minding my own business, trying to be a responsible

---

267 Joseph Goldstein "Investigators Find Hospital Error Caused Mother's Death in Brooklyn" *New York Times* (January 14, 2024).
268 Ann Carrns "Lugging Credit Card Debt Into 2024? Now's the Time to Make a Plan" *New York Times* (January 12, 2024).
269 Pam Belluck "More Women Who Are Not Pregnant Are Ordering Abortion Pills Just in Case *New York Times* (January 2, 2024).
270 Remy Tumin "Ohio Woman Who Miscarried Faces Charge That She Abused Corpse" *New York Times* (January 3, 2024).
271 Stefano Montali "An Overdose Antidote Becomes a Tricky Issue at Some Nightclubs" *New York Times* (December 31, 2023).
272 Proxies are hides, camouflage. A shell company is a hide, designed for the express purpose to cover up the identity of the perpetrator.

person and a kind and loving family member. The Russian government is the Russian Mafia. The Russian Mafia honors no deal the Russian government has made with us, and the Mafia runs the government. Because of multi-decades experience attacked and/or tortured by this iteration of the USSR/Russian government/Russian Mafia, I know it quite well. It's going full bore to unseat President Biden. If it succeeds, we'll be in a world of hurt. The Mafia only sees bombs; the Mafia thinks only in terms of control. The Mafia is pro-slavery. It's hard for Americans to understand because we're culturally anti-slavery but an analogy is if the Russian Mafia were a mammal capable of experiencing an organism, 'slavery' would be it's organism, if the individual in the Mafia were in control of that slave. The Russian Mafia values nothing more than full control over another person or nation. Enslaving Americans is nirvana to the Russian Mafia and Americans need to know that so we can take evasive action or go to war it it's necessary.[273][274]

Since our government is sane and avoids bombing our enemy's unless pushed into it, the Russian Mafia has overrun our nation & our people, enslaved via entrapment unknown millions of us and is preparing to

---

273 Karen Dawisha, *Putin's Kleptocracy*, Simon & Schuster (2015).

274 Craig Unger, *House of of Trump House of Putin*, (Penguin Random House UK 2018) "...Putin and his cronies [...] were allowed to make billions through drug trafficking, extortion, elaborate financial schemes, the sex trade, arms deals, and the like...." Mr. Unger also states "...made Vladimir Putin the richest man in the world, with wealth, according to Hermitage Capital Management CEO Bill Browder, approaching $200 billion."

install one of it's many puppets in the White House. The Mafia could care less about the bombs we have in storage.

We're being stealth attacked by the Russian Mafia. Today's nuclear warheads are far more powerful than those in 1945 so tens of millions of us will die because, unlike America's military who would target the Russian government and avoid bombing Russian civilians, the Russian Mafia will target big cities, our nation's bread basket, our infrastructure, all the things we don't want to lose, the Russian Mafia will bomb.

The Russian Mafia is determined to control the world. The Russian Mafia will never stop stealth attacking us. They can't stop. They don't want to stop. Think of them as a serial killer who lives for that moment of control the murderer feels when they have their prey totally under their control. They are a version of the most entitled of our species. It's not because their skin color is pale, nor is it because they're men. For a combination of reasons I may never know, our species has created this type of apex predator.

They're climate deniers and all they care about is one-upping Americans. It's like a bully sibling who insists on a knock-down, drag-out fight when your house is on fire. They control so many Americans they're extremely confident they'll place a puppet in the White House in 2024 or 2028. We simply won't let them steal our country out from under us. When they finally see that we won't, they'll throw a series of tantrums. In their tantrums all the operatives they've stealth indoctrinated to destroy us, they'll unleash. It'll be bad, many of us will die. But whatever they do, we won't allow them to install a puppet here.

We can try talking to them, in fact I believe we should and we shall, but the Russian Mafia are world class, best in class liars. As I type these words, the Russian government is supposedly bound by the Geneva conventions, anti-genocide treaties, and other rules of war but since 1970-71 the USSR entered my childhood and destroyed it, having signed treaties and agreements decades ago. It's possible the Russian Mafia will suddenly stop attacking our children and civilians but it's highly unlikely because they don't want to stop. They're apex predators and they're proslavery.

If we panic and take a deal with the Russian Mafia, believing their lies because we want to believe them, they'll enslave us then murder us. After the first few years of public reassurances if we allow them to install their puppet, they won't waste their time reassuring Americans they've worked decades to destroy. They'll begin dismantling our government and our defensive capabilities immediately. The Nazis took nearly a decade to get into power, change the laws, and mass ship innocents to the gas chambers. That is not the Russian Mafia – they'll be a lot faster. We will not recover. Notice how the Russian Mafia has done it's setups across countries over decades: (1) Mr. Netanyahu is deeply unpopular but the Israelis, by their laws, are stuck with him. That didn't happen by accident. (2) In Britain, the conservative party is deeply unpopular, but ordinary people can't oust them, by their own laws. Again, no accident. (3) In America, if we allow the Russian Mafia to place one of their puppets into our Presidency, there's no way to legally oust them. Like most of the Russian Mafia's attacks, their coup won't initially appear to be a coup. What will happen is the Russian Mafia will force the tens of millions of Americans they own into the streets demanding that the new president install a new government. The puppet will say that they are obeying the will of the people. Just like you see the GOP Base has been deployed to control it's party's leaders (some of those GOP Base have been compromised, entrapped, and forced to try to entrap their own family and so are slaves – which the Russian Mafia doesn't reveal), just like the Russian Mafia deploys the GOP Base to control the GOP leaders, they'll deploy tens of millions of protesters-slaves to provide their puppet with justification to install a new government. Just like the Russian Mafia had 'homeless' vandals attack post office boxes to provide a believable cover for the removal of 9 of the 11 post office boxes where I live over the last few years. Do you see the pattern?

The Russian Mafia is troubled, something is seriously wrong with them, their psychological and emotional system is all messed up. They don't see people as people. It's like one person looks at a chicken and see's two drum sticks and two thighs, and another person watches the chicken as it hunts for bugs and see's how clever the chicken is, see the chicken is a distinct life and it's life has value. The Russian Mafia see's drum sticks and thighs. We can't expect them to see the world as we do, just like it doesn't expect us to see the world as they does. They're not trying to help us boost our game, they've taken full advantage of Active Measures and the misdirections they've laid for us. This is probably why there's only one group of homo sapiens: 100,000 years ago there were many human species, now we're the last ones. One group of our ancestors likely smiled, plotted, schemed & cheated, as the Russian Mafia does, while our other fore-bearers believed the liars. Then there was a war, perhaps just one side survived. We must be ready for war at all times with this quality of enemy.

The KGB has run it's secret war since at least 1970. I know because in 1970-71 USSR operatives embedded in America attacked me using many of the strategies General Gerasimov discusses in his 2013 article "The Value of Science in Prediction," which appeared in *Military-Industrial Kurier,* an obscure magazine intended for Russia's military. I was 11 when the USSR attacked me. General Gerasimov said the Russian government's new strategies were put into effect in the 21st century, but he lied. The USSR attacked me with many of these very same strategies 42-43 years before the Russian government admitted to using them.

After the demise of the USSR in1991, the successor Russian government which remains in power continued to harass me and expanded the attacks, not only against me but against my country. This book details as many of the USSR and Russian government's attacks as I can recall beginning from the age of 11 to my present age of 64. The attacks remain ongoing. I am an ordinary woman. I worked as a legal word processor/legal secretary in law firms where the Russian government unrelentingly stealth attacked me using my supervisors and coworkers as proxies. I was also attacked outside of my work environment. I am not an operative nor have I been trained in espionage. This is my story.

Chapter 47 - Pre-war

This book is to help you prepare for war, the war you're already experiencing at the hands of the Russian government. That government refuses to tell you they're attacking you because they hate us and refuse to give our civilians or our government any opportunity to defend ourselves. This is a problem because neither ordinary people (aka civilians or our children) nor our government knew the Russian government, and it's predecessor government, the USSR, were/are in America running stealth attacks against ordinary people and our children, running attacks against us in our private lives in the 20th and 21st centuries.

The USSR signed the Geneva conventions, anti-genocide laws, and agreed to other rules of war after the millions of deaths it's soldiers suffered in WWII. But as you'll read in this book the Russian government and it's predecessor government, the USSR, lied when they agreed to abide by those rules of war. They knowingly attacked American families, including civilians and children under 15, against international laws they'd agreed to adhere to, as well as against American's domestic laws, where it is absolutely illegal for an enemy foreign government to commandeer our domestic entrapment system, compromise our law enforcement in charge of that system and aim it to ambush noncriminals, including families, with the intention of destroying our families.

I charge the Russian government, and it's predecessor government, the USSR, with genocide against all people living in America from 1970 onward, or people who lived in America in 1970 and have since died and/or moved out of the country, including Native Americans and all ethnicity's, and including incarcerated peoples incarcerated any time between 1970 to the present.

The Russian government, also known as the Russian Mafia, uses psychological warfare and subversion[275] to attack us because it's cheap and effective and because most civilians have no experience with this type of warfare. Additionally, most people living in modern American have not experienced a coup in America, which coup is what I charge the Russian Mafia is staging here. The Russian Mafia couldn't care less that ordinary people know nothing about psychological warfare and subversion, nor does it consider it their responsibility to warn us. The marked deterioration of our nation over the last fifty plus years is a direct result of the Russian Mafia's, the Russian government's, and the previous USSR government's

---

275 https://en.m.wikipedia.org/wiki/Active_measures. Subversion and active measures form the cornerstone of the Russian government's attacks against Americans.

secret war against all people living in America. Our decline is directly attributable to the success of the Russian government's secret war against us and our total lack of knowledge that we're being stealth attacked.[276]

General Gerasimov, Chief of the General Staff of the Russian Federation said in 2013, "...wars are no longer declared..." "...the very rules of war have changed.'"[277] The General made his comments, which were intended for the Russian military and so were published in an obscure magazine, decades after the USSR and the Russian government had begun attacking Americans. I know because I was 11 when the USSR first attacked me in 1970-71. The strategies used against me in 1970 moving forward over the next 53 years to the present are among those described by the General in his 2013 article. In that article he says the Russian government began these strategies in the 21st century. He is lying about the time frame for the USSR, and later the Russian Mafia, Russian government, used variations of those attacks against me from the age of 11 to the present.

The Russian government and it's predecessor government the USSR was attacking me and my family in the 20th century, illegally and knowingly, in contravention of the USSR having signed the Geneva conventions and the anti-genocide laws the USSR had agreed to, and which they lied.

---

276 https://en.m.wikipedia.org/wiki/Active_measures.

277 General Valery Gerasimov, Chief of the General Staff of the Russian Federation,"The Value of Science in Prediction," *Military-Industrial Kurier* (February 27, 2013). General Valery Gerasimov states: *"...Wars are no longer declared...." "The very "rules of war" have changed.."*

The dysfunction of individuals in our political parties, the fentanyl drug flows, the mass shootings, the homelessness, the drop in student's educational achievements, the dysfunction in our immigration system, the severe shortage in quality, affordable rental housing, and affordable starter homes, and so many more of our societal problems are due to the Russian government's secret war against us, which they've been running since at least 1970-1971.

Chapter 48 - The Russian Mafia exploited my love & respect for my father to influence my career, steered me away from the military and law enforcement when I asked my father's advice. His answer turned me away from those fields since he is a war veteran and retired law enforcement. I'm positive other entrapped law enforcement are used to discourage Americans people looking to serve our nation

My father, my paternal uncle, and both my maternal and paternal grandfathers served our country. Everyone was honorably discharged.

I strongly considered military service and becoming a police officer like my father, but my father talked me out of both. I now know he's entrapped so my theory, since he can't discuss his entrapment with me, is his handlers didn't want me to have the benefit of military training and military friendships with soldiers the Russian Mafia may have struggled to control.

It's not my father's fault he's been entrapped by the Russian Mafia, it's the Russian Mafia's fault. They attacked an innocent man, one of tens of millions, I suspect, given how many American voters the Russian Mafia stealth controls. It's important Americans know that the Russian Mafia entraps veterans, entraps law enforcement, and can manipulate people with espionage training. The Russian Mafia can use compromised veterans and compromised law enforcement to dissuade people from serving our country. These are people our military needs. So if you're considering serving our country, talk to at least two people before you make up your mind.

Chapter 49 - How the Russian Mafia works in America

The Biden Adm is doing everything it can to
help us adapt to the immigration crisis & all
the other crises the Russian Mafia is subjecting
us to this 2024 Presidential election cycle
in their attempt to install their candidate of choice.

Re-building an effective immigration system
while the Russian Mafia[278] attacks the one we
have by: (1) destabilizing immigrant's
home country's political institutions & economy
to make it impossible for workers to earn
a living wage; (2) using political
operatives in America who refuse
to adequately & consistently fund
the IRS, thus depriving our gov of the
billions more we need to effectively
manage our country; & (3) forcing
entrapped Americans to cheat on their taxes
&/or lure/encourage others to cheat on

---

278 "...for decades Russian operatives, including key figures of the Russian Mafia, studiously examined the weak spots in America's pay-for-play political culture...in an effort to compromise America's electoral system, legal process, and financial institutions." *House of Trump House of Putin*," Craig Unger (Penguin Random House UK 2018).

their taxes, thus depriving our IRS of needed funds.

The Russian Mafia is waging a coup here,
by, among other things, controlling how much
money our gov has available to spend.
When the Russian gov controls a government
finances, things don't go well for that government.

This book describes the start and continuation of the USSR's stealth attacks against me when I was an 11 year old African American child living in south western America with my mother and half siblings in 1970-1971. The USSR, and later the Russian government, makes their luck: I've just read General Austin has been hospitalized at Walter Reed Hospital since January 1, 2024.[279] The Russian government demonstrated to me it has controlled our medical system since I was around 48 or so, when, feeling ill, I went to doctor after doctor and none diagnosed my high blood pressure or pre-diabetes, among other health problems.

General Austin, responsible for overseeing parts of both the Ukraine and Israel Wars, essential contests the Russian government intends to win, had a procedure to try to fix a medical problem. There were complication. Because I've been stealth attacked by Russian Mafia controlled American operatives for decades and they've used medical staff including doctors to mistreat me for years (my general practitioner has at least twice tried to entrap me into Medicare fraud, which the Russian Mafia now does because they know entrapment

---

279 Eric Schmitt "Inspector General to Investigate Handling of Austin's Hospitalization" *New York Times* (January 11, 2024).

attacks push me away from getting routine medicare care so they can more easily murder me using natural causes as a label).

Killing General Austin would launch an intense investigation and blow the cover of some of their operatives, which they intend to use against other Biden Team members, so the General will recover, but slowly.

We're far too trusting and behind the curve, not because we're stupid but because, obviously, the Russian Mafia hides it's presence from non-compromised power players. The only reason I know is because I'm not a power player and because the Russian Mafia's American operatives grew so frustrated being unable to entrap me, they began overt attacks against me, believing our legitimate government will ignore my allegations.

We won't survive unless we understand our enemy intends to enslave us, and the Russian Mafia sees the Biden Administration as their last serious threat in this country.

The covert attacks continued for approximately 20 years until the demise of the USSR in 1991, when I was around 30. After which, the next Russian government (currently in power) expanded and intensified it's illegal attacks against me, which attacks continue to the present (January 2024).

I believe the USSR's launched it's attacks against people living in America as part of it's secret war, coup and genocide after World War II, possibly motivated by the USSR's animus resulting from the Cuban Missile Crisis, and other Cold War contests that enraged them. In WWII we were allies. *From post/WWII on we have consistently underestimated the USSR's intense hatred of us, their need to dominate the world and enslave Americans, and their willingness to do anything at all to destroy us that leaves them not bombed.* If I'm overstating anything in the italicized language, it's their hatred of us. Their strongest drive is hubris. Mr. Putin is determined he'll leave Russia as top dog country in the world at his death. If we understand nothing at all, we must understand he'll do anything, take any risk, to ensure Russia is top dog over the world at his death. But to him, a win means we don't bomb the Russian government. Our nation can run down the clock until his death if we install a series of hawks the Russian Mafia believes are willing to risk nuclear war. The problem is the Russian Mafia is so deeply embedded here, we wouldn't be able to get that play past their spies. Even now, I'm suspicious of why they sickened General Austin. Our government will probably never tell us what happened but if Secret Service was with him, I'd be able to take a breath. My theory is he had no protection and the Russian Mafia questioned him. The Russian government made that attack happen for a reason and my theory is the Russian Mafia wanted to know what President Biden has in mind. I read an article recently that says we and Israel have spies in Iran. It's so easy for the Russian Mafia to know what's going on here. We're a sieve, not just us, but many democracies. We have insufficient protections and that's just not good when dealing with the Russian Mafia.

The Russian Mafia doesn't fear our bombs. Rather they don't believe they can recover from being bombed -- there's a difference.

Unfortunately for us, the Russian Mafia has a taste for slavery and we can't break them of that strong preference. They intend to enslave all of us, and we're not okay with being slaves. Additionally, the Russian Mafia is anti-woman. Just in how they've manipulated my brothers to loathe our mother, just by how they've proxy slaughtered Palestinian women and children, in every move they make they shout 'we loathe everyone not us.' Endgame, they intend to annihilate us. All their characteristics are just really bad, yet they insist they're the chosen ones destined to rule our species, when what they really are are a group of hate-filled men who developed an effective weapon they're deployed against the world, killing us and our children.

A new weapon doesn't make the Russian Mafia worthy. Based on their evil behavior, they're the worse possible choice to lead our species. They have no moral compass. And they don't even see that that's a problem. If, for whatever reason we don't take care of the problem, the Russian Mafia will install one of it's puppets, push enslaved dominants into the streets by the millions who demand the puppet install a new government, the puppet says 'the people have spoken,' and the Russian Mafia steals our nation: coup. Worse, far, far worse, the Russian Mafia will annihilate us. Because modern Americans have never experienced a coup nor faced the threat of annihilation, we're like one of those newborn chicks who's struggle to break out of their shell attracts the eye of a hungry snake. Minutes born, it's being eaten. We must stop the Russian Mafia's coup and we do that by: short answer – informing our people and the world about the attack and moving on from there.

Chapter 50 - A zero sum game that's not a game

For us, it's our existence, for the Russian Mafia, it's their ego. They're pro-slavery and refuse to live in a world where they aren't top dog. They hate women and people of color. They've got serious issues.

Is it possible we can play this to a draw? If we convince our people to not vote in the Russian Mafia's puppets, we may be able to hold the line. Mr. Putin and the Mafia control the ground here, control millions here. I'm sorry, that's the truth.[280]

---

280 The Russia government has access to many of our news outlets and can pretty much plant any article it wants. Most of the time it's long gone before it gets noticed by our legitimate government and readers would've been victimized again by another propaganda-slanted article designed to make Americans loathe and fear our legitimate government, when what's really happened is the Russian government has done another slime attack designed to make us lose confidence in our democracy and our government. We're far too slow on the uptake. Unless we bomb the Russian government it's going to destroy us. My government,which I love and respect, is too slow responding to the highly nimble Russian government. My government would say it's easy to criticize from the sidelines, and they'd be right. Bombing the Russian government is what we need to do. Yesterday. The reason our legitimate government doesn't is because many innocents would die and after bombing you can't go back to a pre-bomb mentality. My

The Russian Mafia fears no one. We don't need them to fear us, but we need to understand them. The current Russian government (in existence post-1991), having greatly benefited from the societal and government destabilization achieved by the USSR against Americans by the time of it's demise in 1991, appears to me to be: (1) committed to and quite capable of destroying us, (2) confident they will destroy us without nuclear war, and (3) determined to install puppets into positions of power throughout our nation and throughout our ally's nations.

---

legitimate government wants to be rid of the Russian Mafia, probably more than I do but the cost is too great, the loss of life too great. I think because I've been attacked by the Russian Mafia so very long, and because I'm a woman, and because I see how women are time and again raped and destroyed by bad men, I'm ready to pull the plug on the Russian Mafia. I'm not a politician, and that's not a slur. Political leaders who have to answer to voters is a good thing. But because (1) I'm not a politician, and because I'm a woman (2) tortured for well over a dozen years, and (3) attacked for over fifty, and because (4) so very tired of women being victimized by monstrous men, and because (5) I'm not worried about keeping my party in power, I'd bomb the Russian Mafia, even knowing they'd retaliate. My feelings are beyond hatred. I simply see them. Seeing them, they simply ought not to exist. I think when sentient life forms care nothing at all about the life around them, even killing babies of their own species (as the Russian Mafia is doing to Ukrainian children and by proxy against Palestinian children), they have nothing to offer. Why live, if they live only to bring misery and death to other life forms. To bring only death is unacceptable. To bring only death and cruelty means the perpetrator has nothing to offer. The Russian Mafia is so foolish they will cost us the loss of our only habitat, earth. They believe they're so brilliant because they've learned to manipulate people but that success won't do them any good as they destroy our only habitat by not allowing their allies to work with us to try to prevent the climate catastrophe from worsening. The Russian Mafia is the most clever of us, and the most stupid, simultaneously. They are evil but don't care that they are. They are anti-life, but don't care that they'll remove themselves from life. Do you see? They are us, only foolish. While they believe we are stupid for believing their lies. It seems our species cannot survive itself.

From what I can tell, the Russian government has commandeered some of our law enforcement and some of our law enforcement's entrapment mechanism to aim at noncriminals, prioritizing entrapping a family member. Then the Russian government uses the compromised family member as a weapon to entrap the rest of the family. Example of this attack in action: Two entrapped family members (my sibling and their spouse) told me one of my nephews cheats on his taxes. When I expressed dismay and my intention to talk to my nephew to get him to stop, the spouse asked me not to talk to him, the spouse said they'd helped my nephew with his taxes and because of that, ought not to have discussed the situation with me. This was how the spouse raised the issue of committing tax fraud with me, feeling me out to learn how I'd respond to it without telling me they themselves had committed tax fraud. While my loved ones discussed this subject with me, only I and they were in the apartment, but their handler would've been able to hear what was said. I don't know how but they would've needed to hear the conversation. Had I expressed an interest in lowering my tax bill, the spouse would've offered assistance. Using my family to lure me into committing tax fraud is illegal here but the Russian government couldn't care less about what's illegal or legal in America - their intention is to enslave and/or kills us, steal our country and implement their own laws. They can't wait to crow that they destroyed us without getting bombed once.

Their ego is off the charts and they hate us. They lie like a rug and think we're plain stupid for believing what people tell us.

My loved ones forced to make this pitch are enslaved by the Russian government. They must do what their handlers tell them or presumably they'll be imprisoned. They're not free to warn me to ignore their words. Most slaves cannot free themselves. They've no idea their American handler is controlled by the Russian government. The Russian government has entrapped them as part of the Russian government's coup. There's been no coup in modern American history so it's been crazy easy for the Russian government to commandeer and exploit our entrapment system to steal our nation from us.

The Russian government has established many pro-Kremlin governments throughout the world. Americans are the prime target fo the Russian Mafia but are far from the only one in the Russian government's cross hairs.

That's a problem for us and our allies because the Russian government is the most evil and by far the most effective at stealth warfare our species has ever produced. Any world where they're top dog is a world of people in pain, especially us. Think hate-filled sadists as the top dog and you've got an accurate start.

Many of the negative national events we've experienced in America the last fifty years, were setups and entrapments by the Russian government perpetrated to destroy: (1) our confidence in our government, (2) our belief in our businesses and institutions, and (3) our belief in our democracy.

A current attack is the Israel-Hamas War where protesters blame our President and the Israeli government for the atrocious suffering of Palestinian civilians, including the horrific suffering of their babies is orchestrated by the Russian government. They work in the shadows. Americans get a reputational hit, or worse, the Russian government gains an advantage, no one ever mentions the Russian government is the perpetrator. That's what I mean when I say they work in the shadows.

The Russian government's strategy is discussed by General Valery Gerasimov, Chief of the General Staff of the Russian Federation, in his paper (intended for the Russian military) entitled "The Value of Science in Prediction" *Military-Industrial Kurier* (2/27/2013). The General says: (I bold for emphasis)

"The very "rules of war" have changed…."

"…Among such actions are the use of special-operations forces and internal opposition to create a permanently operating front through the entire territory of the enemy state…."[281]

"The role of nonmilitary means of achieving political and strategic goals has grown…have exceeded the power of force

---

281 General Valery Gerasimov, Chief of the General Staff of the Russian Federation,"The Value of Science in Prediction," *Military-Industrial Kurier* (February 27, 2013). This adequately describes the "friendly" relationship between the USSR and the KKK, as well as other "friendly" relationships between the Russian government and "allies" on the ground in enemy nations.

of weapons...[282]

"...applied **in coordination with the protest potential of the population…." *"….Tactical and operational pauses that the enemy could exploit are disappearing...***[283]

I see evidence in America and around the world the truth of the Russian Mafia's strategy reflected in General Gerasimov's words. Worse, the strategies and behavior General Gerasimov articulated in 2013 I experienced at the hands of USSR operatives when I was 11 in 1970-1971 America, although the Russian government didn't allow the language to be published for public consumption until 2013.

This language is in direct contravention of the Geneva conventions, and anti-genocide laws to which the USSR is a signatory, and which signature binds the present Russian government, no matter it's attempt to shrug off it's responsibilities to civilians and civilian children in it's enemy's nations in times of war or peace.

I witnessed and was attacked by USSR operatives working General Gerasimov's strategies when I was 11 in 1970-71. That General Gerasimov is discussing those strategies in 2013, even though in an obscure military magazine, is ominous. It suggests to me the Russian government believes it is so deeply embedded here, and has seized enough control here, it is 99.99% certain it will publicly own us in the not so distant future, this election cycle or the next. If we don't bomb the Russian government during the Biden Administration, I

---

282 Ibid.
283 Ibid.

believe the Russian Mafia is correct in thinking it owns us. The reason I believe this is the Russian government has people on me 24/7. They know when my legitimate government has operatives watching me because the Russian government's operatives change their behavior.

And when my legitimate government operatives aren't watching me, my enemies begin harassing me again. Indeed, I only know when my legitimate government is paying attention to me because my enemies tell me by altering their attacks against me. The Russian Mafia are all but certain they own the street here and that's very bad new for us.

Chapter 51 - The Russian government is the Russian Mafia

To us, we are Americans and
Immigrants living in America.
To the Russian Mafia we are a competitor.
The Russian Mafia destroys it's competitors.

We continue to try to get them to listen,
but they can't hear us.
They interpret our attempts to live
in peace as a weakness.
Apparently, they believe they're the only
people who can hate, & who can form
hatred into a strategy surviving decades
What they're doing doesn't require intelligence
but only cowardice, ambush,
attacking the innocent, & lying.

They think they're the chosen ones
but all they are are baby murderers.

They understand bombs
which is why they're working as hard
as possible to install their puppets here
so they can control our bombs.

The Russian government is merged with the Russian Mafia. The Russian government overlays the Russian Mafia to provide it a thin layer of legitimacy.[284][285]

It is more accurate to think of the Russian government as a Mafia/KGB hybrid rather than as a government, because the Russian government conducts itself in the world as a Mafia/KGB hybrid. By that I mean it views everyone not them as an enemy or competitor to be either dominated or destroyed.

The Russian Mafia's leader, Mr. Putin, takes the title of president but in the case of Russia, that title is a misrepresentation. An accurate title for Mr. Putin would be Don.

Additionally, and most importantly, the Russian Mafia doesn't differentiate between it's enemy's or competitor's militaries, governments, civilians, or children populations. To the Russian Mafia, civilians and children are legitimate

---

284 Karen Dawisha, *Putin's Kleptocracy*, Simon & Schuster (2015).

285 The Russian government and it's Mafia are one of the most successful illegal drug dealer systems in the world; targeting anyone, including children in West aligned nations is part of the Russian government's & it's allies' strategy to destroy us. According to Craig Unger in his book *House of of Trump House of Putin*, "...Putin and his cronies [...] were allowed to make billions through drug trafficking, extortion, elaborate financial schemes, the sex trade, arms deals, and the like…." Mr. Unger also states "...made Vladimir Putin the richest man in the world, with wealth, according to Hermitage Capital Management CEO Bill Browder, approaching $200 billion." Craig Unger, "*House of Trump House of Putin*," (Penguin Random House UK 2018).

targets.[286][287] The Russian Mafia does not believe it is their responsibility to inform the world of this highly important fact. However, one of their leaders wrote a paper intended for the Russian military audience titled "The Value of Science of Prediction," dated February 27, 2013. In it General Valery Gerasimov, Chief of the General Staff of the Russian Federation, said that the Russian government no longer see's itself as being bound by the Geneva Conventions, or other rules of war. He stated:

> In the 21st century we have seen a tendency toward blurring the lines between the states of war and peace. **Wars are no longer declared** and, having begun, proceed according to an unfamiliar template.[288]

Indeed, civilian and children populations in enemy/competitor nations, including West-allied nations, are highly prized by the Russian Mafia because they are low hanging fruit and are therefore easily destroyed, and are the soft underbelly of enemy/competitor nations.

The Russian Mafia doesn't consider it their responsibility to inform their enemies, competitors, or the civilians and children in enemy or competitor nations that the Russian Mafia is on the ground, pushing fentanyl into the

---

286 Carlotta Gail, Oleksandr Chubko, Cora Engelbrecht, "Ukraine's Stolen Children," *The New York Times* (December 27, 2023).
287 Shaun Walker, Pjotr Sauer and Tom Phillips "Panic and Emotional Pain as Alleged Deep-Cover Russian Spies Vanish," *The Guardian* (April 3, 2023).
288 General Valery Gerasimov, Chief of the General Staff of the Russian Federation,"The Value of Science in Prediction," *Military-Industrial Kurier* (February 27, 2013).

targeted nation,[289][290] destabilizing the government civilians and children depend upon, and entrapping parents and forcing them to entrap their own children, among other evil and genocide attacks.

Chapter 52 - The Russian Mafia breaks our species' hardwired societal contract in order to destroy us

From birth people are taught to understand that if they behave a certain way they'll be rewarded, and if they a behave a different way, they'll be punished. The Russian government flips that on it's head to work it's coup here. It rewards positive societal behavior with negative feedback and unkindness, intentionally frustrating and enraging innocent hardworking people, driving some to self-loathing, drugs, self-harm, and suicide. This is especially harmful to our children and you see our children's decline before our eyes, but, like most adults, children struggle to articulate the overlapping, complex, psychological attacks on them.

The Russian government's basic strategy of giving cruelty as a reward for being good, and their ability to negatively impact most people in America, is one of the

---

289 The Russian government and it's Mafia are one of the most successful illegal drug dealer systems in the world; targeting anyone, including children in West aligned nations is part of the Russian government's & it's allies' strategy to destroy us. According to Craig Unger in his book *House of of Trump House of Putin*, "...Putin and his cronies [...] were allowed to make billions through drug trafficking, extortion, elaborate financial schemes, the sex trade, arms deals, and the like...." Mr. Unger also states "...made Vladimir Putin the richest man in the world, with wealth, according to Hermitage Capital Management CEO Bill Browder, approaching $200 billion." Craig Unger, "*House of Trump House of Putin*," (Penguin Random House UK 2018).
290 Karen Dawisha, *Putin's Kleptocracy*, Simon & Schuster (2015).

reasons our nation is experiencing a mental health emergency. Victims can't explain it because most people don't know the Russian government is deeply embedded here[291], don't know the Russian government's strategy or that they're conducing a secret war with us.[292] Civilians with no knowledge of espionage or the Russian government have no way to figure out that the Russian government is stealth attacking us, using psychological warfare. Our government doesn't know because the Russian government hides it from them. It hugs the shadows and every place our legitimate government doesn't intrude in our personal lives. For example, the Russian government entraps parents and other authority figures in an individual's life, and use them to denigrate and cause psychological harm to that person, including children, all outside the purview of our legitimate government. I was attacked by supervisors and/or coworkers on at least five of my jobs, all outside of the purview of my legitimate government, who had no idea my employers and coworkers were abusing me. Our legitimate government isn't on most people's job. The same when I was mistreated and/or lured by my doctor, or my parents, or lied to by my teachers and got lower grades as a result, and even dropped out of school because my TA accused

---

291 David Smith, "The Perfect Target: Russia Cultivated Trump as Asset for Forty years – ex-KGB Spy," *The Guardian* (January 29, 2021).

292 General Valery Gerasimov, Chief of the General Staff of the Russian Federation,"The Value of Science in Prediction," *Military-Industrial Kurier* (February 27, 2013). General Valery Gerasimov states:
"In the 21st century we have seen a tendency toward blurring the lines between the states of war and peace. Wars are no longer declared and, having begun, proceed according to an unfamiliar template. The very "rules of war" have changed."

me of cheating, even though he said he hadn't saw me cheating but saw another person cheating on my work during a math test. Everywhere in our personal or work life our legitimate government isn't, is where the Russian government hunts us.

When my parents and siblings told me to lie and cheat the government, my legitimate government didn't see that. The Russian government entrapped most if not all of my immediate family and set them on me. Or when I was at a big box store to pick up a prescription, and bought a few groceries, and, unknown to me, the cashier didn't scan all my items, and there were maybe only 10 items in the basket. He was tasked with setting me up to be accused of shoplifting by security when I left the store. My legitimate government couldn't see that because the store security/cashier, unknown to my legitimate government, is owned by Russia-controlled American operatives. The Russian government is hunting us and it hunts us out of sight of our legitimate government, meaning our government can't protect us, and we must protect ourselves, but since we don't know the Russian government is embedded here, it's been a rout on the ground for ordinary people.

Our legitimate government has no idea we're being attacked, often daily, often multiple times a day. And the Russian government doesn't reveal itself to victims. When it had supervisors and coworkers mistreat me, my doctors try to lure me, my parents and siblings try to entrap me, none of those individuals said to me: "we're acting this way because we've been entrapped and our police officer handler is forcing us to behave this way with you." No. What happened was, the Russian government exploited my feelings and relationships with my loved ones, landlord, employers, etc. to make me think the attack was coming from legitimate people in my life, rather than from a foreign enemy government. That's a problem for all victims, but especially for children, or people who are especially dependent on the entrapped authority figure. The Russian government is an evil government. It stealth attacks women, children, and civilians like the cowards they are. What they're doing is akin to kicking a puppy, or slapping a two year old. The Russian government is a monster government not because they covet world power but because they've decided to use the blood of innocents to obtain that power. They fire on civilians and then denigrate the victim as weak for dying. To hear them tell it, our country is deteriorating because of moral weakness and corruption, when they know full well they've enslaved millions of innocent people here by lying to them and convincing them that they weren't really ambushed but were entrapped because they're corrupt. That lie is just one layer of multiple layers of control. The Russian government are master liars.

Destroying a person's happy place by breaking the societal contract, the Russian government rewards excellence with negative feedback, building confusion, frustration and rage in order to herd Americans into overthrowing our own excellent government for the Russian government's pro-slavery model.

It stealth harasses and agitates ordinary people using authority figures and/or important people in their lives, to over time get them to overthrow our legitimate government. Most people don't know that the Russian government is deeply embedded here and manipulates and frustrates millions of us a day by having us sit for hours in traffic, have our grocery store short stock favorite foods to frustrate us[293], who we can reach on the phone[294], deliberately shorting out our kitchen gear and home products to attack us in our apartment, how our landlord treats us, how our employer treats us, if our electric grid works, how medical staff treats us, how our co-workers treat us, etc.

The Russian government does this because they're working a coup here. They hate us, predominantly out of jealousy, and envy. They believe they're smarter than us because they attack us without warning and civilians don't know. They're determined to destroy us by placing a puppet in the White House, and at all levels of politics and business.

---

293 A GOP Base member told me her experience of going to multiple stores and being stuck in traffic for hours just to do their weekly shopping. That's a strategy the Russian government uses to get victims to move out of California. It's determined to make California a swing state and harassing people with traffic flows and not stocking basic foods and forcing them to drive to five stores to buy weekly groceries is one of the ways the Russian government gets people to leave the state.

294 The Russian government has stopped my calls from going through: (1) when I called my doctor's offices (dozens of times); (2) when I called a business for technical help (maybe 10 times); (3) when I called a covid housing assistance program (prevented maybe 5 calls going through; (4) tried to call a longtime friend of our family (I forget how many times I tried calling her; (5) I'm almost never allowed to be put through to the city college switchboard when I call them (maybe a dozen calls), and other calls I can't even remember them all.

The Russian government has broken the unspoken social contract between individuals and society where traditionally individuals are rewarded for good work and good attitude with compliments and salary increases, and instead reward honorable, hardworking, decent people with disrespect and unkindness, and in my case, the Russian government has had hundreds of operatives physically assault me, by kicking me, stomping on me, hitting me with their vehicles, sexually assaulting me, but the majority of the physical attacks were operatives deliberately stealth hitting me with their bags and purses as they got on or off the bus. All this is the Russian government's stealth way of getting us into the streets demanding our government step down, whereby the Russian government intends to have it's puppets ready to step in.[295]

People who are treated cruelly overeat. Distressed, they seek comfort when people are mean to them for no reason.[296]

---

295 General Valery Gerasimov, Chief of the General Staff of the Russian Federation,"The Value of Science in Prediction," *Military-Industrial Kurier* (February 27, 2013). *"The very "rules of war" have changed. The role of nonmilitary means of achieving political and strategic goals has grown, and, in many cases, they have exceeded the power of force of weapons in their effectiveness."*

296 Because the Russian government couldn't entrap me they used authority figures in my life to psychologically attack me, including: a parent to sneer and denigrate me to prevent me writing this book (father); pushed me to file for personal bankruptcy and ridiculed me for refusing (mother); had a partner at my last job yell at me for no reason as I worked for twice the number of attorneys that is usual (boss); had a family member I call on his birthday be very short w/me to the point of rudeness (brother); on my job before last had the temporary supervisor and a coworker ridicule me every time only the three of us were in the room (supervisor, coworker); the job before that had my supervisor and co-workers treat me cruelly, and on and on. I have dozens of examples of bosses and coworkers treat me cruelly. The reason being that after I got settled and established at a company, Russia-controlled American operatives saw no way to entrap me there. Ever hopeful they began,

Some take drugs, others drink, others lose all hope. Most have no idea they're being deliberately attacked by a foreign enemy government on the other side of the world working a coup here.

Most people see themselves as ordinary and can't imagine a foreign enemy government attacking them. The Russian Mafia can't flat out bomb us as they do Ukrainians because we have nuclear weapons, so they developed a different strategy to destroy us, one based very much on a hands-on, dominant by dominant approach. Because dominants are the subgroup who controls our species, and our hard-wired instinct (our genes) force us to elevate their value from birth, and because we depend on them for sounding a species alarm

---

after a few years, having my coworkers and supervisors mistreat me. When that happened I'd find another job, having no idea I had an enemy who is a master of psychological warfare. Time and time again, an entrapped sibling and/or entrapped medical care providers wanted me to say I was experiencing an emotional problem: anxiety, etc. I had problems from overt attacks and the Russian government began overtly, in my face, attacking me after 9/11, when I was 40-41. Because of my decades of experience being psychologically attacked and now understanding why the Russian government was herding me to psychotropic drugs or to use me as a weapon, I need to tell the world. There are so many people experiencing psychological attacks. They must be told they're not crazy, it's not them, it's our evil enemy, the Russian government. I'm not the only one the Russian government attacks this way. It attacks our wonderful president, President Biden. A year or two ago, when the President was doing all kinds of brilliant work, he wondered why his poll numbers were so low. The reason is because the Russian government intends for the President and his family to associate the White House with psychological pain, to encourage them to step down. We need that First Family and the Russian government knows it, so they're doing everything they can to cause them psychological damage. This is flat out truth. So if you're asked to take a poll or leave a comment about our First Family, keep in mind that our primary enemy working 24/7 to get them to step down and say something kindly truthful in appreciation for the sacrifice they're making for us.

about non-overt threats, they were hands-down the obvious choice for the KGB and Russian Mafia to enslave. In my family my mother has five children. The KGB entrapped her and now own all of her children but me. Looking back, I believe the KGB has owned by family since (1) my mother 1972, the year she was destabilized. She beat me around 1975, I believe her handlers forced her to do that under threat that they'd place her children in foster care and imprison her for noncompliance with her plea agreement. The KGB wanted me beaten, they'd deployed three teens to group attack me but I wasn't easily available to those teens. I was easily available to my mom so they forced her to do it; (2) my father, unknown, but for at least the last 30 years (the Russian Mafia made him available for political discussions about the Iraq War and the Bush cabinet when I was in my 30s. They'd failed to entrap me and decided to embed my father more deeply into my life, in an effort to control me. He was a primary influencer, telling me not to enter policing or the military when I told him I was interested and asked his advice (he is a veteran, and retired as a police officer); (3) my brother, since 1975 (they forced him to pick fights with me when I was 16 and he was 15, which ended when he ran away after I picked up a weapon and gave chase. His baiting me was out of character between he and I; he'd never done it before (he ended up picking two fights). Just as my mother's beating me was out of character: she'd never done it before. I think I may have received a swat or two from my mother for misbehavior in my early childhood, but you know, maybe not even that. I have no memory of being spanked during my childhood. I remember my 5 or 6 year old brother stealing $20 from our mom's purse and going to the neighborhood store and buying candy for him and his friends. Ms. Sophie, the kind Jewish lady who owned the small mom & pop neighborhood store, ratted him out to our mother who, unbeknownst to her children, had told her neighbors (Ms. Sophie was one of our neighbors) to tell her when her children misbehaved. I remember he got a whipping and he never stole money from her purse again, and neither did her other two children who were old enough to understand cause and effect at that time. Her beating me for no reason when I was 15 was a first); (4) my sister, 30 years (estimate). She was deployed in the drug mule entrapment attempt, used to get me to her drug dealer father; (5) my younger brother, unknown but estimate 30 years; (6) my youngest sister, unknown but estimate 25 years.

Someone allied with the Russian government knows the shopping patterns of you, ordinary person in America, including immigrants shopping patterns. Anyone who uses a membership card, or who've signed up for a grocery store's sales using the grocery store's card, the Russian government and it's allies know your preferences. They can't frustrate and harass unless they know our favorites.

The Russian government is determined to destroy us, and they have to destroy us one by one so it's faster and more efficient to entrap authority figures like parents, doctors, landlords, employers, etc., and use them to lure, entrap, and harass those subordinate to them. The Russian government isn't trying to entrap 300+ million of us, but they have to control enough of us to deprive President Biden of being re-elected, so that means the Russian government must control around 100 millions voters. To do that, they have multiple strategies, like breaking the societal contract.

Instead of being complimented or financially rewarded for excellent work, a worker is denigrated, sneered at and ostracized by their entrapped boss and/or coworkers; instead of a kind person being appreciated for her/his self-sacrifice, kindness and good attitude within their family, with facilitates family harmony and minimizes drama, entrapped family members treat the targeted person with contempt, with no explanation; instead of a customer being thanked by their entrapped mechanic when the mechanic has kept her vehicle for months longer than originally estimated for repairs, the mechanic informs her that his nephew totaled her car months ago. The mechanic had been letting the customer use his loaner vehicle but months after he finally told her why her car's repairs hadn't yet been done, he asks her to return his loaner car. The customer does so and is cheated out of her own car, nor is she compensated by the mechanic for the loss.

These attacks happened, and not just to me but at least some of them happened to one other person besides me. They are part of the Russian government's strategy to enrage ordinary people here, to prepare them to follow someone who says they understands their pain, to get them to rise up and overthrow our government. January 6, 2021 was evidence of how successful the Russian government has been at enraging ordinary people, we saw thousands of ordinary Americans storm the Capitol and for the thousands who showed up, many other thousands of enraged people stayed home.

Most people internalize the damage from the unrelenting psychological attacks but it's reflected in our obesity rates, our insomnia, our overdose rates, our declining scholastic scores, and our medication needs.

Chapter 53 – The KGB worked to enrage me

Here are more attacks the Russian government ran to deliberately enrage me: (a) I bought an old used car I was told by the seller had been completely rebuilt. I paid premium, her asking price. The engine blew about six months post-purchase and she refused to pay for the repair. I won in small claims court but couldn't collect the $600 awarded me. So I had to pay another $600 to have the engine rebuilt.

(b) I was talked into buying hit and run insurance by a reputable auto insurance company. I was hit and run. The insurance company refused to pay. It cost me $3,000 to repair the car, and I lived paycheck to paycheck. Police witnessed the hit and run and took off after the perpetrator. They found out who he was but refused to tell me. That was the first, but not the last, time I interacted with police who treated me as if they had a problem with me, but weren't willing to tell me what the problem was.

(c) I temped in my late 20s and 30s for extra money and later to gain experience as a legal word processor, legal secretary. After I'd completed a temp assignment at a small law firm, they asked me to work for them directly, not through the agency, for an additional week. I agreed, after which they cheated me out of my week's wages. Again, I lived paycheck to paycheck so being cheated and/or lied to by anyone caused me out-sized financial difficulty.

(d) In high school I worked part time, for necessities and also to contribute to our family. Our family's sole breadwinner was my mother who worked a factory job that didn't pay enough to support our family. We'd been homeless when I was younger and I was determined to go to college. In high school I asked and was told about college entrance requirements, the GPA I needed, the prerequisites I needed, the SAT scores I needed, and how much college might cost. I was told colleges liked to see extracurricular activities. I did all those things, while working. Then I began going to the high school's only guidance counselor asking for direction and advice about how to apply for financial aid. This was around 1975-76, before computers were ubiquitous. She was an operative or owned by the Russian government/KKK gang that ran that city. She helped me the bare minimum, discouraged me from applying for financial aid, and when I kept coming to her office seeking advice she lied and accused me of stealing $20 from her purse I theorize to get me to stop asking her for help. I didn't make a pest of myself, I didn't go to her daily, nothing like that. I'd done everything right but the Russian government has been deeply embedded in our nation for decades and they're anti-woman – you can see that by their facilitating the removal of women's reproductive rights, and their push to more tightly control America's Southern border. They're already anticipating taking control of our country, already dictating what they want here. The Russian Mafia works to destabilize us, our allies, and our neighbors, and doesn't want migrants to have an escape route. You see a similar strategy against Palestinian civilians. Israel bombs Gaza and no country granted bombed Gaza civilians temporary sanctuary, despite the many protests[297] calling for cease fire.[298] Our President pleads for emergency funds from Congress to provide humanitarian aid for Palestinian civilians, but the Russian government controls some of our House GOP. Our

---

297 General Valery Gerasimov, Chief of the General Staff of the Russian Federation,"The Value of Science in Prediction," *Military-Industrial Kurier* (February 27, 2013). "The focus of applied methods of conflict has altered in the direction of the broad use of political, economic, informational, humanitarian, and other nonmilitary measures – applied in coordination with the ***protest potential of the population***."
"*..Tactical and operational pauses that the enemy could exploit are disappearing...*"

298 Gaza Palestinian civilians being bombed would've welcomed an offer of sanctuary but only people with dual citizenship were allowed to exit through Egypt. I don't understand why. Palestinians must have dual-citizenship passports so they can escape bombs. Obviously, we must help Palestinians. I don't understand why they didn't receive more help. President Biden and his Administration pushed hard for civilian protections from Israel, and also for humanitarian aid, which I appreciated, but why didn't we and our allies get Palestinians out of Gaza? I must read up on Gaza. So the next war they won't be sitting there without an escape hatch, with bombs raining down on them. The 2023 Israel-Hamas War left more questions than answers.

Senate works to pass the emergency aid and I pray it gets passed soon.

For sure they don't want women or people of color to advance. The Russian government intends to destroy all of us, including Caucasian Americans, including everyone they've already entrapped. Entrapped police are going to get a nasty surprise if the Russian government's coup succeeds.

I have many more examples of these types of stealth attacks, which were designed to enrage me and make me feel life wasn't fair and that the deck is stacked against me. It is stacked, but by my enemy, the Russian government.

Even President Biden and his Administration are stealth attacked in this way. He, his Administration, our security agencies and our military, are excellent, but you'd never know it by the polls. The Russian government has mastered psychological warfare. It understands people don't work just for money, but to help others, and to be thanked. Instead of being thanked, the Russian government ensures our Administration receives negative feedback no matter how hard they work.

This type of stealth warfare is why Americans attacked the Capitol in January 2021. Before storming the Capitol many were probably stealth harassed to within an inch of their lives for years, but because they're not trained in psychological warfare, they had no way to tie the attacks together. Our Administration has some idea they're being stealth attacked by the Russian government, though I'm sure the attacks still hurt.

Imagine ordinary people with no knowledge of psychological warfare who are financially and psychologically attacked for years. I was attacked in this way. They're being deliberately enraged, maybe for decades. All the Russian government had to do was aim them at the Capitol.

People must be informed and educated as to what the Russian government is doing, or we'll keep being easily manipulated, and used as pawns to destroy our excellent government, while the Russian government laughs and swaps in themselves as a 4th rate slaver alternative.

Chapter 54 - Types of attacks on us by the USSR and the Russian Mafia

The USSR began it's secret war to destroy Americans, our government, and our culture, sometime after the middle of the 20th century, because:

(1) the USSR hated us and was determined to destroy us. The USSR found the outcomes from the Bay of Pigs, Cuban Missile Crisis, unacceptable to their self-image;

(2) our government has nuclear weapons, and in 1945 demonstrated their willingness to defend us against enemy foreign governments who attack us;

(3) the USSR government understood it could not survive a nuclear bomb dropped on it's headquarters, so they chose a secret war to destroy us.

The USSR's secret war is in contravention of our domestic laws, the Geneva conventions, anti-genocide laws, crimes against humanity and rules of war, many of which the USSR agreed to honor but they lied.

By the USSR's end in 1991, it had not destroyed Americans so the subsequent government, the Russian government, continued the secret war the USSR began. Americans can now see clear evidence of the USSR and the Russian government's success. The Russian government facilitated and/or facilitates:

(1) Attack our government. The January 2021 Capitol attack;

(2) Cause overdose deaths. Many thousands of overdose deaths here annually, with Americans in the prime of their lives stealth assassinated by the Russian government flooding our nation with fentanyl and other deadly drugs;

(3) Destroy our government. The increasingly obvious deterioration of our government. In 2023, the House GOP often couldn't even fake being an effective government;

(4) Cause severe under-stock of affordable housing. The millions of unhoused persons who can't find affordable, quality housing because the Russian government and it's operatives have mastered and manipulated our institutions/systems in such a way that affordable, quality housing is in severely short supply;

(5) Force our citizens to vote for puppets controlled by the Russian government. The Russian government, by commandeering our domestic entrapment system to aim at noncriminals and/or noncriminal families, and by entrapping

some of our law enforcement, have built a weapon where, when the Russian government adds influencers, install puppets into positions of power throughout our nation, including placing puppets in the White House;

(6) Commit genocide against us. Out of their hatred for us, the Russian government destroy our families by entrapping parents and force them to entrap their own children, causing a destruction of our families, the cornerstone of our culture, society, and government;

(7) Cause mental health deterioration of our citizens. Entraps parents and force the parents to psychologically and/or physically attack their own children;

(8) Destroys our student's ability to learn. The Russian government's success in degrading our educational system so that, in my personal experience, teachers don't really teach and prepare students anymore but expect students to grasp concepts discussed online. I've had at least two classes where the teachers gave a quiz and midterm but failed to show students how to answer the problem. We, our government, and our people are in a downward spiral because the USSR began a secret war to destroy us which was well in play by 1970, and which the Russian government has escalated;

(9) Destroys our worker's ability to thrive at work. Entrapped supervisors and co-workers and forced them to harass me;

(10) <u>Destroys patient's ability to trust medical care providers</u>. Entrapped medical care providers and forced them to try to entrap me.

Exploiting the Jackson-Vanik Amendment passed in the 1970s, the USSR embedded potentially tens of thousands of operatives, slaves and/or spies in America and some of those people were under order to destabilize us and our nation.[299]

The USSR continued its attacks against me until it ended in 1991. After which the Russian government came into existence and escalated it's attacks against me, using as weapons my supervisors, co-workers, extended family, people I thought were my friends, and strangers, deployed to harass, mistreat, physically, psychologically and sexually assault me, and attempt to entrap me.

When the Russian government found it couldn't entrap me,[300] it worked me to exhaustion and then had my parents and extended family work to entrap me. When the Russian government found that they couldn't entrap me, they decided to stealth murder me by "natural causes." Their attacks continue to the present. This is my story.

---

299 Jackson-Vanik Amendment granted normal trade relationships between the U.S. and other countries, including between the U.S. and the USSR.
300 In my last job the Russian government attempted an embezzlement entrapment I refused to exploit, and at the law firm before that, the Russian government had the accounting department issue me an overly large paycheck which I refused to cash but simply informed accounting of their mistake.

Chapter 55 - The Russian Mafia lies glibly but even w/it's reputation it keeps finding takers around the world - people have a problem sharing power[301]

The Russian Mafia is evil, but it doesn't lead with that while wooing light skinned, male dominant cultures overtly concerned about sharing political power with it's darker-skinned co-citizens.

It's easy to say to people "don't be racist, stop being afraid of other people because their skin is a different color" but that doesn't seem to help. I'm African American and have no fear of Caucasian people even though I should, given that the Russian Mafia has attacked me for 53 of my 64 years of life predominantly using Caucasian people as weapons against me, leading off with two Caucasian boys older and bigger than me, one hitting my 10 year old brother in front of me when I was 11, and the other threatening to be me up, while I'm 11, traveling home on the school bus.

The reason I'm not racist is because I can tell the difference between a male, light skinned Russian Mafia operative actively working to kill me, and/or his operatives stealth working to kill me (which is what normally happened);

---

301 I think it's a hard wiring problem. We're an animal species and animals, apparently, always worry subconsciously that there's not enough food and/or other resources to go around, so are always trying to lock in their share. If it's a hard wiring problem then the Russian government, unless, hopefully, we destroy it, is always going to have an "in" anywhere it seeks an entry. I sure hope I'm wrong. The Russian government has mastered a lot of psychology. It's a bad enemy to have because it cares about nothing but itself, and it's power; it'll throw anyone under the bus, including babies.

and a male, light skinned American minding his own business with no intention of harming me.

The difference is behavior: maleness, ethnicity and nationality have nothing to do with it. But, I was blessed in that the Russian Mafia didn't enter my life until my mother had already socialized me.

If people stopped being racist, the Russian Mafia would be out of business, Palestinians would have their own nation, Americans wouldn't be clinging to our democracy by our fingernails, and our immigration policy would be sorted instead of the mess that it is.

Light skinned male dominant cultures who feel threatened by their darker skinned co-citizens, women, and gay people, keep getting suckered by the Russian Mafia who enter offering KGB type attacks that are highly effective against innocent dark-skinned people but the Russian Mafia doesn't stop with the darker-skinned citizens but end up stealing the country out from under the frightened, dominant, light-skinned, male culture. Currently, in Israel and America, the Russian Mafia is actively working to steal both America and Israel from Americans and Israelis.

Chapter 56 - The Russian Mafia attempts to stealth murder President Biden using overwork & exhaustion

Before President Biden was sworn in, the Russian Mafia had it's operatives attack the Capitol, had them demand and insist that Mr. Trump was the winner despite all evidence to the contrary.

Since President Biden took office, the Russian Mafia has stealth attacked Americans, our government, President Biden, his Administration, our security agencies, and our military (the "Biden Team"), by throwing anything and everything at us the Russian Mafia can think of: forced us and our government to straighten out supply chains, made withdrawing from Afghanistan a nightmare, nearly destroyed on multiple occasions legislation intended to strengthen us and our nation, specialized baby formula shortages, etc.

Currently, the Russian Mafia is attacking us because we're aiding Ukrainians against the Russian Mafia's unprovoked bombings, working to pull us into a wider Middle East War as we aid our ally, the Israeli people, and overwhelming our Southern border with hundreds of thousands of desperate migrants who's nations and governments the Russian Mafia has deliberately destabilized and corrupted.

The Russian Mafia's intention, obviously, is to seize control of the world, and is killing Palestinian babies to unseat President Biden, and anything else they can think of.

In addition, the Russian Mafia is working to murder our beloved President before our eyes by working him and his Team into exhaustion, with the hope that if he collapses, the Russian Mafia will have a shot at installing one of it's puppets.

We see you Russian Mafia. We see you and we understand you.

You demand world dominance while you murder Palestinian babies and blame us, using protesters, cynically,[302] as if we cannot read and understand who and what you are.

You're willing to murder anyone and everyone not you to seize world power, even though all you are is hate, greed, slavery and ambition.

We refuse to allow you to take control of our species.

We see you attempt to destroy our beloved President's joy in providing excellent service. You will not succeed. We see you. We won't accept you. You've only slavery and death to offer.

In their enemy's and competitor's nations, that the Russian Mafia is waging a secret war against them, which goes a long way towards explaining why the United States and our allies have deteriorated so precipitously the last 60 years.

This book is me warning my people and the world about the Russian Mafia threat facing us.

Since, rough estimate, the 1950s, the USSR, and later the Russian government, have embedded operatives in America with the intention of destroying us, and our government. In 1970-71, USSR-embedded operatives, along with the gang they controlled in Southern California, U.S.A. comprised of the KKK, and local police, stealth attacked me when I was about 11. My mother, a 28-29 year old divorced

---

302 General Valery Gerasimov, Chief of the General Staff of the Russian Federation, "The Value of Science in Prediction," *Military-Industrial Kurier* (February 27, 2013). General Valery Gerasimov states: *"...applied in coordination with the **protest potential of the population**...."*

mother of four young children had within the last previous years had graduated from a surgical technician training in the Midwest and been offered a job from a hospital located in Southern California. Accepting the job, she relocated herself and her children and, liking the job, saved her income to make a down payment on a modest house in a neighborhood controlled by the gang.

Our President is excellent. The Russian government is slamming him hard, right in front of us. He is an outstanding President. The Russian government is making him hate the job to sicken him and kill him to make him stand down. We must see we're battling a master of psychological warfare who is determined to destroy us. We must see our reality and defend ourselves accordingly.

Chapter 56.1 – I survived the Russian Mafia in part because my government fed, housed & educated me after the KGB attacked me & my family – feeding impoverished children worldwide helps our species survive

It's excruciating witnessing the Russian Mafia destroy my family and attack innocent people living in America, proxy murder and injure tens of thousands of innocent Palestinian and Israeli civilians, bomb innocent Ukrainians, and attempt to seize world dominance from my government. Our beloved President, his Administration, our security agencies and our military (the "Biden Team") are well aware of most of the many problems the Russian Mafia commits here and worldwide, but are unwilling to wage nuclear war, because, sane and wise, they deem the cost of nuclear war too high to be borne by our people, innocent Russian civilians, and our allies. The Russian Mafia would drop nuclear warheads on high density states, bomb our nation's bread basket states, our fruit and vegetable producing states, our infrastructure, our electrical grids, our water systems, etc. Anything essential for our survival the Russian Mafia would destroy. I agree with the Biden Team's assessment that the cost of nuclear war would be very high and overwhelmingly disproportionately harm innocent Americans and innocent Russian civilians.

Nuclear weapons dropped on Japan in 1945 were far less powerful than those we and the Russian Mafia have today, and given who and what the Russian Mafia is it's unlikely Americans would recover for a few centuries.

I've been attacked by the KGB and later the Russian government in America for 53+ years of my 64 years of life. Their attacks began when I was 11, around 1970-1971.

At that time, the KGB had been embedding operatives for awhile here, with the intention of destroying our government, our nation and our families. I experienced the KGB's attack upon me and my family, being unaware that the KGB was working a secret war against us at the time.[303] Although General Gerasimov didn't admit the Russian government's secret warfare against civilians until 2013 and he implied the Russian government's strategy wasn't up and

_____

303 General Valery Gerasimov, Chief of the General Staff of the Russian Federation,"The Value of Science in Prediction," *Military-Industrial Kurier* (February 27, 2013). General Valery Gerasimov states:
"...Wars are no longer declared ..."

running until the 21st century, he lied. The USSR had their secret war aimed at American civilians and children by 1970 and I strongly suspect before then. I use 1970-1971 as the dates because I know the KGB was attacking civilians and children in America, against our domestic laws and against the Geneva conventions and genocide international laws they'd signed off on approximately decades earlier, because that's when the USSR began their attacks on me and my family.

Most people, and my sane government will try to negotiate peace with a sworn enemy but I say that won't work with the Russian Mafia. They were here attacking families and children at least as early as 20 years after signing the Geneva conventions; their attacks against women and children are illegal even during war, much less in peace. In 1970-71, we weren't at war with the USSR. I argue, because of personal experience, that the Russian Mafia is so blinded by hate and ambition they are unable to negotiate in good faith. Not just unwilling, unable. Their world view is so skewered all they see is themselves: they can see nothing but their need, their compulsion to dominate. Like a rapist can't see the victim as a person with rights but only as a tool to serve the rapist – the Russian Mafia is the same but instead of attacking one woman, the Russian Mafia attacks it's entire species. I'm sorry, I know this is hard to hear, but ordinary people worldwide must be warned.

If the Russian Mafia were a person they'd be rampaging through their society and would need to be imprisoned because they can't be trusted in society. The Russian Mafia can't be trusted in society – the men controlling it are unhinged, so our species must learn how to navigate and defend ourselves against a nuclear-armed, unhinged, apex predator mafia who has figured out how to manipulate human hard-wiring, and with that knowledge is working to enslave the world, after which they'll destroy Americans and steal what we have.

Because modern Americans have never experienced a coup and have never been destroyed, we don't know what it looks it, pre-launch. This book is to warn everyone living in America, and our allies, that we're in pre-coup countdown, set to go live November 2024.

Most people don't know humans are hard-wired so we must be educated on how to recognize attacks to better defend ourselves. First up, everyone worldwide must be warned of the danger. In the Russian Mafia's worldview they are the master race, everyone else is their slave.

The Russian Mafia are functionally insane and we must understand we're dealing with a brilliant, evil, insane, best-in-class mafia with an inferiority complex. If the Russian Mafia were stable they wouldn't have this out-sized need to dominant everyone. Currently, they control over a third of our species via BRICS.

The Russian government is Mafia and Mafia cares only about destroying it's enemies. Expecting the Russian government to think as a government means that we set ourselves up for disaster. It thinks like a mafia because that's what it is.

Yes, the Russian government will sign an agreement with us. It will sign an agreement with Ukrainians, it will sign a negotiated deal, but the agreement, treaty, whatever, will be meaningless because they are a Mafia government and the Mafia thinks in terms of destroying or enslaving it's enemies. To the Russian Mafia everyone not them is an enemy or a potential slave.

Having witnessed and endured attacks by the KGB and the Russian Mafia, having seen them entrap thousands of civilians and weaponize approximately 1,000+ of our children outside of my family, I recommend against making a deal with them unless we accept it as a time-buying strategy. We are a democracy so people must decide for themselves. I am female and a woman of color. I have health problems because the Russian Mafia has attacked me for decades, and my income is modest. Depending on the person listening to this argument, women, people of color, obese people and low income people, any low status person, is far less likely to be believed by a dominant species; plus the Russian Mafia deliberately attacks target's self-confidence, making them doubt themselves, so when low status people have a high quality idea they're too lacking in self-confidence to say so, which cheats our species out of untold millions of technical innovations.

If we fully harnessed our gifts from all our people, all those brilliant children in impoverished nations, all the brilliant children and adults in America, all our quick minded little girls and boys, we'd have a base on Mars by now, and several well-established bases on the moon. Because our hard-wiring prioritizes hierarchy, we're too slow harnessing the innovations from all our people. The Russian Mafia has built it's genocide around our species' hard-wired blind spot but we need to see beyond the immediate threat of the genocide to what we're learning: that our hierarchy-based hard-wiring is working against our species' long-term survival. All the world's children must receive timely, quality education, all the world's children must be fed and protected, live in decent quality housing, be protected in their families from inner-family and out-family threats. We are one species. We need all of our people, all ethnicity's, religions, we need all of us pulling in the same direction. We need all our people worldwide taught to read, taught math, taught science, taught how to plant a garden and to be fed with nutritious, delicious, quality food as they learn. We have no time to lose, we must build up our species. Our long-term survival depends upon it, outside of the immediate threat of the Active Measures genocide.

Let's see the revealed genocide as a wake-up call: we are a hard-wired species who wastes 90% of it's innovative potential because we fail to ensure all the world's children attend a good quality school, have good quality nutritious food, are raised in a stable country, within a stable family. The only reason I had the information to ignore the Russian Mafia's anti-American government propaganda, is because my government demonstrated to me when I was a hungry, homeless child that they wouldn't let me starve. I ate because my government gave my mother food stamps. Because I was kept alive and nourished enough, I've been able to endure the Russian Mafia's attacks and survive long enough to warn my country and the world. As far as I know, I'm the only ordinary person with no espionage training who has remained uncompromised. My people and my government will survive Active Measures but think if I hadn't written this book. We'd not have been warned the way we need to be warned. Our security agencies would've done a Hail Mary over the next several months but we're a government and people of laws and our security agencies hate going rogue, and would've resisted doing so. By the time the dust cleared, we'd have had a puppet in the White House with the next step being compromised slaves in the streets demanding a new government, and the puppet saying 'the people have spoken' – coup. Our legitimate government would've slowed it down but would've been unable to stop it.

My people and my government, our allies, will all survive because when I was a hungry, homeless child my government fed and housed me. All nations need to hear this story. How Americans saved themselves they take care of their children. Being fed and housed allowed me to finish high school. I was poor and wanted to attend university. I was told I could if I worked hard in school, took the appropriate prerequisites, earned good grades and took the

SAT. Even with a compromised school guidance counselor preventing me from accessing the scholarships and grants available to me, I learned enough to apply for a government-sponsored college grant keyed to academic performance. So you see, my government fed and housed and educated me. When I worked for an outstanding attorney who'd become entrapped by the Russian Mafia, he threw me a Hail Mary pass that got past his handlers, an educational package, encouraging me to keep learning so I could present an argument all people can understand.

Entrapped people have no ability to come forward. If he could've found a way, he would've but to him, suicide was his only way out and he refused to be an attack dog aka slave for his handlers. The KGB and the Russian Mafia has done everything they could to stealth murder me and at minimum destroy my quality of life but here I am typing away, warning my people and why am I able to do that? Well, my government owns and controls nuclear warheads which slowed the KGB and the Russian Mafia down. While providing me that level of protection, my government fed, housed, and educated me when I was a child/teenager and couldn't yet do those things for myself. As a teenager, I became a wage earner and a taxpayer and after college I got a full time job and largely remained employed until 2010, when I endured an Exhaustion Warfare attack by the Russian Mafia. I've always been proud to pay taxes, to know my taxes were helping vulnerable children like I was helped long ago. I never spoke about these feelings because nobody ever asked me and my viewpoint is just who I am. The Russian Mafia worked two income tax entrapment attacks against me that failed because I always considered paying taxes an honor and a duty.

Chapter 56.2 - I can't let you hope for a deal with the Russian Mafia without warning you that they won't honor it

In the late 1960s, early 1970s our government was actively working to improve relations with the USSR, a time period known as detente, yet despite that, and despite the fact that we weren't at war with the USSR, that they'd signed the Geneva conventions and anti-genocide laws in effect since the late 1940s and early 1950s, despite all that and more, the USSR was here, attacking and destabilizing civilian families and children at will, with no warning to our government or to the civilians and children under attack. Deliberately attacking civilians is a War Crime and a Crime Against Humanity even in war. In peace, it is evil and illegal. The KGB and the Russian Mafia simply don't care.

One of the groups KGB had befriended by 1970 was the KKK, and some of the local police in the KKK.

A deal with the Russian government, won't work to our benefit because the Russian government never honors deals with slaves it judges as inferior to itself and it judges everyone as inferior to themselves. It is a sadist, slaver government who, like the Nazis, views itself as the master race and every other ethnicity will be forced to bow down to it. The Russian Mafia needs that bow-down and lest you think surrendering the keys to the kingdom will satisfy it, you'd be mistaken. The Russian Mafia has issues. I won't go into detail about what it wants from us because I'm trying my hardest to keep this book readable by children, but I can tell you, you won't like it.

Unfortunately for us, many of our leaders in the mid-to-late 20th century underestimated (1) the Russian government's need to destroy us, (2) their willingness to sign any treaty, and tell any lie to shift our leader's gaze from whatever setup the Russian government was working on, and (3) their increasing mastery of manipulating human hard wiring. Our leaders weren't stupid: many were stealth blackmailed by the USSR. Those who wouldn't play ball, like the Kennedy's, were assassinated. And when the USSR decided they'd committed too many back-to-back assassinations to murder former President Nixon without raising red flags in our security agencies, they unseated Mr. Nixon via old fashioned corruption scandal. The USSR unseated him because he'd refused to allow the passage of the Jackson-Vanik Amendment, which the USSR intended to embed hundreds of thousands of former Soviet citizens here, an unknown number who were spies, assets, and slaves forced to work for the USSR after they'd relocated here.[304]

The Russian government, shielded by their nuclear arsenal which intellectual property they stole from us, has unrelentingly destabilized West-allied governments, and

---

304 Craig Unger, "*American Kompromat*" (Penguin Random House LLC 2021).

worse, is committing genocide against us by entrapping parents here and forcing them to entrap their own children.

Sometime after the end of WWII in the 20th Century, the Russian government (then known as the USSR), began a secret, multi-prong, slow-moving coup against us. Their coup had to be cautious, slow-moving, and secret because (1) the Russian government is determined to destroy us, (2) we have nuclear weapons to defend against predatory attacks, and (3) the Russian government doesn't believe it can recover from a nuclear bomb attack from us and so nuclear war avoidance has, so far, been a through line in their strategy against us.

The Russian government built their initial nuclear weapons using technology they stole from us. You'll find theft, lying, murder and criminality[305] at the heart of the Russian government. The Russian Mafia and old KGB strategies form the foundation of the modern Russian government. It is accurate to think of the Russian government as the Russian Mafia with a thin layer of government atop it.[306] The Russian

---

305 Calder Walton "How the Soviets Stole Nuclear Secrets and Targeted Oppenheimer, the 'father of the atomic bomb.' *The Conversation* (July 24, 2023).

306 The Mafia thinks of it's enemies as targets to be destroyed. Because the Russian government is in fact the Russian Mafia, it thinks of other nationalities as people to destroy and/or enslave. So, heads up, the Russian government will never stop stealth attacking us until it is destroyed. As long as the Russian gov/Mafia hybrid exists, it will work to destroy us. If you read history, you understand relationships between nations wax and wane over time. In 2023 we are close allies of the Japanese government, while 78 years ago our two governments were at war. Because of the nature of the Mafia, which views enemies are targets to be destroyed and/or enslaved, the Russian Mafia has no mechanism in play to declare a truce. It only knows zero sum. Anything the Russian government says is lies to gain an advantage. It see's anyone who forges agreements with them as suckers and weak. Have

government/Mafia hybrid (which I'll refer to as the Russian government) prefer to steal technology or anything of value from whichever person, business, agency, or government they target, rather than develop what they covet from scratch. The Russian government deeply covets the financial strength the United States government enjoys in the world and intends to steal it. As you see from the Russian government's bombing of Ukrainians, when it covets something, things generally don't go well for it's target. The Russian government is a predatory, criminal government that has successfully shielded itself from the repercussions from it's multi-decades long aggressions worldwide by using it's nuclear weapons as a deterrent, as well as by waging secret wars against enemy governments, enemy civilians, and enemy children, none of whom has it deemed to declare it's war against. That means the Russian government has destabilized pretty much whichever government and individuals it targeted because it refused to tell targets they were being attacked. In our case, it's abhorrent the Russian

---

you noticed Mr. Jinping when he and Mr. Putin appear together? Mr. Jinping, no disrespect meant to him, is psychologically dependent on Mr. Putin, was deliberately made so. Anyone in an alliance with Mr. Putin and the Russian Mafia is made into a slave by the Russian Mafia. The Russian Mafia believes they're smarter than anyone else because they've stealth attacked civilians and governments for decades, have seriously destabilized our people and our government, while we believed their words. The Russian Mafia has no mechanism in place to make peace. It doesn't want peace nor will it seek peace. It is confident it is within a decade out from embedding and keeping puppets installed into the U.S. presidency. The Mafia's doesn't make peace. They fight for decades until one side or the other blinks, and whoever blinks loses. I can survive my country losing a war. I cannot survive, however, my nation being enslaved. That's the only thing on offer from the Russian Mafia.

government labels us corrupt knowing it has attacked individuals without warning for, in my direct personal experience, over 53 years.

The Russian government is now a serious threat to West allied nations and to our species because it's unrelenting aggression means our species can't get off the petroleum flywheel threatening earth habitability because, as soon as America makes deals with oil, mineral, and rare earth suppliers of other governments, the Russian government will seize control of those governments. There is no way our species will survive the climate catastrophe with the Russian government in play. It won't tell us because the Russian government doesn't concern itself with truth.

Although the Russian government loves to say we're the problem, it lies. The Russian government is a highly acquisitive, slaver government. It is brilliant and insane, simultaneously. Think of it as a mass murderer who looks normal but who is psychologically unbalanced and suffers from compulsions it can neither control, nor wants to control. This is an accurate description of the Russian Mafia.

What about the model of leaving the Russian government alone? Well, we're doing that. The problem is the Russian government is so acquisitive, egotistical, and insane, it's cannot help itself, nor does it want to stop itself. It has been extremely successful in it's secret wars attacking ordinary people, which has encouraged it's delusion that it's been chosen by god to rule our species. It's like a mass murderer who looks normal but who is mentally ill. It will never believe or accept that it is destroying it's species and the habitability of earth. Not only will it not believe that, it doesn't care that it's destroying it's species, and is in denial.

Think of it this way: the Russian government is Mafia-controlled. The only thing the Mafia cares about is destroying it's enemies and competitors: us. The Mafia component of the Russian government believes that it's veneer government will stop it if the Mafia goes too crazy, but the veneer government has drunk the kool-aid, is convinced that if it allows it's Mafia to have it's head, the Mafia (who cares about risk assessment only if it's about to be bombed by nuclear weapons) will steer the country to victory.

The Russian government veneer is in denial,[307] refusing to believe our species is self-extincting because it doesn't want it to be so. Our species is really in trouble. Hoping the Russian Mafia will steer us out of trouble is to hope for rationality from an insane, power-mad, government who authorized the Israel-Hamas War, knowing thousands of innocents would die. Although the Russian Mafia hides the great majority of it's evil, the small percentage we're allowed to see, like their chemical warfare murdering Syrian civilians, and their unprovoked bombing against Ukrainians, is more than enough to flash bright red warning lights. It is a psychologically unstable mafia buttressed by Active Measures, the KGB genocide bequeathed to them in the 1990s, set to go coup-live

---

307 Mark Fischetti "Our Fragile Earth: How Close Are We to Climate Catastrophe?" *Scientific American* (September 26, 2023).

November 2024. Active Measures has been live for decades but installing a puppet in November indicates the Russian Mafia thinks they have everything in place to launch a coup. The puppet installation isn't the coup – the coup is after the puppet is installed whereupon the Russian Mafia will flood our streets with tens of millions of enslaved Americans demanding a new government. The enslaved dominants are all ready to go, the puppets are primed and ready. Their families have been destroyed and they've been told our legitimate government is culprit running entrapment attacks against innocent Americans, leading to deep hatred against our government that's based on a lie. The Russian Mafia believes lying to victims, like it believes hooking kids on fentanyl, is acceptable behavior. It doesn't care that civilians have no idea they're under attack. To the Russian Mafia, it's the other government's fault for not protecting their people, while the Russian Mafia knows and deploys as a weapon, it's enslavement and manipulation of our dominants which it knows our species are hard-wired to protect. So when the Russian Mafia says 'that's the other government's responsibility,' while it knowingly prevents blackmailed, threatened and coerced compromised plea deal controlled victims to speak, it's a rigged deck.

So, we'll survive a rigged deck. This book is to counter the Russian Mafia's rigged deck.

The Russian Mafia will destroy us if we don't stop them. So we'll stop them.

When a government willingly and knowingly throws children under the bus, that government is insane. It's no use expecting anything other than insanity from it, and it's a colossal mistake to allow that government to control our species, no matter what it says. The Russian government is off the rails. If our species intends to survive, we must better defend ourselves.

There are hundreds of millions of Americans. Let's assume the Russian government drops nuclear bombs on us, targeting high density cities, our nation's bread basket, our water and infrastructure. Many tens of millions of us will die but many millions of us will survive. If we're blessed, our government will survive, as will our military and we'll have a means to defend ourselves.

Centuries ago, England, Portugal, Spain, the Netherlands, etc., were top dog governments who over time stopped being top dog. Top dog governments come and go but it is evil and insane to destroy a people because you envy them their top dog status.

The Russian government plans to install a series of puppets in the White House. Variations of their strategy can be seen in Israel, Hungary, Italy, Brazil, Turkey, Syria, South Africa, France (is up next), etc. In China and India they backed, mentored and protected their choices as the men ascended to power. The Russian government has enjoyed a great deal of success the last decade, meaning it's easier for anyone paying attention to interpret their strategies.

The Russian government is best-in-class at stealing, lying, cheating, and manipulating human hard wiring because that is what they prioritize. I don't include ordinary Russians in my description of the Russian government. I theorize that ordinary Russians are as hardworking and intelligent as any other ethnicity.

Because the Russian government prioritizes manipulating human behavior, and has successfully conned unknown numbers of individuals and governments, it's leaders apparently believe they've been chosen by god to rule our species.[308] That is untrue. The reason the Russian government has successfully destabilized so many nations and peoples, and successfully installed so many pro-Kremlin governments around the world, is simply because they attack without warning, civilians in targeted nations, including children.[309] Any government with a basic knowledge of human behavior can ambush civilians and children, because those people, in general, know nothing about espionage. Even police, reliant on trusted supervisors, hierarchy and chain of command, are easily entrapped by compromised authority figures who've been ambushed by the Russian government.

America has deteriorated so precipitously the last 50+ years because the Russian government is on the ground here,

---

308 Roger Cohen, "The Making of Vladimir Putin" *New York Times* (March 26, 2022).
309 General Valery Gerasimov, Chief of the General Staff of the Russian Federation,"The Value of Science in Prediction," *Military-Industrial Kurier* (February 27, 2013). General Valery Gerasimov states: "...***Wars are no longer declared...***"

stealth destroying families: which form the cornerstone of our culture, society, and our government. To that end, I charge the Russian government with committing genocide against my nation.

I write from direct personal experience. I am a 64 year old African American woman the USSR began attacking when I was 11 and which attacks continue to the present. In this book, I describe how the Russian government is attacking us.

The Russian government believe they are smarter and cleverer than us because they've ambushed us with KGB and Mafia tactics. Because it is Mafia it operates differently than governments. Most governments work to find a medium to interact with enemies but the Russian government works to destroy enemies. The United States is a country and has a government but to the Russian Mafia, it is an enemy to be destroyed. The Russian government will never stop working to destroy us, the Mafia has no mechanism in place to stop attacking us, especially given how successfully it's destabilized us.

The Russian government intends to destroy us, destroy our government, and steal what they want from our government. They have zero problem destroying our families to get what they want. Indeed, the Russian government is incapable of caring about anything but destroying us and stealing the financial power it covets from our government. I charge it with genocide against me and Americans because of the many attacks it has waged against me and my nation for decades.

Chapter 57 - Our children aren't safe, our elderly aren't safe, all of us are under threaten. The Russian Mafia hides it so you won't defend yourself

When someone, no matter how important they are to you, suggests you do something you know will harm you, harm someone else, or may negatively impact our country -- examples include suggesting you break the law, take drugs, or re-open a high-profile investigation, that person is telling you in the only way their handler will allow, that they are compromised. Never ignore that tip for it is unlikely the Russian government will give you another. Don't be distracted by your love or sympathy for that person, whether that person is the love of your life or a most beloved grandparent – the Russian government exploits your emotional attachment in order to destroy you.

Before she died, I'd known my friend for decades, had no idea until the last approximately 12 years of her life she was an operative. In her 90s, in her wheelchair, the Russian government forced her to harass me. I know operatives from my birth, I know an operative from their birth. No matter how much you love someone, if they're entrapped by the Russian government, you're dealing with the Russian government when you interact with them. Friends & loved ones are unwilling slaves but slaves nonetheless. Slaves who cannot free themselves.

The Russian government entraps them to entrap you. The Russian government entraps you to kill you. If you don't believe me, if you forget, or if our government & people don't launch an effective defense, we will be destroyed. (Author's quote)

This book tells the story of how the Russian government and it's predecessor government, the USSR, lied and cheated

Americans in the 20th century, and continue to lie and cheat us (and the world), to realize their idea of nirvana: world dominance and the destruction of America.

Ordinary Americans will be surprised to learn that they and their children are in the cross hairs of the Russian government because of the type of warfare the Russian government has chosen: a coup.

Their version of a coup involves installing people they control into leadership positions throughout our nation, businesses and agencies, including in the White House. They install their people (who I call puppets) into positions of power by influencing and/or coercing entrapped citizens to vote for the puppets and the Russian government does that by ambushing noncriminals (ordinary people) by commandeering our police's entrapment mechanism. Some, not all, of our police are entrapped and entrap ordinary people. Entrapped citizens are herded by Russian government's influencers into the GOP Base, where they are used as a threat against whichever GOP politicians the Russian government is working to control. Entrapped voters don't speak out because they sign confidentiality agreements and because when they're arrested, the Russian government convinces them that they're corrupt, when the truth is they are victims of a world class, best in class Mafia government determined to destroy us and our government. According to Janis Berzins in his paper "The New Generation of Russian Warfare" "...the Russian view of modern warfare is based on the idea that the main battle space is in the mind."[310] They go in hard convincing innocent, ordinary people who have no experience in espionage or entrapment that they're corrupt, they show the victim the proof and the victim doesn't understand, because the Russian government doesn't tell them, that when an authority figure convinces them to lie, cheat and steal, that that's illegal, that they were set up.[311] The confidentiality agreement, the threat of a long prison term is they break their plea deal, their further grooming to criminalize them after they're initially entrapped, and them being convinced by Russia-controlled American handler's that they are corrupt and what you have is millions of who are convinced they're less than ethical, and here's the kicker, none of them know the Russian government is the perpetrator. Just as the Russian government facilitated the Israel-Hamas War and has the world blaming President Biden, it's convinced millions of Americans that they were and remain entrapped by their legitimate government. It's the con to end all cons and if we don't get a clue, fast, the Russian government will destroy us.

---

310 According to Janis Berzins in his paper "The New Generation of Russian Warfare" "...the Russian view of modern warfare is based on the idea that the main battle space is in the mind." Aspen Institute | Prague, *Aspen Review* (March 2014). New Generation Warfare: https://en.m.wikipedia.org/wiki/new-generation-warfare.

311 Russia-controlled American operatives forced my beloved father to try to entrap me by telling me to exaggerate my health problems to Disability after the Russian government had their other operatives work me to exhaustion. Beloved parents are authority figures. My dad worked as a police officer, retired as one. He was an honorably discharged veteran. Using him to entrap me is illegal but the Russian government couldn't care less about our laws, us, our children, or anything else but seizing world power.

This book is largely told from my personal experiences. In it I share: (1) the Russian government is deeply and widely embedded here, (2) how it stealth attacks ordinary people, including our children, and (3) it's warfare model embraces secret wars against it's enemies and their civilians/children.[312] To them, the end justifies the means, which explains the dismaying decline of our beloved nation the last 60 years.

The Russian government embraces cheating: (1) they entrap parents and force them to entrap their own children; (2) they have a cashier they control at a big box store scan half maybe ten grocery items in your shopping cart so store security can accuse you of shop lifting after you've left the store and arrest you; (3) annoyed when you qualify for a covid rental assistance program, they slow-walk your application, forcing you to reapply and kick your completed application out of the completed pile and while you're trying to find out what happened, change the covid rental aid rules without telling you, to set you up to commit fraud if you say you qualify for the new program; and (4) they have a math teacher's assistant accuse you of cheating on a math test and throw you out of the class, even though he says while he's throwing you out of the class, that someone was cheating on you. All the four above stealth attacks I've experienced first hand, and many more, and they're all designed as attacks only the victim and the perpetrator knows about, meaning, Russia-controlled American

---

312 General Valery Gerasimov, Chief of the General Staff of the Russian Federation,"The Value of Science in Prediction," *Military-Industrial Kurier* (February 27, 2013). General Valery Gerasimov states: "*...Wars are no longer declared....*"

operatives who attack you, and you, are the only ones who know you're under attack. I've reported some of these attacks to the FBI and the attacks continue. I'm positive I'm not the only one in this country being attacked in this way and felt I needed to write a book to warn the world.

That's why our numbers of unhoused people, fentanyl overdose, mass murder, plunging student test scores, and other societal problems, are soaring. I believe the Russian government is working similar stealth attacks against citizens in other countries, to increase immigration pressures and to destabilize as many governments it considers it's enemies as it can. The Russian government intends to flip the world order. It strongly prefers the master/slave model, with it as world master and everyone else a slave; (3) the Russian government is responsible for the deterioration of our culture over the last 50+ years as part of it's secret war against us, and (4) the Russian government is committing genocide against us by commandeering our entrapment system, targeting noncriminals (ordinary people), especially parents (and other authority figures), who the Russian government then force (the parent/authority figure) to entrap their own children (loved ones and/or friends). The end goal of entrapping more innocent people is it gives the Russian government more weapons. To the Russian government people are useful as pawns and weapons. It cares about nothing except ascending to world dominance. I believe it began the Israel-Hamas War to flood our media with the horrific images of bombed babies in order to unseat our beloved President. If I'm right, the Russian government couldn't care less about the innocent Israeli's and innocent Palestinians injured, raped, and/or murdered. It's focus is shifting the world's blame on to President Biden to unseat him. It doesn't occur to the Russian government that murdering innocent people, including babies, is unacceptable. It is a very focused and psychotic Mafia government and it's determined to control the world. I prefer that you believe me but if you don't, the Russian government will demonstrate what I've written here all by itself.

The Russian government worked for decades to enrage me. You can see what happens to people the Russian government enrages and aims at a target. The January 6, 2021 attack on the Capitol is an example. We witness the end, the hostility aimed at our institution, but the Russian government hides how they stealth enraged the mob. It's not just grievance politics. To build that rage the Russian government accessed everyone who participated in the Capitol attack, and thousands more who didn't show that day, and had operatives break the societal contract with those people. When ordinary people do good, societal acceptable behavior, from childhood we're rewarded. But the Russian government, deeply embedded here and masters of psychological warfare, rewarded all the people who once were rule followers, with nasty. It's like grooming. At birth we are trained that if we do a certain behavior we get punished and a different behavior gains us reward. To enrage people with the intention of using them to overthrow our government, you must have access to the inner workings of ordinary people's lives and you must use authority figures in their lives to frustrate them. To, day after day, year after year reward the good they do with a psychological kick in the face so that they feel hopeless, that the game is rigged, that they'll never get ahead no matter how hard they work or their best efforts. Then you unleash them. In this book I'll describe how the Russian government worked to enrage me and why it's a very bad idea to let the Russia-controlled American operatives manipulate you.

For the part of their coup where they deliberately destroy our families, which families form the cornerstone of our culture, our nation and our species, I charge the Russian government with genocide against my people.

Chapter 58 - Perfidy against civilians/children in times of peace: the USSR displays it's core

17. Perfidy attack by the Russian government's operative is illegal in peace times and even in times of war under the Geneva Conventions of 1949 which give special protections to women, children and civilian. Because the seducer was a hostile and asset of the Russian government, a enemy foreign government who was stealth targeting a civilian woman and her children, all children under the age of 15, this attack was a War Crime. But General Valery Gerasimov, Chief of the General Staff of the Russian Federation says in

"The Value of Science in Prediction," Military-Industrial Kurier (February 27, 2013) states:[313]

> no more **rules of war,** so to the Russian government this was not a war crime. Now, in 2013 the Russian government is saying this was not a war crime but the crime happened in 1970-1972, when the USSR was in power in Russia and it was a signatory of the Geneva conventions, so this was Perfidy, and was a crime in the laws of America and also because of the geneva conventions.[314]

My civilian mother and her children were all protected under the laws in force in California and America in 1970-1972. The USSR knew what they were doing was illegal, which is why they went in stealth. Also, the USSR was a signatory of the Geneva conventions so even had this been war time, it's attacks against a civilian woman and her children, all the children under 15, was illegal. The attack was especially heinous because the Russian government stealth attacked civilians in peace time, without declaring their intent or identity. The Russian government continues to commit these types of War Crimes to the present day and are not held accountable. People must be informed so they can report these attacks to their governments, in the hopes that their governments will better protect them and/or in hopes ordinary

---

313 General Valery Gerasimov, Chief of the General Staff of the Russian Federation,"The Value of Science in Prediction," *Military-Industrial Kurier* (February 27, 2013). "...**The very "rules of war"** *have changed.* The role of nonmilitary means of achieving political and strategic goals has grown, and, in many cases, they have exceeded the power of force of weapons in their effectiveness...."
314 Ibid.

people better protect themselves.[315] Victims are so shamed by the police, who in my mother's case I believe were part of the gang who attacked her, and the Russian government so experienced in running these types of attacks, that ordinary people, unless they speak up, are left feeling forever diminished and have no idea they were deliberately targeted by a foreign enemy government in a coup attempt

18. Deliberately targeting and making homeless four children under 12 against American law and in contravention of the Geneva Conventions of 1949, to which the USSR was a signatory, which prohibits targeting children under 15 years of age in war time. In peace time the Russian government is operating outside the laws of our country but they don't care;

19. Deliberately targeting and making homeless three siblings, my niece/nephews, as the Russian government targeted my family in it's efforts to destroy me and my family. At least one of these children were under 15 in contravention of the Geneva Conventions of 1949, which prohibits targeting children under 15 years of age;

20. Russia-controlled American operatives take control of security personnel in grocery stores and in big box stores and direct them to harass, stealth attack at least one innocent customer, me;

21. The Russian government works unrelentingly to demoralize and disenfranchise me by covertly and/or overtly

_____

315 Shaun Walker, Pjotr Sauer and Tom Phillips "Panic and Emotional Pain as Alleged Deep-Cover Russian Spies Vanish," *The Guardian* (April 3, 2023).

attacking me pretty much wherever I am. Russia-controlled American operatives go out of their way to demonstrate to me that they control the ground here. They over-prove it, as if compromised American law enforcement believe I'm challenging them when all I'm doing is grocery shopping. Russia-controlled American operatives go out of their way to ensure the society I engage with denigrates me and/or harasses me. A recent example is I sent my mother money I'd promised her for upcoming expenses. On my way to the mailbox, a male, Caucasian operative walked very close to me and belched in my face, as I pushed my mobility device down the sidewalk as it had run out of battery. In whatever way Russia-controlled American operatives can harass me and not get caught, they do so. I talked to my mother and she thanked me for the money but used a racial slur in referencing me as we were hanging up. My loved ones are compromised and Russia-controlled American operatives, their handlers, take every opportunity to use them to insult me or do something they know I won't like, so every time I engage with my society I am harassed by my society. I noticed a variation of this type of attack years ago. I'd do something that benefited someone and the Russian government ensured I'd get a psychological kick in the face for it. Another type of attack is sometimes my mother will smoke when I'm in her apartment when she knows I can't breath well when she does, or I'm in my sibling's vehicle and she'll have her vent open, and an operative in the car ahead of us will not have had their car smog checked and the exhaust will fill my siblings car, which causes me breathing difficulty and headaches. It's important that people understand reading this and believing me that the Russian government deliberately harasses individual ordinary people and often double teams and triple teams targets. They are very hostile and hate-filled and they got a big head start in the 20th century using blackmail and hard-wiring-based attacks;

21. I charge the Russian government with Perfidy against all Americans and migrants living legally and illegally in America, from 1971 to the present. I use 1971 as my charge start date because that was when I and my family were first stealth attacked by a Russian government operative representing himself as a civilian noncombatant. Perfidy is a war crime; and

22. Deliberately destroying families in America by entrapping trusted adults and authority figures like parents, doctors, employers, etc., and forcing them to harass and/or entrap people who love and look up to them is Perfidy, a war crime, and constitutes Crimes Against Humanity because the Russian government intents to not only destroy the American people and our government, but by destroying our families, and preventing us from effectively fighting the climate catastrophe, the Russian government is waging Crimes Against Humanity; essentially throwing our species, and nearly all other earth species under the bus, into extinction.

Chapter 59 - President Biden is excellent, which is why the Russian Mafia needs him gone & we must keep him

You'll notice endless articles about President Biden's age, or anything else the Russian government can think of to get voters to not vote for him. The Russian government doesn't have their operatives explain that they want him gone because he's an outstanding President, and because they can't entrap him. One of the primary reasons the Russian government authorized Hamas to attack Israeli in the 2023 Israel-Hamas War is because they wanted voters to see murdered, bombed, traumatized babies, and protests, all blaming President Biden and Israel in the run up to the 2024 election. Think about it: the Russian government authorized Hamas, a terrorist organization they control through their control of the Iran government, to attack innocent Israeli's, knowing the Israeli government would respond, and because Hamas governs Gaza, the Russian government knew the Israeli government would go to war where civilians, including nearly a million children, live. That's who and what the Russian government is. It facilitates the murder of babies in order to unseat a political rival. Would you do that? Would you kill 10,000 babies and children because you hated a government so much, you'd kill innocent children just to see him blamed for it? The Russian government is demanding species dominance this century. Every time they reach for it, they demonstrate why they're not worthy. If a government facilitates the murder of even one child to political destroy an enemy, that person is unworthy to rule our species.

The Russian government facilitated the deaths of 15,000 children. And counting.

Chapter 60 - The Russian Mafia insures our IRS is underfunded,[316] meaning we can't control our borders, which the Russian Mafia then use as a rallying cry

Brazilian leader President Lula, an ally of the Russian government, implied at a BRICS meeting, which was covered by the *New York Times* and reported on 8/22/2023 that the climate catastrophe is a ruse by our government to control the world. He said: "We cannot accept the greedy neocolonialism that imposes trade barriers and discriminatory measures under

---

316 The Russian government intends to install pro-Kremlin operatives in the White House and throughout our country. Our nation is chronically underfunded which is why we're about $40 trillion in debt. We must overhaul our tax structure, etc. One of the attacks the Russian government enjoys is to destabilize our neighboring country's government and invite a great many migrants. We need migrants, we want migrants but the Russian government, who is tightening it's control over our institutions, doesn't more migrants because, despite what most Americans don't know, the Russian government has a pretty good handle on manipulating people here. That's one way they got so many enraged citizens to show up at the Capitol on 1/6/2021, because the Russian government is running an harassment attack on our people. They even run it on our beloved President and his Administration, exhausting them with overwork, doing whatever they can to make the Biden Team hate the job. I know I keep writing this but we need to destroy the Russian government. Until we do, we need to revise our tax system so our government takes in much more money than it's presently doing. If we were out of the red, we could reconfigure our immigration system and hire all the staff we need and build everything we must put in place to process all the desperate people who need and want to come here. The border, and homeless, are two visuals the Russian government use as weapons to suggest our government is substandard. It's really the Russian government that's substandard, and a slaver government, but they love lying and throwing shade. With the Russian government destroyed, we'd regain control over our government again so it'd actually work for us, and not against us, as the Russian government has it doing.

the guise of protecting the enforcement."[317] He goes on to say in discussing his and the South African government's intention to work together: "...we can work together to combat it's [climate catastrophe] effects, but not on terms dictated by major powers or existing international institutions."[318]

The Russian government uses operatives it controls here to prevent our IRS from being funded as it must be to fund upgrades to our infrastructure and to keep promises we've made to impoverished nations struggling with the climate catastrophe. The Russian governments needs our IRS to be financially weak[319]. Financially weak we can't adequately protect ourselves from the climate catastrophe, can't pay the impoverished nations we've promised who are suffering the effects of the climate catastrophe, because we're out of money. The Russian government has it's operatives continually report that we're a wealthy nation but we're heading towards $40 trillion in debt soon: a rich nation doesn't willingly carry $40 trillion in debt. We must increase taxes just to survive, and change our tax laws to get more money from our people and

317 Lynsey Chutel, "At a Much-Watched BRICS Summit, Putin Tries to Rally Support, *New York Times* (8/22/2023).
318 Ibid.
319 I think this is another reason it encourages entrapped people to cheat on their taxes: it weakens the IRS while gives the Russian government more leverage against victims. I saw this in my family when entrapped family tried to lure me to cheat on my taxes within the last five years. Also, I worked, a lot and hard and by working me exhaustion and getting me on Disability, the Russian government flipped me from a being a employee paying at least 35% of my income to taxes, to me drawing on Disability. I read an article saying there is a high disability rate in England and a bell went off: it's very possible, probably likely, that the Russian government is in England stealth harassing and sickening workers as well.

their companies, and we need more money to effectively fight the climate catastrophe. The Russian government doesn't tell their allies the many ways they're attacking us. We're described as neocolonialists trying to hold onto world power, while we're victimized by the Russian government, and not for the first time. We're being attacked and are barely holding on. We need divine intervention and/or to talk frankly about the Russian government in order to survive the coup they've planned for us. They're going to offer us, probably through their ally China, slavery or war. We're not yet ready for war but we must not accept slavery.

I've seen how the Russian government's American operatives encourage entrapped victims to cheat on their taxes, acquiring ever more leverage to blackmail innocent people, while weakening our IRS with a thousand cuts: even when the IRS gets an infusion of cash, it's rarely fully delivered. The Russian government lies and misdirection in order to blackmail and further destroy innocent people. It wants African Americans[320] and the world, especially it's allies, to see America as some racist outlier, but the truth is the Russian government is deeply embedded in our policing, etc., so embedded I believe they shadow control our nation's policing policies and overall approach to policing our citizens, our prisons and our jails. Were it not for the Russian government's insistence on destroying us, I believe our policing systems

---

320 Jonathan Weisman, Ruth Igirlnik, Alyce McFadden, "Poll Finds Wide Disapproval of Biden on Gaza and Little Room to Shift Gears," *New York Times* (December 19, 2023).

would more closely match that in Nordic nations, where police actually help citizens instead of killing minorities. We must get the Russian Mafia out of our nation else they will destroy us. Even now I'm positive the Russian Mafia intends to assassinate me, just by how the operative upstairs shadows me even when I'm in my hallway (apparently she no longer lives there but the workmen harass and/or shadow me unrelentingly when they are in that apartment). I'm determined to publish this edition of the book before I leave my apartment to pay my rent because I know I may be killed before I return. The Russian government controls my landlord in such a way that the landlord won't allow me to pay two months rent at a time, forcing me to be physically vulnerable to the Russian government's many operatives here. Things aren't looking good for me in terms of longevity but at least I wrote this book. The first edition is a mess, this one is better but I'll work on it more if I survive. I also notice the Russian government has changed the way I can access the New York Times. I truly do smell a hit and I fear it'll mean my death. At least I've warned my people. I must get this book published before I leave to pay my rent. I can pay my rent late, if need be. That's an option. I'll have to pay a late fee I don't have, still, that'll give me more time to get this book out, which is the most important thing.

Our people must be informed and warned about the Russian government or we will be destroyed. I hope people read this book and that it helps them stay safer.

Chapter 61 - The Russian Mafia is determined to install a puppet here

The Russian Mafia is on the clock. Mr. Putin is 71 in 2024 and he's determined to seize world control in his lifetime. He's determined to install a puppet here in the 2024 election, 2028 at the latest. They don't want President Biden or his Team (President Biden, his Administration, our security agencies, and our military) because the Russian government doesn't own them. Operatives are probably on the periphery of the Team, but the Biden Team isn't owned, which is a huge advantage for us.

Bad things happen to citizens when (1) their government is owned by the Russian government, (2) when the Russian government casts a covetous eye on a country (as it's doing against Ukrainians), and/or (3) when the Russian government decides to exploit unprotected people to use as pawns (as it's doing in the Israel-Hamas War by facilitating the murder and/or injury of tens of thousands of Palestinians in order to unseat President Biden and destroy Americans).

The only thing the Russian government fears is our bombs in the hands of a President and Team who (1) generally know what the Russian government is doing here (destabilization, drug flows, etc.) and abroad (destabilization, drug flows, etc); (2) understand it's increasingly likely that, given the dysfunction we're witnessing in various parts of our government, future Administrations will be owned by the

Russian government unless we take action; and (3) refuse to be enslaved by the Russian government.

I've witnessed who and what the Russian government is. Despite the limitations of working within the entrapment mechanism they've set up, and installing puppets, they stealth assassinate our people using drugs, and mass shootings. They commit genocide against us by destroying our families. Even the Nazi's, one of the most evil regimes I'm aware of, didn't force parents to gas their own children. Here, the Russian government enslaves parents and forces them to enslave their own children. They destroy our family unit, and thereby commit genocide against us.

Chapter 62 - The Russian Mafia exploits our species hierarchy & hard wiring to destroy us

We are a hierarchy based species so all the Russian government has to do to destroy, enslave, and psychologically damage millions of us is to entrap one of the family's authority figures, insert a new person into the family and transform him/her into an authority figure, build an authority figure outside the family structure and insert them into the family, etc. Then use that slave to entrap the family. I've seen the Russian government do it in my family. The Russian government is quite evil and quite determined. We must better defend ourselves.

Chapter 63 - Big fish Americans, Israeli's are used as bait to unseat our President

Blaming Israeli's for the excessive deaths of Palestinian civilians, including many thousands of children, is the Russian Mafia's play to make Israeli's feel even more isolated, frightened, ostracized, and vulnerable – to herd them into voting in an even more pro-Kremlin government the next election. This is the Russian government's preferred way of destroying their enemies – coups -- by installing puppets in power positions. Israel is under attack by the Russian government who intends to enslave them. Israeli's vehemently protested Mr. Netanyahu's power grab from the courts. Using Hamas to attack Israel, a move that traditionally see's citizens voting for extremely right wing, pro-defense governments post-attack, is another component of the the Russian government's multi-decades long coup against Israel, their intention, to herd Jews yet again into slavery. Last century the Russian government oppressed the Jews (and other minorities) to such an extent that millions were killed and/or imprisoned. The Russian government is an extremely racist, pro-slavery government who especially loves Jewish slaves, although it will enslave any people it can. This is more than opinion. All you need do is read 20th century Soviet Jewish history to see the blood and despair of innocent Jews was smeared over multiple regimes.

The Russian Mafia is unwilling to let Jews go free. That's a basic truth. My prayer is that Israeli's will slip the noose and over the foreseeable future vote for government leaders the Russian Mafia doesn't control.

In America I theorize[321] Russia-controlled American operatives review the "crime" with the arrested person step-by-step, emphasizing to the target their free will and "culpability" at every step. The Russian government's operatives cherry-pick critically important details like omitting to tell the arrested person that the police who arrested them are compromised and forced by the Russian Mafia to aim entrapments at them.

The Russian Mafia would, no doubt, stress that the target "chose" to make important choices that lead them to being arrested, and will de-emphasize whatever the Russian government dangled in front of them. Russia-controlled American operatives won't disclose that they placed an operative in the target's bed. The Russian government is here illegally, is entrapping our people illegally, has commandeered our entrapment system illegally and that our government, of course, has not authorized an enemy foreign government to entrap civilians.

Nearly everything the Russian government says is based on all lies because they omit key points every at every step. It is illegal to insert an operative into a noncriminal's bed in order to entrap that them. It is illegal to entrap a noncriminal's parents in order to entrap the noncriminal. I believe the Russian government has our legitimate government

---

321 I'm not entrapped and entrapped victims aren't free to discuss their entrapment so I theorize based on my decades of experience being attacked and the behavior of entrapped victims.

believing our American citizens are corrupt when the truth is a foreign enemy government has commandeered our entrapment system in order to destroy us by inserting their puppets into leadership positions here, which is the Russian government's preferred way of working a coup.

So many Americans hate our legitimate government, because the Russian government has commandeered our entrapment system and forces noncriminals to entrap their children, extended family and friends, all under the lie that our legitimate government is working these attacks against it's own people.

Chapter 64 - Why WW3 is preferable to a deal w/the Russian government

The Russian Mafia has entrapped, I believe, millions of our people and possibly tens of millions of people around the world. They were able to entrap so many Americans because the KGB got a head start in the 20th century and hoodwinked, blackmailed and conned our leaders then.[322]

The Russian government won't leave us alone, nor will they respect our laws because they think we're stupid, and that they're smarter than us. The truth is they're not smarter, but they're willing to lie about everything and they don't care that we label them liars. Also, they will kill anyone not them to achieve world dominance. And they inherited an established genocide attack from the KGB, targeting our hard-wiring, so

---

322 The Lacy-Zarubin Agreement (1958), the Jackson-Vanik Amendment (1975).

the Russian Mafia thinks they're all that. Kicking the vulnerable doesn't make them smarter, it means they're cruel. I'm positive they stealth facilitated the Israel-Hamas War because it's to their benefit and it's what they would do: murdering innocent children for political gain. In this case, they're positive they'll seize control of America by turning the world's opinion against us. We have a binding agreement with the Israeli government, who's confidentiality components the Russian Mafia is exploiting to paint us as baby killers when it's actually the Russian Mafia who started this war and who I believe is forcing Hamas to continue waging. Because of a policy implemented here (probably by Russian Mafia controlled operatives) that we don't deal directly with terrorists, our highly competent interrogators can't question Hamas directly, to the detriment of innocent Palestinians who need this war over. Gaza's Palestinian civilians didn't start this war and want no part of it. This is a nightmare for them. Look at their faces: they are enduring physical and psychological agony. Small children having to jostle other small children for a bowl of food. Oh my god.

All this death and horror so the Russian Mafia can seize world dominance. They'll commit any evil so long as they can blame Americans for the deaths of babies[323], unseat our duly elected President, and install one of their puppets.

The Russian government is evil, not clever. They inherited Active Measures, they didn't deploy or develop it. They cheat and then lie about cheating and lying.[324] They don't believe ordinary people will put two and two together to figure out what they've done. However they've been so successful the last ten years, people are noticing, paying attention, and our security agencies are doing a much better job reporting what villains are doing worldwide. The Russian government doesn't care about adjectives and labels: they care only about enslaving everyone but them.

Here's the truth: the Russian Mafia are a sadistic, hate-filled, slaver government. They have brilliant strategists who throw babies under the bus, to ascend to world power.

If we make a deal with the Russian government, a government who hates us and who has been in our country for decades destroying the lives of our children, that government will destroy us. I'm positive they're running the Israel-Hamas War. They care only about power.

We can't ignore them because they refuse to leave us alone.

---

323 General Valery Gerasimov, Chief of the General Staff of the Russian Federation, "The Value of Science in Prediction," Military-Industrial Kurier (February 27, 2013). "The focus of applied methods of conflict has altered in the direction of the broad use of political, economic, informational, humanitarian, and other nonmilitary measures – applied in coordination with the *protest potential of the population*." "*..Tactical and operational pauses that the enemy could exploit are disappearing...*"

324 Ibid.

Chapter 65 – I was driven to exhaustion by the Russian Mafia

I was driven to exhaustion and the doctors I went to for help from world renown hospitals acted like they'd never heard of medical school much less attended one. Russia-controlled American operatives wouldn't allow them to tell me anything useful. I don't know whether I couldn't applied for Disability after serving a very long prison sentence for attempting to defraud Medicare. What I noticed about my family member's arguments was they made no sense.

Because I'd unknowingly evaded entrapment attacks sprung mostly by non-family for decades, Russia-controlled American operatives leaned heavy on my entrapped extended family members in my 50s, 60s to entrap me. My parents, separately, over months tried to entrap me (I think they both did in 2010), my mother tried over years (2010-2020, estimate), and also my siblings (2010-2022), but their arguments made no sense. Now that I know the Russian government is attacking us and how they've structured this phase of their coup, I think I understand. Apparently, this phase of the Russian government's coup is built on entrapment, entrapping as many people as possible to influence (blackmail, etc.) into voting for whoever the Russian government wants in power positions throughout our nation.

At some point I believe our legitimate government signs off on entrapments. I'm guessing they focus on the crime they are told the perpetrator did rather than on how law enforcement obtained the entrapment. Our legitimate government has no idea the Russian government has commandeered our entrapment system and is using it to destroy us.

To get me to lie to Disability, my father told me, unprompted, to exaggerate my health problems to Social Security Disability in order to qualify for Disability. He said: "the government doesn't care about you, you're just a number to them." I remember his words because they were irrelevant to what I'd told him. He said other things but the language I quoted was his strongest argument.

I now know my father is entrapped but in 2010 when I believe we had that conversation I was struck by his reasoning as to why he thought I should cheat the government. I had health problems, there was no reason for me to exaggerate. I'd never applied to Disability so I had no idea they are known to deny people. I theorize the Russian government had my family bring me non-arguments because they were setting up conspiracy charges which apparently require key words.

Looking back it seems to me that Russia-controlled American operatives assumed that, having entrapped my loved ones and forced them to try to entrap me, they believed I'd make illegal and/or high risk choices out of love for my family. I theorize the Russian government and Russia-controlled American operatives know so little about women that they can't see a woman as a thinking person.

Just because I love my family and will give them the shirt off my back doesn't mean I can't do risk assessment. Anyone telling me to lie needs to have an argument ready, and it better be compelling. Having an authority figure that I love, like my mother and father, tell me to lie is insufficient to get me to lie. My father's statement "the government doesn't care about you, you're just a number to them." My mother telling me to lie to Disability, both out of the blue, unprompted by me, was bizarre. My parents aren't bizarre, their attempt to influence me with non-arguments, made no sense.

When my brother asked me to lie for him on an insurance policy (which would've meant I'd have to lie to Disability as I'd already told them I was too unwell to work), his argument was it would allow him to pay a lower monthly premium on a lie insurance policy. That's it. He cared not at all when I reminded him that I'd already told Disability I was unwell and by telling the insurance company I was well (which is what he asked me to do), I would be lying to one of them. He behaved as if he expected me to lie for him, although I'd never lied to him and/or for him in my life. Not only that, although I had no idea about entrapment, I couldn't imagine Disability not investigating my claims beyond the many medical reports I submitted to the law firm handling my appeal. Knowing nothing about entrapment, I'd been a legal secretary for twenty years so I understood it wasn't a good idea to misrepresent in any way to anyone. Legal secretaries rarely get to go to court with their attorneys and most of the time I don't know the status of cases. My job was to timely and correctly file and serve documents. But even not knowing how cases were progressing through the court, I'd xeroxed tens of thousands of pages of discovery and I knew attorneys and paralegals studied them for inconsistencies. To have my brother ask me to lie for him, as if I'd done that before, as if that was normal, was bizarre me. I called my mother while he was in my apartment and she sided with him. I said no. We three didn't speak for awhile.

Now I know by that time, in my early 50s (50-53), the Russian government and it's predecessor government had been hunting me over 40 years. Apparently, because they knew I financially assisted my extended family, they convinced themselves that merely having one of my loved ones ask me to lie, cheat or steal, I'd do it. To help my family financially I worked, one full time job, sometimes a part time job simultaneously, and occasionally I took a temp assignment on a holiday when a legal secretary wanted that day off and her attorney wanted clerical help.

There's a huge difference between working multiple jobs to pay your credit cards off and share your income with your extended family for things and medical care they needed, and lying for them and facing an extended prison sentence. When my brother asked me to commit insurance fraud, Disability fraud, and conspiracy to commit both frauds, I had no sense that I'd be arrested -- at the time he asked me, I didn't understand it was an entrapment attempt, I didn't know he was entrapped. After he tried it I understood he was entrapped but while I was going through it, I didn't know.

I refused out of ethics, common sense, and outrage that he'd demand I risk prison, and be denied Disability, even though I'd told him months ago I felt too unwell to work, and he'd known that I worked the last twenty years. He was willing to risk Disability denying me, willing to risk my being imprisoned because, common sense, Disability investigates claims. I don't know what they do or how they do it, but investigating claims seems like something Disability would do. No one ever told me, I'd never read about it, it just seemed like common sense. Because he wanted to cheat his insurance company, and was too lazy to work a second job, he expected me to risk homelessness, poverty, and prison, after I told him I felt unwell.

I'd worked a second job, on and off, for years. I would've never asked someone who told me they were sick, to lie and cheat for me, and risk prison for me, because I didn't want to find a second job. While he was in my apartment I called our mother to tell her what he asked me to do. She said "do what your brother asks." I said no. My brother left and we three didn't speak for awhile.

The Russian government exploited my love for my brother and my mother, used my love for them as weapons to attack me. The Russian government knew I'd ignore them if they themselves tried to scam me so they camouflaged themselves in the skin of my loved ones.

Exploiting my love and connection to them, exploiting their access to me, the Russian government had nothing beyond that. They had my brother ask me to lie to the government and cheat his insurance company, they had my mother order me to help him but Russia-controlled American operatives had no argument. Why would anyone do such a thing, risk prison, homelessness, being denied Disability to help their loved one fraudulently obtain a lower monthly premium. They know I love my family but when the Russian government weaponized my family to entrap me, and fed them lines to say to me, the Russian government didn't offer an argument. They had my loved ones ask me to lie, cheat and steal from my government, etc., but my loved ones offered no explanation as to why they expected me to do so because their handlers had nothing. They knew greed wouldn't work with me because I gave many thousands to help my family.

Chapter 66 - The Russian Mafia murders targets using 'natural causes'

The Russian Mafia weaponizes stress to stealth attack targets, raise their blood pressure, kill them by "natural causes." My theory is, after decades of attacking me covertly and overtly, Russia-controlled American operatives decided to stealth kill me by harassing me to death, but another operative who's attacked me for nearly 15 years, the woman upstairs, has apparently died, grown sick, or perhaps she now lives in a nursing home or with her adult daughter, I don't know. I suspect she's dead and I suspect they killed her. She was over

80 and she was a person who, had she been questioned by a solid interrogator, would've eventually answered questions the Russian government wouldn't have wanted her to answer. They prefer to kill people via natural causes so if they murdered her, they would've made it appear natural.

They had my GP within the last six months try, again, to lure me into cheating Medicare. They already know I'd say no, they've hacked my phone (I believe) and my whatsapp discusses my GP's prior entrapment attempt as I complain to my mother. Russia-controlled American operatives have no fear of being caught. Since they can't entrap me, they'll harass me to death. They know I've no idea what officers are running these attacks, which agencies, etc.

The Russian government is working a coup entrapping people using entrapped law enforcement in America and forcing entrapped people to entrap their own children and to vote for puppets the Russian government wants installed. It's been easy for them because the Russian government inherited the weapon already up and running from the USSR, and has fine-tuned it, made it even worse. Our overdose rates reflect the Russian government's success as it stealth destroys hundreds of our people per day.

Ordinary people have no idea they're being emotionally slaughtered by the Russian government. I've no idea what compromised law enforcement think. I don't know the names of Russian operatives embedded here, and/or names of any entrapped law enforcement except my father. My immediate family are entrapped. I'm hoping my government reads this book and opens an investigation. We need a national conversation or the Russian government will install a puppet here, and destroy us.

Now I know my brother was forced to try to entrap me when I was somewhere around 51-53, estimate, then I didn't know.

Russia-controlled American operatives camouflaged themselves using my loved ones as hides, but even after they enslaved my loved ones, they offered no argument. Just because my loved ones asked me to do something, didn't mean I would do it if it meant breaking the law. But time and again, from when I was 50 to my 60s, the Russian government scam attacks using my loved ones but with no argument. Loved ones asked me to lie, cheat and/or scam and offered no argument as to why anyone would do such a thing.

My sister recommended a law firm to handle my appeal and that law firm accepted my case. I thanked my sister. Her handlers had her tell me to apply for Unemployment, even though she knew I'd applied for Disability. I listened to her and said nothing.

Russia-controlled American operatives offered no argument as to why anyone would apply for Unemployment and Disability simultaneously, even though a claimant is telling Disability they can't work, and telling Unemployment they're available for work. They used my loved ones to tell me to make self-harming choices. While I'd never been on

Disability or Unemployment, I understood the basic concepts. Russia-controlled American operatives apparently assumed because they had my loved ones tell me to do something, that I'd do it, even if it was illegal, unethical and/or self-harming.

I began to understand, because Russia-controlled American operatives repeatedly used my extended family to try to entrap me and lure me, that I couldn't' trust most, if any, of my loved ones. The pain is huge. I don't blame them. They are slaves even though Lincoln freed the slaves long ago. I'm 100% positive there are many more ordinary people enslaved by the Russian government in America. I intend that this book help free them.

Russia-controlled American operatives had one of my loved ones tell me to apply for a program anyway, even though I told her I didn't qualify. She told me to let the program decide if I qualified.

Love isn't weakness, love is sharing what you have to make your people stronger. The stronger your people, the stronger you are. The stronger your nation, the stronger you are. The stronger your species, the stronger you are. The stronger all life on our planet is, the greater our chances of becoming space explorers and a long lived, wiser people. The universe is our oyster *if* we learn to respect the gift of life, to internalize that all life has the right to exist and experience peace, joy and a decent quality of life. Better we learn here, on this planet, if we're capable of learning it, rather than while we're trying to colonize a moon.

My father knew I'd worked for decades, sometimes multiple jobs, and paid my own way. He knew I'd given thousands of dollars worth of needed items to my extended family and that I'd earned that money sometimes working multiple jobs. To help my family gain marketable skills I bought them new computers decades ago, I helped them pay rent, spent thousands to buy extended family members groceries, to help them attend college I bought and/or gave extended family members three cars, going without a car myself to gift mine to one of my nieces as a fulfilled promise: if she went to college, she'd get a car to help her attend college.

My father knew many of these things. He knew I don't feel my government owes me anything, so I'm guessing his argument was forced on him by his handlers. I'm guessing entrapment requires that key words be used. Because the words are the opposite of who I am and what my life experiences are, the Russian government was unable to entrap me with them.

I've never lied to my dad and he's never behaved as if he suspected I've lied to him, so for him to prompt me to lie was strange. I was 50.[325] The USSR attacked me from the age

325 I believe the reason the Russian government is having entrapped
   parents begin grooming their children immediately after the parent
   accepts a plea deal is because the Russian government has found that as
   people get older their parents have less influence on them. The Russian
   government used my parents to try to entrap me when I was over 50,
   both telling me for the first time in our parent child relationship to lie. I
   said no. Children who've been groomed by their entrapped parents
   won't say no. Children of entrapped parents will have begun being
   groomed as soon as the parent was ambushed, entrapped and accepted a
   plea deal. I've witnessed entrapped parents forced to use their
   newborns as beards. I've experienced 9-10 year old children working as

of 11. I theorize I'm one of millions they were attacking here, and assume they didn't start fine-tuning their entrapments and other attacks against me specifically until they'd entrapped many in my family and I remained unentrapped. When I was around 30-31, with the fall of the USSR, is when the Russian government began to intensify their attacks, to try to entrap me and enrage me simultaneously. As I aged, they increasingly had their operatives physically assault me including hitting me their vehicles twice as I had the right of way on my mobility device, stomping on my foot, kicking me hard, nearly kicking me in the face, using their packages and purses hundreds of times to hit me as they passed me on my mobility device on the bus, calling me the n word, having my family call me the n word (which when I was raised my family didn't apply that word to family members).

The USSR was here illegally, the Russian government is here illegally, and the police attacked me extra judicially, so I'm guessing for a long time no one wrote reports, not wanting to leave a paper trail. Long before I knew who was attacking me, in my 40s and on when I knew I was being attacked, I noticed there was a lot of repetition in the attacks. I knew police were involved because they harassed me and ordered operatives to harass and attack me, and didn't care that I saw them do it.[326] I didn't know a foreign enemy government was

_____

co-harassers with their entrapped parent to harass me. Recently, within the last year or two.

326 I witnessed a uniformed policeman force a young Caucasian man in his early 20s to harass me: he intimidated me and veered into my sidewalk space as if he intended to walk into me. Before he'd seen the police car

involved. Because these attacks are illegal and stealth, it makes sense that neither the USSR, the Russian government or local police would leave a paper trial.

I'm the only person who knows all the attacks I've endured under order of the USSR, and the Russian government in the 20th and 21st centuries and I want to write them down so people will have an idea of how the Russian government is attacking us in the 21st century. For most of that time, about the first 30 years (age 11-41), I had no idea I was being hunted because the USSR camouflaged it.[327] When the Russia-controlled American operatives did show me I was under attack (when I was around 40-41 or so), they used my entrapped father to gas light and misdirect me. I didn't tell him or anyone about the attacks for the first several years because I didn't know what they were.

The first attacks were at a high end grocery store I liked, and a big box store I'd shopped at for years. Mostly plain clothed Caucasian male strangers got in my space (and

---

and officer monitoring him, the young man walked a reasonable distance from me but after he saw the police officer driving slowly behind him, watching him as the young man approached me, he put himself closer to me as if he were going to walk into me. I saw that the policeman had forced him to attack me, and the police officer didn't care that I witnessed his behavior.

327 The Russian government is big on camouflaging itself. The Israel-Hamas War is a proxy war facilitated by the Russian government to destroy us and to unseat our President. I've read nowhere that the Russian government facilitated that war but it's obvious to me that it did. Red flags the Russian government is possibly involved in an evil occurrence: (1) a poor and/or impoverished people are being killed or significantly harmed, (2) Americans or our government is being blamed, (3) the Russian government is not being blamed, and (4) the attacks promote the Russian government's interests of dominating the world.

sometimes uniformed security did in stores as well), and stealth harassed me, cut in front of me in line, etc. The first several years hundreds of mostly Caucasian people, men and women, some with children, on the bus, sidewalk, in grocery stores, etc. harassed me, forcing me off the sidewalk and into the street when a male operative nearly walked right into me on the sidewalk as I walked to or from the bus stop to work, forcing me to step down into the street; over the years, hundreds of walkers walked nearly into me; hundreds of skateboarders nearly skateboarded into me; dozens of bike riders rode too close to me; dozens of joggers nearly jogged into me, etc. I began to be maced on the bus by operatives with something that wasn't regular mace but a lighter version of it, misting me. The mace like attack used a product that had a scent, burned my eyes, caused me to have re-occurring sinus infections and migraines from the attacks. I was deliberately made sick for about a week when an operative put what tasted like dirty dish water in the chocolate glaze of a yeast donut I bought at a donut shop. A work friend who I guess was an operative picked me up from the bus stop and offered to drop me off at my destination, a few miles away at a favorite donut shop. Apparently his car was bugged and Russia-controlled American operatives had their operative in the donut shop put dirty dish water in my donut. I was violently ill and believe I've not returned to that donut shop since.

I was sexually assaulted by an operative on a packed bus on my way to work. I've been sexually assaulted by four

Russia-controlled American operatives from my teens to within the last several years. When I told my mother about the most recent attack, she gas lighted me and said it didn't happen. You'll get that from entrapped authority figures in this coup. Victims report the attacks they experience to authority figures and the Russian government has the authority figure dismiss their claims. It's a way the Russian government influences victims to not report the attacks they experience. It's especially glaring given that my mother received unwanted sexual advances from her mother's boyfriend when she was 12-16. If anyone should believe someone who tells them they've been sexually assaulted, it's a victim who's experienced sexual assault. But the person who sexually assaulted me is an operative and it's likely my mother was told how she must respond if I told her. My mother is up in age. I think she believes she'll be sent to prison if she breaks her plea deal. Gas lighting is a variation of psychological warfare. The Russian government has had both my parents and one of my brothers use it against me.

I think I told my father about the attacks around my mid-40s. It was difficult to accept anyone would want to attack me. I denied it not only because my father misdirected me but because I didn't understand it. I worked, a lot, I slept, I read, tried many diets, exercised sometimes, occasionally took classes, and visited my family. There was nothing in my life that should've lead to a deep-pocket enemy who controlled our police.

I had no enemies I knew of, certainly no enemy who could control hundreds of harassers on the bus, in grocery stores, on sidewalks, and the police. When I finally decided to tell my father, he was used to turn my gaze away, gas light and misdirect me and he was effective. I'd considered writing down every attack but didn't because I believed what he said about shoplifters. You'll notice with people entrapped by the Russian government, influencers always have a believable explanation, believable to people untrained in espionage, or who don't understand the Russian government's overall strategy.

I finally disregarded my father's advice to ignore the harassment in stores when it became overwhelmingly obvious that I was being harassed nearly daily and it had nothing to do with stores being concerned about shoplifters, because most of the attacks were now occurring when I as not in a store.

In my personal life the two men I loved were both in committed relationships, so were unavailable to me. I was raised to respect people, and especially to respect women. I'd never break a woman's heart by having sex with her man, and so experienced unrequited love for decades. I decided to view my feelings for them as the blessing it was -- I couldn't have them in my life but I'd met them, liked them, and respected them.

The grant I'd lied to get to attend electrical engineering school I'd repaid long ago to the kind family who gave it to me, so even in my personal life, I'd done my best to hurt no one.

By my late 40s I knew I had enemies, powerful ones but not who was directing the police to attack me or why. The attacks progressed outside of stores, and largely on the bus and/or on sidewalks, etc. They weren't talking beyond harassing, attacking, assaulting and/or torturing me as I moved into my 50s and 60s where I experienced entrapment attacks by my parents and extended family, torture by the operatives upstairs and in surrounding apartments, etc.

Chapter 67 - my mother's 180 degree ethics shift shocked me; although operative strangers had harassed me on the street for the past decade+ by 2011, I was over 51 the 1ˢᵗ time my mother told me to lie. That's why the Russian Mafia now force America's entrapped parents to take their young children on harassment attacks – it grooms the parent to stop being a parent & start being an operative, even w/their children

My mother really shocked me when she suggested I lie to Medicare to obtain benefits, as she drove me to her apartment, I think it was pre-Christmas 2010, so I could spend the holiday with her. I was shocked her words were the exact opposite of how she'd raised me. My father hadn't raised me but he'd never told me to lie about anything and I was 50 when he recommended it. Since he'd worked as a police officer, and he was probably retired by the time he recommended it, I was just very surprised. I would've understood had I asked his opinion on whether I should lie to the government but it never occurred to me to lie so I listened politely to him vent. His position was just so strange and out of the blue.

Like my father, my mother offered no argument. She told me to lie. I reminded her that she'd raised me to not lie. She was happy I remembered that. After that conversation in her car, where a police car paced in the fast lane and to the left of us and which exited the freeway after I responded to my mother, suggesting to me that my mother's car was bugged, when a month or so later I reached the conclusion that my beloved mother was entrapped.

My father told me to exaggerate my health problems to get "whatever you can" after I told him the attorney at the free legal advice place advised that I apply for Social Security Disability when I explained my situation to the attorney. I think I called my dad and told him what the person said. My father's argument: "the government doesn't care about you, you're just a number to them." I had health problems and I didn't feel well. I knew nearly nothing about Disability, I'd never thought I'd need to file for it. My father had retired, I believe by the time I was 50, and had this conversation with him. I hadn't asked him whether I should cheat the government, it had never occurred to me to cheat the government, so I had no idea why he brought it up. I felt unwell, there was no need to exaggerate anything.

My father's handlers had his pitch ready but it made no sense. There was no reason for me to exaggerate health conditions I'd finally been able to get diagnosed. Several years later I went to get tested for sleep apnea. I suspected I had it because I woke myself up snoring, woke up many times throughout the night, was obese, etc. The Russian government rigged that test.

As the sleep tech was wiring me up, operatives pumped a sleep inducing gas into the room that had no scent but which effects I felt immediately. The only reason I knew something was being pumped into the room is because I have sleep disorders due to Russia-controlled American operatives deliberate destruction of my natural sleep cycle beginning when I was around 51-52. Operatives would wake me up after I'd finally fallen to sleep: they'd drop heavy things loudly in the apartment above mine, drag furniture across their floor. Operatives would appear outside my window day and night, they'd have workmen operatives arrive early in the morning, laughing and talking excessively loudly. When I asked them to stop, they ignored me, some smirked. I took pictures of their license plates. They expressed no concern law enforcement would question them.

When I feel sleepy, most of the time I can't sleep. Tortured by the Russian government's operatives for years, my natural sleep cycle is gone. So when I began to feel sleepy, something extremely rare for me, it really frightened me. I think I was maybe mid-50s by then, and so had been overtly attacked for nearly 15 years. I didn't doubt my reality. I was so frightened that even with the sleep inducing gas, I couldn't sleep so the test was a fail. I had to return another time and when I did, no gas was pumped into the room. I believe Russia-controlled American operatives wanted to freeze me obtaining diagnoses. The more diagnosis I obtained the harder it was for them to push whatever lie occurred to them.

I was uninterested in exaggerating my health problems. When a person feels unwell it doesn't occur to them to exaggerate symptoms. For example, say a person is diagnosed with a treatable cancer and goes through chemo. Chemo, I've read, is usually a miserable experience. A person enduring chemo and applying for Disability wouldn't think it necessary to exaggerate their symptoms. A person enduring chemo who had a loved one tell them to exaggerate their symptoms to get Disability to believe them would find that a very strange suggestion. While I've not been diagnosed with a health condition where I have six months to live, I use the chemo example to highlight how my father's suggestion struck me as strange. By the time I was 50, I'd been overtly harassed for nearly 10 years. When my father tried to entrap me he was the lead off family member -- I believe he tried to entrap me sometime around mid-2010. My mother didn't try to overtly entrap me until December 2010 as she drove with me to her apartment, where I'd spend Christmas with her.

At 64 I now know the Russian government destroyed my family as part of their coup. I'm positive they're doing similar attacks against other families here.

What I experienced during about the first ten years of overt attacks was that hundreds of strangers, including uniformed police and plain clothed police, overtly harassed me for no reason I was aware of and no one bothered to tell me why.

I didn't immediately draw a connection between the overt attacks I'd begun experiencing in my early 40s, to when I was 50 and my father suggested I exaggerate my health problems.

Many of the overt harassing attacks by strangers when I was in my 40s were swarming attacks, strangers getting in my space, suddenly appearing out of nowhere to deliberately startle me, plain clothed security already at the products I often bought, blocking my access for an inordinate amount of time to force me to engage with them, entrapped parents used their children as beards to block my way, strangers on the bus were too friendly, an operative began pushing my shopping cart as I pulled it (that may have been a harassing attack in my 50s). I was sexually assaulted on the bus in my 40s, and the dirty dishwater in the chocolate glaze on my yeast donut to sicken me occurred, I believe, in my 40s.

The attacks were bespoke, aimed specifically at me, designed to the casual eye to look like normal behavior by the perpetrator. A child blocked my path as I entered the post office, her mother several feet from her, watching my reaction. A 5 year old little girl in a post office doorway triggers nothing from a casual observer. I tended to be harassed by children, my guess is because the Russian government forced entrapped parents to begin grooming their children as soon as the parent took a plea deal (if I'm right about this coup works). My parents weren't forced to overtly try to entrap me until I was in my 50s. In the way Russia-controlled American operatives attacked me in my 40s and beyond, even though I didn't know the Russian government was the perpetrator until 2021, I noticed the heavy use of children in the harassing attacks against me. Because of the wide use of children to harass me, and because in my 50s I learned my mother and some of my siblings were entrapped, though not by who, by the time I was 58 or so I knew I had to write a book warning children.

I was terrified for my mother when I realized when I was 51 that she had somehow been entangled. She was unable to entrap me because I'd long ago embraced the ethics she'd taught me as a child, but she was used a lot throughout my 50s by her handlers to try to entrap me. Watching the babies, young children, older children and teens used to harass me and/or used by their entrapped parents or teachers as the parents and/or teachers harassed me, I knew I had to warn the children, years before I learned who the perpetrator was.

When my mother first tried to entrap me in the car, which I believe occurred in December 2010, I didn't know that she'd been forced to try to entrap me. I didn't do as she suggested for the reason I told her, that she'd raised me to not lie. After that, she wasn't forced by her handlers to get me to lie to Disability. Instead they used her to browbeat me into filing for personal bankruptcy -- for maybe four or five years in my 50s. Russia-controlled American operatives had also used an employee in the HR department at my last job to encourage me to file for bankruptcy. I looked into it but refused, my argument being that I'd borrowed the money with the intention of repaying it and I didn't want to get more debt before I repaid what I owed. My mother used former President Trump as an example of an intelligent businessman who'd filed for bankruptcy often, and he was a billionaire she argued. She implied I was stupid for refusing to take advantage of bankruptcy laws which, she said, were put into place for people like me who'd developed health problems and could no longer repay the debt. I wanted to repay my creditors. I felt bad about borrowing money I couldn't pay back. For the first year I was unwell I made small payments to my creditors, explained everything to them, sent them medical reports.

The Russian government pushed me hard to file for personal bankruptcy so somehow they've gotten control of that system.

They also forced my mother to try to entrap me using other frauds. In my early 50s my brother pushed me hard to commit (1) insurance fraud, Disability fraud, and conspiracy to commit both. My mother worked with my brother to commit fraud. I refused.

That was the second time in my life, in my early 50s, that my mother told me to lie. I no longer trusted her advice but I didn't learn what entrapment was until I was 54-55, maybe. And I didn't learn the Russian government was on the ground working a coup until I was 61. I endured multiple entrapment attacks before I knew what was happening and who was doing it. Because we have nuclear weapons, and because the Russian government intends to destroy us without us bombing them, they developed this coup built around entrapment. They inherited the entrapment scam from the USSR, and expanded it. Currently, they're committed to the entrapment model because they've enslaved so many of use (and probably many millions of innocent people worldwide) using it. And they've attacked millions of us by starting a war against us without telling us. Anybody, but especially civilians untrained in espionage, can be ambushed if their enemy doesn't tell them they're attacking them. None of this is our fault. However, we better get on the same page fast or the Russian government will destroy us.

Because I'd been overtly harassed since my early 40s, I thought all of it might in some way be related but it wasn't until sometime in my mid-50s I learned about entrapment. I feared for the children, that, like my mother had been forced to do, their parent(s) would be forced to entrap them.

Remember, I had been deliberately exhausted by Russia-controlled American operatives for the last several years, at my final law firm legal secretarial job, and the Russian government wouldn't allow the doctors I sought medical help from, to help me, so my high blood pressure when I collapsed was undiagnosed, my insulin resistance/pre-diabetes was undiagnosed, my sleep apnea, insomnia, migraines, peripheral neuropathy were all undiagnosed when I collapsed. I'd been worked to exhaustion, sometimes working for six attorneys, instead of the usual 2-3. My beloved grandfather was suddenly gone, and my family needed my financial support.

When my father told me to exaggerate my symptoms to Disability, I'll estimate we had that conversation maybe the middle of 2010. By that time I'd begun to be diagnosed. I'd switched medical care providers before the Russian government disabled me and some of those doctors began to diagnose me. I don't remember how I felt in mid-2010 but since I hadn't worked since the end of December 2009, I probably was feeling more rested, but it was a highly stressful time as I had no idea when I would work again or how I'd financially survive. I went to a Santa Monica shelter to ask if I could stay there but they said they had no openings.

My father's advice seemed strange to me at the time, which is why I remember it. I listened politely to his advice, but didn't agree with it, which is why the Russian government couldn't use it to entrap me. I didn't know my father was compromised, I didn't know Russia-controlled American operatives were feeding him lines to entrap me. His suggestion made no sense to me, so I said nothing, just politely and respectfully listened. Born in the 1950s, I was raised by my mother and grandmothers to treat everyone with courtesy and respect, especially my elders, so I listened to my father and didn't interrupt him out of my love and respect for him. I had zero idea I was experiencing an entrapment attempt. I had zero idea the Russian government was exploiting my relationship with my father to use him as a weapon to destroy me. Now I know, then I didn't.

If someone in your life suggests you lie, cheat, or steal, whether the target of the theft is a person, business and/or government entity, never, under any circumstances, agree with them, in writing, verbally, or by nodding your head. If you do, the Russian government will enslave you, your family, your nation, and destroy everyone and everything you care about. It is deeply embedded here, has commandeered our entrapment system and entrapped some of our law enforcement. It will destroy you and everyone you care about. This book is a warning.

Because I listened politely on the phone to my dad's entrapment attempt but didn't respond to it, and because Russia-controlled American operatives needed me to respond in order to entrap me, they forced my mother to try to entrap me in person. They had her come collect me, I believe right before Christmas 2010, and while she drove on the freeway to her apartment where in another city, she told me to lie to Disability.

I was shocked because she'd raised me to not lie. I hadn't asked her whether I should lie to Disability, I wasn't interested in lying to anyone. My father told me exaggerate my symptoms I believe earlier that year, my mother told me to lie right before Christmas. They'd been divorced since I was a few years old, had remarried, divorced and had other children. It didn't occur to me that they discussed me unless they told me they had.

My mother raised me to be an ethicist, to not lie, cheat or steal. That conversation, which I believe took place when I was 51, was the first time in my life my mother had ever told me to lie. Confused, responding quietly, I said "but you told us not to lie." Her response, which made no sense to me at the time was "Yes! I knew my kids don't lie!" Now I understand she knew law enforcement officers were listening to us in the car; then I didn't know that. I was genuinely confused, both by her advice and her response.

Because police had been overtly harassing me for about a decade by then, I noticed as I sat in the front seat, turned slightly towards my mother during our conversation, that a police officer had been driving near us, but in the fast lane, while my mother drove in a slower lane. The officer had been driving a couple of car lengths behind and to the left of us, but after my response to my mother the officer suddenly crossed multiple lanes to exit the freeway. I noticed the officer's exit because, overtly harassed and attacked by police-directed operatives for about a decade by then, I understood police in my vicinity might mean they were there to attack me.

But it wasn't until January or February 2011 that I understood my mother was involved in whatever was being done to me. I was terrified for her and blamed myself. I thought she'd been entangled because I hadn't warned her of the overt attacks I'd experienced the past decade. I hadn't told her because (1) I hadn't wanted to worry her; (2) I didn't know how to explain the attacks -- I was attacked but I didn't know why; (3) I had no idea the attacks against me were part of wider attacks against my family and my country; and (4) my father successfully gas lighted me (misdirected me) for several years in my 40s, convincing me to interpret the store attacks as shoplifting related. Through it's weaponization of my father, the Russian government controlled my *interpretation* of the attacks. They constructed a frame, a misdirection, around the attacks that was a lie. By the time my mother was forced to try to entrap me, I knew the attacks weren't shoplifting related, but I didn't know what they were. I simply had insufficient information to reach a conclusion. I knew I had a serious problem, that men who wore guns were attacking me, but not why.

When my mother was forced to make her first overt entrapment attempt, I didn't know I'd successfully evaded my father's first overt entrapment attempt earlier that year, when he told me to exaggerate my symptoms and I didn't respond. I didn't know my parents were owned by my enemy.

I knew I had an enemy but not who my enemy was or why they were my enemy. This book is to help people worldwide connect the dots faster than I did. The Russian government works in the shadows, uses people who have access to your life as weapons to destroy you. They're working a secret war here, a coup. They double, triple and quadruple team victims. This book is to help you survive.

In approximately January or February 2011 when I understood my mother was in danger, weeks after she'd been used to try to entrap me, I became terrified for her. I was terrified for me.

That Christmas when I was 51 at my mother's apartment, I didn't know I'd just evaded another entrapment attack. I didn't learn about entrapment until sometime in my mid-50s after I read an article in the New York Times discussing conspiracy charges and entrapment attempts. That is when I finally had enough information to begin connecting the dots to entrapment attempts I'd evaded decades ago.

Christmas 2010, I didn't know the police officer pacing along with us on the freeway would've pulled my mother over immediately to arrest me had I answered her in any other way than I did.

Although I didn't know who was ordering police and strangers to attack me or why, I thought about my mother and my pre-Christmas conversation sometime in January or February 2011, and I understood she was somehow captured by whoever was attacking me. I suddenly understood the officer on the freeway pre-Christmas, who exited as soon as I responded to my mother, would've arrested me had I answered her in any other way. I knew I would've been arrested but I didn't know why I would've been arrested. I knew nothing about entrapment.

I responded to my parents' entrapment attempts, and to nearly all the other entrapment attacks against me over the decades, by refusing to lie, cheat and steal. By my mid-50s, I'd chosen to live an ethical adulthood since around my mid-20s. I hadn't discussed it with anyone because: (1) no one had ever asked me, (2) the subject hadn't come up in casual conversation, (3) I hadn't known I was being hunted until Russia-controlled American operatives began to overtly harass me, out of the blue, in my early 40s, (4) I didn't know the Russian government was working a coup built around entrapping ordinary people, and (5) I didn't know my lifestyle protected me from a human apex predator government.

The USSR first attacked me when I was 11. I was 64 before I understood the attacks on me during my childhood were under order of the USSR. I learned that when my mother's handler's were working another scam to attack me. During a phone conversation the summer of 2023, my mother was allowed to tell me how she met the faux boyfriend she doesn't know was owned by the USSR and inserted into her life to destroy her when she was 28/29.

She'd been set up by a palm reader she went to see and by one of her new California friends, both owned by the USSR/KKK police gang who ran the neighborhood where we lived in 1970. Her handlers the summer of 2023 were fishing to see if they could scam me into visiting a palm reader.

I didn't know any of that during our conversation. My mother told me her "meet" story with the USSR controlled operative inserted into her bed and life in her late 20s. I thought through our conversation after the call ended.

What initially got me thinking was, she began her story in a specific way, told me something particularly disgusting that I responded to naturally, not initially understanding her conversation was a fishing exhibition by her handlers. My mother chided me for interrupting her story (with my yuck reaction). Most people would've responded the way I did but my mother isn't trained in espionage; she was forced to do this chore. Her handlers told her to get to the palm reader part. We didn't have a mother/daughter conversation, the Russian government stole my mother from me, I don't know when they entrapped her. They exploit my mother's access to me in order to destroy me. What they're doing to me, they're doing to you. My mother is a slave in her own country, even though America outlawed slavery over a century ago.

Reviewing the call later, I understood she was ordered to hit specific parts of the story, and that her handlers hadn't prepared her for my realistic reaction. I didn't know during the call that it was a set up and fishing expedition. So when she was short with me for no reason, I noticed.

Because Russia-controlled American operatives wanted to know my thoughts about palm readers, my mother was allowed to share helpful information as to how she met the operative used to destroy her quality of life and set her on the road to slavery, and to be owned by the Russian government.

For forty years (age 11 to 50) the Russian government (then the USSR) worked to destroy me, using operatives from individuals, in businesses, school employees, police, and my extended family which to camouflage themselves, to use as hides. Had Russia-controlled American law enforcement not gone loud on me when I was 40-41, had not begun to overtly attack me to my face, I would not have known I was under attack, and would've been unable to connect the dots when the Russian government forced my parents, siblings, and other operatives to try to entrap me. Although it is excruciating to be overtly attacked, I've learned enough to warn you.

Although I'd worked about 20 years as a legal secretary, I mostly worked for law firms who focused on business law, not criminal law. I knew nothing about entrapment until I was, rough estimate, in my mid-50s, after I read an article about it. Because of that article I began to understand to piece together the random experiences I'd endured and understand that they were entrapment, enraging, grooming and luring attacks, but I still didn't know who or why.

So it was not until I was about 55 that I understood I'd evaded multiple entrapment attacks, no one told me. The Russian government doesn't tell victims and my loved ones were entrapped and prevented from warning me. So I went through decades of entrapment attempts and attempts to enrage me, just as I now believe other people in America (and probably elsewhere) did.

Living through the many entrapment attempts, enraging attacks, and/or hostility aimed at me on a near daily basis by a foreign enemy government fronting as my society, my employer, my grocery stores, my police, my family, my educational systems, etc., and evading them by the grace of god (and my mother's & great-grandmother's teachings), means I know how the Russian government is destroying us and our nation, and how we will fight back.

Everyone must be told. Knowing a foreign enemy government is on the ground in our country (and probably in countries allied with us), waging a secret war against ordinary people[328] gives people a heads up. Describing how the Russian government has attacked allows people to access their memories and relationships to see if they've been or are under attack, and if they haven't been, telling them helps them be on their guard. The information also helps our government better defend us.

Because the Russian government has a deep knowledge and mastery of how our institutions work, forcing my dad to tell me exaggerate symptoms was weird and although I didn't learn until 2021 that he is entrapped, I remember his conversation and now understand why he encouraged me to lie and put myself at risk of imprisonment.

I believe no American government agency would set up a noncriminal to lie and cheat a business, and the government simultaneously. The Russian government tried to entrap me to cheat the state and federal government, simultaneously. I have no why they'd think anyone would do that: perhaps they kept

---

328 General Valery Gerasimov, Chief of the General Staff of the Russian Federation,"The Value of Science in Prediction," *Military-Industrial Kurier* (February 27, 2013). General Valery Gerasimov states: "…. *Wars are no longer declared …*"

throwing anything at me, knowing they'd not get punished for harassing me to death. idea what they were thinking. I love my family, have spent a lot of money on them over the decades but that was, except for my credit card debt, my money. I didn't steal to help them, I worked, sometimes multiple jobs, paid my own money to support and/or financially them. Just because you love your people doesn't mean you'll self-harm for them. Loving your family doesn't mean you'll cheat for them. That makes no sense. All the scams the Russian government forced my family to bring me made zero sense and since I had no idea who was attacking me, I was very perplexed with their behavior.

The Russian government tried to get me to defraud Medicare. A doctor worked to get me to lie to Disability, twice. No one not entrapped, no one except someone who is being forced to try to entrap you, will, in their professional capacity, offer to help you cheat Medicare. The Russian government tried to get me to commit immigration fraud. I worked at a West Los Angeles law firm and out of the blue one day, after I'd worked there maybe three years, I got off work and went to the garage to drive home. A stranger approached me, an attractive African American male. Now I know he was a plains clothed police officer but then I didn't know. He asked if he could talk to me a few minutes. I said yes, standing a few feet away from him. He explained he had a friend he was trying to help out, a man in a foreign country who was being persecuted in his home land and who needed to escape. The victim in the foreign country needed a green card, needed someone willing to marry him for a brief period of time and then there'd be a divorce. The marriage would be unconsummated, I'd be paid. I said I was so sorry the man was being persecuted but that I couldn't help him. I then got in my car and left. I felt bad for years after that, that I couldn't help a migrant in need. I honestly believed the man although it was weird to be approached by a stranger out of the blue. I had no idea I was experiencing an entrapment attempt. I was in my, maybe mid-30s. The reason I was nonresponsive, aka unentrapped was because I was concerned I'd have to lie to my government, which didn't interest me. Knowing nothing about green cards or marrying an immigrant I assumed when an immigrant married an American that somewhere, on some form, my government would ask me why I'd married the person. I was unwilling to lie, and assumed it was illegal for an American to marry an immigrant for money, despite the fact that the operative offered me money. The man trying to entrap me explained none of this to me. I didn't know it was an entrapment while I experienced it and for years later felt bad that I couldn't help a migrant in need. I considered the offer over a minute or two then politely declined, got in my old car and left.

I was the person who'd reported her game show prizes worth $1,000 to the IRS, just in case the IRS might want to know. A person paying taxes she's not sure are owed is not a person who'll lie to her government about why she married someone. I bought my family computers to help them learn technology faster and get jobs, who sometimes worked up to three jobs to pay down credit cards and help my family with things they needed like help them paying rent, buying food, needing surgery. A person who shares her income like that doesn't have a problem with greed. But the Russia-controlled American operatives in charge of entrapping me ran entrapment after entrapment, repeating many of them. After they sickened me in my 50s, they launched multiple entrapments. I think Russia-controlled American operatives are paid by the entrapment attempt, that's how they acted. They weren't dinged for failing, apparently they are compensated for the attempt.

One of the Russian government's go-to weapons here, based on their entrapment attacks against me over decades, is their belief that most people are greedy. I don't believe most people are greedy, I believe most people have insufficient income for their needs and wants, and that most people are incurable romantics who long to be rescued and so reach for the too-good-to-be-true scams the Russian government dangles in front of them.

The Russian government is the villain, the victims forever hopeful. An analogy is when you cook something and your dog smells it and approaches you, hoping you'll share a bite or two. People are like that, forever hopeful. The Russian government exploits people's hopeful nature, use it against them as a weapon. The Russian government is evil and masters at manipulating people. I'm positive the Russian government manipulated Hamas into attacking Israel using lies. I don't believe Hamas, who are mostly Palestinian, and who were once Palestinian children, would've attacked Israel if they'd understood 20,000 Palestinian civilians would be murdered in ten weeks.

People must better understand who and what our species is and for sure, as many people as possible need to understand who and what the Russian government is or our species won't survive.

The Russian government has destabilized our nation to such an extent that our tax system is insufficient for our nation's needs. They've used operatives to influence us into outsourcing and dismantling much of our manufacturing base. We've got to rebuild and it's possible there'll be a hot war involving us and at least one of the Russian government's allies. That's because the Russian government intends to control the world. Forget what they say, follow what they do.

At that time, before 9/11, the Russian government worked to entrap me using entrapments they could claim to our legitimate government were on the up and up. None of the entrapments were on the up and up but our legitimate government wasn't going to be told the details by entrapped American law enforcement. At that time, because they needed to have their cover straight as our security agencies hadn't been distracted by Middle East terrorists, my theory is the Russian government was here, working their stealth attacks, but those attacks had to fit specific parameters: the legitimate government had to be lied to and/or cheated, or the target had to be a drug dealer.

Chapter 68 - The Russian Mafia promotes cruelty against Americans in America to form the victims into weapons; then Russian Mafia aim & trigger them

I've seen a little video and/or photos of abuses in the Israel-Hamas War. I've seen video of Americans attacking the Capitol. I believe I can tell the people who've been victimized the Russian Mafia because those people are extremely brutal and trigger-happy.

The Russian Mafia deliberately abuse people because all it need do to aim them and get them to do violence is to trigger them. People are socialized to not be aggressive. Before the Russian Mafia can use people as weapons, the Russian Mafia has to abuse them and one thing the Russian Mafia loves to do is to abuse people living in America and our allies. I speak from direct personal experience being abused, harassed, attacked &/or tortured.

Her half siblings were brutal to Amber and now she's a weapon. Cruelty is how the Russian Mafia destroys people, from childhood to their deaths. I've witnessed newborns with their entrapped parents as their parents harassed me, and I knew a 93-94 year old slave of the Russian Mafia and a year or so before she died she was still being forced to harass me from her wheelchair. No lie.

Before the Russian Mafia can trigger a human being and form them into a weapon, the Russian Mafia ensures they are horrifically abused, physically, and/or psychologically.

The younger the Russian Mafia can begin attacking victims the better they like it because that means children have been socialized less and are easier to manipulate. Rampant cruelty on the innocent by the Russian Mafia using their proxies here is why our nation is experiencing (1) a mental health crisis, (2) an obesity, high blood pressure & diabetes health problem, (3) an overdose epidemic, (4) declining life spans, and (5) increased mass shootings.

The Russian government is stealth attacking us, their secret war is illegal and immoral, so they can't come here in person to force a targeted person to take a specific action. So they use their entrapped slaves here as spurs and whips to psychologically attack individuals the Russian government want to feel emotional pain and/or to do one thing or another, depending on what the Russian government desires.

For over fifty years Russia-controlled American operatives have forced entrapped people with access to me to abuse me, including an older boy in the neighborhood I lived in during my childhood, someone I didn't know, threaten to beat me up when I was 11; three teenager girls in high school I didn't know threaten to beat me up for no reason; and my high school career guidance counselor accuse me of stealing $20 from her purse to stop me asking her advice on how I could afford college.

Russia-controlled American operatives had my heretofore amazing mother beat me, out of the blue, for no reason when I was 15; and had my beloved brother inexplicably pick two fights with me when I was 16.

At my first full time job post-college, Russia-controlled American operatives had my supervisor physically assault me. After I transferred out of that department they had my next supervisor sexually assault me and deny me a promotion I qualified for and hoped to get even after telling me I was doing a good job. In the next department I transferred to, they had operatives mistreat me for no reason. I bore it as best I could.

When I left that company to return to college to earn an electrical engineering degree, Russia-controlled American operatives had a teachers assistant they owned accuse me of cheating on the math test, even though he said a student (who was likely entrapped) was looking over my shoulder, and cheating on me as I took the exam.

In, rough estimate, 90% of my post-college jobs, Russia-controlled American operatives had my supervisors and co-workers treat me cruelly after I'd been there a few years (apparently Russia-controlled operatives used that time trying to figure out how they could entrap me at the company). The operatives wanted me to move on to another company, one where they hoped to be able to entrap me. They didn't care that I enjoyed working for that company and that I enjoyed the work.

They especially love weaponizing authority figures like a supervisor, a parent, or someone the target particularly admires to attack them when the target is in a subordinate position. They had my father sneer and ridicule me for the first time in my life, in my late 50s, or I might have been 60. He told me to not write this book, and strongly implied people would think me crazy. He focused on the fact that I didn't know who was attacking me when I decided to write the book. At the time I discussed my intention to write the book I didn't know he was a slave. The Russian government didn't want the book written or published and used my father to attack me emotionally to discourage me to not write it.

They had my mother sneer and ridicule me (also a first), and try to browbeat me into filing for personal bankruptcy, strongly implying that I was stupid to not take advantage of bankruptcy laws, given that smart businessmen like then-President Trump had filed for bankruptcy multiple times, and look, he was president and a billionaire.

I think my aunt was killed on order of Russia-controlled American operatives via deliberate overdose about twenty five years ago in order to emotionally destabilize me. By that time, somewhere in my 30s, they'd been hunting me for over 20 years with no success entrapping me. In the car and at the airport was the first time my aunt ever sneered, ridiculed, or mistreated me. I'm positive her handlers forced her to do so because that wasn't who she was with me, and, looking back,

Russia-controlled American operatives had already developed a pattern using entrapped people to attack me emotionally.

I approached the gate and turned one last time to wave to her. She stood there watching me and I could tell she regretted snapping at me. We exchanged waves and that was the last time I saw her alive.

I suspect but cannot prove that Russia-controlled American operatives made her an addict with the intention of entrapping me after they learned she was my favorite aunt. An addict is easily controlled by the Russian Mafia since addictive drugs are their bread and butter.

Having an addict with access to me must've had them salivating. I believe they tried to addict another relative of mine as well. There's a reason the Russian Mafia floods our country with fentanyl and other addictive drugs: an addict is a slave even before the Russian government entraps them. Slavery is one of the Russian government's things. The Russian government is all about control, has zero value for lives not their own, and is so morally warped they thought I'd prefer being a drug mule to working for a living.

The Russian government tried to get me to lie on a federal complaint. My supervisor, who'd previously been cordial to me, began treating me cruelly. I didn't know Russia-controlled American operatives had been working to entrap me for years, and at that time didn't realize I'd been pushed out of jobs. There was no way I could know, given that Russia-controlled American operatives used the people at the specific job site, and there was no way I could know there was an evil puppet master working the shadows to entrap and destroy me.

I'd worked at the law firm as a word processor maybe three or four years, did great work because I like doing a good job, and proofed my work. Out of the blue my supervisor began treating me cruelly. I said nothing, did my work. I was accused of doing shoddy work, and I denied it. Since my boss' attitude had inexplicably, negatively and radically changed towards me, and I began to be accused of doing shoddy work which I knew wasn't the case, I began to make a copy of documents I worked on before I began them and after I completed them. The next time I was accused of shoddy work I presented my evidence, which enraged my attackers, but it stopped their work-quality-based attacks.

After, rough estimate, six months of mistreatment from my supervisor one of my coworkers, a woman of color, approached me and asked if I wanted to join a complaint charging racism against the manager and I think it may have been against the law firm as well.

This was a setup. I suspect but cannot prove the supervisor was entrapped and forced to be mean to me by her handlers, and the co-worker was to give me the opportunity to retaliate, using a charge of racism. Even the complaint already existed. The manager had hired me, and other people of color in the department. I didn't think racism was the problem but I didn't know what the problem was since my supervisor wasn't allowed to tell me.

The Russian government give targets the opportunity to retaliate. Russia-controlled American operatives create the problem, then provide the opportunity to retaliate. If the target takes the bait, the Russian government tightens the noose. I didn't know that then. I didn't join the lawsuit because I didn't think the manager was a racist. She has since died and she died relatively young. It's likely the stress her handlers subjected her to by forcing her to abuse a good worker, sped up her demise.

Because the Russian government has stealth attacked me for 53 years, and overtly attacked me for about 23 years, and because they deliberately sought to enrage me, it's highly likely some of the people who stormed the Capitol in 2021 experienced cycles of psychological abuse, were deliberately enraged, and deliberately frustrated when they sought relief.

Russia-controlled American operatives use authority figures to psychologically abuse me. This is one of their go-to strategies. It works because it increases the target's emotional pain. The reason the Russian government needs us to feel agitation, rage and frustration is because it is grooming us to overthrow our government. Contented people don't overthrow their government. The Russian government is building a destabilization weapon, forming ordinary people to destroy our government, and con us into installing one of their puppets. If we do, we will most assuredly regret it.

Remember the large employee turnover rate a few years ago? That was caused by the Russian government. It's burrowing deeper into our country and into the psyche's of our people, to shred and destroy them. Ordinary people are being stealth abused at work and in their personal lives by Russia-controlled American operatives, in order to emotionally destabilize them, push them into self-harm, into seeking comfort in drugs, alcohol and/or any way that's dangerous to them.

The Russian government hates us. It's hard for most ordinary people in America to believe the Russian government hates us but only deep hatred, jealousy, and ambition explains their willingness to destroy not only our quality of life, but any life they believe they can use as a pawn to destroy us. The Israel-Hamas War is an example of who and what the Russian government is. It owns Hamas. You won't read that anytime soon in the news but it owns Iran, Hamas, the Houthis, and any other terrorists Iran owns. The Russian government intends to keep the news cycle focused on our President, who they are determined to destroy and replace with their puppet. To focus the news cycle on Gaza's being horrifically bombed, and the loss of life requires civilians to die in high numbers. The Russian government authorized the war and is ensuring the war continues. I can't prove it but it's what I believe because it bears the hallmarks of previous stealth attacks by the Russian government: (1) high loss of life of innocent people, (2) Americans and/or our leaders blamed, (3) the Russian government benefits (in this case the Russian government intends to install a puppet in the White House); and (4) the

Russian government isn't publicly blamed for the loss of life and there is no paper trail tying them to Hama's attacks and/or the Israeli government's war strategies.

Chapter 69 - The Russian Mafia intends to destroy us

The Russian Mafia won't declare war because we're much easier to kill when we're not warned, and believe me, the Russian government intends to kill us. The attacks on ordinary people usually take place out of sight of our legitimate government because the Russian government is here illegally and while it couldn't care less about out laws, it's, as I said, easier for them to kill us when we have no idea they're hunting us.

The Russian government tried to get me to shoplift by having cashiers not ring up all your cart items, or even everything you put on the conveyor belt.

The Russian government tried to get me to steal recently released movies on an app I didn't pay for, put on my phone by an entrapped family member without my permission.

The Russian government tried to get me to fraudulently apply for a federal program for low income people offering low cost or free wi-fi and low cost or free telephone serve I don't qualify for. When I told my loved one I don't qualify, she said apply and let the program decide if I qualify. Had I followed that advice, that's fraud. The Russian government would love that.

The Russian government tried to get me to use a fire stick in an illegal way. Had I done so their operatives would've used that to push me into other illegal acts.

The Russian government tried to stealth encourage me to embezzle my employer. I was tasked with buying corporate gifts on my boss' behalf, told I was responsible for approving purchases on the credit card, that only I would review the statement, and that I'd have no oversight. By that time, around 47 or so, I'd been harassed in the street, in stores, on the bus and by co-workers overtly at my last job the last 6 years, so I knew something was going on, but not what. I knew the police were involved (they harassed me), but not why. The Russian government isn't big on disclosing it's attacks to targeted Americans because they're here illegally, and the entrapped police they own harass extra judicially. The Russian government hates us and loves to confuse us but their North Star is they intend to rule our species and kill as many Americans as possible. Understanding that truth will help you better see their moves. They're not big on transparency.

Chapter 70 - The Russian Mafia attempted to entrap me w/a covid rent aid program

The Russian government tried to get me to defraud a federal covid program by slow walking my application so that it wasn't reviewed timely; then tweaked the program mid-program, employees telling me to submit a supplemental application. When I completed the supplemental application, employees of the program told me my supplemental application had been received and was awaiting review. That went on for multiple weeks.

After my calling about the status of my completed supplemental application, which Russia-controlled American operatives forced me to fill out by having their operatives slow-walk my original application (blaming it on the fact that the covid program prioritized applications of those facing immediate eviction, the Russian government always has a believable lie handy).

Russia-controlled American operatives are quick on their feet when attacking innocent people, figured out how to entrap me using the program. Operatives, mid-program, changed the program rules on the supplemental application, removed my completed supplemental application from their completed pile, and allow employees to tell me my application hasn't actually been submitted, even though it had been, and the employees I'd talked to for over weeks told me my completed supplemental application had been received and was awaiting review by my case worker. After weeks where I was told all was well, Russia-controlled American operatives sprung their trap. When I called, I was told my supplemental application needed to be resent again, and that, great news, the supplemental program now allowed for six months of rental aid, not the three months in the original program. All I needed to do was press the send button. Having no idea it was a setup I went to the library to use a computer there, since when I sent the supplemental application on my phone, apparently, although the webpage said I'd send it, now I was told it hadn't been sent. At the library I pulled up the supplemental application and pressed sent, and left, needing to pick up my prescription from the pharmacy. I hadn't read the supplemental form because I'd not been told there was new language on it but right before I pressed send again, my eyes had noticed new language just above the send. I hadn't really read the language, just skimmed it, because I hadn't been told there was new language.

On the bus on the way to the pharmacy, I began to be concerned that new language had been added and that I hadn't been told. At the pharmacy I called the program and for the first time in a couple months, even though I called with the program was open, my call wouldn't go through. I'd get a voice message but my was not put through. This was a deliberate attack on my phone by Russia-controlled American operatives. I called, I don't know, five time or so, and my call wasn't put through.

After paying for my prescription I took the bus home and in my apartment called a different number for the program. This number wasn't blocked and I explained my concern. The woman who answered the phone assured me the program rules hadn't changed but I knew the supplemental language was different. On the new supplemental application, just glancing at it briefly, and trying to remember what it said, but the changes appeared to mean I was no longer eligible.

The worker reassured me but I asked her to transfer me to the main covid rent aid number. She was able to put me through but Russia controlled American operatives attacked my phone by putting a horrible echo on it, the worse one I've ever heard. That is how I learned that operatives control who I can reach on my phone. The woman I spoke to reassured me the program rules on the supplemental application hadn't changed but I believed they had been and asked that she cancel my application. She lied and said she couldn't do it and I said, okay, I have the opportunity to return the money and cancel my supplemental application then. She was laughing at me, as if she were an operative but what I cared about was not applying for a program I didn't qualify for.

The next day was Christmas Eve and the program was closed, the day after that was Christmas, Saturday, and the program was closed, but the program operated on Sunday the day after Christmas so I called then, reached an employee, explained my concern. She told me a case manager was already working on my supplemental application but I asked her to cancel my supplemental application and she did, no problem.

Every worker I spoke to after I resubmitted the supplemental application told me there was no new language added but new language had been added. Had I known I was being set up I would've take a picture or somehow made a copy. As I said, I used the library computer, and I didn't know

I was experiencing yet another entrapment attack by the Russian government.

I'm careful about programs I apply to. My income is modest but I don't qualify for extremely low income programs. If the language I glimpsed briefly when I re-submitted my supplemental application had been on the original supplemental application, which I read carefully to ensure I qualified, I wouldn't have submitted my first supplemental application, I would've known I didn't qualify for the program.

Russia-controlled American operatives attempted to set me up and like with most of their entrapment attempts, I didn't anticipate it but reacted in real time to whatever the situation is. Sometimes I know it's an entrapment attempt, sometimes I don't know until later.

The Russian government cheats. It's American operatives hated that I qualified for the covid rent aid program, cheated me out of a month of rent aid (I received two months of rent aid but should've received three months total), and cheated me out of financial assistance for my electricity bill (my electricity bill is high so I would've appreciated help with it although I did receive my gas bill aid request for over $200 (workmen operatives, I believe but cannot prove, deliberately blew the grid and destroyed something in my oven that cost me $200 for the part). I couldn't use my oven for about a year after that until I paid to have it repaired. By slow walking my original application, illegally removing my completed supplemental application from the completed pile on the website, then changing the rules on the supplemental application, Russia-controlled American operatives lied and cheated to entrap me. When I called the program to rectify the problem, operatives prevented my phone from reaching the program by jamming my phone in some way. There are multiple ways to prove this. I still have the phone and phone number I used to place the calls.

That was how I learned that (a) the Russian government has access to our computer systems and can manipulate an American's application in a federal covid program, and (b) that was where I learned the Russian government can block calls by Americans on our smartphones. I've also learned the hard way that the Russian government can and does cause outages and surges of our grid, so unless you have something that must be plugged in all the time, you might rethink leaving your items plugged in that are difficult and/or expensive for you to replace.

I'd been a legal secretary at law firms that for the most part practiced general business law, not criminal law. I'm not a criminal although the Russian government has worked for decades to try to make me one. As recently as within the past six months Russia-controlled Americans operatives forced my GP to try, yet again, to lure me into cheating Medicare by offering to lie on my behalf for the weight loss drugs I don't qualify for. Diabetics qualify, pre-diabetics don't.

Chapter 71 - The Russian Mafia's attacks me on my final job

The Russian government was unable to corrupt me so they attacked me in the following ways at my final job.

The Russian government made me sick by: (1) having my employers overwork me to the point of exhaustion. Normally a legal secretary works for 2-3 attorneys/paralegals. At my last job I was assigned to 5 attorneys, and asked to assist additional attorneys if their secretary didn't come to work on a particular day and the attorney needed something filed or served, or a letter to go out on a specific date. It wasn't unusual for me to work for 6 attorneys; twice what is normal in law firms; (2) Russia controlled American operatives harassed me unrelentingly on the street, in stores, on the bus, etc. They had operatives on the bus mist me with a light version of pepper spray, which caused me to have reoccurring sinus infections where I called in sick and was forced to use all my vacation time as sick time; (3) Russia-controlled American operatives had my senior citizen mother and extended family tell me they had insufficient food to eat, needed transportation to attend school, needed dental care assistance, and other needs, and asked me for financial help. When I asked my three other half siblings to help me assist our mother, they refused (I later learned most if not all my half siblings are entrapped, (4) Russia-controlled American operatives had credit card companies flood my mailbox with credit card invitations, (5) I was close to my grandfather. I believe Russia-controlled

American operatives killed him to emotionally destabilize me. We'd been discussing our planned RV trip to the Grand Canyon. My grandfather told me, out of the blue, that he'd bought an insurance policy leaving me as the beneficiary. After he died unexpectedly, I believed his adult children stole my inheritance, and sent me a money order for $75. However I've learned of a way I can verify if my grandfather left me an inheritance and I'd be very happy to be wrong in thinking that my relatives stole it from me. I was unable to attend his funeral because I had no funds as my credit cards were maxed out to assist my extended family, and to buy flooring and a stove top in my apartment, as the entrapped landlord who owns the apartment where I live refused to make basic repairs. I had no vacation time as I'd had to use my vacation time for sick days deliberately made me sick by operatives on the bus spraying me with a mace-like mist, and harassed me on sidewalks when I walked home. My father, who's father had just died, did not offer to fly me to the funeral. He has represented to me over the years that he had low funds, which I believe within the last few years my mother says he and his wife own property she saw on Facebook (a problem is my loved ones are compromised and the Russian government has tried for years to get me into Facebook. I don't visit the site because the Russian government uses anything and everything as a weapon and they make my life miserable as it is); (6) the Russian government had one of the partners I worked for treat me cruelly, yell at me, as I worked my guts out for the attorneys I worked for; (7) When I went to doctors to try to find out why I felt so ill, the Russian government had the doctors it controlled not diagnose me. It turned out I have high blood pressure, I'm insulin resistance, obese, suffer from migraines, sleep apnea, severe insomnia, peripheral neuropathy, etc. None of those health problems were diagnosed by the several doctors I sought help from before collapsing; (8) The Russian government had the landlord where I live lie to me and say they'd received my rent check, even after I saw my rent check hadn't cleared. It turns out, distracted by grief over the unexpected loss of my grandfather, I'd mistakenly written my rent check on the wrong checking account. When I called the rental company, several times after seeing my check hadn't cleared, they assured me there was no problem, only to later serve me with a 3 day pay or quit notice on my door. I was confused as to why they hadn't told me when I'd asked them, several times, why my rent check hadn't cleared. (9) The Russian government had me go through a process with them where I paid them my rent, their legal fees (as it was my mistake that I wrote my rent using wrong checking account). The Russian government used that opportunity to protect the real estate business that owns this building, had me agree that our dispute would be under seal so the owners abuse and lies wouldn't be easily accessible to anyone who cared to look into the practices of this particular landlord.

The Russian government knows our systems backwards and forwards.[329] At that time, I knew I had an enemy, I couldn't help but know since I was attacked on the street, bus, etc., but the only hint I got who it was was that uniformed American police were involved. Starting from around 2001 or so, when I was 41 I believe, I was attacked unrelentingly by people I didn't know, most of them Caucasian males, with several Caucasian females too, and maybe 3,000 adults over two decades, and maybe 1,000 children over those decades. One of the groups harassing, intimidating and/or forcing others to attack me were uniformed and plain clothed American police. I witnessed a uniformed policeman force a young Caucasian man in his early 20s to harass me: he intimidated me and veered into my sidewalk space as if he intended to walk into me. Before he'd seen the police car and officer monitoring him, the young man walked a reasonable distance from me but after he saw the police officer driving slowly behind him, watching him as the young man approached me, he put himself closer to me as if he were going to walk into me. I saw that the police officer had forced him to attack me and the police officer didn't care that I saw what happened. At my last job, California experienced a police not strike but the police didn't

---

329 "...for decades Russian operatives, including key figures of the Russian Mafia, studiously examined the weak spots in America's pay-for-play political culture – from gasoline distribution to Wall Street, from campaign finance to how the K Street lobbyists of Washington ply their trade – and, having done so, hired powerful white-shoe lawyers, lobbyists, accountants, and real estate developers by the score, in an effort to compromise America's electoral system, legal process, and financial institutions."*House of Trump House of Putin*, Craig Unger (Penguin Random House UK 2018).

go to work for a day or so. For one day I was not harassed on the streets, on the bus, on the sidewalk as I'd been just the day before. After the very short walk out, the police returned to their work schedules, and I was again harassed nearly everywhere I went. At that time, the only place I wasn't harassed was in my apartment. After the Russian government made me disabled and had the operative in the apartment above mine begin to harass me when I was around 51 or so, even my apartment was no longer a haven for me. The Russian government tortures me by having it's operatives deliberately interrupt my sleep cycle to the point where I no longer have a natural sleep cycle. They know I hate when my GP tries to entrap me and have him do it to make me so uncomfortable I visit the doctor as little as possible. They work to facilitate my death as quickly as possible from "natural causes."

I learned conclusively the police were attacking me when I witnessed one directing an attack against me, and when the police staged a walkout and I wasn't harassed by strangers for that day but was harassed when police returned to work. My own police, who I pay my taxes to employ, were managing and forcing attacks against me.

Chapter 72- Shoplifting entrapment attempts

I didn't have many groceries in my shopping cart. On my mobility device I'd gone to the big box store to pick up my prescription. I didn't have much money but was hungry and so put maybe ten food items in the cart, basics like peanut butter, etc. I didn't' add up the items in my head because I was tired and hadn't planned on buying anything, just collecting my prescription. I hadn't eaten, but I didn't think the food cost more than I had. I waited for the cashier to tell me what I owed him, gave him the money and left. I had no idea I'd just been setup, that the cashier hadn't scanned nearly half of the relatively few items in my cart.

I'd spoken to the store manager at the store about some issue, I think I'd made a return on a previous visit and been given too much on that visit, and so told the manager and gave him the extra money I thought I owned, I don't remember the precise details but I believe I paid the manager money I'd been wrongfully given the last time I was there. As I exited the store, the store manager, an African American man, watched me. I nodded good bye to him.

Exiting the store after the exit person handed me back my receipt I glanced at it, then glanced at the food in my cart. I stopped rolling (I was pulling the shopping cart as I rode my mobility device). The cart wasn't loaded and I'd never had a cashier not scan everything in my cart (I'd had items double scanned but not entirely omitted), so I wasn't familiar with this type of entrapment, which was a conspiracy involving the compromised cashier and compromised security personnel. The Russian government was determined to kick me into the criminal justice system. This attack happened during covid.

Seeing I hadn't been charged properly I immediately turned around, against the flow of exiting customers, and returned to the cashier. I told him I hadn't been charged for all my groceries. He asked me how I'd known, asked if the exit person told me. I told him no, I'd noticed and returned to pay him. He seemed flustered.

He scanned an item and I pointed out that he missed two more items I hadn't paid for. It took him three tries to ring up maybe 10 items. That was bizarre but I didn't' know it was a setup then. What saved me was ethics. I didn't think the free groceries were a blessing and keep going, only to be surrounded by security guards ten feet out the store. As soon as I saw I hadn't been charged, I returned to pay what I owed. Even when the cashier left off two items after I returned to him and had to ask him a second time to scan all the items, I didn't know it was a setup. I understood I'd been set up as I ate something outside the store, and plain clothed officers stood staring at me.

The Russian government cheats. I'm not a shoplifter so they had a cashier they owned not scan everything in my shopping cart. What would've happened is I would've told security I went through the line and paid and had no idea why the food wasn't scanned. Security would've roughed me up because this was a Russian government entrapment attempt. I would've protested and been charged with resisting arrest. I'd be yet another dead citizen. People protest after an innocent person is killed but the victim is still dead. This happened during covid so it may have been after President Biden was elected but before he was sworn in, I forget exactly when it occurred.

I'd written President Obama when he was in office and, before I saw how former President Trump interacted with Mr. Putin in Finland, I'd written Mr. Trump asking for help with the unrelenting stealth attacks I'd experienced over years. It's possible, and likely, that the Russian government didn't want me to write President Biden requesting help the way I'd written the two previous presidents. I believe Russia-controlled American operatives made this shoplifting play after President Biden had been declared the winner but I don't' remember for sure.

After that entrapment attack failed Russia-controlled American operatives had several more cashiers at other stores do something similar. The reason Russia-controlled American operatives want me in the criminal justice system is so they could beat and rape and destroy me. The Russian government is deadly serious in it's intention to destroy us. I've no idea what they tell American law enforcement they've entrapped but, given that these officers are owned, the Russian government probably lies freely.

I read an article where rape is prevalent in California prisons and of course, the Russian government and it's American operatives know exactly how prisons are run here, I suspect Russia-controlled American operatives run some of our prisons and/or jails. In *The Guardian* article by Sam Levin dated 10/25/2023 referenced in the footnote below, a serial rapist guard in a California prison had 96 abuse charges, and those were from women victims willing to come forward to lodge complaints.[330]

Chapter 73 - Don't accept a plea deal that insists you aid law enforcement for the rest of your life to stay out of jail/prison – that's a Russian Mafia entrapment. They target families. They'll know if you have a deadline. Remember: masters of psychological warfare

Once the person is entrapped and taken into custody, I theorize police show them the evidence and recordings they've amassed,[331] then offer the victim a deal: no prison or jail time, and in return the entrapped person will assist police when asked.

I had no idea I was being hunted before the Russian government went loud and began attacking me on the street, etc. I'd been stealth attacked since I was 11. Although it's true I had no training, most ordinary people have no training in espionage or psychological warfare and so it is, I suspect, extremely easy to convince untrained people that they are corrupt when the interrogators are specifically trained to con them, as the Russian government would ensure the arresting law enforcement are.

---

330 Records and interviews suggest the California prison system let Gregory Rodriguiz get away with rampant sexual abuse, while his victims were punished. Sam Levin "One Prison Guard, 96 Abuse Charges: Women Say 'Serial Rapist' Targeted Them Over a Decade," *The Guardian* (10/25/2023).

331 Janis Berzins "The New Generation of Russian Warfare" "...the Russian view of modern warfare is based on the idea that the main battle space is in the mind." Aspen Institute | Prague, *Aspen Review* (3/2014). https://en.m.wikipedia.org/wiki/new-generation-warfare.

My father, a law enforcement officer, was entrapped. I know because he was used to try to entrap me. The Russian government and their American operatives are masters of psychological warfare, whether the target is a police officer like my father, or an ordinary person like myself. It's critically important if this coup is to work, and it is working, that victims are made to believe they themselves are corrupt. A person who believes they've been set up, deliberately targeted, will forever seek exoneration, and I've detected nothing from the people in my family that they are fighting or resistant of whatever's been done to them. I'm the only one I see who's resisted and it's probably because, as the oldest, I bought into ethics more, had more time being trained by my mother because I'm the oldest, while my younger siblings were too young to really get ethics locked in as young children. So when they were entrapped by the Russian government, they were compromised. Before the Russian government's handlers went loud on me, I had zero idea I was being investigated. I knew nothing. I didn't like the "friends" the Russian glued to my side but I had no idea they were operatives; their behavior was odd.

But if you'd have asked me were they operatives I wouldn't have had a clue what an operative was, and by that time I'd been stealth hunted for about 30 years. The Russian government are masters. The only reason they went loud on me then, is because I was laughing with the women operatives and told an off-color joke and the plain clothed officer sitting next to us was offended, and grew annoyed, and lost his temper. I noticed his annoyance. He probably was annoyed that I'd gotten another job, away from his operatives. Maybe he would've gotten a big bonus had he entrapped me, maybe that's why he was annoyed, there's no way I can know. But had he not lost his temper, I would've never learned I had a Russian government enemy. When you don't know you have an enemy that dangerous, things don't end well for the target.

I think they'd been hunting me so long they thought I knew they were there, but I didn't. I'd evaded multiple entrapments by then but I had no clue I had. I'm just an ethical person – a little sensitive to people's feelings around me, but nothing beyond that. Ethical people exist.

Most of the people who've tried to entrap me haven't been to jail or prison that I know of, so I assume part of a deal must be no prison/jail time. And since I've known these people for decades and haven't heard of them being arrested, I assume that in entrapment attacks ordered by the Russian government victims are arrested away from their family, friends and job, so the Russian government can use them to entrap their friends, family, and work colleagues.

I suspect the victim is told they can't discuss their arrest and deal with anyone but the charging officer or his/her designee, and that if they break the deal, even if discussing their arrest with an unauthorized person, including their spouse, parent, another law enforcement officer, the deal is voided, and the person will be immediately imprisoned for the original crime, and the crime of breaking the deal. There has to be a reason why millions of entrapped people aren't publicly complaining and that reason must be because the Russian government uses employed, uniformed law enforcement to work these scams. It would never occur to an ordinary person that they're being attacked as part of a coup by an enemy foreign government.

Once the victim is entrapped and takes the deal, I believe Russia-controlled American operatives groom them to commit more crime/attacks. They're also rewarded when they follow orders. I've witnessed entrapped victims receive affordable housing even in tight and/or expensive housing markets, and/or a well-paying job, and I've witnessed extended family who are entrapped facilitate tax fraud and attempted to lure me into committing tax fraud within the last several years. The key to why entrapment works is the Russian government convinces the victims that they're criminals. Conned into believing so, the Russian government then grooms them to criminals. I theorize their Russia-controlled handler manipulates and encourages them to commit, to more easily blackmail them, to entrap more people in our family and to financially weaken our government. I've witnessed parts of these attacks on a job, where I was stealth harassed by two entrapped people to get me to quit a job; and I've witnessed family victims who appear to have been encouraged to commit tax fraud, and those victims seemed to me to facilitate tax fraud in other, younger family members. No one tells me: I've had to piece the behavior together over years while evading stealth entrapment attacks by extended family members and unknown police controlled by the Russian government.

I believe, based on approximately 1,000 attacks by strangers with their children, and another approximately 2,000 attacks by individuals with no children, plus the entrapment of some of my own family members, and evading multiple entrapment attacks, that the Russian government prioritizes attacks on families. After a parent is entrapped and takes a deal, the Russian government coerce the parent to groom their children, and/or try to entrap their child (both my parents tried to entrap me, which is how I know this part of the coup exists); both my parents were used to gas light me; I believe they were forced. I've witnessed children being used by their entrapped parent as beards and/or as a co-attacker, working in unison with their parent to harass me. I've evaded Russia-controlled American operative's entrapment attempts for approximately 53 years (most of the time unknowingly); so now they harass, threaten, rev their engine as if they're going to hit me with their vehicle as I cross in front of them on my mobility device, and/or torture me (with sleep deprivation) on a near daily basis.

Russia-controlled American operatives weaponize the children of entrapped parents and use them to bully, befriend, harass, lure, and/or entrap other children in families the operatives want destabilized, or to attack an adult Russia-controlled American operatives want harassed (I've been harassed by Russia-controlled skateboarding, bike riding, swarming children for decades).

I also believe compromised teens and young adults are used by Russia-controlled American operatives to gain access to teens in other targeted families, to get teens or young adults to take and/or buy illegal drugs, to the detriment of the targeted person. The Russian government camouflages itself inside a compromised loved one or trusted friend, lies to the targeted child or young adult, and the compromised friend or loved one being used as the hide isn't allowed to warn the target to run.

I believe it's very easy for the Russian government to stealth attack our children and young adults in this way, and I believe Russia-controlled American operatives are doing so. My nephews and niece when they were children (9 to 13 or so I think), were set up by Russia-controlled American children, their school friends, and asked to stash stolen school laptops in their apartment. My nephews and niece were told by their school friend that they'd be given one of the laptops if they helped their friends out. Setup, when they were children, by their school friends. They lost their public housing and became homeless; my nephew went to juvenile hall. None of them finished high school; one got his GED. They'd not participated in the theft, had no knowledge of it, had no idea it was a setup, but the Russian government destroyed their childhood and they've not recovered. I didn't find this out until some time later; Russia-controlled American operatives were attacking me at the same time, and the operatives in my family weren't allowed to tell me.

Russia-controlled American operatives have entrapped, I believe all of my family. From what I've seen from other families in Santa Monica, most of them Caucasian because Caucasian people are the largest ethnic group in Santa Monica, the Russian government is making huge inroads into destroying my entire nation. I've been attacked by thousands of Caucasian people over decades, including being physically attacked, kicked, stomped on, called the n word, etc. I'm positive all those people weren't racist; some were, the majority weren't.

When a parent brings their days old infant on an harassment attack, you know there's a problem. Most parents are protective, especially of their infants. I was the target in these attacks so I witnessed the parent's behavior. No way is an non-compromised parent, someone who recently gave birth, going to get out of bed and take their newborn on an harassment attack unless that parent is forced. People must be warned about how the Russian government works on the ground to destroy our families.

Chapter 74 - The Russian Mafia has stealth commandeered some of our police

I believe the Russian government commandeered our police's entrapment weapon and has turned it against ordinary noncriminals to reinstall slavery in America. To entrap ordinary Americans on the scale I believe the Russian government is doing, they would need to have entrapped many, though not all, of our domestic police.

Even stating that some of our police are entrapped and controlled by the Russian government, and are being used to run entrapment attacks against ordinary Americans is a claim fraught with danger for police. For people who believe me, I don't want them attacking our police officers, many of whom are innocent civil servants doing their job and minding their own business. Entrapped police are victims, just as many of us are victims of the Russian government. Attacking victims of the Russian government would lead to a national blood bath, which would delight the Russian government, but destroy us. That's unacceptable. We must find a way to tell the truth without being further victimized by our enemies.

There's so much hatred of police in America, which the Russian government facilitates in order to frighten, stress and make police feel isolated; to make ordinary Americans feel this is not our country, not their government; to make police and ordinary Americans feel unappreciated, encourage everyone to take a "them versus us" attitude, encourage everyone to see ethnicity instead of a citizen, deserving of respect and courtesy. All of this plays into the hands of the Russian government who is working us.

Chapter 75 - The Russian govt turns victims into mean attack dogs

The Russian government deliberately makes ordinary Americans, including people they entrap, into attack dogs: makes them mean, short tempered, irritable, stressed. Torments and/or irritates them, has them always on a simmer, unhappy with their life, and easily triggered by a Russian government operative. The reason the Russian government aggravates and agitates ordinary Americans, including people they've already entrapped, is because it's much easier to shape enraged people into a weapon than it is to manipulate a calm person. From their entrance into my life, the Russian government has worked very hard to enrage me, using a variety of strategies, so I have a good sense of how they're enraging ordinary Americans.[332]

332 I worked at a law firm in my 30s as a legal word processor, had gotten the job by going to the law firm, demonstrating my skill set and requesting a part time weekend job, while I worked a full time job at a different law firm. I eventually quit the job I'd worked at in Century City, California and worked full time at the West side law firm. I was on good terms with my supervisor. At some point, maybe I'd worked there a few years, my supervisor began to treat me cruelly. I didn't know why. My work quality was good. Law firms I worked at didn't have unions, so a worker did good work all the time or they could be fired; there was no worker advocate to explain a worker's point of view. After a few years, out of the blue, my work quality began to be questioned and after which I began protecting myself and made a copy of the document before and after I worked on it. In word processing departments, multiple word processors usually work on documents. The next time I was accused of subpar work, I presented my before and after, demonstrating where I worked on the document was quality. My proof enraged my supervisor. It demonstrated that I wasn't the culprit but the fact that I'd thought of a way to prove my innocent made my supervisor and her assistants really angry with me. I didn't know at the time but Russia-controlled American operatives aimed my supervisors at me. At that time, in my early 30s, I had no idea many of the experiences I had were stealth attacks by Russia-controlled American operatives. My supervisor at this particular law firm treated me cruelly. Her behavior was a very distinct change, as she'd hired me into the department. She was mean and I just worked and did my best. After several months of her apparently disliking me, another word processor, a woman of color, came to me and told me she'd filed a complaint of racism against our supervisor and would I like to join it. This was a setup but I didn't know at the time. I'm African American so Russian-controlled American operatives had a woman of color approach me with the offer. Our supervisor was Caucasian (she's since passed away). The setup was to get me so enraged by my supervisor's cruelty that I'd join a federal lawsuit in retaliation, accusing her of racism. I declined my co-worker's invitation because I didn't think my

The reason our policing systems, and many of our jails and prisons, are stuck in the post-slavery, Jim Crow era is because the Russian government needs us stuck there so they can more easily destroy us. You put a citizen in a Texas prison in summer, without air conditioning, with cell temperatures regularly at 115F (and sometimes topping 149F) and, if the citizen survives, they're going to be a very different person post-prison than they were pre-prison.[333] Most imprisoned people have families and/or friends, and most are released from jails/prisons at some point. After guards or other inmates have beaten and/or raped them, after they've been slow roasted in the summer and frozen in the winter, the citizen released back into our society is a traumatized, easily triggered bomb, and you better believe the Russian government will trigger them. That's the whole point of the Russian government's strategy in forming our jails, prisons, and police into horrors: to incite rage and agitation, to get us to overthrow our own government, and to get us to accept a puppet controlled by the Russian government installed in the White House. Grievance politics is

---

supervisor was racist. She had a problem with me but I didn't think racism was it. This is an example of how the Russian government creates rage and then tries to use that rage to attack it's target, in this case, me. Had I joined the federal complaint, when I knew my supervisor wasn't racist, I could've been charged with filing a false federal complaint. The supervisor knew I was African American when she hired me. Maybe 35% of the word processors in the department were African American and/or Hispanic, I couldn't see racism. I was mistreated, but Russia-controlled American operatives hoped to enrage me so I wouldn't think through the argument, look at the ethnic mix in the department, and conclude that there was no racism argument there.

333 Michael Schwitz and Michael Winerip, "Kalief Browder, held at Rikers Island for three years without trial, commits suicide," *New York Times* (6/8/2015).

a real thing, it's a strategy they've worked on for decades, and it's nearly perfected as evidenced by the Russian government's attack on the Capitol, using enraged Americans as weapons.[334]

Based on the Russian government's unrelenting efforts to entrap me, I have a very strong feeling it controls some, if not all, of our jails, prisons, prisoners, and/or staff. We must reform our policing and prison systems. I strongly suspect our enemy controls those systems and are attacking and destroying our people and/or destabilizing them so when they get out, it'll harder for them to assimilate into society and easier for the Russian government to trigger them into violence against their loved one's and society.[335]

---

334 The Russian government is deeply embedded in America and controls many of the systems, businesses and institutions most ordinary people rely on. For example, where I live in Southern California, the Russian government controls most, if not all, of the grocery stores within a five mile radius of where I live. It has security and workers harass me in stores where I've shopped for decades. In a big box store where I've spent thousands of dollars for needed items for extended family over decades, their operatives try to lure me to theft by having an operative leave her unattended purse in a grocery cart right in front of me while she wanders down the isle; or the Russian government will have a cashier fail to ring up all the items in my cart even when I have only about ten items. To get me to shop at a store the Russian government prefers I shop at because it's operatives have an easier time harassing me there, they had one store stop stocking several of the groceries I prefer and had a different grocery store stock two of those items. Most people don't understand the Russian government controls the ground in America. Until we all know what to look for, the Russian government isn't going anywhere.

335 Ed Pilkington, "People Are Dying: Texas Prisoners Say Heatwave Turns Cells into Ovens," *The Guardian* (7/31/2023). In Texas animal shelters are required by law to not allow ambient temperature to exceed 85 degrees for over two hours. In Texas prisons where approximately 100,000 citizens have no cooling system – they are **subjected to torture** with temperatures rising as high as 149F, and regularly above 115F. About 70% of it's units have partial or no cooling systems. Air conditioning legislature has been introduced to the Texas state legislature since 2019 but each time it's come up for debate the

Chapter 76 - Jackson-Vanik Amendment seeded spies here

The Russian government has had operatives on the ground in America since at least the 1950s, probably in the 1940s. Because the Russian government has embedded operatives here the last 60+ years, and embedded many more by exploiting the Jackson-Vanik Amendment in force in 1975, it has embedded an unknown number of spies and Mafia, possibly hundreds of thousands: the exact number is unknown. Approximately 600,000 Soviet immigrants migrated to America under the Jackson-Vanik Amendment. Not all of them were spies or Russian Mafia but some of them were.[336]

The Jackson-Vanik Amendment is no longer in effect but the USSR, and later the Russian government, exploited that amendment to seed our nation with a large number of operatives.[337]

My theory is the Russian government installed operatives into positions of authority as well as lower level positions, in police departments, unions, and throughout the country in various businesses, institutions, etc. According to Yuri Shvets, a former major in the KGB who came to the

---

Republican-controlled state senate stymies it. This year the Texas House passed a bill to invest more than half a billion dollars in air conditioning. The senate opposed the idea and despite the state having $33b in reserve, it opted instead for $0 investment. "When inmates are begging and pleading for their lives in the searing heat, that is a problem," said Carl Sherman, a Democratic state rep who has sponsored cooling bills but who says state senators are stuck in the old mindset.

336 Craig Unger, "*House of Trump House of Putin,*" (Penguin Random House UK 2018).

337 Ibid.

United States and now lives outside Washington DC "The KGB is very patient. It can work a case for years."[338] Over time I believe Russian government operatives set up entrapment attacks targeting police. I believe that because I've been extra judicially harassed by hundreds of police over decades, uniformed and/or plain clothed. After more time, and for sure in Southern California where I live, the Russian government was able to take control of some of the police's entrapment mechanism to target ordinary Americans. Based on how I've seen the Russian government exploit our media when it wants a particular type of story run,[339] it's strategy is not to trumpet it's access to our systems. If it did that, it'd be outed. Instead, the commandeered system works nearly the same as it normally should, except for the times when it's hijacked, re-purposed. That too is done stealthily, to make the commandeered system appear to work as it should, but it's not quite normal because, although the Russian government works stealthily, their agenda is to destroy us. I'm not saying or suggesting specific journalists are owned by our enemy. I'm saying the Russian government has found a way to get articles in our newspapers that serve their interests, not ours. When they commandeer one of our systems, they use that system to hurt us and to help them. Again, in using a specific article, I'm not saying the reporter is owned, I'm saying the Russian government has found a way to get stories into our newspapers

338 Craig Unger, "*American Kompromat*" (Penguin Random House LLC 2021).
339 David Leonhardt "China's Problems are Real," *New York Times* (August 25, 2023).

that serve their interests and hurt our own.[340] Based on how I've seen the Russian government exploit our media, I theorize entrapped law enforcement work legitimate entrapments, as well as entrapments aimed at noncriminals, which serve the Russian government's intention to destroy us. Some of our compromised police will have to come forward and describe how the scams work in detail, if they can.

My father is a retired police officer, a senior citizen, entrapped. He is not free to discuss his entrapment with me, nor are any of the entrapped people I know free to discuss their entrapment. I only know they're entrapped because they tried to entrap me. To anyone looking at them, they're ordinary people, workers, grandmothers, grandfathers, aunts, uncles. I believe several of them earn good salaries. None are homeless, the Russian government financially rewards people they've entrapped. The homeless people in my family, now in their late 20s, and early 30s, were attacked by the Russian government as children, and hooked on drugs and/or otherwise destabilized.

I learned my father was entrapped in 2021. Before that, when I trusted him, he said something I remembered after I understood he is owned by my enemy. We were talking, I might have alluded to a recent attack against me, and he said something like "you wouldn't want me to die, would you?" I said no, of course not, something like that. My takeaway, if I'm right that my father was referencing his entrapment, is that entrapped people are told they will be immediately imprisoned

340 Eswar Prasad, "The Problems with China's Economy Starts at the Top," *New York Times* (August 26, 2023).

or killed if they discuss the terms of their entrapment with anyone other than their handler. I'm assuming their handler is a compromised American police officer, either retired or working. My father was used by the Russian government to try to entrap me using Social Security Disability fraud, as was my mother and a couple of my siblings. I was 50 when he tried it, so he would've been retired by then, as he's a generation older than me. I theorize that compromised retired or employed police are used in entrapment attacks, based on my personal experience with my father, and based on the many entrapment attacks aimed at me but I don't know what happens when the Russian government entraps noncriminals with the intention of enslaving them.

Chapter 77 - From entrapment to placement in the GOP Base – how the Russian Mafia runs it's coup against Americans

Citizens are influenced and/or coerced to vote how the Russian Mafia wants them to vote. I personally know two of the GOP Base. I know they're entrapped because they've tried to entrap me many times between the two of them. Based on my personal experience, I believe the Russia Mafia entraps individuals then use a variety of means to get them into the Base. I've been quadruple teamed by entrapped family members (to get me to stop using a flip phone and start using a smart phone which the Russian Mafia can more easily control and access), so it's no problem for the Russian Mafia to have compromised law enforcement entrap/ambush targets, then influencers in the new church the Russian Mafia pushes victims to join, or have an operative the Russian Mafia places in victim's bed to groom victims to do whatever the Russian Mafia wants. After noncriminals are entrapped their lives are never their own. I've no idea exactly when my mother was entrapped by this past summer, 2023 she was used by her handlers to see if they could get me to a palm reader. Apparently they've worked that scam on at least one of my family members, one or more.

The Base members I know who admitted they're part of the GOP Base are my mother and youngest half sister. From what I can tell they were entrapped and at some later point flipped into the Base by influencers or someone. The Russian government double, triple and quadruple teams targets. They'd have no need to set off entrapped citizen's red flags by insisting they vote for a particular candidate. Instead I suspect victims are flipped into the Base by the Russian government's other influencers.

I've had compromised migrant workers who work next door to where I live torture me with sleep deprivation for three, four years. They've finally stopped now but I begged them to stop a couple years ago and they wouldn't. A few smirked at me. The Russian government loves to make people beg. To them, it's music.

Chapter 78 - The Russian Mafia attacks me using our grid

The Russian Mafia and/or it's allies control our electric grid. I know because Russia-controlled operatives have harassed me over the last approximately 14 years by deliberately shorting out multiple household appliances including my oven, fan, blenders, food processors, vacuum, microwave, all of which I've kept as proof. On September 29, 2023 at 5:40am, Russia-controlled operatives believing I was in the process of writing a book outing them, attacked the grid in Southern California and later surged the power back on, targeting the circuit I normally use to charge my laptop – with the intention of destroying my laptop after they learned I was using it to write this book.

My computer wasn't plugged in but they couldn't tell. They have eyes on me illegally in this apartment. When they caused the power outage, my plug strip had chargers plugged in but they couldn't tell that the laptop wasn't charging because of my belongings on the bed. When I woke to go to the bathroom, the power was out and, not knowing the Russian government planned to trip the circuit, I unplugged the chargers. I believe Russia-controlled operatives killed the power in the apartment. The operative upstairs couldn't tell I'd unplugged the chargers when I went to the bathroom - nor did I think about that. I unplugged the chargers out of concern that when the power came back on, that it not destroy the chargers, although I didn't know if it would or not. I unplugged them out of an abundance of caution.

If your government is aligned with the West, expect Russia-controlled operatives will harass you by turning off your power, likely in the heat of summer, when you need your fan or air conditioner. Have backup plans in place. And if you use electricity for heat in the winter, have a backup then as well.

Chapter 79 - The Russian Mafia controls some of the grocery stores & a big box store in Southern California. I know because their operatives harass me there

Russia-controlled American operatives control our grocery stores. I know because to get me to stop shopping at one store Russia-controlled American operatives stopped stocking the affordably priced low carb foods I used to purchase there, and had a store they wanted me to shop at begin to stock two of those items. Most stores in America have memberships or special offers customers can access using the grocery stores card for a special deal. In this way and others Russia-controlled American operatives know people's favorite foods. To frustrate and harass ordinary people, Russia-controlled American operatives have grocery store employees in charge of ordering products, under-orders favorite products. Remember the Russian government believes in exploiting and using people as weapons. Contended people aren't weapons grade; frustrated, pain-filled, enraged people are. I've spoken to one of the entrapped GOP Base about this and she's experienced it too, although she doesn't know the Russian government is behind it. To frustrate ordinary people, to make them unhappy and stressed, and encourage them to move from California (which the Russian government is determined to make a swing state), to another state, Russia-controlled American operatives use under-stocking favorite foods as an effective weapon. Forcing families, targeted and mistreated at work, poorly served by their landlord, by any authority figure, forcing them to sit in traffic for hours while they travel from store to store hoping to buy needed foods, is one of the many ways the Russian government frustrates people and makes them a ticking time bomb.

Grievance politics doesn't encompass the evil totality of how the Russian government stealth attacks people. Well-intentioned, ordinary people like myself give their best at work and to their families. In retaliation, the Russian government ensures those people are psychologically kicked in the face, and/or exploited by operatives the Russian government controls. And understand, the operatives in my family don't know the Russian government controls them. They do what they're told to do by their handlers.[341]

---

341 I bought my sister a car for school, gave it to her. I have health problems and no driver's license. To grocery shop I sometimes use a transportation service. My sister gave her car to our mother to use. While I likely would've agreed with her choice, she didn't' offer it for my use. I've since requested that if our mother no longer drives it, I'll have access to it. I bought a artificial Christmas tree and my sister liked it. I told my family it was a family tree, so they knew they could use it too. I left it at my mother's in the box, my mother didn't want it put up because it was too big for her space and I had no way to get it to my apartment. My sister, without asking me, took it out of state, probably because her handler told her to, in order to get me to visit her during the holidays, where the Russian government intended to attack me. Russian government operatives don't tell people they control what they intend

To victims in Ukraine who are being bombed by the Russian government being stuck in traffic for hours over the course of two days to buy groceries (a chore that shouldn't take more than four hours), sounds much better than being bombed, and it is. But stealth attacks day after day, year after year, and in my case decade after decade, causes health problems.

Palestinian civilians have their homes and apartments bombed, and their babies and loved ones killed. Nothing is worse than that. But in the Russian government's secret war against Americans, the Russian government destroys families by forcing entrapped family members to work for the enemy government, and use the family member to lure the target and/or obtain information from the target. In both hot war and covert war, innocent civilians bear the brunt of the attacks.

Chapter 80 – I believe the Russian Mafia controls traffic flows on the freeways in Southern California & uses traffic jams to encourage people to leave CA; it's determined to turn CA into a swing state

I believe the Russian government controls traffic flows in Southern California by placing operatives they control in infrastructure, road expansion and/or road planning positions, and then having some of those operatives make choices that increase traffic jams. I believe this because Mr. Berzins

---

to do to targets they're luring, in this case I'm the lured target. I've been attacked by the Russian government for over fifty years now and nothing good comes from the Russian government's exploitation of your loved ones to get you (the target) from point A to B. Only evil comes from that government, especially to those like me it's working to destroy to prevent me ever having a frank conversation with American security agencies.

described how the Russian government manipulates citizens to rise up against their own government[342] and because of what one of the GOP Base members told me, describing their intense frustration with once a week grocery shopping and how it has inexplicably become a multiple day chore, swallowing up the limited relaxation hours a family has at the weekend. The Russian government's whole thing is to seize world power by destroying America. To get ordinary Americans to rise up and overthrow our government, the Russian government makes us as miserable, exhausted, and annoyed as possible, and then has one of it's many influencers point the figure at our legitimate government. As you saw in the attack on our Capitol in 2021, the combination of triggered and enraged citizenry, aimed at a target by their influencers, resulted in the deaths of at least ten of our law enforcement officers.

Ordinary people are easily made miserable because the Russian controls millions of entrapped people here. Work supervisors are owned, so it's no problem to force an entrapped supervisor and coworkers to mistreat another worker (this type of attack has been done against me many times; and is how the Russian government got me to leave companies and find a job elsewhere, starting with my first job, post-university). Doctors are employees, entrapped doctors are easily controlled by their supervisors and handlers. Presumably Russian controlled law enforcement references Medicare fraud and Medical Centers

---

342 Janis Berzins "The New Generation of Russian Warfare" "...the Russian view of modern warfare is based on the idea that the main battle space is in the mind." Aspen Institute | Prague, *Aspen Review* (3/2014). https://en.m.wikipedia.org/wiki/new-generation-warfare.

likely tell their employed doctors to cooperate with security, and there you have it: my doctor deployed by the Russian government to entrap me into Medicare fraud. The last time they tried it was at my 2023 doctor's visit.

An entrapped doctor tried to steer me into taking psychotropic drugs after I complained to one of my doctors that police were stealth harassing and attacking me. One of the Russian government's most effective weapons here are parents and doctors because ordinary people don't understand they're under attack by an enemy foreign government who has entrapped their beloved parent and/or trusted doctor. Entrapped parents and entrapped medical care providers cause a lot of harm to ordinary people in America. Unneeded hysterectomies, unneeded C-sections, etc., the Russian government makes medical centers happy by having doctors recommend expensive services the Medical Center will be well compensated for, whether the target needs the service or not. Someone in my family was subjected to 7 surgeries in 4 years. The Russian government's intention was to hook them on pain medications to make them more willing to entrap me, similar to the play I believe the Russian government made against my favorite maternal aunt. Another family member, a senior citizen, had a vaginal prolapse problem and was forced to get a hysterectomy in order to get the prolapse repaired. Not even lying.

One of my doctors stopped treating me because the few times I mentioned I was under attack, she simply didn't care.

The Russian government controls my phone number so controls which calls my phone completes. When I tried to get prescription refills and make followup doctor appointments, the Russian government wouldn't let my call go through, put me through to voicemail, in the middle of a work day when the doctor's office was open.

The Russian government has bosses and/or coworkers of targeted individuals mistreat them, have their landlord perform the cheapest repairs possible, and deliberately frighten the tenant by making loud noise outside the targeted person's apartment, has the target's car by a hit and run driver and have the police not only not tell the target the name of the perpetrator, which name the police know, but have the target's car insurer refuse to pay the claim, even though the target had hit and run insurance. No lie.

There's a thousand ways to stealth harass and enrage ordinary people and Russia-controlled American operatives know all of them. Because of their expertise and years of practice destroying the quality of life of ordinary people in America, it was no problem to get enraged citizens to storm our Capitol in January 2021.

Chapter 81 - The Russian Mafia controls who gets affordable housing in Southern California

I've seen various entrapped family get affordable housing. I was lured to this rent controlled Santa Monica apartment over 35 years ago by Russia-controlled American operatives who, unknown to me at the time and not until decades later, had been working to entrap me for years. I didn't know at the time. I know from direct personal experience Russia-controlled American operatives control access to affordable housing here.

Chapter 82 - The Russia Mafia controls how often tenants are harassed by their landlord & which tenant gets decent versus shoddy repairs

My landlord stealth harasses me, and has for decades. Most recently the manager of the building where I live, his family owns the building and manages it, noise harassed me and startled and alarmed me to such an extent, my heart elevated to 200 bpm and I needed to call 911. After my ER visits, when I went to the Santa Monica Police Department to request to file a report, the desk officer asked me to describe the problem. After listening, and he was nice, he said the Santa Monica Police Department didn't accept complaints of harassment. He said I could file a lawsuit, if I wanted to go that way. My GP told me my 200 bpm heart rate wouldn't have lead to my death which is what I believed, when I described the experience to him in the summer of 2023. Because he's used to try to entrap me with Medicare fraud, I've no idea whether he was lying or not.

Chapter 83 - The Russian Mafia controls who gets disability

Russia-controlled American operatives control who gets approved for disability. In America, and I believe in other West aligned countries like England, the Russian government is in the process of sickening, disabling[343] or getting hooked on prescription or street drugs, as many working age people as possible so that if the Russian government and/or it's allies make their move against America (and our allies), we will have many fewer able bodied people to serve in our militaries. I read that England is experiencing an uptick in disabled people.

Chapter 84 - The Russian Mafia's IRS strategy is to starve our nation of funds we need to thrive & to prevent us from fulfilling our promises to nations needing climate catastrophe assistance

After the Russian government saw that it couldn't corrupt me on the job, it had operatives at the law firm I last worked at overwork me to exhaustion. After which it set multiple family members to trying to entrap me. By switching me to a hard working tax paying citizen, contributing to my nation's coffers, the Russian government flipped me to a person financially reliant upon my government to the detriment of me, my family and my government. The Russian government has worked at financially weakening America for many years, and has had people in the GOP it controls stop our IRS from being adequately funded for years. This is another way how the Russian government is working to defeat us. Presently, our nation needs many hundreds billions of dollars to pay for infrastructure to transition to an electricity powered nation.[344]

343 Robert Booth, Michael Goodier "Surge in Young People declaring disability in England and Wales," *The Guardian* (February 8, 2023).
344 Russian government operatives repeatedly shorted out my expensive mobility device batteries until they no longer charged. I'm concerned they'll easily short out batteries in electric cars. I hope manufacturers have a protective barrier around batteries to make it more difficult for

The Russian government has not only flipped people like me from being a tax payer to needing financial assistance, it has, I believe, groomed some members in my family to cheat on their taxes[345] in order to more easily blackmail the victims in my families, and I suspect my family are not the only ones Russia-controlled American ops have groomed to cheat on their taxes over the last ten years, thus not only allowing the Russian government to more easily blackmail victims here but also to financially weaken our nation. When poorer nations call on us to pay them for climate catastrophe mitigation, we're tapped out financially, which the Russian government and it's allies

_____

the Russian government to disable electric vehicles. The Russian government deliberately destroyed the batteries in my first mobility device. Some years after that they had my mother ask her then neighbor to sell me her neighbors used mobility device, a clunky, heavy device I bought for $500. Had the Russian government's operatives not destroyed my lithium batters on my first device, I'd have been uninterested in buying the device I'm presently using. As I've said, the Russian government creates their opportunities and then use their weapons to destroy the target. My mother didn't tell me she asked her neighbor for years to sell me the spare mobility device, my entrapped sister told me. This is an example of triple teaming: the seller, my mother and my sister, as well as destroying what I had that worked, to get me to buy a needed, but subpar item and to make me feel beholden to one of the Russian government's slaves, my beloved mother. Additionally, *most people won't report entrapment attempts and setups against them by their parents and/or loved ones out of misguided loyalty. The Russian government exploits family love and loyalty to destroy us. Love and loyalty are highly effective weapons in the hands of an enemy who loathes us, which is one of the reasons the Russian government entraps family members and unleash them against their extended family. I've experienced the Russian Mafia using my brother in law and sister to see if they could interest me in income tax fraud. I suspect they may have been used to entrap our nephew.*

345 After entrapping my loved ones I believe Russia-controlled American operatives sicced them onto our wider family, to more easily groom and/or lure the extended family into entrapment and criminality. I believe I've witnessed such a scheme when my loved ones felt me out to see how I'd respond to income tax fraud.

use to spread the lie, the propaganda, that we don't care about impoverished nations. We and our government do care. The money isn't forthcoming because the Russian government has manipulated our nation in such a way that we have relatively little disposable income.

I was initially denied for disability and went to a shelter in Santa Monica, the city where I live, to see if they had room for me. They didn't. I was trying to figure who I knew had an old trailer I could live in but no one came to mind. I appealed because a senior citizen cousin of my mother's had apparently worked in a related field decades ago and told me to appeal. Had she or someone else not told me, I wouldn't have appealed as I knew almost nothing about Social Security Disability. The Russian government needed me to, at minimum, appeal because some of their subsequent entrapment setups using my family as a weapon against me involved me committing Disability fraud. Had I not appealed, they'd have been no potential to charge me for Disability fraud.

Then Russia-controlled American operatives had family member after family member try to entrap me, try to get me to lie and cheat. It was a nightmare. I asked my father if I could live with him. He said no, because he had health problems and his wife took care of him and he didn't want to overburden her.

My father told me to exaggerate my health problems to Social Security Disability. I believe he was forced to do that. He didn't offer me financial or any other assistance. A few years later, I needed a couple hundred dollars to pay my rent one month and asked him for a loan, and I asked my neighbor, neither whom I knew were entrapped when I asked them. They both agreed to financially assist me and I immediately repaid them the next month and never again asked either for financial assistance. The Russian government, to dissuade me from ever again asking my neighbors for a loan, had them noise harass me including, following me from room to room, dropping items on their floor wherever I happened to be in my apartment, had them drag their furniture at any time day or night while I was trying to sleep, had them come stand over me in their apartment when I typed complaining comments about the Russian government in the *New York Times* articles, had them take a shower when I was, which caused the water to scald me, and many other stealth attacks. It's very hard when your loved ones and people you thought were your friends are entrapped. Very, very hard. I've been a good daughter to my family and a good neighbor and citizen to my country but that's the opposite of what the Russian government wanted from me so they've destroyed my quality of life. The only upside to that is I've survived long enough to warn you. If you use the warning to effectively defend yourself, then that's a win for decent people worldwide.

The Russian government is focused on destroying Americans and seizing world dominance. They're going to get a very nasty surprise with the climate catastrophe. They've refused to work with us, refused to allow their allies to work with us, so the next decade or two they're working to destroy us would've been the time they ought to have been working with us. By the time they can no longer ignore it, it'll be too late. They'll have their scientist slaves worldwide try to stabilize the climate, but the time to stabilize the climate is passing us by. A decade from now, when they're able to focus on it, it'll be too late.

Chapter 85 - The Russian Mafia controls public transportation strategies, including using bus drivers to ask disabled people which stop is their exit when they enter the bus

Russia-controlled American operatives control what bus drivers say to riders, and how bus drivers treat riders, kindly or rudely. I use a mobility device. To make it easier for Russia-controlled American operatives to harassers targets the Russian government, had bus drivers on Santa Monica and Los Angeles buses about five years ago begin to ask me which stop I intended to get off on. In this way, the operative on the bus listening could tell their co-conspirators where I was headed position operatives to harass me in that store.

Chapter 86 - The Russian Mafia harassed me on my jobs using compromised supervisors and co-workers as proxies

Russia-controlled American operatives had at least six of my supervisors mistreat me over the decades, so ordinary people should know that the Russian government will use their boss to harass them at work. Russia-controlled American operatives would let me get settled and established as a valuable worker, and after finding no way to entrap me there, or seeing that I evaded entrapment, Russia-controlled American operatives would force entrapped supervisors and/or coworkers to stealth harass me until I applied for work at another company. The Russian government attacked me in this way throughout the majority of my work life, from my first full time job post-college, to my final job at the end of December 2009. When the Russian government wants me to not do something, I've figured out they'll have an employee at the business and/or school be mean to me and have noticed I tend to avoid that person and institution.

By studying my responses to their attacks over decades the Russian government learned to use abuse as a weapon against me. They began when I was 11 and have never stopped attacking me. That is how I know them, and that is how they know me. When I tell you the Russian government is incapable of honoring any agreement they sign, and that the only thing they care about is world domination and enslaving everyone in the world around them, that's the honest truth. Hoping that I'm wrong and that they'll treat you better won't work. They'll kill and/or enslave you, your family and your country.

I survived to warn you. Use my warning to save yourself. It's not that the Russian government admires strength. They don't. When they encounter strength, they attack the strong individual.

When I worked in offices, Russia-controlled American operatives would say something to my supervisors and/or coworkers, and the quality of my work life degraded without explanation or warning. I was told nothing, no explanation given me as to what had happened or why. But it was Russia-controlled American operatives who outed me job after job, as they'd outed me from my family home when I was 12. These unrelenting attacks using proxies against a defenseless family and children is who the Russian Mafia is. Proxy killing Palestinian civilians to seize world dominance is who the Russian Mafia is. Although the KGB attacked me the Russian Mafia was even more brutal.

They always pushed me to a new company, one where they hoped, apparently, to entrap me. They cared not at all that I liked where I was, enjoyed my work, or that the stealth attacks traumatized me. They cared not for my trauma, but only about entrapping me, illegally, extra judicially. Law enforcement were in charge of these kinds of attacks, entrapped law enforcement who had no monitors. I've not a single name or department to tell our non-compromised law enforcement. My parents are entrapped and I hope our security agencies question them.

I've lived a nightmare life for approximately 53 years. I know the perpetrator government so I can warn you against believing their lies, but just know, you're dealing with the Mafia. The Mafia is not a government, it is labeled a government in Russia but it thinks like the Mafia. To the

Mafia, they murder competitors and West-allied nations are competitors.

Against our laws the Russian government attacks and hunts ordinary people. You can believe, if you so choose, the Russian government will honor a deal it makes with you but any deal you propose they'll consider as a sign of weakness by you. I've been attacked by entrapped, mostly Caucasian people for decades because I've worked and lived in predominantly Caucasian businesses and neighborhoods for the majority of my life. Moving to Southern California when I was 9, I was young enough to accept any ethnicity and so feel comfortable with any ethnicity. It also helped that my mother wasn't racist when I was a child and raised me to respect all people of all ethnicity's.

To be entrapped and used to harass me, targets were studied, hunted and attacked. The Russian government is working Slavery 2.0 and this time, everyone here will be a slave. For people who believe the Russian government values light skinned Caucasian Americans, most of the victims I've been attacked by are Caucasian Americans. Skin color means nothing to the Russian government: they see themselves as the master race and everyone else is a slave, no matter the skin color. If you're Russian, the Russian government might value you, but like as not, they won't. Russian operatives have sought asylum here for decades. As far as I know, not one I've read about told our government about the stealth warfare aimed at ordinary people here. It's hard to believe that none of them knew about it

Chapter 87 - Drug wars

Several hundred people a day die in America from drug overdose,[346] most of those deaths are attributable to fentanyl and other opioids. Fentanyl precursors come from China. Fentanyl is used legally by our medical system but for the most part that's not being sold on the street. What's being sold on the street are precursors shipped from China to cartels, most in Mexico, and the precursors made into fentanyl, which is killing us.

Over a hundred thousand of our people are dying annually from drug related deaths. The Israeli government has gone to war because about 1,200 of it's people were murdered and a little over 200 innocents were taken hostage.

Because the Russian and Chinese government are killing us with drugs, they've figured out a loophole to prevent us from bombing them in retaliation as they stealth murder our people. They're killing hundreds of us daily but because our people don't know the Russian government has launched a secret war against us, because they refuse to tell ordinary people,[347] murder by fentanyl, like murder by mass murder gun

---

346 Josh Katz, Margot Sanjer-Katz, Eileen Sullivan "Some Key Facts About Fentanyl," *New York Times* (October 23, 2023).

347 General Valery Gerasimov, Chief of the General Staff of the Russian Federation,"The Value of Science in Prediction," *Military-Industrial Kurier* (February 27, 2013). General Valery Gerasimov states: ".... **Wars are no longer declared** …" "...*development of the armed forces over the long term*…." "**The role of mobile, mixed-type groups of forces, acting in a single intelligence-information space** *because of*

violence, isn't framed as the stealth attacks they are by an enemy foreign government but, using our media as a weapon against us, the Russian government labels the mass murders and overdose deaths, which murders they re-frame as evidence that we're in decline when in actuality, we're being attacked by the Russian government, the Chinese government, and their Mafias.[348] Understand we're at war and our enemies are using drugs to kill us because if they bomb us, like the Russian government is bombing Ukraine, we'd bomb them back. The only thing the Russian governments it can't recover from is us bombing the Russian government.

Chapter 88 - The Russian Mafia doesn't want more migrants here at this time; it's easier for them to attack victims spread out per country

The Russian government is working a coup against us and controls millions of us. It's more difficult to maintain control of a people when thousands arrive every week. Additionally, the Russian government is determined we don't pivot into become a manufacturing powerhouse nation again, as such a nation is harder for them them to convince we're in decline. This is all a scam to the Russian government but it's key that we believe the scam. Immigrants who come here are hard workers, want to work, want to prove their value to us. In short, hard working immigrants make us stronger. For the same reason the Russian government sickened me and put me on

_____

the use of the new possibilities of command-and-control systems has been strengthened…."

348 Anton Troianovski and Adam Satariano "Putin Denounces the U.S. as a Fading World Power," *New York Times* (June 17, 2022).

Disability, instead of having me work two or more jobs paying taxes: it is determined to financially weaken our government and more hard workers strengthens us and our country, which the Russian government doesn't want. The Russian government is seizing control of our nation. When it has various states illegally halt and/or slow immigration, that reflects the Russian government's strategy of weakening us and the Russian government needs us weak. I've been a student off and on for much of my life. The Russian government now has it so, it is my experience, teachers don't teach optimally in any of the programming and/or accounting courses I attempted. In the city college I attend basic math isn't taught. Everywhere, from local libraries to city college, to entering the country, the Russian government is working as hard as possible to control us and our country. Because of the KGB's success last century in embedding hundreds of thousands of it's operatives here by exploiting the Jackson-Vanik Amendment, and the Russian Mafia inheriting that success, the Russian Mafia is more than confident they will destroy us and before that, they're determined to stop migrant flows into our country, pound our students into believing they can't learn anything, and generally con us into believing whatever lie the Russian Mafia is currently hawking. We must better defend ourselves by warning our people.

Chapter 89 - The Russian Mafia tried to hook me on drugs

An operative I didn't know was an operative stayed the night with my aunt while they were visiting Southern California. In the middle of the night the operative sexually assaulted me. When I later told my mother, she gas lighted me, said it never happened. My mother has known this particular operative for about 55 years. The Russian government tends to use the people in my family they've entrapped to gaslight me: my father turned my attention away when I told him I was being harassed by security in big box stores; my brother told me I wasn't insulin resistant when I first became disabled; and my mother told me the woman who sexually assaulted me hadn't sexually assaulted me. My theory is, given I know my mom, dad and brother are entrapped, the Russian government orders them how to behave with me and they do so. Psychological warfare, to get the target to doubt her own reality. My sleep cycle is messed up so I was sleeping 12/26/23 around 11am and an operative upstairs made noise in such a way as to noise harass me. When I awoke later that evening around 6-7pm, I didn't hear the operative so maybe they've gone between Christmas and the New Year holiday, hopefully. When I told my father I intended to write this book, he told me no one would believe me and that people would think I'm crazy. The Russian government definitely doesn't want any book I write to see the light of day. I'm positive they stealth attacked the electric grid in Southern California on September 29, 2023 to destroy my laptop. Fortunately it wasn't plugged in at the time, just the chargers were plugged in. Had I been charging them when the Russian government surged the electricity back on, the surge so strong it tripped the circuit I normally use to charge the laptops, the laptop would've been destroyed and I wouldn't have been able to write my book on this laptop.

I also believe the Russian government gave Hamas the okay to attack Israel to distract our security agencies from the book, which was a nasty surprise to the Russian government as it discusses quite a few subjects they don't want well known. I must get this edition published. There's quite a bit people need to know.

When my maternal aunt and her friend left, they left a drug. They could've thrown it away on the way to their vehicle but they made a point of letting me know they had it.

It was yet another set up by the Russian government to expose me to drugs. They've tried for decades to get me to take drugs, despite consistent contra indications from me that I'm interested in drugs, and despite my approaching senior citizen age.

During their visit my aunt and the operative smoked multiple times, going away from me, outside, to do so. They seemed to enjoy the experience. The Russian government had their operative leave a drug with me in the hope I'd be curious and partake. If you know anything at all about the Russian government know this: they are evil and hate everyone not them. They've targeted and destroyed people of color in

America, including Native Americans and African Americans, like they've worked to destroy Palestinians. Whichever minority the dominant culture wants destroyed, the Russian government is all over that, anything, everything, they can do to make their bones with the dominant culture to destroy that culture. Currently, the Russian government is running similar plays against Americans, using people of color, against Israelis, using Palestinians, and against the French, using people of color in France. They've used a variation of racism against so many countries it makes you wonder why no one has caught onto the scam.

To put it another way, the Russian government will destroy whatever ethnicity the dominant culture wants destroyed because they use that as a way to ingratiate themselves into the dominant culture to destroy the dominant culture. The Russian government are acquisitive. They're collectors. They're slavers. If you want to be enslaved by the Russian government, be racist, sexist and classist, open your country to them, and have something they covet: then you too will be their slave.

While minding my own business and put-putting on my mobility device, the Russian government had an African American operative suddenly, out of nowhere, position himself in front of me, blocking my path with his wheel chair, and forced me to engage in conversation with him. He asked me to go out drinking with him. When I politely declined he acted like there was something wrong with me.

Think of yourself as a woman walking along, and suddenly out of nowhere a man you don't know jogs up to you, positions himself in front of you blocking your path, and asks you to go drinking with them. These attacks are harassing attacks, forced engagement attacks.

Russia-controlled American law enforcement operatives harass me on the street using people I don't know and don't want to know. When one of their operatives harassed me to such an extent I had to go to the ER the summer of 2023, Russia-controlled American operatives wouldn't allow the paramedics transferring me to take me to the ER hospital I requested, they had me taken to an ER further away. While there, Russia-controlled American operatives had maybe a dozen entrapped operatives stealth harass me, while I experienced a medical emergency, anything to stealth kill me and hurry my death. No exaggeration.

The African American female sexual assault operative with my maternal aunt acted surprised when I didn't take her up on her drugs. She'd sexually assaulted me in the middle of the night and then expected me to her drugs. I've never taken drugs, which my aunt quickly told her. When they left it, I threw it out.

The Russian government had a drug dealer approach me and offer me any drug I wanted for free within the last several years. I politely declined his offer. I've had operatives offer me a free trip to Vegas, been invited to a bar, etc. Men I'd never seen before and don't want to know invited me. A handsome operative at a store looked at me with interest but you must be warned, anyone entrapped by the Russian government is a slave. If you sleep with a slave the Russian government will enslave you.

In my experience the Russian government focuses accessing me through people I used to trust, my family. They've taken my loved ones, camouflage themselves with my loved ones flesh and attempt to control me. After 53 years I now understand that when my family suggests something they know doesn't interest me and that I know involves risky behavior from me, it's the Russian government working me to self-harm. But children don't know their parents are slaves and I'm writing this book to warn victims who've no idea what's happening in our world now. We're under attack. The Russian government would just as soon bomb us but if they did, they'd be bombed back and they know they can't recover from that. So instead of bombing us, they entrap our people and force our loved ones to stealth attack us. The sexual predator operative who visited with my aunt couldn't have gained entrance into my apartment on her own. She used her friendship with my aunt, and probably drugs, as they both smoked drugs during their visit. Drug addiction is an easy-in to a target, which is why the Russian government has worked for decades to hook me despite my disinterest. They keep trying because it's easy for them and, apparently, no one files reports because entrapped law enforcement is running these attacks extra judicially. Reports would be a paper trail Russia-controlled

American operatives would have no easy answer for. So they keep doing the same kinds of attacks. This book is necessary to warn people.

For years Russia-controlled American operatives had lesbians approach me to see if I was interested. The Russian government uses sex, drugs, anything and everything as a weapon. They have all people living in America in their sights. I suspect people in West allied nations are getting swarmed in a similar manner.

The Russian government is a mass murderer, with, easily, the longest string of murders over decades of any other known mass murderer. However, because the Russian government has entrapped and controls individuals in our media, no news article ever frame them for the mass murderers they are.

Currently, the Russian government is facilitating the Israel-Hamas War, knowingly murdering thousands of babies. No one is outing them. No one is calling them baby murderers but me. Protesters blame the Israeli government, which is a pro-Kremlin government, and President Biden, the leader the Russian government intends to unseat in 2024. No one is blaming the actual perpetrator, the Russian Mafia. That's why you see Mr. Putin with this smirk: he and his government murder with impunity, without even being outed. When villains commit mass murder with impunity, things don't end well for our species. I'm warning you so you'll at least have an idea of what's going on.

Chapter 90 - The Russian Mafia used my landlord, neighbors &/or workmen to stealth harass me for decades

Since maybe 12 years ago the operative upstairs, who I hadn't even known was an operative until she and her daughter began noise attacking me, began following me room to room and deliberately interrupting my sleep, has been used by the Russian government to stealth harass me. Apparently she's not living upstairs, meaning she's died or moved elsewhere. If she's died the Russian Mafia made it appear to be from natural causes or something easily explained. The Russian Mafia didn't want her questioned by law enforcement should my legitimate government investigate my charges. Since she's apparently no longer available, the maybe thousands of stealth attacks she and her adult daughter committed against me, can be confirmed, although her daughter, I believe, is still alive.

Chapter 91 - The Russian Mafia tortures me with sleep deprivation as recently as February 2024

I dropped my Spring 2023 class, because I was convinced Russia Mafia American operatives intended to attack and/or kill me and that I needed to complete this book before they did. When I leave my apartment to pay rent on January 5th, 2024, I'll do my best to get something readable and helpful to my people but I don't like not being able to perfect and get a good edit. I don't have the money at this time, and every minute counts with the Russian Mafia stealth attacking us.

My book is getting better but I've only been writing for four or five months and I'm often stealth noise harassed by the Russian Mafia. The Russian government doesn't want any version of my book in the world, and can remove the threat by killing or disabling me before this edition is completed, so I must get it out before I leave my apartment to pay my rent. I've definitely noticed that the Russian government has told the owners of this building to not accept early rents, the one or two times I tried to pay in advance meaning the Russian Mafia can depend on my leaving my apartment to pay my rent monthly.

Chapter 92 - The Russian Mafia controls the quality of medical care I receive in Southern California

1. The Russian government prevented an ER doctor from adequately numbing my finger before stitching it up when I cut it maybe 10 years ago. In my 50s, after the Russia-controlled American operatives had exhausted me to illness and prevented me from getting quality medical for years in my 40s, I was disabled and cutting chocolate in the kitchen when the knife slipped and I cut my finger. The bleeding wouldn't stop so I went to the ER. After I explained the problem to the ER doctor he was called away for a few minutes. When he returned he got a needle and gave me a shot in my finger to numb the pain of the stitching, only I could feel every stitch and asked for another shot. He gave me another shot and continued sewing the wound but there was no pain numbing medication in the syringe, my theory is it was just saline. When he was called away I suspect, since doctors being called away

after I'd explained the problem" was something Russia-controlled American operatives had been doing to me for years. Because Russia-controlled American operatives always had eyes on me, they knew where I was headed, they knew when I cut myself because they illegally monitor me inside my apartment. I know because the operative upstairs has noise harassed me and followed me room to room, in real time, for years. When I cut my finger I didn't know the Russian government was my enemy. I knew I had an enemy and that the police organized and oversaw physical and psychological attacks against me but I had no idea who would be ordering them to do that. I didn't learn the Russian government was the hostile until January, February 2021, so I was 61. In my 50s I'd been harassed by a fathom hostile since I my early 40s, 40, or 41. I think by 42 I already knew. I was working at a law firm downtown when I was 42, I'd left the Westside law firm I'd worked at for some years but it was at the end of my employment at that firm when Russia-controlled American operatives went loud on me, harassing me in the street, bus, most places.

2. A sleep doctor I went to at a major hospital in West Los Angeles tried to get me to see a doctor who dispensed psychotropic medications after I told her I was being harassed by police. She couldn't have cared less. Her doctors report was inaccurate and when I asked her to correct it, she refused. I asked the medical center and they refused. I actually filed a report with a California department where patients can file

complaints but nothing ever came of it. The medical center in West Los Angeles said that because a correction wouldn't change her diagnosis, the doctor needn't correct her notes. Why even bother giving patients a copy of the doctor's notes if the doctor won't correct them when asked.

3. At the above-referenced medical center I was tested for sleep apnea. The test was rigged, or more specifically, the room was rigged by Russia-controlled American operates. I was hooked up with wires. It wasn't late, I don't remember the time. I've experienced severe insomnia for years because I've been harassed for years. It's highly unlikely for me to feel sleepy enough to sleep but suddenly, at the sleep study, I began to feel extremely sleepy. I hadn't felt normal sleepiness for years and suddenly I knew gas was being piped into the room to make me sleep. I was terrified. The terror pushed away the sleepiness and though I stayed the entire night, the sleep study was a fail because I was unable to sleep. I had to return and this time, no gas was piped into a different room and my sleep apnea was diagnosed. Russia-controlled American operatives conspired with a major medical center in Los Angeles County to misdiagnose me to prevent me from being diagnosed. I'm not naming the hospital because I don't want to be sued, and there's no way I can prove the allegation except with my test results but they wouldn't show the gas being piped into the room or which gas it was.

4. I had a doctor sneer at me when I asked him to examine me. Not even exaggerating. The man shouldn't have

been a doctor because he had no problem actually sneering at me. I don't know if it was because I'm African American, obese or female. My regular doctor was sick I think and his replacement was the sneering doctor. He didn't even try to be professional. Before medical appointments I make sure I wash, so it wasn't because of body smell. Some compromised people really are racist and this man had all kinds of issues. Still, he didn't try to entrap me which my current general practitioner does.

5. Before the Russian government made me disabled I went to multiple doctors trying to find out why I felt so sick. I'd never felt unwell before, I knew there was a problem. But doctors I saw said they found nothing. I collapsed months later. After I collapsed other doctors began to diagnose me.

Chapter 93 - The Russian Mafia controls who I can reach on my phone

Russia-controlled American operatives control who I can reach on my phone. People they don't want me talking to, like some of my doctors, the switchboard operator at the city college I attend, a tech business I was trying to hire for my company, and a covid rent aid program they were trying to entrap me with, those calls the Russian government refused to allow me to reach, even as recently as Fall 2023 when I tried to get through to the city college switchboard operator.

Chapter 94 - The Russian Mafia forces the general practitioner I see to try to entrap me using Medicare fraud

The current doctor I see has encouraged me to lie to Medicare on two occasions, to receive benefits he said I don't qualify for. I qualify for glucose test strips, a battery for my mobility device but Russia-controlled American operatives refuse to allow me access to those benefits, even though I pay extra for Medicare and for those benefits, even when I have prescriptions to obtain those benefits, even when I have Medicare on the line with me trying to get a medical equipment store to install a battery I was trying to obtain. It's surprising how much power compromised law enforcement have over targets. Deeply embedded here the Russian government knows just how they can have their law enforcement operatives push, which is why they target some of our law enforcement for entrapment. My father is retired law enforcement and he's entrapped.

If our legitimate government understood just how compromised the Russian government has made our nation, our security agencies, who have seen everything, twice, might be surprised. To me, all this is shocking so I've no idea what information they'd find most problematic. It's not that they don't care it's that they, I think, respond to fires. I know the Russian government is swarming and stressing our legitimate government to break them, to break us. I want to protect and defend us. The Russian government is so deeply embedded here the only way I know of to get them out is to bomb the

Russian government, but that's something our legitimate government believes would cause more harm than good. My argument is the Russian government is enslaving us. Slaves take their freedom when they bomb their oppressor.

Chapter 95 - The Russian Mafia had an operative attack me w/come kind of handheld device in the grocery store after I complained about a doctor's subpar care. Now I tend not to complain about subpar care or that my doctor tries to entrap me. Stealth attacks are how the Russian Mafia controls here. The only reason I'm allowed to complain is that I've evaded entrapment

I wrote a letter complaining about a doctor, I wrote him a letter and left it at his office. To retaliate a Russia-controlled American operative attacked me with some kind of handheld weapon he had in his pocket. After the doctor's office I went to Ralph's to buy groceries. I was the only person in line except for the man behind me. He got very close to me and all of a sudden I felt very unwell, as if I was about to pass out. He was starring dead in my face, he knew he'd made me sick. I was on my mobility device. It didn't matter to him as he stealth attacked me.

I cannot describe the depth of hatred the Russian government has for Americans and for migrants living in America. Their hatred is real. Even though they use American operatives in attacks, their operatives are hate filled. Some smirk, some delight in frightening me. No lie. I had a skateboarder operative, young adult male. Used as I am to skateboarders harassing me. I paused, pulled over to the edge of the sidewalk, as close as I could get without my mobility device falling off the street. And he stopped skateboarding and starred at me. Seeing he wasn't needing to pass by me, I began moving and so did he, swerving nearly into me. That harassment was within the last few years, I was over 60.

Chapter 96 - The Russian Mafia had two surgeons pressure me to get an unneeded hysterectomy

I've had two benign tumors in my uterus and though my menses should've stopped by now, it hasn't. I was going every three and I think every six months to get tested to ensure cancer wasn't growing. The surgeon I'd been seeing left and the surgeon who replaced her leaned on me hard to get a hysterectomy. If you've had a hysterectomy you know they change your body and not in good ways and the only reason to get a hysterectomy is if you're in pain or have cancer.

I was not in pain nor was I diagnosed with cancer but hysterectomies are a money maker for hospital and because the Russian government pushes, as best I can tell, all business they can influence to embrace greed (as part of the Russian government's strategy to corrupt and destroy us), I've experienced medical centers pushing me to have procedures not in my best interests, like an unneeded hysterectomy. I asked the surgeon to allow me to continue the monitoring and he refused. He said I needed a hysterectomy. When I said no, he refused to allow me to continue getting intravaginal ultrasounds, even though my insurance paid for them and even though I'd had two benign tumors removed.

Chapter 97 - I've found most doctors don't want to help their patients & when I call 9/11 & paramedics arrive, this summer I had to them twice. I have insurance. The second call, the Fire Dept (which I didn't call) tried to talk me out of going to ER, told me I'd have to wait in the Waiting Room, I told them that as fine. Then I suspect they intervened in my medical care. I'd been working on a program for my Final & the building manager came to my front door & noise harassed me, startling me. Hours later my heart sped up inexplicably to 200 bpm. Before his stealth attack I'd been studying and writing my Final. Then to get me to file a lawsuit & forget writing this book, the Russian Mafia had operatives at the grocery store fronting that they were attorneys. The Russian Mafia is deeply embedded here

I had one endocrinologist I told I was being harassed and she just didn't want to know. Her staff stopped answering my calls, and Russia-controlled American operates kept putting me through to voicemail.

Chapter 98 - Access to upper division math & programming classes are out of reach when lower level math is not on offer as the Russian Mafia stealth closes off avenues to middle class earnings for impoverished people

Too much of our country in inaccessible to poor people and immigrants. When upper level courses are not available to the poor, because lower level basic classes aren't on offer, poor people can't access the middle class.

A free, open-source operating system, intended for poor people, can't be made using a library computer because library computers and city college computers are set up in such a way to not make them accessible to be useful to people who need it. Poor people attending the city college I attend can't start at remedial math and work their way up to calculus, they can't take and understand difficult programming classes because entry level math courses are no longer on offer at the city college where I live.

Meaning poor people must own a computer to access a free operative system, we have a problem. Libraries must have at least one computer from which a patron can download a free operative system or other program. Chromebooks aren't helpful in downloading an operating system. If you need to download a free ISO (an operating system), something difficult to do on a chromebook, you can't access unless you already own a computer. From what I've seen affordable computers, chromebooks, didn't allow me to make a bootable flash drive.

The Russian Mafia blocks access by impoverished Americans to middle class income. At the city college I attend lower level math courses aren't offered, so to access the upper division classes you'd have needed to taken lower level math in high school or you need to attend a different city college to take lower level math and then you can return to the city college I attend to take high level math. The problem is, it's just so easy for Russia-controlled American operatives to complicate your life. When I was attending a city college in

Los Angeles I was harassed a lot by operatives. Strangers, mostly men, would jaywalk across a major boulevard to walk into me as I stood at the bus stop. Russian-government American operatives still order their people to harass me in this way. I'm doubly harassed when I leave the apartment.

Chapter 99 - News papers everyone in the country needs access to have price barriers, making them inaccessible to poor people

The New York Times is one of the most important papers for people to read here but it costs. I didn't know for years I could get a free temporary subscription. Although I've seen the Russian government use the New York Times to spread lies, at this time (December 2023) most of the articles I see aren't propaganda, so everyone in this country should read that paper. It should be free, it is that important a paper.

Chapter 100 - The Russian & Chinese Mafias have found a way to kill tens of thousands of us annually using drugs

The Russian government has figured a way of killing several hundred of us a day without triggering us bombing them. They use fentanyl. The Russian and Chinese governments, and their allies, their mafias and multiple cartels use their nuclear weapons as shields to prevent us bombing them while they stealth murder hundreds of us daily.

This is not sustainable. Almost no one in America wants a hot war with the Russian and Chinese government but we must defend ourselves. Although the Russian and Chinese governments kill hundreds of our people daily, we don't kill any ordinary Russian or Chinese people.

Chapter 101 - The Lacy-Zarubin Agreement & the Jackson-Vanik Amendment: how the USSR gained access to us in the 20th century

To take control of us, the Russian government has inserted itself into our families, institutions, businesses and systems but before that, it had to have access. The Lacy-Zarubin Agreement and the Jackson-Vanik Amendment gave them access.

To gain access into our country in the 20th Century the Russian government (then known as the Soviet Union) got us to pass the 1958 Lacy-Zarubin Agreement which gave them access to our technology, education, science, medicine, film, dance, music, tourism, and scholarly exchange, etc. In 1975 the Soviet Union got us to pass into law the Jackson-Vanik Amendment, where, as best I can tell, we supposedly got the Soviet Union to allow immigration of marginalized groups like Soviet Jews, evangelical Christians and Catholics to migrate to America, about 600,000 since 1975. About a million migrated from the Soviet Union to Israel, etc. To reward the Soviet Union for allowing these oppressed groups to migrate, we formalized a trade partnership with the Soviet Union.

The Russian government cheated us in both these agreements. From access granted them by the Lacy-Zarubin Agreement, they stole access to our tech so that in 2023, the Russian government and it's wide network of mafia and hackers extract information from the majority of computer systems they target in the world. It's likely that only computers with the highest security aren't compromised and even those, the Russian government probably has in it's sights. With it's theft and dominance of technology that our government, meaning we, paid for, the Russian government is hunting us and our children in our own country. Everything is tied to computers. The Russian government applied old KGB attacks, I theorize, long before Lacy-Zarubin was signed it was put forward to some of our politicians to get them to sign the agreement after Soviet spies saw they couldn't access the all the intel they wanted.

With Jackson-Vanik the Soviet Union placed possibly hundreds of thousands of their spies here. If you're American or have lived in America the last forty years, you'll have noticed our country has noticeably declined in standard of living for Americans. The Soviet Union granted permission to individuals to migrate and I've read that the Soviet Union placed conditions on migration – in some cases the migrant had to agree to help the Soviet Union once it resettled here. There is no telling how many of those 600,000 migrants were spies, but some of them were, and the decline of our nation tracks with this seeding of the Soviet Union of it's spies in America. The Soviet Union seeded it's estimated million migrants and Israel's population is under 10 million. I theorize some of those immigrants were spies and as you see, Israel's government in 2023 is pro-Kremlin, and there's a very strong ultranationalist vein in Israeli politics. Ultranationalism is a red flag that the Russian government is at play in a targeted nation.

We've been played by the Russian government for over 60 years on so many levels, it's a wonder we've survived. We're barely hanging on. Our government is old school, trying to protect us from the bad news but we need to hear it, if only to stop so many of us from trying the poison fentanyl, which is a stealth weapon the Russian and Chinese governments, with the help of their mafia, kill us where they don't have to worry we'll bomb them. Israel launched a war against Hamas when it murdered about 1,400 innocent Israeli's and others, and kidnapped over 200 hundred people. Hamas has no nuclear weapons, which is the advantage Israeli has in being able to retaliate against an enemy who attacked it. China and Russia's mafias murder our people with fentanyl and because they have nuclear weapons, and because we prefer to have a nuclear war because we're sane, we don't have the luxury of retaliating against them as we'd like to. That luxury would cost a great many innocent lives.

So we take these evil gut punches: the stealth murder of our innocent. We must warn our people. Fentanyl is meant to reduce the number of military age American soldiers and one of the reasons the Russian Mafia wants to close the border is to stop the endless supply of children who'd be willing to serve in our military. The Russian Mafia is about degrading our defense.

Chapter 102 - The USSR & the KKK

When the Russian government stealth enters a country it directs some of it's operatives to seek out like-minded people and institutions.[349] Its operatives then embed themselves by ingratiating themselves with the intention of absorbing the people and destroying their government, to be replaced by the Russian government. You see their strategy against Americans, Israeli's, the French, and Italian peoples, to name a few. In their allies' countries, the Russian government choose the like-minded candidate they prefer, and remove the rivals of their preferred candidate. Mr. Putin, Mr. Jinping nor Mr. Modi face a serious challenge from an opposition candidate because the Russian government and the Russian Mafia, aided by it's allied mafias in many countries, remove opposition candidates. Our species is facing the world's most powerful Mafia, concerned only with growing it's power and wealth. The Russian government see's America as a competitor to destroy, not as a government and people to respect, which is why they're waging a secret war against us.[350] Their secret war is

---

349 General Valery Gerasimov, Chief of the General Staff of the Russian Federation,"The Value of Science in Prediction," *Military-Industrial Kurier* (February 27, 2013). General Valery Gerasimov states: *".... Wars are no longer declared ..." "...development of the armed forces over the long term...." "The role of mobile, mixed-type groups of forces, acting in a single intelligence-information space because of the use of the new possibilities of command-and-control systems has been strengthened...."*

350 General Valery Gerasimov, Chief of the General Staff of the Russian Federation,"The Value of Science in Prediction," *Military-Industrial Kurier* (February 27, 2013). General Valery Gerasimov states: *".... **Wars are no longer declared** ..." "All this is **supplemented by military means of a <u>concealed character</u>, including carrying out actions of informational conflict and the <u>actions of special-operations forces</u>...**"*

against the law here but the Russian Mafia cares nothing at all about our laws. If we keep playing by rules that work during peace but are disastrous during war, the Russian government will destroy us, and smile while doing it.[351] I've written this book because I don't want innocent people in the world, or their children, to be destroyed by a mad-dog government determined to rule the world with nothing on offer but blind ambition, cruelty, slavery, and death.

Chapter 103 - Russian Mafia attack strategies: New Generation Russian Warfare

The Russian government calls it "New Generation Warfare." It's based on Sun Tzu's idea that "supreme excellence consists of breaking the enemy's resistance without fighting."[352] This strategy isn't new. The USSR used a version of it against Americans as early as the 1970s, probably earlier. I use 1970-71 as the time frame because I was 11 then, when the USSR first attacked me.

---

351 "Mr. Jinping has led a well-publicized crusade against corruption, but it has been mainly a purge of rivals," according to U.S. National Security officials and Chinese dissidents. In fact, they said, "Chinese intelligence services have quietly expanded their ties with Chinese mafias, known as triads, for mutual benefit." …. "There is no question there is interconnectivity between Chinese organized crime and the Chinese state," said Frank Montoya, Jr., a former top counterintelligence official at the Office of the Director of National Intelligence. "The party operates in organized crime-type fashion. There are parallels to Russia, where organized crime has been co-opted by the Russian government and Putin's security services." (Sebastian Rotella and Kirstein Berg "How a Chinese American Gangster Transformed Money Laundering for Drug Cartels," *ProPublica* (10/11/2022)).

352 It is impossible to break the enemy's resistance without fighting. The Russian government flat out lies to enemy governments and to enemy civilians.

In 2024 the Russian Mafia will attempt to ambush nearly everyone in America, because it is working a coup here and determined to embed puppet after puppet moving forward. Everyone is in play. Even our beloved President is being targeted in order to deliberately sicken, exhaust, and overwhelm him: make him hate his job in order to force him to stand down. I call it assassination-by-overwork. The Russian Mafia tried a version of it against me in 2009 after they saw they couldn't corrupt me. They then decided to force my loved ones to have a go at entrapping me. I felt devastated. But if I can help our President and my nation survive, that pain will have been worth it.

A parent, authority figure, dominant and/or service provider is ambushed and entrapped by American law enforcement, who themselves are entrapped slaves. There's no way I can know how many of our law enforcement are entrapped but not every police officer is, so, if you believe me that entrapped police are ambushing and entrapping ordinary people, please don't assault a police officer.

I believe millions of people in America are entrapped. The 74+ million who voted for former President Trump, and the unknown numbers of the GOP Base who say they believe in him. I don't know who flips entrapped people into the GOP Base. Because I've been quadruple teamed by entrapped family members,[353] and triple-teamed by medical care

---

[353] The Russian Mafia swapped out the flip phone one of my compromised brothers had lent me and swapped in a smart phone using my sister, brother-in-law, mother and that same brother.

providers, the Russian Mafia doesn't have to rely on compromised law enforcement to flip voters into the Base. It's possible handlers (compromised law enforcement who entrapped victims) "encourage" them but the Russian Mafia has so many influencers here entrapping officers aren't required to get people into the GOP Base. It'd be good if our legitimate government investigated what I'm writing in this book so they can see for themselves how victims are entrapped and flipped into the GOP Base. It is being done, I don't know the mechanism because entrapped victims aren't allowed to discuss their entrapment with me. As best I can tell, an entrapped person believes they will be killed/imprisoned if they discuss their deal. As best I can tell, entrapped people are sicced on their extended family and forced to entrap their loved ones.

At least five of my immediate family members are entrapped, two of my entrapped family are part of the GOB Base. I know because they told me they're part of the Base and my mother and sister have tried to entrap and/or lure me multiple times. They are my beloved family but the Russian government controls them, and the rest of my entrapped family. They aren't free to discuss this with me so I've had to put two and two together based on their behavior, the attack behavior of acquaintances and people I thought were my friends, and the behavior of several thousand strangers, some of them children, who've harassed and/or attacked me over decades. I estimate I've been stealth harassed and/or attacked by about three thousand people, nearly all of them unknown to me. About a thousand of those people were children, mostly with their parents, but quite a few were older teens and/or adolescents who harassed me at the direction of, I believe, compromised law enforcement. The Russian government doesn't declare itself: you know you're being attacked when someone is attacking you, no heads up, no warning shot. The Russian government has had operatives it controls deliberately destroy my sleep cycle. They have eyes on me in my apartment, illegally. They monitor my phone, illegally, all of which I know because the operative in the apartment above mine has followed me from room to room, in real time, when I type a comment in the New York Times outing the Russian government that operative often comes and stands over me in their apartment, and drops something or noise harasses me in other ways. In addition to the upstairs operatives, the Russian government had Hispanic workmen harass me for maybe the last five years, now not so much, but over the previous four years or so they deliberately interrupted my sleep. I asked them to stop, explained I have sleep disorders. They ignored me, under orders from the Russian government camouflaged in the skin of American police the Russian government owns.

Mr. Trump says the system is rigged and he's right, but it's rigged by the Russian Mafia, not by our legitimate government. The Russian Mafia has commandeered our entrapment system, lying to both our government and our civilians, similar to the lie the Russian government is perpetuating to American educators that students are cheating. Because our educators believe we're cheating they've, in my experience, stopped teaching. It's another scam the Russian government is running here to stagnate us, prevent us from learning, and discourage us. In any system in America where the Russian Mafia can get between two groups, it's default is to push us away from each other, to destroy us.

I've taken more than one city college class lately that was not conducive to a student learning but rather designed to discourage students. Especially in programming classes. We need programmers and it serves the Russian Mafia's interests to discourage Americans and migrants in America interested in the field.

According to Janis Berzins in his paper "The New Generation of Russian Warfare" "...the Russian view of modern warfare is based on the idea that the main battle space is in the mind."[354] The Russian government is a world class, best in class liar and has learned from conning people around the world that once they've gained their target's confidence, it's over.[355] The Russian Mafia lies to people in ways big and

---

354 Janis Berzins, "The New Generation of Russian Warfare," Aspen Institute | Prague, *Aspen Review* (March 2014). New Generation Warfare: https://en.m.wikipedia.org/wiki/new-generation-warfare.

355 Have you noticed how Mr. Jinping behaves when he's with Mr. Putin? Mr Jinping is leader of over a billion people but when he's with Mr. Putin he behaves as if he highly trusts Mr. Putin. To get that trust, Mr. Putin communicated to Mr. Jinping in a way Mr. Jinping found helpful and believable. Mr. Jinping values Mr. Putin's friendship so much that he doesn't care that the world see's his esteem for Mr. Putin. I say this not to denigrate Mr. Jinping but to point out that Mr. Putin made it his business to cultivate Mr. Jinping's friendship and trust. Mr. Putin is old KGB, he's a master of psychological warfare and control. Mr. Jinping is smart, capable and savvy but he's not a trained in psychological

small. The truest thing they'll ever say is they intend to control the world, but they couch their intentions in complaints about America's financial power, using words like "hegemony," or they couch their attacks in phrases like "new generation warfare," knowing most people don't understand what they mean.

New generation warfare," according to Wikipedia is a Russian theory of unconventional warfare which prioritizes the psychological and people-centered aspects over traditional military concerns, and emphasizes a phased-approach of non-military influence. In this type of warfare, there is always an enemy, and the Russian government has chosen West aligned innocent people. Wikipedia goes on to say this type of warfare is "dominated by information and psychological warfare….[..] morally and psychologically depressing the enemy's….civil population." That's us, people, the civil population. In other words, the Russian government's war strategy prioritizes lying to us, and convincing our legitimate government, and us, that ordinary Americans are a pack of liars. Had the Russian government succeeded in getting me to lie to Social Security Disability by exaggerating my symptoms, as the Russian government forced both my entrapped parents to tell me to do, the compromised American law enforcement the Russian government controls would not have told my legitimate government that: (a) the Russian government deliberately

---

warfare. Mr. Putin is, and it shows. When you see Mr. Putin engage with Mr. Jinping or with Mr. Trump, you can tell who has mastered psychological warfare. No disrespect meant to Mr. Jinping or Mr. Trump but Mr. Putin has mastered psychology.

worked me to exhaustion by giving me twice as many attorneys to work for than is usual for legal secretaries; (b) had their operatives harass me nearly everywhere I went on foot and/or when I took the bus, and even prevented me from voting in the 2008 presidential election; (c) had their operatives routinely mist me with a mace-like spray when I took the bus, causing me to endure re-occurring sinus infections that forced me to use all my vacation time as sick time, and prevented me from attending my beloved grandfather's funeral; (d) had my extended family ask me for financial help where I spent several thousands of dollars on them, which put me under a great deal of financial strain; (e) had my landlord refuse to make needed basic repairs, putting me under even more financial strain when I paid for the needed repairs out of pocket; (f) had the doctors I went to trying to learn why I felt so ill, and who were owned by the Russian government, refuse to diagnose me or to help me in any way; (g) had one of the partners I worked for yell at me for nothing I did, but merely to cause me emotional distress to precipitate a collapse.

For most of my work life, beginning with my first full time job post-university, to my final job at the law firm where the partner yelled at me, the Russian Mafia's American operatives had my supervisors and coworkers mistreat me after I'd been at the firm awhile and they saw they couldn't entrap me there. Russia-controlled American operatives had people be rude to me for no reason in the hope that operatives would succeed in entrapping me at the next job. Before I was mistreated, I'd liked the businesses I worked for, the jobs and my coworkers. The Russian Mafia couldn't have cared less. They wanted me entrapped and destroyed my work life to make that happen. At my last job they'd overworked, exhausted, and maced me to such an extent that I collapsed, although pre-collapse I'd updated my resume in preparation of finding another part time job, because I was concerned my credit card was far too high. At that last law firm I worked for up to six attorneys, and yet I prepared my resume to find another part time job. Because of the abuse the Russian Mafia heaped on me, I recognized their attacks on President Biden and did what I could to warn our government so they could protect him; (h) had operatives flood my mailbox with credit card invitations at the same time the law firm I worked at reduced my salary (it was 2008 and they said they reduced everyone's salary due to the economic downturn, although there is no way I could verify that); and (i) after the Russian Mafia's American operatives attacked me to the point of collapse, they had both my parents, unprompted, tell me to lie to Disability and exaggerate my symptoms. I refused.

Had the Russian Mafia succeeded in entrapping me I don't believe they would've informed my legitimate government about the relevant information referenced in (a) – (i) because my legitimate government would've recognized them as illegal entrapment attempts. Although my case is usual because I'm resistant to entrapment, the fact that any of these attacks were used suggests that the Russian Mafia has commandeered our entrapment system and exploit it to (a) install their puppets here, and (b) enslave our people -- both of which are illegal here during times of war or peace, and both of which are illegal under international law, which the Russian Mafia well knows and which is why they work these scams under the radar, well out of sight of our government. Whatever plea deal the Russian Mafia coerce people to agree to is illegal, because according to the Geneva conventions, people can't give up their rights. The Geneva conventions state that people cannot be forced to give up their rights so whatever plea deals the Russian Mafia, using compromised law enforcement as proxies, offers victims is illegal as it forces noncriminals to assist law enforcement on order for the rest of their lives. My mom was destabilized and collapsed when she was around 29. In July or August 2023 I believe was the month, her handlers were forcing her to fish to see if I had any interest in palm readers. On the assumption that she behaved the way she did with me was because her handlers forced her, and using her collapse as the year they attacked her, I believe my mom has been under the USSR and/or the Russian Mafia's control, unknowingly for 55 years. She is a slave. My theory is she believes her handlers, the people giving her orders, are American. They are, but the government controlling them are the Russian Mafia.[356]

_____

356 Geneva conventions, Article 8 – **Protected persons may in no circumstances renounce in part or in entirety the rights secured to them by the present Convention…."**

Chapter 104 - Stealth attacks in contravention of the Geneva Conventions, War Crimes, War Crimes using Perfidy

The Russian Mafia studied our systems decades ago according to Mr. Unger[357] and that's a real problem for ordinary Americans and our legitimate government because the Russian Mafia believes they only have to keep doing what they've been doing and they'll win a coup for the Russian government. That can not be allowed because it'll mean the Russian government destroyed our families.

How these types of attacks looked on the ground, in action, in the early 1970s: my then 28 year old, divorced mother of four young children, a trained surgical technician, first time homeowner and our family's sole breadwinner, was set up. A Russia-controlled American operative was slipped into her bed (by a palm reader she went to for a reading, and a "friend" she'd met since she'd moved to Southern California were controlled by the Russian government, unknown to my mother). The palm reader and my mother's "friend" steered her to a man, I'll call him Peter. Within a year or so of meeting him my mother had lost her career, her home, she was sick, she'd been in jail I think for the first time in her life: she was devastated. She had no idea she'd been set up a foreign enemy government. How could she know that. Few people knew the Russian government was on the ground in Southern California in the 1960s, embedding itself in the local KKK and police to

---

357 Craig Unger, "*House of Trump House of Putin*," (Penguin Random House UK 2018).

form a gang. My mother didn't learn KKK families were neighbors of ours until she'd bought her modest home. We'd moved from the Midwest, not from the South. We were an ordinary family, gave kindness and respect, and expected to receive it, but the KKK families didn't see it that way so their allies, the Russian operatives, and the domestic police in that city, dealt with us.

I think my mother would've committed suicide except that she loved us, and she saw that we loved her unconditionally, and needed her. I was her oldest around 12, her youngest was 4. She refused to abandon us. She got to her feet after the Russian government sucker-punched her and carried on as best she could. But they destroyed her momentum and she never got it back. She never fully explained what happened to her children. She'd flown us back that summer so we could get some grandparent love when the Russian government set up my mother. I don't think my mother fully knew what happened. The Russian government isn't big on explaining attacks to victim. It attacks and the victim finds out when they're on the floor. She was entrapped at some point but I don't know when.

This type of unconventional warfare is illegal but our government has to catch the Russian government doing it and must have people come forward to press charges. Entrapped people like my family are prevented from coming forward plus targets are lied to so convincingly the Russia Mafia has victims believing they're corrupt.[358] I'm coming forward because I evaded entrapment, thanks to my beloved mother who taught me ethics when I was a small child.

In unconventional warfare Special Forces, inserted deep behind enemy lines are used unconventionally to train, equip, and advise locals."[359] In our case I theorize the KKK and local police were already commingled in Southern California before Russia's operatives arrived. Different operatives are charged with different infiltrating tasks and the KGN was in our country working to destroy us from the inside out.

My and my family's strength attracted them because their task was to destroy America and any sign of strength they observed which suggests to me that the Russian Mafia has been here decades destroying the strength of individuals in America since at least the 1970s. The Russian operative came up with the idea of slipping "Peter" into my mother's bed, using a palm reader and a "friend" to have her accept and look for Peter. My mother was beautiful, had an abusive childhood. A man nice to her and her children was welcome. I don't know when he told her he was a heroin addict. It's hard to believe he would've led with that. As far as I know, my mother had no experience with

---

358 If you've seen video of Mr. Putin with either Mr. Trump or Mr. Jinping, it makes me suspicious Mr. Putin had engaged with both men before they met in front of the cameras. I don't mean this disrespectfully, but have you ever seen a really outstanding dog trainer, someone who can get inside a dog's mind and sooth him or her. That's what Mr. Putin reminds me of when he meets with these world leaders and it's not natural, Mr. Putin is showboating to the world that he has these men in the palm of his hand. He pursues them, gets them to trust him, but he's playing them, putting on a show at their expense.
359 https://en-m.wikipedia.org/wiki/unconventional-warfare.

drug abusers. When she wasn't working and parenting us, she had a social life she kept separate from us. Had not the gang targeted her, she would've continued on her course, paying down her mortgage, raising her children, enjoying her friends. Instead, as part of the Russian government's long running coup my mother knew nothing about, her life was destroyed and she was left to pick up the pieces. Instead of leaving her alone, the Russian government entrapped her, using compromised American police, and they forced her to try to entrap at least one of her adult children: me.

Special forces Russian operatives also "spread subversion and propaganda."[360] There are several clues who was involved in sucker punching my mother, and who made me and my family homeless when I was around 12. And who has hunted me ever since and entrapped most, if not all, of my immediate family.

Chapter 105 - The Russian Mafia using parents, influencers & proxies to misdirect & spread discontent among the population

Mr. Berzins goes on to explain how the Russian government works their stealth wars: "...this includes creating discontent with national institutions among the population." The Russian government creates much discontent among the population and uses it to unseat the legitimate government.

In my experience, they seized control of some of the businesses I've worked for and forced my bosses to treat me

---

360 Ibid.

cruelly; wealthy and high earners are entrapped by the Russian government, I believe at least one billionaire is entrapped. They've seized control of some of our institutions, systems, and in my experience, destroyed essential relationships (like those between parent-child, teacher-student, doctor-patient), and use those we trust as weapons to destroy us. Both my parents have tried to entrap me, groom, gas light and lure me. They're not free to tell me why they would do such a thing: entrapped people aren't free to discuss their entrapment with anyone but their handlers, I theorize. They've tried to get me to break the law in ways that would get me arrested with a very long prison sentence: defrauding the government and conspiring to defraud the government would be a few of the charges. I am my parents first born child, I'm middle age. I've been a loving daughter to them both. The only reason I can think of they'd both, separately, within about a year of each other tell me to lie and cheat the government is that they were forced to do so.

Chapter 106 - My father

My father is a retired police officer and war veteran, honorably discharged. I sought his advice when I was considering joining the military and the police force. He talked me out of it. I now know the Russian government has been attacking me since I was 11. I understand why they wouldn't want me to be a professionally trained soldier or police officer, that would've made it harder for them to destroy me.

My father was forced by his handlers to harm me, lie to me, misdirect me, he sneered at me doing what he could to get me to not write this book. But a kind thing he did for me was when I was disabled and had no money to pay my rent, I asked him to loan me $300 and he did. I repaid him the next month. I sincerely appreciated the loan.

Chapter 107 - How the Russian Mafia forced my mother to completely change from the parent who raised me. From ethicist to someone encouraging me to steal; from non-racist raising her children to racist

My mother was a hardworking, tax paying worker who raised five children on her own. Most parents don't advise their children to cheat the government, unprompted. I didn't say to my parents: "mom, dad, do you think I should lie to Disability to get a monthly check?"

The Russian government's control of my mother was especially glaring because, although I didn't know who controlled her, and in fact, I didn't know anyone controlled anyone in my family when I was 50 (although, by that age, I later learned I'd unknowingly evaded multiple entrapment attacks by living ethically), but when she was first used to try to entrap me, I had no idea it was an entrapment attempt. Indeed, it wasn't until a month or so after that that I understood my beloved mother was entrapped. I didn't use that word. I just understood, thinking about it a few months later, that her behavior must mean someone forced her to do what she did. I was terrified for her and for me. To know a person you love is entrapped is a horrible feeling. At that time I had no idea the Russian government was the perpetrator.

Another way the Russian government is exploiting systems to entrap innocent Americans is by having a doctor lure them to lie. One of the doctors I go to has tried that, twice. The doctor I went to before my current one, the Russian government wouldn't let me get though to make appointments or to get prescription refills so I had to choose another doctor. The most recent doctor helped me get my high blood pressure under control but when his medication caused my legs to swell, and I told him, he didn't tell me to stop taking the medication, that a side effect is leg swelling. I took it for maybe two years longer before he seemed to realize the high blood pressure medication he prescribed caused leg swelling. Now my legs are swollen, not a good thing. The Russian governments wants me dead so bad they can taste it, but they're not far enough along in their coup to shoot me in cold blood so they're stealth assassinating me to make it appear that I died of natural causes.

They had their drug dealer offer me free drugs, of my choice, I declined and when they had their op leave a drug at my apartment I threw it away, so they see drugs aren't something I'll do. Drugs to the Russian government are a highly effective weapon of death they use often. Anyone who abuses drugs, if they possibly can, get off them.

To entrap a target, anyone the Russian government believes might be an authority figure or a specially loved loved one to the target, the Russian government will entrap.

According to Mr. Berzins the Russians have placed the idea of influence at the very center of their operational planning and use all possible levers to achieve ....<u>deception operation</u>." Meaning the Russian government is deceiving whoever they can, their enemy government, the enemy's American operatives they control, the targets they intend to entrap, anyone, everyone.

Mr. Berzins says "The 3rd phase objective is to make the opponent's legitimate political and military leaders support measures against their own country..." While Mr. Berzins doesn't reference police, it is our domestic police who, in my experience, the Russian government has entrapped and turned into a weapon against it's own people. It convinces our legitimate government that those entrapped are legally entrapped, especially those brought to them by law enforcement our legitimate government doesn't know are compromised.

Based on the entrapments the Russian government has waged against me, our legitimate government will sign off on entrapment for: (1) thefts committed against the government, federal/state; (2) drug crimes; and apparently (3) people filing personal bankruptcy.

Chapter 108 - The Russian Mafia destroyed many of my personal belongings. First they destroyed my childhood, then stole my family, then my career (engineering), then my replacement career (legal secretary), then they stole my country from me. I can't go grocery shopping w/out being stealth harassed by the Russian Mafia's American operatives

"Destruction of real and personal property is forbidden."[361] I've had many personal items stolen and/or destroyed by the Russian government's operatives, thousands of dollars worth of valuables illegally destroyed.

Chapter 109 - Fourth Geneva Convention, Article 47-62

Occupying power shall ensure and maintain food, medical supplies, as well as the medical care to the population of occupied territories.[362]

As to the Fourth Article, I was denied medical care when I was hunted and harassed by Russia-controlled American operatives in my 40s. In my late 50s, I was denied access to medical care when Russia-controlled American operatives blocked my phone dozens of times from making doctors appointments and requesting prescription refills. When I changed to another doctor, that doctor not only worked with his handlers to to try to entrap me via Medicare fraud, he also prescribed a blood pressure medication that has caused my legs to swell and continue to swell, even though I pointed out to him the swelling during multiple doctors appointment, and not

---

361 Fourth Geneva Convention, Articles 47-62.
362 Fourth Geneva Convention, Articles 47-62.

for quite a while later did he take me off the high blood pressure medication.

At the same medical center I requested a weight loss diet pill and the doctor I paid out of pocket failed to warn me of the heart palpitations side effects, which palpitations I continue to suffer with to the present.

A surgeon who replaced a previous surgeon who allowed me to monitor my unusual intrauterine condition refused to allow me to continue monitoring my condition via regular intravaginal ultrasounds. He said I needed a hysterectomy, though I had no pain, uncontrolled bleeding or cancer and when I refused the hysterectomy, he said he would not authorize the continued intrauterine checks, even though I'm still having periods at the age of 64, which is an unusually long age and which puts me at risk of cancer.

I cut my finger in my 50s and going to the ER, Russia-controlled American operatives called the ER doctor out of the room and when he returned the shot he gave my finger was of saline; Russia-controlled American operatives forced me to get stitches with no pain numbing, out of spite and cruelty.

Chapter 110 - Several of my loved ones illegally entrapped by the Russian Mafia received subpar medical care by order of the Russian Mafia to hook them on drugs

(1) My aunt, who I believe was murdered by Russia-controlled American operatives when she failed a second time to entrap me, was denied effective medical care and endure

multiple surgeries to hook her on opioids to make her more amendable to try to entrap me, as I believe her handler's demanded. When she was forced to get out of her sick bed to get me to a game show, even though she was in remission from cancer and nearly too weak to stand, her handlers forced her to attend the game show to get me there, where they had me win a prize in their attempt to entrap me using tax fraud. I charge Russia-controlled American operatives with cruel and unusual punishment in forcing her to travel across country when she was obviously unwell, and I charge her Russia-controlled American operatives for causing her to overdose; they used her death to try to emotionally destabilize me. In murdering her, they stole from her perhaps a decade of life, stole from her children and grandchildren their beloved mother and grandmother, hooked her on drugs and made her into a junkie against her will, all so they could use her as a weapon to entrap me, a target they'd been working to entrap for over 20 years at the time of her murder.

(2) My relative was not provided adequate medical care for a female related condition and was forced to get a hysterectomy to fix the actual condition she had.

(3) Another relative was forced to have repeated failed surgeries, many over a few years, to get her hooked on opioids to make her more amenable to working to entrap me. After years living in excruciating pain, she now does whatever her handlers tell her to do. Within the last several months she was used to get from me my most recent prescription medications, I

believe so Russia controlled American operatives can swap out prescription meds for compromised medications. I've had an experience of either Russia-controlled American operatives or a thief, stealing half of a sleep aid medication I once took. When I reported the shortage, half the supply of ambien was stolen, the pharmacy did nothing, didn't replace the stolen pills nor allow or suggest I fill out a theft report. Apparently they had nothing in place when a customer had part of their prescription stolen before receipt.

Chapter 111 - I've witnessed at least 1,000 children illegally forced into a stealth child army by their entrapped parents, school minders, etc.

"Several provisions of the Geneva Conventions grant a special protection to children under 15." When Russia-controlled American operatives first attacked me and my family we children were all under 15: At 12, I was the oldest. My brother was 11, my sister was 9, my baby brother was 4. Later, when our last baby sister was born, she too was harassed and attacked beginning around the age of 7 or 8, and continuing for much of her childhood and adulthood. Additionally, at least one of my nieces and several of my nephews were set up to receive stolen property by school friends controlled by Russia-controlled American operatives; and one of my nephews, who is now addicted to illegal drugs, I don't know exactly which operative(s) influenced him, but we, and the children I list, were all under the age of 15 when Russian controlled American operatives entered our lives and began to destroy the quality of our life in contravention of the Special Protections for Children under the Geneva Conventions.[363] The Russian government's stealth warfare is especially heinous as it covertly attacks with no war declared between our two nations, targeting the youngest, most vulnerable of us. I've witnessed entrapped parents forced to bring their newborns on harassing attacks with me as the target.

Chapter 112 - Illegal attacks on me, a civilian woman, by the Russian Mafia

Humane treatment is granted by the common article 3 of the Geneva Conventions: It forbids: "violence to life and person (including murder and torture);" I charge that two Russia-controlled American operatives deliberately drove into me while I crossed in front of them on my mobility device when I was in my 50s. Both clearly saw me, I had the right of way. In one of those attacks a marked police card was already at the scene, facing me, with I suspect camera rolling, to capture the attack on me. I was pinned underneath a small pickup truck. I believe it was a setup, another attack, by Russia-controlled American operative police officers, forced by their handlers to harass and attack me both physically and psychologically. In addition to being deliberately struck by two Russia-controlled American operatives, I've been harassed and intimidated by multiple Russia-controlled American operatives who deliberately rev'd their engines as I passed in front of them on

363 Deyra, Michel (1998). Droit International humanitaire. Paris: Gualino Editeur. p. 109. Children are protected as victims of the armed conflict.

my mobility device. One operative didn't stop when I had the right of way; another operative, when I rode my mobility device one night to return a library book to the drop box, drove at approximately 2am with his headlights off down an alley as I approached the alley to cross it on the sidewalk. Only the fact that I heard the engine prevented me from being hit by that operative.

I've been kicked, stomped on, called an ethnic slur, deliberately hit with hundreds of bags and purses by operatives as they pass me on the bus, no matter how far away I lean, Russia-controlled American operatives find a way to hit me with their belongings. Russia-controlled American operatives cut in front of me while I wait in line at the pharmacy, at a restaurant, cut in front of me as I reach for products at the grocery store, nearly swerve into me on sidewalks as they jog nearly into me, walk nearly into me, skateboard nearly into me. All of which has been ongoing for decades.

Chapter 113 - The Russian Mafia's operatives prevented me from voting in the 2008 election, decades after my right to vote was guaranteed

In 2008 I was prevented from voting by Russia-controlled American operatives who harassed me when I got off work to vote. They terrorized me so much, one at minimum appeared to be an off duty police officer, that I signed up for mail ballots and have voted by mail ever sense. Inside my apartment, operatives come stand outside my window any time day or night and make sudden loud, startling noises. On May 31,

2023, the manager of the apartment building where I live suddenly appeared outside my front door and made a sudden loud noise that frightened me so much my heart rate reached 200 bpm and I was forced to call 911. The operative in the apartment above mine follows me room to room in real time and drops things on their floor, loudly, to startle me. When I'm trying to sleep they drag their kitchen chair repeatedly, loudly, across their kitchen floor to interrupt my already light, poor quality sleep. All this and more are in contravention of the Geneva Conventions.

Chapter 114 - The Russian Mafia forced my entrapped supervisors to harass and degrade me at work in most of my full time jobs

I charge Russia-controlled American operatives with "outrage upon personal dignity, in particular humiliating and degrading treatment," when the Russia-controlled American operatives had supervisor after supervisor harass and insult me and be cruel to me, from my first time position as a clerk in Human Resources when my supervisor put her finger on my forehead and insulted me, and the next supervisor I had in another department after I transferred, sexually assaulted me and passed me over for a promotion when I didn't respond to his sexual advances. From my first full time post college job, to my last full time legal secretarial job where I was overworked and exhausted to such an extent I collapsed after one of the name partners yelled at me for no reason, while I worked as his secretary and legal secretary to four other attorneys, plus helped attorneys who needed a filing done and their secretaries were absent. I was assigned to five attorneys, and it wasn't unusual to work for six. A normal work load for a legal secretary with the twenty years legal experience I had was 2-3 attorneys and/or paralegals, at the most. So at the last job I worked at I had twice as many attorneys to work for. I collapsed from exhaustion. Not only was I overworked but when I went to doctors to get help for feeling unwell, it was as if they'd not gone to medical school, Russia-controlled American operatives ordered them to not help me in any way, so the doctors didn't and I got sicker and sicker. After I collapsed, Russia-controlled American operatives had my family work to try to entrap me. I refused entrapment. It wasn't until I was 51, I want to say January or February 2011 that I realized my mother was entrapped.

It wasn't until after January or February 2021 that I realized my father was entrapped. He's a retired police officer so he'd know more than my mom on how to cover his actions. I honestly had no idea. After I understood my mother was entrapped and being used to try to entrap me I was heartbroken for her and for me. I never would've seen the connection except I'd been harassed in the street since I was around 40, 41 and when my mother told me to lie, that was the first time in my life she'd said that to me. Still it took me a month or so after that to understand she was entrapped. I didn't know what entrapment was. It wasn't until I read an article in the newspaper several years later, when I was maybe 55, that I understood so many things that had happened to me were entrapment attempts I'd evaded because I committed to living ethically when I was 25.

Chapter 115 - Attacks I've endured at the hands of Russian Mafia's American operatives in contravention of the Geneva Conventions which the USSR agreed to in 1948

The additional protocol II completes the article 3 of the Geneva Conventions adds several forbidden actions and grants to protected persons the rights to "respect for their person, honor and conviction and religions practices."

According to the Fourth Geneva Convention "the rights of protected civilian persons are absolute and uninalienable... Protected persons cannot renounce their rights," so whatever deal Russia compromised law enforcement made to victims is void. The additional protocol I "prohibits indiscriminate attacks or reprisals against the civilian persons, their objects, and objects necessary to their survival."[364]

Humane treatment – protected persons are "entitled, in all circumstances, to respect for their persons, their honor, their family rights, their religious convictions and practices, and their manners and customs." They shall be protected against acts of violence, intimidation, insults and public curiosity.[365][366]

"Equal treatment – protection should be provided without discrimination on the base of race, nationality,

---

364 Additional Protocols to the Geneva Conventions of 1949.
365 Fourth Geneva Convention, Article 27.
366 Deyra, Michel (1998). "Droit International humanitaire. Paris: Gualino Editeur." p. 106.

religion, pinions. Health, rank, sex, and age distinctions are accepted."

Security – protected persons could not be used as a human shield. Corporal punishments, torture, murders, collective penalties and experiment are forbidden."[367]

Chapter 116 - War Crimes done against me by the Russian Mafia, including sexual assaults

The Russian government broke the Geneva Conventions' special protections for women when they attacked and harassed me for most of my life

The Geneva Conventions grants special protections to women in all circumstances, which Russia-controlled American operatives disregarded and failed to honor. The Geneva Conventions say "Women shall be especially protected against any attack . . . or any form of indecent assault."[368] Two Russia-controlled American operatives sexually assaulted me. One visited with my maternal aunt and sexually assaulted me around 3am as we all three lay in bed, them at the foot of the bed, me at the head. My crotch was groped by the female sexually assaulted me. When I told my mother about the assault, she gas lighted me, told me it didn't happen.

One of the many psychological attacks by the Russian government is when they have beloved and/or trusted family not believe you, when prior to those loved one's entrapment they believed what I said. Russia-controlled American

---

367 Fourth Geneva Convention, Article 29.
368 Fourth Geneva Conventions, Article 27.

operatives behave as Russian-government American operatives, not beloved family members even though they are beloved family. Their behavior is so odd and disconcerting because they are being forced to be a hostile to an individual they once had a loving relationship with. The Russian government doesn't care what their relationship once was, as long as it provides them access to the target. After that access, the Russian government forces the entrapped person to carry out the agenda the Russian government demands, and that agenda doesn't fit in the history of the relationship the target had with the entrapped person, so it's pretty glaring that something is wrong but the entrapped person isn't free to disclose the problem. That there is a problem is clear, so the stealth entrapment the Russian government is working towards is still ongoing but the oddness, even to an untrained person like myself, is jarring. Even knowing nothing about entrapment I noticed something was off.

The Russian government is working a stealth war, it hates every American and every person living in America. Even for people who are entrapped, they must kick themselves after the fact when they look back and say "why did I ignore that?"

It's largely that targets are inexperienced in psychological warfare, and they're not being warned, that's given the Russian government their edge in attacking our innocent ones. After people are warned, the Russian government won't be able to as easily entrap targets because targets will know the scam and won't take the bait.

I charge the Russian government with all the above-referenced breaking of the Geneva Conventions.

Chapter 117 – I charge the Russian Mafia with Genocide against me, my family, and my people who are the American people and everyone living in America from 1971 to the present, the approximate years I've been attacked by the KGB and the Russian Mafia, minding my own business in my own country

I charge the Russian government with genocide against me, my family, the American people and everyone living in America, starting in 1971, when the Russian government first attacked me and my family, to the present.

Genocide is the intentional destruction of a people in whole or in part. The 1948 United Nations genocide convention defined genocide as any of the five "acts committed with intent to destroy in whole or in part, a national, ethnical, racial, or religious group These five acts were: (1) Killing members of the group, or (2) causing them serious bodily or mental harm; (3) Imposing living conditions intended to destroy the group, (4) preventing births, and (5) forcibly transferring children out of the group. Victims are targeted because of the real or perceived membership of a group, not randomly.[369][370]

---

369 United Nations 2019: office of the United Nations Special Advisor on the Prevention of Genocide 2014; Voice of America 2016.
370 Convention of the Prevention and Punishment of the Crime of Genocide art. 2, 78 (UN.T.56 277, 9 December 1948).

Chapter 118 - Specific attacks the Russian Mafia's American operatives ordered done to me are war crimes

Killing members of the group, including acts of extreme violence, such as torture fall under the 1st prohibited act.[371] Russia-controlled American operatives have deliberately prevented me from sleeping on many occasions. Deliberate sleep deprivation is designated as torture. Torture is the deliberate infliction of severe pain or suffering on a person for reason such as punishment, extracting a confession, interrogation for information, or intimidating third parties. Psychological torture includes exhaustion by sleep deprivation. Sleep deprivation is inadequate duration and/or quality of sleep. Complications include obesity and cardiovascular disease. I suffer from high blood pressure, obesity and other conditions related to sleep deprivation.

Causing serious bodily or mental harm to members of a group, 2nd prohibited act Article II(b).

When a partner of a law firm yelled at me, after I was working for up to six attorneys, five assigned to me and any attorney who's legal secretary was out and the attorney needed sometime file with the court or served, not long after I collapsed from the stress. In all I estimate at least eight of my supervisors over my work life of about 30 years have been entrapped and/or compromised by Russia-controlled American operatives and forced to psychologically abused me at my job.

---

371 Prosecutor v. Semanza, Case No. ICTR-97-20-T, Trial Judgment, para. [320], 15 May 2003; Prosecutor v. Ntagerura, Case No. ICTR-99-46-T, Trial Judgment, para. [664], 24 February 2004.

From my first full time job when I was 22 or 23, to my last full time job ending when I was 50, Russia-controlled American operatives have used my bosses as weapons to psychologically harass me in contravention of the Genocide Act.

In addition, Russia-controlled American operatives forced my parents to abuse me emotionally when I told my father I intended writing a book describing the torture and harassment I've endured. He sneered at me and said people would not believe I was mentally unstable. His handlers had him treat me overtly unkindly for the first time in my life. It hurt. I didn't know at the time I confided in him that he's entrapped.

Chapter 119 - The Russian Mafia had my mother bully me to file for personal bankruptcy – I refused

When I refused to file for personal bankruptcy, which Russia-controlled American operatives had pushed me to do using my mother to tell me to file, and an HR employee at my last employer's, my mother sneered at me and said it was stupid for me to not take advantage of the personal bankruptcy laws to write off my debt. I didn't want to write off my debt and I explained why to my mother but her handlers had her push me hard, over years. She held up Mr. Trump when he was president as an example of a smart businessman who took advantage of the laws of the land to write off debt and he was lauded as a smart businessman. That the Russian government used my mother to extol Mr. Trump, their preferred candidate. By the time she pushed me really hard to file personal bankruptcy I knew she was compromised. Once you understand someone is entrapped, you view everything they say through the lens of their enslavement. One of the doctors I see is entrapped, I suspect, because he's tried to entrap me on more than one occasion. So when he suggests something to me I view his suggestions through the lens of his enslavement, meaning I don't have a doctor I can trust. The Russian government is to blame for the subpar medical care I receive.

Chapter 120 - My once beautiful family has been stolen from me by the Russian government and it's Mafia

Forcibly transferring children of the group to another group. I didn't know where to fit this into genocide but I know it's part of genocide. The Russian government entrapped my parents and half siblings I was raised with and forced them to either try to entrap me or influence me to my detriment. In essence, the Russian government stole my family from me. My parents and siblings live but they are owned by my mortal enemy. I can believe nothing they say and everything they say I must view through the lens of how the Russian government is trying to sucker punch and kill me. I have no illusions about the Russian government. It is ruthless, evil, cruel. It cares only about world power, it's throwing our planet's habitability under the bus by preventing it's allies/slaves from helping us to seriously mitigate climate warming.

The Russian government and it's Mafia stole my family. My family is alive but they're slaves of a monster government who means me to die as quickly as possible, preferably without writing another edition of my book. The Russian government didn't steal me from my family, the Russian government stole my family from me. This aspect of genocide is a flip of the stealing a child from their family the Russian government is perpetrating against Ukrainian children and their families. In my case the Russian government stole my family from me. I remain the same, my family are enslaved by my enemy. It is a theft, but a twist on the theft of children, in my case the Russian government has stolen my family from me. The twist mean the War Crime doesn't exactly match when a child is stolen from their family: I chose it because I am my parent's adult child and my family was stolen from me.

Russia signed this genocide pact when it was the USSR. Researching I found it interesting and telling that the USSR refused to sign the genocide document unless political and social group oppression was omitted from genocide document. That suggests to me the USSR was already using political and social genocide strategies against groups.[372]

Killing members of the group (Russia-controlled American operatives I believe killed my beloved aunt, my beloved grandfather, and my beloved friend, people they knew I love deeply, who they murdered to emotionally destabilize me. Killing members of a group (article 1) includes torture, and Russia-controlled American operatives have tortured me for over decade with deliberate sleep deprivation.

---

372 Stanton, Gregory H., "What Is Genocide?"Genocide Watch.

Chapter 121 - I charge the Russian government and it's Mafia with 3rd Prohibited Act – deliberately inflicting on the group conditions of life calculated to bring about its physical destruction

I was worked to exhaustion at the last law firm I worked at. In this prohibited act "deaths are not immediate but rather create circumstances that do not support prolonged life. Courts must consider duration of time before actual destruction would be achieved."[373]

Chapter 122 - The Convention on the Prevention Punishment of the Crime of Genocide, put in force on 12 January, 1951; signed by the USSR on 12/16/49

All signatories to the CPPCG are required to prove and punish acts of genocide both in peace and in war time. I charge the Russian government with these crimes, as it's predecessor government signed off on these laws and agreed to be bound by them.

Chapter 123 - Our people's national self image has been attacked by the USSR since the 1960s; the Russian Mafia has made us doubt our stretch because they intend to destroy us

The Russian Mafia controls part of our media and use that control to further their interests and to destroy us. Last month an American internationally known newspaper ran repeat articles about how China's economy was going downhill. It was odd because there were daily reports for maybe five days

---

373 Stanton, Gregory H., "What Is Genocide?"Genocide Watch.

in a row. One article, okay, two, okay, but more than that, saying basically the same story, was strange.

The Russian government was behind it. The Chinese government is an ally of the Russian gov. The Russian government is running the lie to Chinese officials that Americans and our government wish the Chinese people ill. The reason the Russian government is working that lie is because it intends for Chinese people to back it in the Russian government's war with us, and after the Russian government encourages the Chinese military into a war, which the Russian government is presently doing. The Russian government has gotten the Chinese military all jacked up about becoming a world leader and taking on America. China is already a world leader. The Russian government intends to use China's military and the Chinese people as cannon fodder, the same way it uses it's own people as cannon fodder in the Ukraine War.

For that to happen, the Russian government must convince the Chinese people that Americans and our government are disrespectful of China. We're not, but the Russian government lies and to get Chinese people willing to be used as cannon fodder, the Russian government must lie to them about America and what we think about Chinese people.

Over a decade ago when I went on prepper forums to learn about prepping the Russian government was pushing a very strong strain of anti-China racism. It did that so it could show China's military leaders and hackers that "see,

Americans are so racist against Chinese people." We're not. What we are is a country deeply embedded by our enemy, the Russian government, who hates us, spreads lies about us, and intends to destroy us. The Russian government fails to disclose to the Chinese military that it is they who are the originators of anti-China sentiment here.

The Chinese military can disbelieve me if they want to, but I'm telling the truth. The Russian government intends to rule the world. It's the most racist government I've ever seen and it believes no one is it's equal. It'll befriend Chinese people, India's people, whoever and whatever it needs to do to destroy Americans and America because world power is their idea of nirvana.

Chapter 124 - The Russian Mafia intends to use heat as a weapon against us in the summer

Heat as a weapon I think I've already covered. One of the reasons the Russian government has taken control of our electric grid is if we're at war with the Russian government or one of it's allies, that war will occur in summer. The Russian government will destroy our grid and watch us die and laugh. The Russian people wouldn't laugh at our pain, but the Russian government hates us and will laugh.

Chapter 125 - Earth habitability

To rule the world, the Russian government isn't allowing it's allies, governments in BRICS, or anyone the Russian government influences, to work with us to fight the climate catastrophe. The Russian government is actively stopping our species from effectively fighting the catastrophe, believing in it's denial, that if the problem is that severe, after it takes control of the world, it'll deal with it. The Russian government believes in slavery. It will remake the world in it's own image and you better believe any Americans who takes a deal with the Russian government will regret it. I believe it was the Brazilian president who said, during BRICS, he believed we were manufacturing or exaggerating the climate catastrophe to maintain world control.

First, we don't have world control, that's a lie the Russian government loves to spread. Second, if you believe whatever the Russian government tells you, you're a moron. The Russian government lies like a rug. When it speaks the truth it's a thin coating of truth surrounding a huge lie, usually a lie of omission of something the Russian government has done.

Chapter 126 - Ambushing targets using their loved and trusted ones, exploiting hierarchy, & stressing targets are some of the Russian Mafia's favorite strategies

By studying people and exploiting human hard wiring, the Russian government observed that victims tend to do what their supervisors, elders, priest, parents, police, authority figure, tell them to do. They further developed that to target victims when the victim is in a subordinate position, using our hierarchy against us.

A person like me, who the Russian government has worked to entrap for most of my life, after decades of trying and failing, they decided to use my parents to tell me to cheat the government. I wouldn't. But the Russian government used them against me. My parents insulted and sneered at me, following orders from their handlers. My father was used to dissuade me from writing this book, told me people would think I'm insane. I'm being as specific as possible because people are still alive and if our legitimate government questions us, people will either lie or tell the truth. If the Department of Justice questions us, it'll be interesting how my parent's handlers will direct them. I can't imagine Russian-controlled ops will give my parents or anyone in my family permission to speak the truth. But when my family are prevented from telling our legitimate government the truth, how will their handlers maintain the lie that our government has entrapped them. If there's anyone who can lie their way out of anything, it's the Russian government.

My mother was used to pressure me into filing personal bankruptcy. When former President Trump was in office, she used his experience of filing for bankruptcy and said that that was what intelligent people did, that bankruptcy filing is there for people to use when they have financial problems. What I doubt her handlers told her was that they'd set me up to become over-extended. The Russian government informs on a need-to-know basis and it doesn't believe slaves and/or victims need to be told anything except when given orders.

It takes a certain type of evil to force parents to entrap their first-born child. Russian-controlled American ops have no problem using my parents as weapons against me. And, based on the approximate thousand, mostly Caucasian parents forced to harass and/or attack me with their children in tow over the last couple decades, Russian-controlled American ops have no problem ordering American parents to corrupt their children by bringing them along to harass a woman using a mobility device. Nor did Russian-controlled American ops have a problem getting operatives to threaten me using their vehicles, and hitting me with their vehicle twice as I rolled past them on my mobility device. Or kicking me, stomping on me, calling me the n word or lifting their leg and trying to kick me in the face on the bus. All these attacks, and worse, were done to me by Russian-controlled American ops.

Chapter 127 - The Russian Mafia ensures the worse possible thing happens to you, usually out of the blue – their strategy is to make people as miserable as possible, then trigger & aim them at whatever the Russian Mafia wants destroyed. In America it's our government, so the Russian Mafia aimed our people at the Capitol building on 1/6/2021

The Russian Mafia makes attack dogs. You see it in some of their operatives, you see it in ordinary people who've been entrapped, you see it in entrapped law enforcement who are cruel, and it's not just here. Police in France have a horrific reputation for targeting innocent minorities. As do police in England. Israeli law enforcement and military have horrific reputations for targeting Palestinian children. Police in India have a horrible reputation for targeting or failing to protect Indian Christians and Muslims. Hindi's are the ethnicity in power with Mr. Modi. And many of us have read about Iran's and Afghanistan's horrific abuse against it's female citizens and against anyone who dares to protest.

In my family my half brothers are entrapped. They weren't raised in an abusive home, I lived with them until I was 17. But they became abusive after the Russian government got a hold of them. Russian-controlled American ops sic men or anyone it can, as I said, the Russian government makes attack dogs. Attack dogs in our jails and prisons are brutalizing our people in those jails and prisons.

Chapter 128 - Never trust the Russian Mafia or the Russian government

The Russian Mafia and the Russian government are our enemy, not the Russian people. They've always been our enemy except for a brief time when Nazi Germany attacked them, then we formed an alliance with them in War World II. The reason some of our political leaders got confused in the 1960s and 70s is because the Russian government, the KGB, blackmailed them and made them confused. In the 1950s, 60s, 70s, etc., the people largely controlling our country were Caucasian men. Like many people, when someone is your sex and your skin color, it's easy to believe they see the world as you do. The Russian government sees the world very differently from how Americans do. That's fine, people worldwide have different belief systems, religions, marriage rites, etc. The mistake some of our leaders made in the 50s-70s was believing Russian government officials saw the world as they did, and under estimating the USSR and it's KGB. The reason our then-leaders underestimated the USSR and the KGB is because the Russian government wanted our leaders to underestimate them.

The Russian government is a slaver government. It believes it is the master race and everyone else is it's vassal. It won't come out and say that, it needs it's allies as cannon fodder when the Russian government launches it's hot war against us, but, based on my experience being attacked by them in my country, they believe they're a superior being. The reason they're so sure they're the brightest is because they're thieves and will still anything not nailed down. Because they're skin is Caucasian and because their leaders are male, and because they stole their people's money and so was able to present to the world they are successful and wealthy, the truth is they are criminals and thieves, and their government is a mafia government. What that means is, when the Russian government wants something to happen in an enemy country, they use their mafia to make it happen. It's all stealth. The Russian people follow blindly because if they don't they'll be squashed like bugs. The thing we need to do is to bomb the Russian government. The problem with that is if we do they'll bomb us back and many of our innocent people will die, and out infrastructure will be destroyed, and many of us will starve because nuclear weapons are much more powerful than they were in the 1940s, when a country could survive them. You may have noticed our newspapers are filled with destroyed infrastructure of innocent people, and that's part of the Russian government's psychological warfare. When it makes it's pitch to our government and likely it will through the Chinese government, it's counting on all the Americans seeing the destroyed towns and villages the Russian government and it's allies have strewn in multiple countries, and it's counting on that, depending on that, so when our legitimate government comes to us to tell us the deal the Russian government has hatched for us, the Russian government is ensuring ordinary Americans will accept it. If we accept any deal proposed by the Russian government, we will regret it. Exhibit A: Ukraine accepted a deal where they gave up their nuclear weapons and you see they're being bombed unrelentingly. We can listen to any deal the Russian government proposes, but we must always, always look at what they've done. Their behavior is them. The Russian Mafia is them. They hide nearly all the evil they've done but they aren't able to hide everything, because our security agencies helped us see, and because brave heroes whose names we'll never know, got essential intel to the world. You trust the Russian Mafia to your peril.

Chapter 129 - The Russian Mafia has set a trap for us – pardoning millions of innocent people is how we can evade the trap. We have a lot of people to talk to about their entrapment & how it came about

To escape this trap, we're going have to free ourselves from it. People taking polls aren't free to explain they're entrapped. We need to assume they are and come up with a solution before next year's election. My vote is President Biden and as many states as we can get to back him, pardon specific types of crimes, in the millions. People who actually deal drugs, beat up people, not them. I'm talking about where people are charged with conspiracy to defraud, especially defrauding a state and federal entity. The Russian government pushed me hard to cheat on my taxes, Medicare, Social Security, etc.

Instead of waiting to catch a break from the Russian government who is working to destroy us, we need to accept the Russian government has worked decades to enslave us and we'll catch no breaks. We must create our own breaks. Pardoning people will work.

Because people in America don't know the Russian government is working a coup, because people in America don't know the types of stealth warfare being used against them, we're all of us being attacked separately. Our government doesn't know and we're not protected. I'm listing as many destabilizing attacks I've experienced as possible to share why I believe ordinary American people attacked the Capitol, and why I believe so many people are fed up. When people follow the rules and are spat on, they become embittered. The Russian government knows this and deliberately frustrates victims.

The Russian government knows there are millions of ordinary people on their best behavior, trying their best. When those people are rewarded, content, they don't overthrow their government. The Russian government deliberately agitates and frustrates people who follow the rules because they intend to use us to overthrow our own government.

I read that President Biden wondered for a year or so ago why his poll numbers weren't reflecting his achievements. He too is being stealth attacked by the Russian government.

Chapter 130 - Stealth murders of my loved ones

I believe but cannot prove the Russian government murdered my beloved aunt in my 30s after she was unable to interest me in becoming a drug mule; killed my beloved grandfather in my 40s after they saw how close we were and knew I'd turn to him after they drove me to exhaustion; and killed my beloved mentor in my 60s in their attempt to have me fail my programming Final. (The Russian government pushes really hard to have me fail or drop out of school. (1) Since I was around 15 (when their guidance counselor op accused me of stealing $20 from her purse to stop me coming to her office seeking help on how to pay for college), (2) When I was around 25 or so, and the TA op in my math college class accused me of cheating on a math test and threw me out of the test even though he said he saw another student cheating on me over my shoulder (I had no idea at the time I was being hunted and I dropped out school, devastated); (3) as a beginning part time programming student the Russian government has had their operatives noise attack me, during Finals, during quizzes, I had to go to the ER Spring 2023 Final because an operative frightened me; the attacks are really noticeable. Because of the school-related attacks I've experienced from the age of 15 until my recent programming classes in my 60s, I'm convinced the Russian government is behind the decline in student's test scores in America.)

All three of my loved ones were civilians I believe the Russian government stealth murdered to emotionally destabilize me because that's who the Russian government is. The Russian government has covers for these deaths. They always have explanations and the covers sound believable. To generate suspicion from an non-compromised investigator, the officer would have to know me, have to know what the victims meant to me, and when and how they died to understand they are hits. The Russian government only cares that I know; to the casual glance they have believable covers. The Russian government continues to harass me to the present day November 2023, age 64.

Attacks by an enemy foreign government against civilians are illegal in war and in peace. The Russian government doesn't care. And with the Israeli's government's possible War Crimes against innocent Palestinians, the Russian government loves muddying the water. There's a reason the Israel-Hamas War began on October 7, 2023, after I'd published my first edition on October 5, 2023: the Russian government always muddies the water, always reaches for cover, didn't want our security agencies to be free to study this book uninterrupted. Even as I type these words at 3:25 a.m. the operative upstairs is standing in their hallway monitoring me as I type in mine. The Russian government had to okay Hamas' attack on Israelis. The Russian government controls Iran and Iran controls Hamas. When Russia-controlled operatives saw they hadn't destroyed my laptop after they surged the electricity in my apartment on 9/29/2023, the Russian government authorized Hamas to attack, for multiple reasons. Hamas attacks on Israeli's are evil. Hamas is a terrorist organization with a history of attacking Israeli civilians. The Russian government authorized this war, is causing the horrific suffering to unseat President Biden in 2024.

The Russian government has declared no war against America and attacks Americans illegally, taking control of our entrapment system and some of our police to destroy our families, to commit genocide against us. The Russian government knows it would lose in a fair fight so it attacks women, children, and civilians. I've witnessed with my own eyes, entrapped parents in America used to harass me with their newborns in their arms. The Russian government uses our newborns in their war against us. This cannot stand.

While the Russian government worked to destroy me, and successfully destroyed my family as it once was, I saw how the Russian government's operatives stealth destroyed other children in my extended family, and because I am apparently the only American the Russian government has been unable to hook on drugs, entrap, or emotionally destabilize the last 53 years, it's American operatives overtly and/or covertly harass and/or attack me, using any means necessary to harass me. Because they've thrown so much at me, I've witnessed Russia-controlled American operatives including police, force ordinary Americans to harass me. And I've witnessed entrapped parents being their newborn infants on harassment attacks where I was the prey. Most people with newborns aren't willingly taking their babies anywhere which leads me to conclude that at least some parents are forced to bring their children with them to use as beards (cover) as the parent harasses me and/or attempts to engage me. I've also witnessed parents using their young children (so under the age of 15), as a co-harasser, meaning children are being forced into warfare, which is a War Crime during war and is definitely a War Crime when war hasn't been declared.

I witnessed the Russian government in the process of destroying at least 1,000 other families, and at least 3,000 other people's lives as Russia-controlled American operatives forced thousands of strangers stealth attack me over multiple decades. Based on the approximately 3,000 mostly Caucasian people who I've experienced stealth and/or overtly harass me the last decades, I believe the Russian government is running a highly successful entrapment scam where they're enslaving people using an entrapment mechanism police long ago, I believe, aimed at criminal suspects but which is now being used to target and entrap innocent people, to destroy families and to force people to vote for whoever the Russian government wants seated in power positions.

Chapter 131 - The USSR & the Russian government illegally discarded the rules of war

Deliberately attacking children under 15 during war is illegal and is a War Crime under the Geneva Conventions of 1949. It is also illegal for a foreign enemy government to attack children in it's enemy's country when there is no declared war, or even when there is a declared war, but the USSR did so anyway, breaking the law and destroying my childhood. The USSR knew, and agreed to abide by the Geneva Conventions but it flat out lied. Below I describe their attacks on me when I was 15 and under.

I was a child of approximately 11 in 1971, a 6th grader living in a Southern California city with my mother and three half siblings, when a gang consisting of an unknown number of USSR operatives, KKK, and local police attacked me by conspiring to: (1) have an older Caucasian boy I didn't know threaten to beat me up at the bus stop after the bus driver dropped us off after school; (2) have a USSR-controlled African American op who was fronting as my mother's boyfriend (I'll call the op Peter) meet me at the bus stop to question me; (3) have Peter drive me home and exaggerate to my family about what had happened at the bus stop (introduce the concept to my family that I was a "tough" kid when before that I had been known as a responsible kid in my family); (4) set my mother up as an accomplice to a crime Peter committed. My mother was set up in such a way that she became unemployed, we lost our modest home, my mother became ill, and we became homeless; (5) at some point the USSR or the Russian government entrapped my divorced parents, and my half siblings. My family doesn't discuss their entrapment with me so there's no way I can know the exact date individuals in my family were set up by the perpetrators; (6) had my mother beat me, out of the blue, for no reason, when I was around 14 or 15, which precipitated my decision to leave home to attend university when I was 17 (leaving my youngest half sibling to be abused by multiple operatives, including children operatives controlled by the USSR); (7) had three teenage girls at the high school I attended threaten to beat me up and attempt to do so.

Chapter 132 - The Russian Mafia illegally hunts civilians in America (and very likely many other countries)

The Cold War never ended. Nor was the Cold War strictly cold because the USSR, and later the Russian government, covertly attacked civilians, including families with children in America, which is illegal under the Geneva Conventions and other rules of war. I was one of the children the Russian government attacked in the 20th century and I'm one of the many adults the Russian government is attacking in the 21st century, although they're still attacking children. President Zelensky says the Russian government has stolen tens of thousands of Ukraine's children which is one reason Mr. Putin and another government official face arrest warrants. The Russian government takes control of children's minds and make them weapons. You'll have noticed the huge spike in mass murders perpetrated by young people here. That's courtesy of the Russian government, who's found a way to commandeer our entrapment system, entraps all kinds of people including parents, and forces those parents to treat their children how the Russian government wants them treated. I know because the Russian government's American operatives used both my parents to do things the Russian government wanted or didn't want me to do.[374] The Russian government wants monsters created so they can aim them at innocent people in the grocery stores, bus stops, etc. The Russian government is everywhere our legitimate government isn't, meaning it's in our families, in our doctor's offices, in the work place ensuring vulnerable workers are being harassed, making sure landlords don't do reasonable repairs, etc.

The Russian government has found a way to entrap parents (and other ordinary people) using compromised

---

374 Both my parents are entrapped by the Russian government but believe they're controlled by our legitimate government. When the Russian government wanted me to not write this book, it had my father discourage me from writing it by denigrating me. And when the Russian government wanted me to file for personal bankruptcy it had my entrapped mother tell me it was stupid for me to not file, denigrating me. I was in my 30s, 40s, 50s and 60s when my parents were used by the Russian government to control my behavior. Imagine what it's like for young children when their parents or beloved caregivers tell them to do or not do something. Children are easier to control than adults but the Russian government will tell you, adults aren't especially difficult when the puppet master is unscrupulous enough, and the Russian government is nothing if not unscrupulous.

American law enforcement to entrap them. The Russian government then forces the parents to treat their children however the Russian government wants them treated. Our legitimate government isn't in children's homes but the Russian government is, forming and shaping mass murderers because the Russian government absolutely loves turning death and destruction on innocent people in America. This book is intended to help ordinary people better defend themselves and their children against a monster who hides in the shadows: the Russian government.[375]

The Russian government intends to rule the world, and I'm pretty sure, based on BRICS and other aggressive, effective maneuvers, that it expects to control the world within Mr. Putin's lifetime. He's about 71 now so he expects to place a puppet in the White House in the 2024 election or within the next decade. He's the world's richest man because he's head of the Russian Mafia. He and his allies control drug cartels and use those cartels to flood our people with poison. For people like me who eschew drugs Russia-controlled American operatives send drugs and/or drug dealers to me and/or use compromised loved ones to coerce me to meet with their drug dealer relatives, or the Russian Mafia will have an aunt visit me and she'll bring an operative along and they'll leave drugs with me. For people uninterested in drugs, the Russian Mafia's American operatives make sure targets have plenty of drug opportunities, so imagine the flood for people who are actually

_____

375 Ibid.

interested. No wonder we're in an epidemic of overdose deaths.

According to Karen Dawisha author of *Putin's Kleptocracy*, Mr. Putin gets a cut of every scam the Russia Mafia works[376] so every time a child is sold into prostitution by the Russian Mafia or a child is forced into a gang by the Russian Mafia, Mr. Putin gets a cut. Plus he has this huge spider web to keep him entertained: the enslavement of all Americans, 21st century American slavery, Russian-style.

The Russian government intends to seize control of the world. Most Americans have no idea they are in the Russian government's cross hairs. In this book I explain why we are being hunted, and how to better defend yourself.

It's hard for most people to understand why anyone would want to control the world but some people have serious ego issues and think world power is the best life has to offer. The Russian government is one such entity and unfortunately it's formed a hybrid with it's Mafia to make the Russian government's dreams of world domination come true. You've heard of people with compulsions, like shop until you drop, or eating disorders, or sex addictions. The Russian government is a group of like-minded people who are determined to enslave the world at any cost. They won't stop, they can't stop.

They facilitated Hamas' attack on innocent Israeli's on 10/7/2023, knowing the Israeli government would drop bombs on Palestinian civilians in Gaza. They don't care about the

_____

376 Karen Dawisha, *Putin's Kleptocracy*, Simon & Schuster (2015).

murder of innocents. All they care about is destroying Americans, destroying our President, and flipping the world order so the Russian Mafia is top dog. Once top dog, they'd run our species on the slavery model because that's the system they're most comfortable with. Because they're back-stabbers and bold faced liars, they're unable to trust anyone. Slavery is their first choice, with them as the world's only master. They see themselves as the master race. The Nazi's did too. Something about entitlement and racism drives some people insane.

Because they've attacked civilians and and children with no warning, and have gotten away with it for decades, they think they're smarter than everyone else. But what they actually are are cowards, killing and/or harming innocent people. Not just America, but our species. They're off the rails.

Mom hadn't raised us to be racists so it was hard to understand people might hate us because our skin was darker. That made no sense, but we accepted this was truth because our mother told us. By the time she warned us we were already friends with kids in KKK families, and with most of the many kids on our block.

Chapter 133 - Nearly everything I've learned about the USSR/Russian Mafia I've experienced through trial by fire. I don't wish that for you

I've learned about the Russian government's attacks on me through trial by fire. Throughout the book I list time and again specific attacks by which I hope to help you defend yourself against an enemy who hides in the shadows, and lies. Who will use your trusted loved ones as camouflage to manipulate you into making life-destroying choices. Your loved ones won't be free to warn you to ignore them. After they take a deal from the Russian Mafia's American operatives, your loved ones are slaves, 21st century American slaves. The Russian government loves this, making Americans a land of slaves, forcing them to destroy their own children.

Chapter 134 - To the Russian government everyone in the world is a slave except them

Under the Geneva Conventions of 1949, and other generally accepted rules of war, it is illegal for a foreign government to covertly or overtly attack civilians of it's enemies.

The USSR signed off on the Geneva Conventions, and the present Russian government is legally bound by those laws. However, the Russian government is the Russian Mafia, and mafias care nothing about laws that protect the innocent, so the Russian government has stealth attacked us for decades in ways it hides from our legitimate government. This book is intended to shed light on the many stealth ways the Russian government is destroying our nation to help us better defend ourselves.

Because the Russian government stealth attacks ordinary people in ways designed to not trigger our legitimate government's attention, and because most civilians have no training in psychological warfare, it's been a blood bath here, figuratively and literally, as the Russian government stealth attacks us, and slaughters over a hundred thousand of us annually via fentanyl and other drug poisons.

The huge upswing in mass shootings over the last decades, I believe that's the Russian Mafia. The deterioration in grades of our kids, again, the Russian government. The voters in the GOP Base in swing states who in poll after poll insist say they'll vote for Mr. Trump, that's the Russian government entrapping people and forcing them to vote for whoever they're told to vote for.

The Russian Mafia and it's government hate and loath us and is determined to destroy us, our families, and our government. It's hard for most people in America to believe that a government far away from us would care about us, much less hate us, but that's what I'm telling you and I'm not lying to you.

Chapter 135 - Russia's agenda - proxy wars in distant lands, & stealth attacks on civilians using proxies, including weaponizing parents against their children. Attacks against ordinary people in America; destroying us & our government

This book goes into detail about multiple Geneva Conventions, articles, prohibitions and other War Crimes but if you want to know some of the current plays the Russian government is working against us and how to evade them, then this chapter is for you.

I believe the Israel-Hamas War was started and is being run by the Russian government, to remove the Biden-Harris Administration. By flooding our media with babies and their families suffering daily bombings and horrific deprivation, witnessing the despair and heartbreak of Palestinian civilians who are being used as pawns by the Russian government, and the nearly constant worldwide protests demanding a cease fire, with nearly everyone blaming our President, the Russian government intends to have this war continue until after election day November 2024. After which, if they're successful, they'll have their puppet take office, whichever puppet it is, and seemingly effortlessly, end the war.

Mr. Putin inherited his access to our country and he's exploited it. There can be no mistake what the Russian government's intentions are. Most people in America have no idea we have an arch enemy who's name is the Russian government and that's not by accident. The Russian government exerts a fair bit of control over our media. Most ordinary people here will not believe that their shopping habits are known to the Russian government, but they are. To destroy a more powerful enemy you're determined to destroy, and, if you're determined to survive and to not be bombed, requires decades of stealth attacks, attention to detail, and a stealth war model. The Russian government entraps dominants like parents, landlords, teachers, older siblings, etc., and use them to attack subordinates: children of entrapped parents whether adult or underage, renters are stealth harassed by landlords, students setup by teachers, younger siblings abused by older siblings, are just a few examples of how the Russian government exploits human hierarchy to stealth attack and destabilize whoever it wants destabilized. I can't tell you how many supervisors have mistreated me under orders of the Russian government.

I believe Mr. Putin has made it his mission to destroy us in his lifetime. The Russian government's first choice, because it prefers to not have nuclear weapons dropped on it's government, is to install a puppet here. Any pro-Kremlin U.S. president is acceptable although, given Mr. Trump's publicly stated rage and intention to remake parts of the government, I believe he's the Russian government's favorite choice of the top three GOP candidates.

I published the first iteration of this book on October 5, 2023 and two days later Hamas attacked Israel. The first edition is a mess because I hadn't planned to publish it, it was the first draft, but while working on it I saw evidence the Russian Mafia intended to kill me, was trying to lure me out of my apartment to kill me. Because the Russian Mafia murders just about whoever they want, I assume they'll be successful and that I needed to get something out, so that, even though it was a mess, I hoped people might wade through it to find helpful information.

When I was typing parts of the book on September 29, 2023, Russia-controlled operatives attacked the grid in Santa Monica and my power went out. The blackout was to cover the power surge which was the actual attack aimed at me, to destroy my computer, which the operative in the apartment above mine, and who had eyes on me, saw my charger was plugged into when the person was monitoring me in the early hours of 9/29/2023. Fortunately, I hadn't plugged the laptop to the charger. I woke up during the black out, and unplugged the chargers. Because the apartment was in darkness no one, unless they were watching with night vision goggles, could see that I'd unplugged the charger. Some time later, while I slept, the actual electricity surge attack happened, which the power outage was a cover for, and negatively impacted the one circuit out of the nine circuits in the apartment I usually use to charge the computer. Had my computer been charging, it would've been toast, and the book on it.

I'm convinced the Russian government has put a hit out on me and I was determined to get something published, but within it is quite a bit of high value intel the Russian government didn't want our security agencies paying attention to, so they approved Hamas' murder of innocent Israeli's knowing Israel would attack and kill thousands of innocent Palestinians. Our government would have to respond to prevent a wider war, and it's done so. It'll take months for our government to stabilize the situation, just what the Russian government wants. The Israel-Hamas War is an attempt to unseat President Biden, and a running down the clock, putting our President's focus on the Middle East. November 2024 will be here fast and stabilizing the region will take years.

While the real threat to our people and our government is the Russian Mafia and the tens of millions of voters (and growing) I believe it controls here, voters coerced and/or influenced in a chain of illegal accesses to the voter, to get the voter to vote for whoever the Russian government wants seated. Instead of our legitimate government being able to focus on the low poll numbers, the Russian Mafia has our government focused on stabilizing the Middle East while the Russian Mafia runs down the clock attack, and the world protests. General Gerasimov said in "The Value of Science in Prediction," that the Russian government uses protesters as weapons to attack us politically.[377] People who are rightfully protesting the Israel-Hamas War would find it helpful, I believe, to understand that this war is another example of the Russian government's extreme cruelty and indifference to the suffering of ordinary people, in this case Palestinian civilians and their children, they use as pawns to seize world power. Anyone considering installing a puppet of the Russian government needs to think twice about who they're voting for.

Chapter 136 – Don't ignore the polls

Months ago I thought the polls would self-correct, now I understand that won't happen. Entrapped people are slaves and slaves can't free themselves.

I'm positive our security agencies, military and Biden Administration see it, but it'd be very helpful if Americans saw

---

377 General Valery Gerasimov, Chief of the General Staff of the Russian Federation,"The Value of Science in Prediction," *Military-Industrial Kurier* (February 27, 2013). *"The focus of applied methods of conflict has altered in the direction of the broad use of political, economic, informational, humanitarian, and other nonmilitary measures – applied in coordination with the **protest potential of the population**."*

the war for what it is: the Russian government's effort and intent to install their puppet.

To free our people, slip the noose, we offer pardons, millions of pardons, on both federal and state charges, precision pardons. Not pardons of violent criminals but targeting the type of conspiracy crimes the Russian government favors to entrap the innocent: when victims had someone they trusted advise and/or tell them what to do, including an authority figure, like a parent, a doctor, a teacher, etc.

Pardoning on this scale takes time and currently our focus is on the innocent Israeli's and innocent Palestinians being murdered, held hostage, and/or used as pawns. The Russian government is determined to dominate the world by any means necessary, as long as they're not bombed. Bomb avoidance has shaped the Russian government's strategy. It's the only thing that government fears. That doesn't mean we threaten to bomb them. It means we understand they have a strong preference that has shaped their stealth warfare model against us, over decades.

Chapter 137 - Specific illegal attacks I've endured since 11

The Russian Mafia has, through it's Russian operatives, it's American operatives, and it's non-American operatives: targeted, hacked (computer/phone/apt), harassed, tried to entrap (many times), physically assaulted (kicked, stomped on, hit with vehicles), sexually assaulted (twice), tried to kick in the face, called the n word, repeatedly sprayed with a mace-like mist on many buses to cause me to get repeat sinus infections, and/or tortured since I was 11.

My book details many of those attacks and why I and my nation are under attack. The Russian government is working a coup against us, is destroying our families, and is working to destroy our government, motivated by hatred and jealousy of us, and it's hubris.

The Russian government stealth destroys our families by ambushing and entrapping parents and forcing them to entrap their children. I speak from direct personal experience.

My father, a retired police officer and an honorably discharged veteran, told me to lie to Social Security by exaggerating my health problems after the Russian government worked me to exhaustion when I was 50. He was also used to gas light me, influence me, misdirect me, and groom me over decades. The Russian government works to destroy families in America, using family authority figures to order and/or advise at least one adult child, me, to self-harm. Had I done what I was told to do by my parents, I would've been imprisoned a long time or, if like my parents, I'd opted to take a deal to stay out of prison, I would've been enslaved by the Russian government, forced by that government to entrap other innocent people and/or I suspect, forced to vote for whoever they told me to vote for.

I know two GOP Base members who are entrapped. I know they're entrapped because they tried to entrap me. This leads me to believe that some of the millions of GOP Base are being influenced, coerced and/or blackmailed into voting for pro-Kremlin candidates under secret order of the Russian government, who control compromised American law enforcement handlers who I believe manage entrapped people here.

My mother, who raised me to be an ethicist, told me to lie to Disability, to exaggerate my health problems. She also was used to gas light me and misdirect me for years.

My sibling told me to lie to an insurance company, to tell them I have no health problems when I know I do, so that he could get a lower monthly payment for an illegally sought life insurance policy. He gas lighted me by telling me I wasn't insulin resistant when I told him I am.

To take control of voters in America, to get us to vote for whichever candidate the Russian Mafia wants seated, the Russian government destroys our families. The deliberate destruction of families by an enemy government working a covert and illegal war is genocide. What the Russian government is doing is against the law even in times of war, and is especially heinous when no war has been declared. But the Russian government cares only about seizing control of our government and our nation: destroying families in America means nothing to the Russian government.

The Russian Mafia hides behind the compromised American law enforcement it owns. No one who's tried to entrap me in my family, or those who've tried to entrap me outside of my family, have ever told me they're entrapped, forced to do what they do against me under orders. The Russian government commits the war crime of Perfidy, and multiple other war crimes, illegally using civilians as fronts when the civilians are actually enslaved operatives of the Russian government which that government uses to attack civilians. The Russian government cannot directly order Americans. To go around that, it has entrapped American law enforcement, force some of our law enforcement to run entrapment scams targeting families, among others, and force entrapped family authority figures, and entrapped siblings, to order, influence and/or browbeat an innocent family member(s) to break the law, thus enslaving and destroying families. I know entrapped grandfathers and grandmothers which the Russian government force to weaken and/or destroy their families. The deliberate destruction of families by an enemy government is genocide.

Another sibling told me to apply for Unemployment after I had applied for Disability. I'd been denied for Disability but had been told to appeal, so I did. While awaiting Disability's decision, my sibling told me to apply for Unemployment. Had I done what she told me to do I would've been charged with Unemployment fraud, Disability fraud and conspiracy to commit both. She was used to give me bad advice after she'd helped me find a law firm that agreed to handle my Disability appeal. What I've found with the Russian government is they mix it up: they'll have an operative do something nice or helpful (like bring you groceries when you have little money, or recommend a law firm to help you with a legal matter; then the Russian government will use that operative to try to get the target to self-harm, as when my sibling asked me to lie to an insurance company, or a different sibling recommended I apply for Unemployment even though she knew I'd already applied for Disability. Applying for Disability and Unemployment simultaneously means you're defrauding the government.

My mother pushed me, hard, for years, to file for personal bankruptcy, using former President Trump as an example of a successful businessman who'd filed for bankruptcy, and who was (then) president. She insulted me when I refused, told me it was stupid for me not to apply.

I have more examples of immediate family members pushing me to do illegal acts and/or gas lighting me. All of my family who tried to get me to break the law are entrapped. I know because they tried to entrap me. Because my immediate family told me to do illegal acts, with no prompting from me, I understand the Russian government controls them and has destroyed my family as it once existed. When I talk to my parents and/or my siblings, I am talking to the Russian government and the Russian government is always working to destroy me. In addition to my immediate family, my general practitioner has tried to get me to lie to Disability. He's offered multiple times to lie to Disability on my behalf, unprompted by me. It is alarming when a doctor who went to medical school offers to throw it all away by lying to Medicare for a patient he barely knows. The Russian government doesn't declare itself so my family nor my doctor can tell me why they're telling me or facilitating my breaking the law. The Russian government hides behind my parents, siblings and doctor, uses their face and my history with them, use them as camouflage to order me to self harm.

I'm positive I'm not the only person in America pushed by the Russian government to break the law but it's possible I'm one of a few who have evaded entrapment for the following: (1) unemployment fraud/conspiracy (suggested by a sibling); (2) Disability fraud/conspiracy (suggested by both parents and two siblings, on four separate occasions); (3) social security fraud/conspiracy (suggested by both parents and two siblings, on four separate occasions); (4) insurance fraud/conspiracy (suggested by my brother and my mother); (5) bank fraud, (6) tax fraud, (7) immigration fraud/conspiracy (suggested by a plains clothed police officer who approached me); (8) filing a false federal complaint/conspiracy (compromised law enforcement got my boss to be cruel to me and a woman of color co-worker to approach me and offer me retaliation by filing a false complaint); (9) drug mule/conspiracy (my aunt and sister were used in this setup); (10) shoplifting (big box cashier didn't ring up my purchases in my cart even though I had only maybe 10 or so food products; (11) Covid rental assistance (the Russian government manipulated the covid program, slow-walked my application, pushed out my completed application and prevented me from getting through to the program by manipulating my phone. When I was finally able to speak to someone, an operative told me that submitted the application she couldn't stop it. I think told her I'd simply return the funds upon receipt as I had that option. She was surprised I knew that and was part of the conspiracy, though I can't prove it. A few days later I called the program again and the woman told me she would cancel the submitted application. The Russian government controls too many of our systems, institutions and too many of our people); (12) federal wi-fi/phone low income programs/conspiracy (my sister and mother tried to get me to apply even though I told them I didn't quality for the program); (13) embezzle employer (I was given a business credit card I was put in charge of, including the statement, and asked to make purchases of gifts for business clients that appealed to me. By that time I'd been harassed in the street, stores, buses by strangers since I was around 41, so I was suspicious when I was given apparent carte blanch with a company credit card because so much weirdness was going on and no one was bothering to explain why); (14) illegal movie phone apps/conspiracy (my sister, without my permission, put two apps on my phone where I could view the latest movies for free. I removed the apps); (15) amazon fire stick illegal use/conspiracy (my sister and brother in law tried to get me to use a fire stick in an illegal manner. My brother in law later told me the fire stick could no longer be used inappropriately);

(16) medicare fraud/conspiracy (both my parents and my doctor (twice), tried to entrap me using medicare fraud, so on four separate occasions the Russian government attempted to entrap me to commit medicare fraud by exploiting authority figures in my life). Every time conspiracy is listed in the 1-16 entrapment attempts, someone, an immediate family member including a parent and/or sibling, and/or my doctor, unprompted, recommended I do an illegal action I had to decline doing. Given these types of attacks I'm positive the Russian government, using compromised law enforcement to entrap parents and other authority figures, force those authority figures (dominants) to order subordinates to break the law.

In addition to the above entrapment attempts, the Russian government used: (a) drug dealer to offer me free drugs of my choice, a man I didn't know who approached me in a parking lot near my apartment as I made my way home. I had no working battery, the Russian government had destroyed my expensive mobility device batteries so I was propelling myself forward on my mobility device using my feet; (b) used a sleep study doctor to influence me to take psychotropic drugs after I told her police were harassing me, and (c) the medical center she worked for (a world renowned medical center) rigged a sleep study I participated in to try to ensure I wouldn't be diagnosed with sleep apnea, they pumped a sleep inducing gas into the room which I couldn't smell but felt the results of and terrified me; and (d) the Russian government used an operative who accompanied my aunt during her visit. The operative not only sexually assaulted me but left a drug for me to try, which I threw away without sampling. I'm anti-drug, especially with an enemy like the Russian government who use drugs as weapons to kill.

To harass me stealthily, the Russian government has used operatives in the apartment above mine, in nearby apartments, workman working on the apartment next to mine, and more stealth and overt attacks than I can remember as I type these words. They used a college TA to accuse me of cheating on a math test and I was so traumatized I dropped out of that program. The TA told me someone else was cheating, looking over my shoulder unknown by me, but he threw me out of the test, despite my protests. I was maybe 26 and the program was challenging. The Russian government's TA's attack was the last straw. I had no idea how to appeal such a finding. I'd never been accused of cheating and was very distressed. The Russian government goes for distressing targets when at all possible.

My high school counselor tried to get me to not apply for scholarships and grants and accused me of stealing $20 from her purse to get me to stop coming to her asking for help. I've endured attacks from thousands of people controlled by the Russian government, which is why I know they're not only working a coup against us, not only working to destroy our government, which is unacceptable, but working a genocide attack against us as well, to destroy our families. Everything they're doing here is illegal. They couldn't care less.

My family are alive but they aren't my family as they once were. The Russian government has destroyed my family as a group of trusted loved ones, and reformed them into a foreign enemy government intent on destroying me. I cannot trust them and I'm positive the Russian government is the reason why. I'm also positive the Russian government is why my doctor's handlers have him offer to cheat Disability on my behalf. Behind my doctor, and behind my family are compromised people I infer are there because my family and doctor are puppets, with their puppet masters hidden from my sight. Behind the compromised American police puppet masters is the foreign enemy Russian government.

Chapter 138 - Why this type of coup? The Russian Mafia inherited it from the USSR & since the USSR got far along in destabilizing us the Russian Mafia decided to continue but ramp up the intensity. Mr. Putin intends to be the Closer

The Russian Mafia has designed a slow-moving stealth coup because we have nuclear weapons and they strongly prefer to not be bombed. If we had no nuclear weapons, we'd have long ago experienced what innocent Ukrainians are currently enduring. They found a way to weaponize our police's old entrapment mechanism that used to be aimed at criminal suspects. The Russian government commandeered it, repurposed it, entrapped some of our law enforcement, and now aims that weapon at ordinary people in America, in order to build a stealth army it is using to destroy us, our families, and our government. My book explains how the Russian government is doing this.

The Russian government intends to rule the world. To that end, it has made it it's mission to destroy Americans, all people living in America, our government, and our nation. Ordinary people, including office workers, restaurant workers, bus drivers, police officers, bank tellers, garbage collectors, postal workers, caregivers, working in the home moms, everyone who works and lives in America is a target to be destroyed by the Russian government. Ordinary people in America believe no foreign enemy would bother with them. They're wrong. I am an ordinary woman. I was an office worker. The Russian government destroyed my quality of life and negatively impacted my health, flipped me from a worker paying thousands of dollars in taxes per year to my government, financially strengthening my government, to a disabled person financially supported by my government.

Decades ago the Russian government, then the USSR, sent operatives to America to study us, our businesses and our system of governance. Since then it has since refined that intel and is using it to destroy us. This book describes some of their attack strategies. Because the Russian government deliberately destroys our families, I charge them with Genocide against the American people, War Crimes, War Crimes including Perfidy, Crimes Against Humanity, and other crimes because the

Russian government illegally targets civilians and our children.[378]

Chapter 139 - Entrapment is the Russian Mafia's preferred strategy, but will change after people are warned & believe

The Russian government is working a multi-decades long coup against Americans. A key component of the coup is to entrap by ambush ordinary Americans. It has been unsuccessful in entrapping me despite their many attempts. Because of the many ways they've tried to entrap me (using tax fraud immigration fraud, unemployment fraud, defrauding Social Security, defrauding Medicare, insurance fraud, making false statements to a federal law enforcement officer, bank fraud, drug mule, embezzling, shoplifting, defrauding a covid federal program, defrauding a federal wi-fi and telephone program for low income people, movies theft, grooming, luring, etc.), I was forced to try to learn some of their attack strategies. They threw a lot of crime at me over decades, which attacks I'll describe in the book. Because the Russian government has attacked me so many times I've come up with a theory as to how they're entrapping ordinary Americans. I know they've

---

378 "...for decades Russian operatives, including key figures of the Russian Mafia, studiously examined the weak spots in America's pay-for-play political culture – from gasoline distribution to Wall Street, from campaign finance to how the K Street lobbyists of Washington ply their trade – and, having done so, hired powerful white-shoe lawyers, lobbyists, accountants, and real estate developers by the score, in an effort to compromise America's electoral system, legal process, and financial institutions."*House of Trump House of Putin*, Craig Unger (Penguin Random House UK 2018).

entrapped many people because they force those victims to harass, attack and/or torture me.

Chapter 140 - Fentanyl & the Russian Mafia

The Russian government and the Chinese government, and their Mafia's, are after us. It's not merely to replace us, the Russian government, especially, hates us, and is the ring leader. I know we'll adapt because we'll either adapt or we'll die. I understand why we're not declaring overt war on the governments, despite the fact that they're largely responsible for killing over 100,000 of us annually. Our political system is both an advantage and a disadvantage. The disadvantage is the party who declares war will be held responsible for the war which we're not prepared, yet, to fight. And the Russian government uses it's media ops to flood our paper with images of devastated people in war so when they try their pitch with out government next month, they're hoping their psychological seeding will have taken root and that Americans will go along with whatever the Russian government has planned for us (China will make the pitch), which I can tell you straight out, anything the Russian government asks will mean enslavement for us. And if you take nothing else away from this book, please take away that the very last government we should partner with is the Russian government. It will mean enslavement and/or death for us. I vote we fight to the death rather than be enslaved by the Russian government. In the end, to be defeated, the target has to accept defeat. If we don't accept defeat they'll be hot war but at least we'll rid ourselves of the Russian government who, insofar as I can tell, hates us and has no agenda except our death or enslavement. If you trust them, you will regret it.

The Russian government and it's allies have played us for decades. The good news is that our security agencies understand. The bad news is that ordinary people don't. If ordinary people in America understood what was really going on with the Russian government and Chinese government stealth killing so many of us with fentanyl, no one would be taking fentanyl. If we flooded our people with the intel that oxycodone, and the other feel good medications are being used as weapons to kill them, used to tack fentanyl and other opioids onto, far more people would get their pleasure medications from a doctor. Assume anything you buy from a dealer has fentanyl in it. Just assume it.

Understand that we are at war. Our enemies are determined to destroy us and most of us haven't got a clue that some of our police have been compromised. We must tell our people so they won't be mowed down like so many of us have been.

Our government is in an extremely difficult predicament. We're being stealth attacked by an enemy who is determined to destroy us, and they've got aggressive allies. We've got nuclear weapons, plenty of them, but nuclear war is excessively destructive and only a last ditch-survival option when it becomes obvious the Russian Mafia is about to destroy us.

We must warn our people and the world and we must pivot to become a completely different nation, our people made aware of how we're being attacked. Embedded here, the Russian Mafia and it's American operatives and other operatives, are preventing us from adapting. Exhibit A: the House GOP. My book is intended to help us pivot.

Chapter 141 - How entrapment appears to me from the outside looking at people who are entrapped

From what I can tell, given that (1) entrapped victims aren't allowed to discuss their entrapment, and (2) I've been attacked using the strategies in this chapter but remained unentrapped, so I'm describing them from the outside looking in, and as such, I don't know all the beats of the attacks the way an entrapped person experiences them.

A basic entrapment the Russian government uses is to have it's compromised American law enforcement target an ordinary person, a noncriminal, and sets them up to cheat for gain, specifically, to cheat the American government for gain.

The information I relay to you in this chapter is based on my first hand experiences being stealth and overtly attacked by Russian government controlled operatives, who used one of my parents as a weapon against me, and an HR employee at my former job.

I theorize the Russian government sets up innocent people in our country and has our legitimate government be the target because at some level, compromised American law enforcement must show the names of entrapped ordinary

Americans to our legitimate government. It is much easier, is my theory, for the Russian government's American operatives to get our legitimate government to sign off on an entrapment, when the Russian government's American operatives demonstrate a person tried to cheat our government. Another easy sign off, I believe, is when the Russian government's American operatives can show the accused person is in the drug trade. And yet another easy sign off, based on how hard the Russian government's operatives tried to push me to file for personal bankruptcy, is for an ordinary American to file for personal bankruptcy. Given that the Russian government worked for years to get me to file for personal bankruptcy. I believe personal bankruptcy filing is another way the Russian government lies to our legitimate government, by convincing them that people who file for personal bankruptcy are scammers. When the Russian government found they couldn't corrupt me, they pushed me hard to file for personal bankruptcy, which suggests the Russian government can convince our legitimate government that people who file for personal bankruptcy are cheating the system.

My entrapped mother's Russian government controlled American handlers forced her to pressure me to file for personal bankruptcy; they also forced an entrapped HR employee at my last job to advise me to file for personal bankruptcy, unprompted by me. The Russian government also used my extended family to tell me they were in great need, lacking food and an inability to pay their rent, while the

Russian government had it's operatives flood my mail box with offers of credit. That's how the Russian government manipulates our people, government, destroys our families, and destroys our nation.

The key for the Russian government's stealth coup and building it's secret army of entrapped Americans, including grooming and forcing our children under 15 to attack targets the Russian government wants attacked, is to get our legitimate government to believe that entrapped ordinary Americans are liars and thieves, when in reality they are innocent victims the Russian government is setting us up while it lies to both ordinary Americans and to our government, while they work a coup to destroy us, our families, our government and our nation.

After the Russian government's compromised American law enforcement show our legitimate government that an ordinary American was entrapped for good cause (which is nearly always going to be a lie when the Russian government is involved), the Russian government can then do whatever they want with the entrapped American, including killing them with fentanyl or by any other stealth means, including forcing and/or influencing them to vote for whichever candidate the Russian government wants to install here.

Many ordinary Americans are entrapped, that's how the Russian government got a hold of them, but our legitimate government has no idea that the people they've been told are criminals and thieves are victims of the Russian government.

What Russia controlled American operatives aren't telling our legitimate government is that they're entrapping, setting up, influencing and targeting ordinary Americans, including our children, in ways that are illegal, unethical, and immoral, and are part of the Russian government's coup against us.

In working their coup against us, using the weapon of family destruction in their undeclared war against the American people, the Russian government is committing genocide against us by destroying our family structures. When they force our parents to entrap their own children, their own families, that's committing genocide against a people. People in ripped apart families are then forced into the Russian government's underground army, where they destabilize and attack whoever the Russian government aims them at. I've personally witnessed individuals in the underground army and it includes American children under the age of 15. Some of those children have been used to harass and intimidate me under orders of entrapped parents, entrapped adult minders of children on student field strips, and, I believe, under order of compromised American law enforcement controlled by the Russian government.

Chapter 142 - How the Russian Mafia flips victims into the GOP Base

I doubt entrapped voters are told to vote for a pro-Kremlin candidate by their police officer and/or retired officer handlers but honestly I don't know how victims are flipped. Instead, the Russian government double and triple teams people. So influencers get victims to transition to the GOP Base, not law enforcement. That prevents some entrapped law enforcement from suspecting the Russian government has entrapped them and it helps prevent ambushed noncriminals who've been entrapped from making the connection between being entrapped, being in the GOP Base, and the convenient coincidence the Russian government just so happens to prefer the candidate influencers persuade voters to choose.

Chapter 143 - Entrapped people are no longer your friends & loved ones but are slaves of the Russian Mafia

No matter how much you love them, entrapped people are slaves. Help them and yourself by reading this book and helping our government get our people free of the Russian Mafia.

No one should be asking you to lie, steal and/or cheat for them and if they are, something is wrong.

People who have a job but who sneak into a theater and tell you about it are on a fishing exhibition by someone controlling them.

Chapter 144 - The battlefield is the mind

The victim must be convinced they really are corrupt by the entrapping officers. The victim must believe they weren't lead and entrapped. To that end, when arrested I suspect entrapped police go step-by-step through the crime with the victims to convince the victim that they were entrapped fair and square. This is essential for the entrapment to work. People who believe they've been set up will be eternally seeking exoneration, which the Russian government cannot afford.

Because victims felt greed and reached for the cash or whatever the Russian government dangles in front of them, they believe that means they're corrupt when what it really means is they've been lied to by world class liars.

When a person is entrapped by their lover, authority figure and/or an operative of the Russian government, it means the victim was targeted, and is innocent. The Russian government won't tell the victim that though.

Chapter 145 - Tips to stay alive

Chapter 145.0 - The Russian Mafia prefers to attack victims when the target is in a subordinate position. I've been attacked by my parents, some of my doctors, cashiers, my landlord, various supervisors & coworkers. I've been manipulated by bus drivers, etc. In these attacks I was subordinate to the attackers, a psychological attack.

Chapter 145.01 – The Russian Mafia exploits the doctor-patient relationship to destroy and destabilize us – the Russian Mafia double, triples & quadruple teams targets, using a variety of compromised slaves to position the victim into doing what the Russian Mafia wants them to do

In a doctor-patient relationship where you're the patient, you're in a subordinate position. The doctor is the dominant in most doctor-patient relationships and from birth we're taught that doctors, who we meet a birth, are one of our trusted circle, to be obeyed. We're taught to ignore pain and allow our doctors to give us shots, from infancy. Parents, grandparents, and doctors are among the first group of dominants a baby knows.

Your doctor can be forced by the Russian Mafia to offer to lie to a government agency or other entity for your benefit. I want to take a weight loss drug but being insulin-resistant, not diabetic, I don't quality under my insurance. The doctor, unprompted, offered to lie to my insurer as recently as the summer of 2023 to help me out. Before I went to see the general practitioner, an ER doctor advised me to see a general practitioner after I'd experienced a medical emergency, and one of the ER nurses at a hospital I'd requested not to go to but was taken to because the paramedic said the closest hospital's ER I'd requested was closed to more ambulance arrivals that day, that ER nurse advised me, unprompted, to ask the general practitioner if I qualified for one of the weight loss drugs. The Russian Mafia owns a wide range of people here. In the summer 2023 entrapment setup, those owned included (1) the operative who noise harassed me to raise my heart rate to 200bpm which necessitated a trip to the ER; (2) one of the paramedics who repeatedly said he'd seen multiple people suddenly experiencing medical problems, to draw attention from any of the other paramedics thinking it strange that my heart rate would suddenly accelerate to 200 bpm; (3) the general practitioner who offered to lie for me; and (4) the nurse who recommended I ask the general practitioner for one of the weight loss drugs. I mention these four players because they're all compromised and were activated by the Russian Mafia, but likely believe lead to believe their legitimate government is their handler. It's important for people to know that the Russian Mafia double, triple and quadruple-teams victims, including children. I know because I was 14-15 when they sicced three teenage girls I didn't know to beat me up. When those girls failed to beat me up, the USSR had my mother, in a bizarre twist, beat me. The four operatives in my 2023 medical emergency were coordinated to get me to the doctors and ask him about weight loss shot. Additionally, that same doctor was used to push me to see physical therapists affiliated with the ER hospital I was taken to in 2023. When the doctor who attends you offers to lie for you, don't take any of his subsequent recommendations in terms of non-urgent medical assistance, like physical therapists. Given that he's compromised, the Russian Mafia will continue to exploit his dominant position to push you to more of their compromised slaves, all without warning you.

Chapter 145.02 – How the Russian Mafia weaponizes newspaper accounts of a drug, and the operatives in a target's life, to get them to take injectibles – why injectibles are an attractive way to murder targets the Russian Mafia wants murdered

Currently, the Russian Mafia is stealth working to get me to use one of the weight loss drugs. I suspect they will then murder me by installing something else instead of the weight loss drug. The Russian Mafia knows I read the New York Times and knows I want to lose weight, so they ensure there are plenty of articles about the drug in the NYTs as well as ensuring operatives communicate to me their satisfaction with the medication – this weight loss drug isn't the first injectible medication the Russian Mafia has attempted to stealth assassinate me with. The Russian Mafia is hot to kill me. Whatever works that gives them deniability. I know the Russian Mafia can access prescriptions before they're collected because when I was taking a sleep aid, half the prescription was stolen before I picked it up. When I noticed and tried to get the other half, I had no proof and had half my prescription stolen with no way to prove to the pharmacy that I'd been victimized.

Chapter 145.03 – There are legitimate entrapments by our legitimate government but they're very different from Russian Mafia orchestrated stealth entrapments: here are some of the differences

When our legitimate government has worked an entrapment and decides to arrest you, they'll arrest you before the world. If you're at home they'll arrest you there, if you're at work, they'll arrest you there, if you're at church, at school, at the grocery store, wherever they believe it's best to arrest you, they'll do so and couldn't care less about the witnesses as long as the witnesses don't interfere with the arrest (although they may avoid alarming young children by avoiding arresting their parents in front of them).

On the other hand, the Russian Mafia entraps people out of view of the target's family, employer and friends because the Russian Mafia intends to use that victim to destroy their family, their co-workers and their friends. The Russian Mafia believes in ambush. Family, friends and co-workers might be wary of someone who's been arrested.

Our legitimate government won't tie you to any plea deal that forces you to take your children on harassment attacks against noncriminal civilians.

Our legitimate government won't demand you help all law enforcement in any manner they request into perpetuity. From what I can tell, my mother was destabilized by the USSR when she was 29. This past summer 2023, in her 80s, she was forced by the Russian Mafia to try to get information from me the Russian Mafia sought in a draft of another setup they hoped to launch, one featuring palm readers. So they forced my mother to tell me her palm reader story, where I learned the USSR had weaponized a palm reader they owned in Southern California, and a supposed "friend" of my mother they also owned, plus the male operative conspire to destabilize my family. As I've said, the Russian Mafia triple and quadruple teams and the KGB has been working that way against us since at least 1971, probably earlier. That's how the USSR got one of their operatives into my mother's bed, and the Russian Mafia wanted to see if I'd go for that same attack so they used my mother as a weapon because she has access to me, probably the only reason they've allowed her to live this long. They're hot to destroy me. If they kill her they'd cause me immeasurable pain, which they'd love, but they'd burn that access to me. The Russian Mafia values maintaining access to targets, to fight the war another day.

Our legitimate government won't insist you entrap your own children, parents, or extended family. The Russian Mafia insists that you do. They won't tell you that though. They'll groom and criminalize you, get you into such a position of dependence, psychological destabilization, physical pain, and blind obedience, victims do whatever they're told. I've seen the Russian Mafia put a woman through six or seven extremely painful knee surgeries in four years, until all she knew was pain, for years. Now she does what she's told. This type of attack is available to the Russian Mafia because they've seized control of our medical care system, and it's extremely easy to quadruple team people, convince them to choose surgery because our people don't know we're under attack and that the Russian Mafia uses compromised medical staff to stealth/proxy destabilize us. This book is meant to at least sound the alarm that ordinary people are under attack by a best in class, world class mafia determined to destroy us. Our security agencies aren't in ordinary people's lives and ordinary people haven't been trained in espionage so the Russian Mafia has had a field day for decades here, and before them, the USSR.

Americans need to know that we're not somehow deteriorating, we're being attacked by a foreign enemy government who's had access to us and our nation for decades. And our people need to know who's doing the pushing, and why.

Chapter 145.04 – The Russian Mafia exploits the cashier-customer relationship to get victims charged with shoplifting, to place targeted individuals into the criminal justice system by fair means or foul. Notice the Russian Mafia exploits any relationship where you're the subordinate

In a cashier-customer exchange where you're the customer, you're in a subordinate position. The cashier is the dominant in most cashier-customer encounters and in any dispute, the store's management, store security, and police will automatically believe them in a suspected shoplifting.

Customers automatically assume that when we wheel our carts to cashiers to scan, they do so, then charge us. But the Russian Mafia entraps cashiers and deploys them in shoplifting entrapment attacks. I barely evaded the first such attack, I think it was sprung on me in late 2021, and I endured several more after that within the next four months or so.

In the first one the cashier scanned maybe 6 of my 10 items. I wasn't paying attention, hadn't figured up the cost fo the items in my head as I usually do, because my purchases were modest, peanut butter, peanuts I think, and several other low cost staples. I'd gone to the store to collect my prescription and hungry, decided to buy a few items. I didn't have much money so I bought affordably priced basics. The Russian Mafia hadn't used this specific type of attack against me previously so I had no way to anticipate it. The cashier scanned my items, I hadn't added them up, I paid him, having no idea I had just been attacked. Exiting the store having shown the exit employee my receipt I glanced at my receipt, stopped moving on my mobility device, glanced at my shopping cart, down at my receipt, saw that multiple items hadn't been scanned, turned immediately against the flow of traffic, re-entered the store, approached the cashier, showed him the discrepancy. Flustered, he asked me if the exit employee had told me. No, I said, I noticed and returned to pay. He was so flustered, he scanned I think two more items but still left one out. I pointed it out to him. He scanned it, I paid and exited the store.

For many years that store's security had harassed me in innumerable ways but sadly for me, 99% stores harass me after the Russian Mafia put me on an "please harass her" list. I live in an apartment, don't have the ability to grow my own food, am dependent on grocery stores for essential foods, stores the Russian Mafia controls.

Had I not noticed the omissions by the cashier, store security would've approached me and asked for my receipt, which I would've surrendered. Then they would've pointed out the discrepancy between the items in my cart that hadn't been scanned. I would've been surprised, told them I'd just gone through the line and had paid the cashier. They would've charged me with shoplifting because this was a shoplifting entrapment. I would've protested, the police, already there in plain clothes (I saw later as I ate a bun-less hot dog), would've arrested me and charged me with resisting arrest. They would've slammed me to the ground by yanking me off my mobility device. I'd have been sat on by four or five men insisting I'd resisted arrest. (Protesting your innocence is considered resisting arrest in the law.) In the melee the Russian Mafia would've made sure I was killed, or be made a vegetable. There'd be protests in front of the store repudiating American police brutality, when the attack was actually orchestrated by the Russian Mafia (who is responsible for our law enforcement's excessive use of force) as a stealth assassination against a target the Russian Mafia had tried and failed to entrap for decades. It would clearly be shown by store video that I'd gone through the line and paid what was asked and left the store. It would be established that I'd not shoplifted. My family would receive a settlement, probably from the store and the cities of the employee police in Southern California. I wouldn't be alive to write this book. I believe this attack happened in the end of 2021, after I'd established my company RussnCoupAttpt Publishing Co.

How I unknowingly evaded that entrapment: (1) I was fortunate enough to glance down at the receipt a few feet after exiting the store and noticed the discrepancy; and (2) I immediately returned to the cashier and paid what I owed.[379] It was not until my second exit, when I saw so many plain clothed police staring at me, that I understood I'd just evaded another entrapment. I'm 64 in March 2024 so I estimate this entrapment attack occurred near the end of covid, rough estimate October to December 2021.

----

379 That day, while collecting my prescription(s), I'd reported to the store manager that during my last visit I'd received more for a return than I should have and I insisted on repaying the store – it wasn't that much, I don't remember the amount. That store has a history of harassing me in all manner of ways so I made sure any interaction I had with them was as accurate as possible. The manager accepted my repayment. As I left the store the first time, I noticed the manager staring at me. I nodded at him a goodbye. Now I believe he'd been given a heads up that something was about to go down on me. This is just me theorizing but I remember he was staring at me as I left the store and he didn't appear to be happy that I'd insisted on repaying the store for the overpay I'd received.

Chapter 145.05 – The Russian Mafia exploits the bus driver-passenger relationship to obtain destination information from targets which allows them to attack the target at every point in their journey

In the bus driver-passenger relationship where you're the passenger, the bus driver is the dominate and you're in a subordinate position. We're trained from birth to accede to the requests/demands of dominants when we're in a subordinate position in relation to them. The Russian Mafia see's people as weapons and targets, as dominant and subordinate relationships. When you want something from a target, put them into a subordinate relationship with a compromised dominant or a dominant who's under orders by management to behave a certain way. Bus drivers needn't be entrapped by the Russian Mafia to gather intel in this way, they need only be employees and their bus company began a new policy the driver is tasked to follow. The Russian Mafia exploits so many levels, to get the Los Angeles Bus line I usually take and the Santa Monica bus service the Russian Mafia had people with authority in those bus agencies initiate this policy. Even after I leave this area, die, the Russian Mafia will keep this policy in place for future targets they're hunting who use mobility devices.

The Russian Mafia had Santa Monica and a bus line I take to doctors through Los Angeles change their policy to have bus drivers ask me my stop. The intel is gathered only on people using mobility devices and was targeted specifically on me as the Russian Mafia's American operatives sought to attack me everywhere I go. Now when I enter the bus most times the driver asks me my stop. It took me a while to learn to say, I'll just ring the bell. Bus drivers will accept that but they don't tell you that.

Chapter 145.1 - Avoid Russia-controlled sex partners

You probably won't be able to spot them. In general, they'll probably be into you unnaturally fast, but the Russian Mafia's operatives read too, so after they read this book they may change their modis operandi. Just overall be careful. We live in a world where the Russian Mafia intends to slave you. Don't make it easy for them.

I'm not entrapped but my family is. My theories on how it works are based on what I've observed from them and what I've observed of other entrapped operatives who've attacked me. Multiple family members seem to have had an operative placed in their bed. Everyone who has had an operative placed in their bed is entrapped, so that's a highly effective strategy for the Russian government. Be careful who you let into your life and your bed.

Some of my relatives I've never heard were arrested or got into legal trouble but I know they're entrapped (because they were forced to try to entrap me), so if you're surprise-arrested and the arrest happens outside of the purview of your family, friends or coworkers, and you're later offered a deal with no jail or police time, you may be being set up by the Russian government.

The Russian Mafia entraps noncriminals by grooming them, luring them, or just plain setting them up. **<u>Don't accept anything you haven't paid for</u>** – <u>that is the beginning of their grooming/luring you</u>. The Russian Mafia had cashiers at four different stores over several months not ring up all the items in my shopping cart, etc. Check your receipt to ensure the cashier scanned all your groceries. ***Don't think you've been given a gift from god if you receive something you value, including words***. Russia-controlled American operatives were so pressured by the Russian Mafia to entrap me, especially right before President Biden took office, they began having cashiers not scan all my purchases. Had I not noticed, they would've finally got me into the criminal justice system, where the Russian Mafia rules.

I believe uniformed police must be arresting targets and taking them to processing buildings or jail to make sure entrapped people believe our legitimate government is arresting them in a sting. Innocent people are being targeted so this is a set up but the Russian government's mafia does their homework. They'll know if you need to be out of jail to pay your rent, pick up your kid, they'll know the concerns on your mind and they will exploit your fears and worries. That's one of the reasons the Russian government targets families: most parents will do anything to avoid sending their kids sent to foster care, avoid yanking them out of school, avoid pulling them out of their home.

Because of their love for their underage children, parents are easy to entrap and I theorize will take a deal because they refuse to distress their young children to that extent. A great many families live paycheck to paycheck and few parents will willingly send their child to foster care, have them lose all their belongings because the parent is stuck in jail and can't get out to oversee packing, the logistics of putting their children somewhere safe, with little money: many parents take the deal. They aren't told when they take the deal that they've just begun the process of entrapping their own children.

When the Russian government entraps one parent, they get the whole family. It takes a certain kind of evil to groom a parent to entrap their own child: that's the evil we're dealing with, and most of us don't know. Please believe me. I would never make this up. If there's something I'm unsure about I'll say I'm theorizing, extrapolating or believe. If I'm not using those words, it's because I've seen enough evidence on the ground to reach these conclusions.

Some of our police are entrapped. I don't know what those police say to innocent people they've set up. I doubt it occurs to victims to question they've been entrapped by their government. When you're arrested by a uniformed police officer you automatically believe they represent the government. No one, I believe, would think that their sting was part of a coup by the Russian government.

From my decades of experience being attacked by Russia-controlled American operatives, and evading entrapment, they spent many years working to entrap me and failed. Russia-controlled American operatives harassed me repeatedly, doing the same type of attack, year after year. The police were doing these attacks, I knew they were extrajudicial attacks but how to complain. I complained to the FBI, to the last three presidents (I wrote Mr. Trump early in his presidency before I understood his situation), before his meeting in Finland with Mr. Putin. When I wrote presidents, complaining about the attacks, local police stepped up their over harassment against me. Uniformed police did all kinds of stealth harassment against me, the operative(s) upstairs, operatives who were workmen, it was really bad. Still, I kept trying.

They used my parents to tell me to break the law, and I refused. When first my father, and later my mother, told me to lie to the government, I had no idea that they were entrapped and being forced to harm me. I refused, quietly, their directive because what they were saying was opposite to who I am. I couldn't conceive of doing what they told me to do. My mother shocked me because she'd raised me the opposite of what she was saying. My father surprised me because I had no idea he felt the way he did, about our government. When people are entrapped, they lie. Their loved ones must understand entrapped people cannot be trusted because the Russian government can't be trusted. Russia feed victims a script. Victims are slaves and must do what they are told. They do not have the option to protest. As best I can tell, if they don't do as they're told they face immediate imprisonment. I suspect, but cannot prove, the Russian government controls at least some of our prisons. Entrapped people sent to prison for breaking their deal are unlikely to survive.

The Russian government must be laughing their heads off at the number of families in America they own. They probably will try to manipulate President Biden's Administration but our President, his Administration, our security agencies and our military know what the Russian government is. I pray they don't take a deal.

We are at war, undeclared war. The Russian government is illegally stealth attacking us, deliberately targeting civilians during peace time. What they are doing is illegal even in war. So they care about nothing but destroying us. And world power.

Chapter 145.2 - Don't be a Drug Mule

The Russian government tried to get me to become a drug mule. My family is from back east and at that time my grandparents were alive and I visited them every few years. On one trip, I was in my 30s but I don't remember my age, the Russian government wanted and tried to set me up as a drug mule. They had my sibling plead with me to visit her father, who'd been my step father when I was a toddler. I had no childhood memory of him but my sister's handler's pushed her hard to beg me to see him. Not knowing he was a drug dealer, I agreed.

After I'd agreed, my aunt I think it was, who I now understand was entrapped, told me the horrible news that the father of my half siblings was a drug dealer. I couldn't image much worse than learning your father dealt drugs. She adored her father, both my half siblings let me know they deeply loved their father. Because him being a drug dealer was a rumor, something someone told me but which I myself didn't know, I decided not to break my sibling's hearts by telling them such an awful rumor. It was all a setup I later learned, I was actually the only one who didn't know he was a drug dealer. Certainly my sister knew she'd sent me to a drug dealer's home, putting me in danger if he was raided during my visit, if there was a shootout, etc. She is entrapped and entrapped people are slaves. I hope my government investigates while my parents are alive so we can free everyone who lives in America enslaved by the Russian government, or by anyone else.

He'd been my mother's second husband when she was in her early 20s and I was a toddler. I barely knew him, and wasn't into drugs, so there was no reason I'd know he was a drug dealer when I went back east to visit my extended family. Russia-controlled American operatives needed me to know he was a drug dealer because of their entrapment scheme, aka, their attempt to make me his drug mule.

I flew back east and visited my loved ones. Because this was a set up by Russia-controlled American operatives they monitored me, which I didn't know at the time, and knew

I'd asked no one to take me to visit him. I couldn't be entrapped as a drug mule if I didn't meet the drug dealer.

I think it was the last full day of my visit, maybe the last night. My flight out was the next day and I hadn't mentioned the drug dealer. Russia-controlled American operatives had my sister call and plead with me to visit her father, told me he missed me. That was so odd that I remember it. I'm not his child, there's no way he could miss me but my sister was forced to say whatever worked to get me to his house. I agreed to go. I couldn't think of a way to tell her the rumor without devastating her and as you'll notice, it's a reoccurring theme in this book how the Russian government use my loved ones as camouflage to lure me to locations or try to get me to do something illegal they as the Russian government would've been unable to get me to do.

It's important to understand how the Russian government thinks: they have zero problem harming you if it benefits them. They approach the world that way. In Syria, in Ukraine, in America, in the U.K., in Central and South America, in Gaza, in Israel, etc. They demand dominance over our species but they think of people only as pawns, as weapons, they can climb over to get where the Russian government wants to go. They respect no one's laws or rules. They're running an illegal war against ordinary people in America and couldn't care less that they're breaking our laws, the Geneva conventions, that they're destroying our families and committing genocide against us. They simply don't care about anyone but themselves and their ambition. They see themselves as the chosen ones and everyone else as slaves. They've got nothing to offer their species or any other species, but death, despair and slavery, and they're determined to make our species accept their dominance.

Russia-controlled American operatives got me to a drug dealer's home, a place I'd never ordinarily go. They'd enslaved my sister, which I didn't know, and used her as a weapon against me to get me to a drug dealer's home.

I asked my aunt to take me to visit him on the way to dropping me off at the airport the next day and she agreed. We visited him maybe a few hours, then we left. On the way to the airport I said little about him, expressed no interest in the drug trade in response to my aunt's few comments referencing him. Now I know the Russian government's American operatives were listening to the conversation in my aunt's car, but at the time I didn't know that. I had no interest in drug dealers and I didn't know him so I had no reason to sustain a conversation about him with my aunt. I worked, sometimes multiple jobs. If I wanted a product I worked for it. I'd never consider harming innocent people by selling them drugs. I've read the way some people become dealers is they were initially users, and became addicted. When they could no longer afford to pay for drugs, their handler's repurposed them into dealers. From what I can tell based on casual observation of people who use it, it appears addictive, a person needing to smoke every approximately 20 minutes or so. I'm concerned weed is augmented with addictive substances to get people hooked. Another of my aunts, who visited me with her friend, who is owned by Russia-controlled American operatives and who sexually assaulted me during their visit, left a supposed weed joint at my apartment after they left, no doubt the Russian government hoping I'd be curious enough to try it. I think I was 60 or so. I threw it away. Anything the Russian government gifts a target is poison.

My favorite aunt, who I didn't know was entrapped, an opioid addict, and controlled by Russia-controlled American operatives, became inexplicably annoyed with me out of the blue as she drove me to the airport. I theorize now that, given I expressed no interest in the drug dealer, and since my aunt her handlers wanted her to facilitate my becoming a drug mule, she became testy with me, anticipating that her handlers would blame her, and/or they'd told her to be mean to me.

Chapter 145.3 - Don't let anyone get you to do something illegal; the grooming will start small. Once you do one illegal thing, my suspicion is that person will harass/tease you into repeating the act – the Russian Mafia forced my parents to sneer at me to force me into behavior the Russian Mafia wanted me to do but which I wouldn't

One way to stay alive is to understand and believe the Russian government is running a multi-decades long coup here, and that you, an ordinary person, are known to them and are targeted. Most people find it hard to believe the Russian government knows anything about them but that government has been here for decades, embedded hundreds of thousands of slaves, ops, victims and/or criminals beginning in the 1970s. Some of those people entrapped others and now, I believe there are millions of entrapped slaves controlled by the Russian gov, living not only in America, but worldwide. Almost no entrapped person living outside Russia know they're owned by the Russian gov. That government uses local police and the nation's government and has compromised, entrapped slaves believing they've been entrapped by their own government. My guess is 99.9% of the people the Russian government owns worldwide have no idea the Russian government controls them.

Chapter 145.4 - Pay attention if someone appears well groomed or poorly groomed for their income

Tips someone may be an operative are if they're too well dressed or too poorly dressed. Too well-groomed for their salary, in other words, they have disposable income to spend on high end grooming products, which are crazy expensive. They buy very nice quality clothes that quietly say "quality," and drive a luxury car, but more telling, they buy and use high-end skin and hair products so their overall appearance says "wealth," while their job title says "middle class." Some people have sources of income other than their job so you can't assume someone who works in intelligence and looks posh has flipped but I noticed a former security agency official who was illegally assisting our enemy and he had a polished appearance. His clothes looked expensive to me.

Operatives the Russian government have used to harass me many times have appeared to be homeless. You'd be surprised how often the Russian government uses "homeless" operatives. Housed people don't look at homeless so homeless operatives tend to be ignored.

Chapter 145.5 - Be kind to yourself & take care of yourself

It will help you survive if you love yourself, if you're kind and psychologically and emotionally generous to yourself, if you don't talk to yourself sarcastically. You can tease yourself but not be cruel. Operatives have special training, we don't. If you can be kind and polite to others, you can be kind and polite to yourself. You're worthy of politeness, and gentleness.

Chapter 145.6 - Don't ignore odd behavior; don't ignore when people lie or cheat in small ways – the Russian Mafia puts out feelers using slaves they own; their slaves are your family, friends, people who have access to you. The Russian Mafia uses them as camouflage. I'm sorry. That's the truth

Chapters 145.7 – 145.9 - Your people are no longer free to warn you. President Biden isn't free to fully explain his backing of the Israeli government because confidentiality clauses in our alliance agreement prevent government officials from discussing key aspects of that agreement – which the Russian Mafia is exploiting to get Americans to not vote for him. President Biden is hoping to help Palestinians in a more long term way by helping them build their nation. Irregardless, the Russian Mafia has created this situation. Stealthily. Mr. Biden is U.S. President & the Russian Mafia is manipulating him & us. The Russian Mafia are world master of psychological warfare. It doesn't matter how smart you are (Mr. Biden is smart), how powerful you are (Mr. Biden is powerful), the Russian Mafia works to destroy people they want destroyed. That includes you, me & Mr. Biden. We won't survive unless victims share what they've experienced to try to help those being targeted.

Chapter 145.10 - Russian Mafia use some of our American law enforcement to entrap noncriminals. Your entrapped friends have no idea the American officers who entrapped them are controlled by the Russian Mafia

The Russian government is here presently, entrapping people. That'll change but as far as I can tell, they're still entrapping people to build their voting base of compromised voters. People with access to you will be used by their handlers to see how you feel about lying, cheating, drugging, in minor ways because currently the Russian government is entrapping people here.

When someone introduces the subject of cheating on taxes, or someone they know who cheat on their taxes, or sneaking into a movie theater, or putting an illegal movie app on your phone, which you didn't ask for, that's not normal. Most people who do those things won't discuss them, even with friends and/or loved ones.

The Russian government usually won't have ordinary people suggesting people in their circle rob a bank, but they need to prod for ease of corruption, so it'll be something small, a small cheat. Don't ignore it. If you don't want to call them on it, that's fine but understand, it's not normal for people to sneak into a movie theater when they have a job, it's not normal to put free movie apps of the latest movies on your phone when you've not asked for it. It's not normal for someone you know to try to convince you to apply for a federal program after you tell them you don't qualify for it.

If you have friends or loved ones, they're likely telling you they have a problem in the only way they can, in the only way the Russian government will allow them to tell you. The Russian government is running scams here; they're hunting Americans and migrants living in America. The Russian government can't come right out and tell you they're here, so they use people with access to you as a weapon, force them to suggest illegal acts to you; use your loved ones as "hides."

When someone you know does that, it's not normal, and they're doing it for a reason. Usually, they're saying it because they've been compromised and are being forced to feel you out about criminality.

The Russian Mafia entrapped those people using entrapped American law enforcement, so, to your loved ones, they're being forced to entrap you under order of American law enforcement, so they've probably been led to believe our legitimate government is running these scams: that's a lie.

Although I can definitely be wrong, it's harder for me to believe my loved ones and American law enforcement knowingly entrap me so the Russian Mafia can blackmail me. Knowing the Russian government as I've been forced to understand them, it's far more likely the Russian government is running an entrapment scam than that my legitimate government is running one.

Expert interrogators questioning entrapped law enforcement using lie detector equipment would have more tools and expertise than I do to learn what entrapped law enforcement know and/or don't know. The Russian Mafia, I suspect, stole a copy of the first edition of my book when I was trying to figure out how to email it, and has definitely read the previous drafts of this book, and already have their strategy on how they'll move forward if the book ever comes up with American law enforcement interviewing Russian Mafia, which is unlikely to happen.

Russian Mafia spies who've sought asylum here will likely say they knew nothing about my allegations in this book. And even though I've been attacked, harassed and/or tortured by operatives of the Russian Mafia for decades, I've no idea who gave those orders to have me attacked. I know where I worked and who my supervisors and co-workers were. I know the name of the operatives in the apartment above mine. I know the names of the people in my family who've tried to entrap me. I ask my legitimate government to question me, and them.

But my security agencies will have to use their expertise to find out more. In this book I am as specific as possible, in case the Russian Mafia murders me before our security agencies question me and, needing to rely on the book, they have some kind of starting point.

If the Russian Mafia continue working these attacks the way they've worked them against me since they made me ill when I was 50, initially they pushed my parents and siblings to entrap me with Disability fraud, Unemployment fraud, insurance fraud, conspiracy to commit the frauds, and they used my mother to push me hard to file for personal bankruptcy.

Then, they had operatives I didn't know offer me drugs, and alcohol; and they went hard denying me quality medical care, had doctors they control lure me into committing Disability fraud. Before that they had a doctor encourage me to get on psychotropic drugs after I told her police were harassing me. When I tried to talk to my doctors, those doctors tended to drop me and the Russian Mafia, after they gave me the smart phone, stopped my number from being able to reach my medical care providers on multiple occasions, which I didn't understand at the time the problem was caused by the Russian Mafia.[380]

It's impossible for me to know how the Russian Mafia will adjust to the information in this book being available in the world. It's my understanding that a million books per year are published so it's easy to have even important books get lost in the pile. The Russian Mafia will likely monitor if this book is noticed and then decide how to adjust their strategies.

Chapter 145.11 – Behaviors I noticed compromised people displayed

Below is what I noticed about loved ones who are entrapped by the Russian Mafia and who have been used by the Mafia to fish to find out if I'd be vulnerable to entrapment. The Russian Mafia used one of my most beloved family members to try to entrap me using the following:

1) If your loved one has a job but sneaks into movies for free, and tells you they do so;[381]

2) If your loved one (and/or their spouse), make a point of discussing personal taxes with you and let you know that someone in your extended family cheats on their taxes. When

---

380 I didn't know it at the time, I just knew I had to call the hospital and ask the switchboard to transfer me to one of my doctors.

381 Telling you is necessary because that's how their handlers (who are listening) ascertain whether you're interested in cheating or lying to obtain something of value.

you say you're going to speak to the tax cheat to get them to stop, your loved ones ask you not to because, they were speaking out of turn and had helped the tax cheat in the past with taxes, and shouldn't have mentioned them because they'd helped the tax cheat prepare taxes. When you ask your loved one to talk to the tax cheat to warn them, the person says they'll warn the cheat, but because you're not allowed to speak to the tax cheat because you've been told all this in confidence, there's no way to know if the cheat has been warned or not;

3) Your loved one, without your permission, put free movie apps on your smart phone,[382] on two separate occasions. These movie apps display movies currently in theaters, and you have access to them for free. When you ask them about how the apps are paid for your loved ones tell you they're not paying for them. You remove the apps;[383]

---

382 This is your first smart phone, purchased for you by your loved ones a few years earlier, and you know little about the technology which became widely available after you became disabled, and which you initially know little about beyond how to make phone calls, watch Youtube videos, and send Whatsapp and email messages.

383 Free movie apps are the thin edge of the wedge. If you accept anything of value for free that you know should be costing you money, the loved one will then browbeat you into accepting items of ever more increasing value until their handler has entrapped you. I don't know how I know. I just know. This particular loved one, in my early 50s before I learned my loved ones were entrapped, recommended a law firm that helps disabled people appeal their denials from Social Security Disability. After the law firm agreed to accept my case, I called this loved one and thanked her. She was then promptly weaponized by her handler to tell me to also apply for Unemployment, even though she knew I had applied for Disability. I knew nothing about Disability or Unemployment so I didn't challenge her suggestion but I thought that couldn't be right: applying for Disability, where you're saying to the government you're unable to work, while simultaneously saying to the State, that you're available for work. I thought to myself, "that can't be right." I didn't challenge my loved one, I simply thought through the situation and reached my own

Pay attention to yourself, to your loved ones and your friends. If your friends or loved ones will sneak into a movie even though they can afford to pay, notice. Let them know you're not comfortable with that. People who feel it's okay to lie or cheat in small ways are more likely to lie or cheat in big ways. Don't trust anyone blindly, no matter how much you love them.

The Russian government is deeply embedded here and although you don't believe they know who you are, they do. They're always testing the water, seeing what makes you uncomfortable. If you're with friends or family and something they do makes you uncomfortable, tell them. If you don't want to make a big deal of a small indiscretion, understand this: entrapped people aren't free to tell you're they're entrapped, and if they are entrapped, more likely than not, their car, phone and apartment/home is too. Don't let anyone talk you into

_____

conclusion. Had I done what she said, I've have been charged with Social Security Disability fraud, Unemployment fraud, and conspiracy because my loved one discussed them with me. I am a quiet natured person. When someone suggest I do something, especially something I know little if anything about, I try to find out a correct course of action. If I disagree with them, I don't challenge them, I just decide their suggestion isn't for me. **But, importantly,** if someone tells me to do something I later learn would've caused me harm, including criminal jeopardy, I put an asterisk by that person's name in my mind. If they are my loved one I still love them, but I no longer trust them. Because the Russian Mafia has exploited most of my immediate family to use as weapons to harm me, I no longer trust my immediate family. I love them, but I don't trust them. That's because the Russian Mafia intends to murder me and uses my loved ones as camouflage to get close enough to launch the Russian Mafia's attack. So by the time my loved one began putting free movie apps on the smart phone she gifted me towards my later 50s, she'd in my early 50s been used to try to entrap me with Social Security Disability fraud, Unemployment fraud, and conspiracy to commit both.

lying and/or cheating. Don't think you'll get away with something, that no one will know. We have an enemy government in our midst and that government gives no warning shots. You're one day living your life and the next day, out of the blue, you're under arrest at the police station. Don't think entrapments are over because I wrote this book. The Russian government is here to steal and destroy. It hates Americans and people living in America. Don't forget.

Chapter 145.12 - Never agree to lie, cheat &/or steal either verbally, in writing, or by nodding

The way the Russian Mafia has worked entrapments for decades here is they try to get you to lie and/or cheat for gain or for someone you care about's gain. Don't do it. Never agree, verbally or in writing, or by nodding your head that you agree when someone brings a scam to you. It'll be someone with access to you, a friend, an acquaintance, a family member, a lover, an authority figure, someone you won't want to challenge, someone who's feelings you won't want to hurt, someone you don't see as a threat.[384] If someone tries to get

---

384 Someone I know was harassed for decades and is now an operative. The person was used to suggest scams to me and the Russian government made sure the person told me all the emotional attacks they'd experienced over the years, to make me very hesitant to say anything to hurt their feelings. Under this: the Russian government targets individuals, they know my history, they'll know your history, they'll target psychological attacks based on their intel. For instance, in my case, the person who endured horrific psychological abuse is someone I love. Loving someone and knowing they've been deliberately attacked is heartbreaking. That combination of my not wanting to hurt that person's feelings meant the Russian government used that person in all kinds of way to harm me. President Biden is an empathic person. An attack aimed specifically at him is the Israel-Hamas War. Not only is the war designed to unseat him politically but

you to lie, steal or cheat, say nothing if you can. If you're forced to say something, say no. Say I don't feel comfortable doing that, and then say nothing else.

I go to a doctor because I need a doctor and the Russian government ran me off from my previous general practitioner, wouldn't allow my calls to go through, etc. The doctor I have now has offered on at least two occasions to lie for my benefit to Medicare. These were offers unprompted-by-me, in other words I asked a question about a prescription and as part of the doctor's answer he offered to lie for me. When someone you don't know but who you see in their professional capacity offers you in their professional capacity to lie and cheat, that person is working to entrap you. His handlers have told him to make those offers to you. Think of it this way: if you spent years studying and working to become a doctor would you risk throwing it all away by defrauding Medicare? No. So if

---

it's designed to destroy him psychologically because, no matter what the Israeli government and Hamas say, the Russian government who is running this war, intends to have tens of thousands more Palestinian civilians, many of them children, killed between now and November 2024. The world will blame President Biden and our President, although he is strong, and has been working nearly constantly to persuade the Israeli government to move to precision strikes and better protect civilians, won't want to be president if tens of thousands more children are killed. Their deaths would break his heart, which is what the Russian government is going for. Know this: the Palestinians injured and killed, including 10,000+ children, and the Israeli's injured, raped and killed is facilitated by the Russian government working it's coup against us. The Russian government authorized Hamas to start this war, knowing it would be fought in Gaza, among civilians. The Russian government didn't care. The Russian government doesn't leave paper trails. They create situations advantageous to them, even use protesters, and limit pauses, all General Gerasimov said in "The Value of Science in Prediction, *Military-Industrial Kurier*, (February 27, 2013), to defeat enemies. The Israel-Hamas War is another proxy war conceived by the Russian government to destroy us.

someone offers you something for your benefit that you know requires them to lie to and/or cheat the government (or anyone else), decline the offer. If you can say nothing, say nothing. If you're forced to say something, decline their offer. Say you feel uncomfortable doing what they offered.

The Russian government is deeply embedded here. The Russian government is Mafia. Mafia lives and breathes blackmail. It's hard to break away from them once they've compromised you. Best to understand our nation and we are under attack, and probably much of the world has been compromised by the Russian gov, the Russian Mafia, and it's allies. Protect yourself as best you can, say nothing to any offer you'd have never thought of asking for.

Chapter 145.13 - Avoid lying

If you lie casually, no big thing, work to get yourself out of the habit. Avoiding that slippery slope can save your life at this phase in the Russian government's coup. This phase will change so you will adapt. Give yourself small, easily achieved goals. Don't lie for a day. Then expand it out. When you slip, put a dollar in your anti-lie jar and start over. If you can stop lying for a month solid, you get to take the lie jar money and buy yourself something you want. Just keep at it. If I hadn't lied, and regretted it, and thought about lying and what I wanted and needed from me in my life, I would've been entrapped by the Russian government long ago. When you make a mistake, talk honestly and kindly to yourself, recommit to improving, forgive yourself and move on. No one is perfect but if you have a problem lying, work to get out of the habit.

Chapter 145.14 - Don't do drugs – drug deaths are a cheap and easy kill for the Russian Mafia trying to steal your country from you and murder you

The Russian government loves to get people here hooked on drugs, either street and/or prescription. Avoid drugs. Avoid pot. Avoid anything you know makes you easy to manipulate. The reason is because the Russian controls that market in America, uses drugs to stealth assassinate us. If you absolutely must have pot, try medical pot, or grow your own. But I strongly suggest no drugs, don't even try it. If you try it you'll like it and you just don't want to go there. The Russian has stealth assassinated so many people worldwide using drugs. They had a dealer offer me whatever I wanted for free. Would you take rat poison if it made you feel good for five minutes then killed you? I doubt it. Fentanyl is similar. Don't.

Chapter 145.15 - The Russian Mafia uses kindness as a weapon

The Russian government loves to use "beholden-ness" as a weapon. First they study you and your life in order to harm you. I'll describe four attacks where they used "behold-ness" to try to set me up to destroy me.

Chapter 145.16 – Luring targets to another company

(1) <u>Lured me to a company in order to more easily attack me.</u> I worked at a downtown law firm as a legal word

processor in the word processing department, and as a legal secretary on various desks, from around the age of 41 to 46, rough estimate. When I worked there as a legal secretary, I worked a 3 attorney desk, and a 2 attorney desk for attorneys I liked.

One of them committed suicide. I now suspect he was entrapped and being blackmailed to harass me. He'd been a teacher before he became an attorney and he was big on education, gave me educational dvds as a staff Christmas present. He was a kind, very high quality person. I think he refused to harass me and follow other orders the Russian government demands of their slaves and I suspect, but have no proof, that he was hazed. Based on how I was mistreated by entrapped supervisors and coworkers operatives before that job, and after leaving the company, my theory is he was harassed while working on an important case where he'd hoped to make partner. He'd gone out of town with a group of attorneys for a trial. At first he hadn't arranged for me to go with him, and then later tried to get me to come but I couldn't go (my desk I believe supported 2 attorneys then. I think going out of town meant the other attorney I worked for wouldn't have his secretary to file/serve documents. It was no problem for someone to take my place, but the other attorney would've had to okay it, and HR. The other attorney would've had to walk a temp through court filings/servings, which most attorneys don't ordinarily do, so aren't familiar with). My kind boss had no legal secretarial assistant dedicated to him on an important case while he was trying to make partner. I don't think it was an accident -- his lack of clerical support set him up to fail. Legal secretaries were there, but they would've prioritized work from their own attorneys. It's possible, though I'll never know for sure, my boss may have been hazed, with a little extra spitball from Russia-controlled American operatives working to psychologically destabilize him. Not long after, he committed suicide. These are my theories, largely based on my own experiences. My boss didn't confide in me. He was a very nice man.

I eventually transferred full time to the word processing center. I loved the job and supervisor. He got sick with cancer and had to take medical leave. I was maybe 45 or 46, and believe I was 41-42 when I was hired by that law firm.

Unknown to me Russia-controlled American operatives had two African American co-worker sleeper operatives in the word processing department, and the Russian government put them in play to harass and ridicule me, out of the blue, when just the three of us were in the word processing department. I think the male operative was hired after I'd been at the firm awhile and the female operative was a longer term employee already there. Before she was used to harass me, she'd made a sex-referring comment in my presence to note my reaction, and she'd asked if she could attend an Octavia Butler reading with me when I said I planned to go during a meal break. The downtown library hosting Ms. Butler's reading was close enough to walk to from the law firm.

These sleeper operatives were unleashed on me to get me to leave the firm. I didn't know that. The Russian government isn't big on disclosure because they're here illegally, and entrapped police managing these attacks also operate them illegally. Victims aren't told. I'm writing this book to warn people.

I was stunned when the operatives began ridiculing me, had no idea what was going on. Before that I'd thought they were co-worker friends, had danced at the Christmas party with the married male operative. At Ms. Butler's reading, I noticed she looked completely miserable, and not long, she died. The lady operative was the person who told me, noting my distress.

I've noticed that Russia-controlled American operatives pay very chose attention to my response when they're involved in the death of someone I love and/or who's work I like. An operative told me Octavia Butler had died, an operative told me Michael Jackson had died, and when my three loved ones had died, operatives told me.

Russia-controlled American operatives have attacked me for so many decades, and I've evaded them for so long, they're always looking to see how I'm doing it: whether I see them, anticipate them, or evade them in some other way. The truth is, until my 50s when the Russian government began using most of my family to try to entrap me, I had no idea I was evading entrapments. I live ethically, that's my secret. For the most part, I didn't anticipate their attacks but reacted to the set up in real time.

Chapter 145.17 - Unusual offer of a company credit card

The only entrapment attempt I noticed at the time, when I was maybe 47.[385] My boss, a partner at the law firm before I was removed from his desk, gave me his company credit card, and asked me to purchase gifts for customers on his behalf, that I'd be the only one who'd review the statement, and that I'd approve the purchases. I'd worked for partners before, been permanently assigned to at least one, but none had given me their company credit card, asked me to buy corporate gifts on their behalf, and told me I'd be the only one to approve the statement. By that time, I'd been harassed on the street and in grocery stores, on the bus, etc., since I was 41-42. My entrapped father who I trusted and believed, was used to misdirect me so I didn't know I'd been lured to the company who worked me exhaustion, or that coworkers at my previous jobs were operatives deployed to harass me into leaving that

---

385 By the age of 47, I'd in my 30s unknowingly evaded an income tax fraud setup, a drug mule setup, an immigration fraud setup, and fraudulently filing a federal complaint setup. I'd also unknowingly evaded luring and grooming attempts from my mid-20s. At the downtown law firm I'd been lured from, I'd been paid some exorbitant amount on my paycheck, which I reported to payroll, having no idea it was a setup. Russia-controlled American ops use operatives embedded into firms or have those operatives hired to join the firm after I'm there, to harass me out companies when their entrapments fail. I didn't know that. I didn't know that receiving a paycheck with a big money amount, far more than I was supposed to receive, was a setup. I'd never experienced it before and I thought it was an accident and so reported it to payroll. Seeing my response, the Russian government harassed me out of that job and lured me to a new job. All that is illegal, but the Russian government cares nothing about our law, about human rights, about nothing but destroying Americans.

job. I loved that company but the Russia owned the ground there.

At 47 I noticed the company credit card offer because it was unusual and strange. But because of my entrapped father's gas lighting and misdirection, I didn't immediately tie the credit card embezzlement setup with me being harassed on the street, or my previous jobs where my supervisors and coworkers had harassed me, or when my paycheck at that job reflected a huge amount that I reported it to accounting, thinking it was an accidental overpayment. Because I was so unrelentingly harassed on the street (bus, grocery stores), I had become sensitized to unexplained weirdness so I noticed when my boss gave me his credit card and explained I'd be the only one to review and approve it, but I didn't know it was part of an enemy government's coup-related attacks on me since I was 11.

When Russia-controlled American operatives stealth murder someone I love and/or who's work I enjoy, they do their best to have an operative tell me, to get my reaction. They're always trying to see if I know they're in my life or on the ground here. In 2023 they killed a mentor I'd loved since I was in my 20s. Operatives made sure I knew he'd died; they wanted to hear my reaction and my pain. That's who the Russian government is. They're like wasps but vindictive. No one but the Russian government would facilitate a war they know will potentially murder hundreds of thousands of children. The Russian government is facilitating the Israel-

Hamas War to unseat President Biden and couldn't care less how many Palestinian and Israeli children are murdered. This government demanding world dominance is the government least worthy to rule our species.

I believe but cannot prove that Ms. Butler was murdered by Russia-controlled American operatives. I believe Michael Jackson was murdered by their operatives. I believe Carrie Fisher was murdered by the Russian government's operatives. I believe my maternal aunt, my paternal grandfather and my mentor were all murdered. None of the murders appear to be murders. Of the six people I've listed, the Russian government knew I loved and/or admired their work. Michael Jackson said someone was trying to kill him, but he died in such a way that nothing could be traced to the Russian Mafia. Ms. Butler was visibly miserable when she promoted *The Fledging* at the downtown library and I believe she was entrapped because of her body language and the behavior of the people who surrounded her during her reading. Carrie Fisher I believe was lured out of the country to work and swarmed with drugs. It's evil to offer an addict who struggles with substance abuse drugs, but it's reported Ms. Fisher had drugs thrown her way. She barely made it back to us alive. I believe she was murdered in such a way that it can't be traced to the whoever ordered her death. And they killed her mother, the beloved Debbie Reynolds, who couldn't bear to live without her.

In my family, my aunt had failed to turn me into a drug mule as the Russian government wanted and was killed to emotionally destabilize me, my grandfather told me he took out a life insurance policy and died not long after, although he'd sounded fine our last conversation. And my mentor died the end of Spring semester 2023, and an operative made sure I was told to get me to fail my Final and drop the class.

There's a reason I believe the Russian government owns the ground here. It's not because I want it to be true but because they've gone out of their way to show me, over decades. After they saw they couldn't entrap me they've concentrated on destroying my quality of my life, stealing my family from me, and my country from me, forcing me to witness as they use babies in harassing attacks, and gloat at me in triumph as I work to understand who my enemy is and how they're destroying my people and my nation.

As I type these words on 12/21/2023, the operative upstairs noise harasses me in their hallway above. Now, 2/7/24, apparently that operative is dead. Although a senior citizen, if she has died instead of going to live in a nursing home or with her daughter, my theory is the Russian Mafia killed her and made it look like natural causes. If my government investigates my allegations they would've spoken to her. Now, if she's dead, they can't and that's the way the Russian Mafia likes it. It's not funny how things go their way and innocent people like Palestinian civilians are slaughtered because the Russian Mafia think they're worthy to rule our species. They are not.

By the age of 45-46, I'd been harassed and attacked in stores, on buses and on the sidewalk for maybe 5 years by that point. Police were involved because they were some of the harassers but I was a fifteen years away from understanding the Russian government was embedded here, controls some of our police, and had their operatives attacking me. I talked to my father about being harassed in stores and he'd gas lighted me, convinced me to ignore it, so in my mid-40s I was in father-influencer denial, and didn't piece together the harassment by strangers on the street, my boss' suicide, and the verbal harassment by previous supervisors and co-workers. To me, I was an ordinary office worker who'd done nothing to make enemies, especially enemies who by that time had hundreds of strangers harassing me on the street, bus, stores.

Looking back now, I'd been harassed into leaving most of my jobs since graduating from college twenty years+ ago, but I worked a lot and slept. Working at the downtown job, I worked full time as a word processor, and part time at a second law firm.

At the downtown job, I worked late hours in word processing and men I now believe were plain clothed law enforcement, as well as uniformed police in their cars, stared at me, a lot. I didn't know the plain clothed men were police. I talked to my father about the problem and called the Santa Monica Police Department to ask about gun ownership, because I live in Santa Monica. I explained why I wanted a gun. The officer said it was unlikely I'd be given a hidden carry permit, that those permits were given to body guards and private investigators, etc. My father, when I asked his advice, told me a perpetrator would probably wrestle the gun from me and shoot me with it. I didn't get a weapon.

Police were harassing me downtown in my mid-40s, but I was a few years away from actually witnessing a police officer direct a young man to harass me (that happened when I was 48-49, estimate), or observing that during a brief police strike, no one harassed me, but as soon as the strike ended, I was harassed again (that happened when I was 48-49, estimate).

I knew police were involved in harassing me but I didn't know in my mi-40s that police were extra judicially attacking me and conspiring to attack me. I wasn't shown that information until my late 40s.

At the downtown job I worked full time in a word processing department at a law firm. I loved that job and my regular boss but he was out on medical leave due to cancer. I also worked a part time job at another law firm. I was double teamed by my temporary supervisor and coworker who had been sleeper operatives in my life before they were used to harass me, and partners at the law firm who lured me to work for them as full time legal secretary, who were sleeper operatives who lured me and later worked me to exhaustion. First, my temporary supervisor and coworker began harassing me, suddenly, out of the blue, when no one else was around. Then my part time job, which was a few miles from my apartment, offered me a full time job and said I could name the salary. I said $60,000, given I had nearly 20 years of legal secretarial and legal word processing experience. The luring law firm cut my salary 10%, they said they cut everyone's salary due to the 2008 economic downturn but I have no idea if that's rue. I was offered a legal secretarial position with one of the name partner at that law firm and I'd temped for them them off and on for maybe five years in the past. Before they'd exhausted me by overworking me, one of the name partners yelled at me for no reason. The Russian government is big on getting authority figures to yell at victims. All a setup.

Chapter 145.18 - Pay attention and proceed with caution if someone is inexplicably financially generous with you or your children or family

I don't remember much about the USSR operative put in my mother's bed. As I've said, my mother kept her social life, including her sex life, away from her children. I knew she liked him because he spent the night sometimes and she rarely let men sleep over. I only remember him sleeping over in terms of him being her lover and sleeping over, although I didn't know these words then. I occasionally heard sounds which I now know were sex sounds. Mom had male friends that occasionally spent the night, if they'd been out late and he didn't want to drive to the base, but those were platonic friends, not lovers. I wasn't playing close attention. Sex hadn't really registered with me by 11 or 12, it wasn't part of my life at that age, so people's sex life probably went right over my head unless someone had a crush on a boy.

What I do vaguely remember is the operative was financially generous with mom and us kids. I don't remember what he may have gotten me for Christmas but I remember he got my little sister a pretty coat. Since I was 11 or 12, she was 8 or 9. She may remember the coat. We lost most of our belongings due to him (my opinion) so I've no idea what happened to the coat.

If someone enters your life spending money very freely on you and your children, proceed with caution. When I was 11 and 12 the USSR placed an operative in my mother's bed and I believe he exhibited that behavior. The main thing I remember about him was the one day he showed up at my bus stop and drove me home, after which he exaggerated my behavior at an almost fight.

Chapter 145.19 - How the Russian Mafia exploits your loved ones to destroy you

The summer 2023, my mother's American handler's controlled by the Russian Mafia were working another attack against me. During a phone conversation, my mother was allowed to tell me how she met the faux boyfriend she doesn't know was owned by the USSR and inserted into her life to destroy her when she was 28/29.

Chapter 145.20 - How the USSR embedded an operative into my mother's bed to destroy her quality of life

Some of mom's ex-in-laws were from the South and when she was in her early 20s living in the Midwest, she heard some of them talking. On at least one occasion she went to a palm reader in the Midwest and believed what she was told. My mother has been used to tell me that palm readers are accurate and a singer who's music I love has said she had a believable experience with a seer. No coincidences, not with the Russian Mafia. They're working to lure me to someone they control because that's who and what they are.

In California, mom, a 27 year old young woman with a new job and a new career, and with four young children she was solely responsible for, no doubt felt some anxiety, and apparently went or was perhaps lured by the USSR to see a palm reader. I didn't know. I was 9. Mom didn't discuss her spiritual beliefs with me when I was 9.

My suspicion is, open as she had to be in a new environment to meeting new people, and based on my half a century being attacked by the USSR and Russian Mafia, and seeing how they routinely pushed me from job to job, in order to try to better position me for entrapment, my theory is that as soon as mom told her new girlfriends (at least one of whom was compromised), that she'd visited a palm reader or seer in the Midwest, the USSR seized on that as the easiest way to insert a "romance" operative into her life. After a "romance" operative is installed, I've yet to see anyone escape entrapment. Bed mate attacks work because ordinary people don't know the Russian Mafia use them as a go-to. I've been targeted twice with "romance/mentor/sex based" attacks over the decades. One was brutal because I fell in love with the man. What saved me was ethics: I loved him but he was married. He was my supervisor, I left that job. After I left I told him why. He didn't know until I told him and I never saw him after I quit. At the time I had no idea my feelings were a result of an attack by the Russian Mafia; most people won't know. The USSR and the Russian Mafia are best in class, world class criminals and masters of psychological warfare. His being married was a blessing, I couldn't take it further because he belonged to another person. I refused to destroy that person's relationship. I think of a male spouse as belonging to his wife/husband. His marriage saved me from sex-related entrapment. A different operative was pushed on me years later but I wasn't interested in him. I didn't' know he was an operative but I knew something was off. The Russian Mafia pushed lesbians at me to see if I was interested but I'm straight. The USSR and the Russian Mafia also had three operatives sexually assault me, one was my supervisor when I worked at a studio, one is a friend of my aunt who my aunt brought along with her during my aunt's visit, and another operative assaulted me on the public bus.

The Russian Mafia was why I never married, not a lack of desire on my part. After 30, and after the Russian government took control of Active Measures here, the attacks against me were much more precise. This after the many proxy attacks against me by the USSR starting when I was 11. I wanted children but I was under attack. Even though the attacks were via proxy, many financially negative things happened to me, plus I was attacked in school. Attacks, negative financial outcomes, being stopped from achieving educational goals aren't conducive to marriage, making babies and building a family. Again, I'm positive I'm not the only American being proxy attacked in this way, just look at the many Americans attacking the Capitol. While probably several thousand showed up, I estimate millions of people in this country are stealth/proxy attacked but were unwilling or unable to travel to the Capitol. The USSR used proxies to stealth attack me so there was no way I could know an enemy foreign government was hunting me, but when the USSR attacks you, there's a problem. The Russian Mafia is even more precise and they're growing their numbers here.

The reason I don't have a family is because I've been attacked for the majority of my life by an enemy foreign government. I'm positive I'm not the only one here, in fact, I know I'm now, which is why I'm writing this book.

I believe mom was set up by the palm reader she went to see and by one of her new California friends, both owned by the USSR and/or their affiliate the KKK and/or local police.

In the summer of 2023 the Russian Mafia was fishing to see if they could scam me into visiting a palm reader the way they'd scammed my mother and apparently other family members. The Russian Mafia has destroyed my natural sleep cycle and was able to get the names of my new prescription from me by using my sister as a weapon. I was too slow to understand and gave her the information. I've learned a lot the hard way and I'm not trained. I make mistakes. I'm sharing what I've learned to help you better protect yourselves.

Mom told me her "meet" story, and by comparing it with the many stealth attacks with which the Russian Mafia has aimed at me over decades, I pieced together how the USSR illegally (using perfidy, a war crime) inserted their operative into her bed and life in her late 20s, to destroy her life, to destroy her family, in order for the Russian Mafia to run their coup here.

I thought through our conversation after the call ended. Of the nearly 400 million people living in American, nearly none understand we've got a world class, best in class enemy destroying us, our children, our families, and our government. People must be warned.

Chapter 145.21 - Pay attention when behavior doesn't make sense

What initially got me thinking through the conversation with my mother, a stealth hunting attack by the Russian Mafia, is she began her story in a specific way, told me something particularly disgusting (someone put human waste in someone's food) and she told me because one of her ex-in-laws told her. I responded to that story beat naturally, just as you would. (ugh) But, not initially understanding mom was ordered to tell me this story by her handlers, and for them the money was the palm reader not the human waste, when I responded to the idea of human waste being put into someone's food, mom chided me for interrupting her story (with my yuck reaction). Most people told that story would've responded as I did but my mom is a slave in 21st century America through no fault of her own, through no fault of any of my loved ones, or anyone who is entrapped: the perpetrator is the Russian Mafia. The operative who used to live upstairs, if she'd dead, I'm suspicious she was murdered so if our security agencies read this book and believe me and go to her to question her, she'd dead. That is how the Russian Mafia would behave, even though she attacked me under their orders for over a dozen years. Entrapped people everywhere need to know, the Russian Mafia hates people living in America. If you're entrapped don't think you're of value to the Russian Mafia. The only thing the Russian Mafia cares about is themselves and world domination. Oh and destroying Americans: that's their idea of nirvana.

I believe Mom's handlers ordered her to tell me the meet story so they could judge how I'd respond to her visiting a palm reader, to see if showed any interest in visiting one. The Russian Mafia is always looking for attack entry points.

Mom is elderly now and she's a slave. Her handlers force her to behave in specific ways. They've lead her to believe they're American law enforcement, they've not told her, obviously, that they're controlled by a foreign enemy government working a coup here.

From mom's point of view, she's being given an order from our legitimate government and that's her job. She believes her handlers are legitimate American law enforcement, not slaves compromised by an enemy foreign government working a coup here. She's been enslaved for decades and now does what she's told.

I wasn't thinking any of that during our conversation but only afterwards did I see her behavior was odd for what she was doing. If a nonentrapped person told that story, they would've told it for the yuck payoff, but to her handler's the human waste wasn't the money, the palm reader was the money, and they'd communicated to mom what they needed from her.

Mom isn't trained in espionage; she was forced to do this chore. Her handlers told her to get to the palm reader part. We didn't have a mother/daughter conversation, *the Russian government stole my mother from me and use her body and exploit our mother/daughter history to use as a weapon against me, in order to entrap and/or destroy me*. I'm positive the Russian Mafia uses similar attacks against millions of other innocent people.

Reviewing the call later, I understood she was ordered to hit specific parts of the story, and that her handlers hadn't prepared her for my realistic reaction. I didn't know during the call that it was a set up and fishing expedition. So when she was short with me for no reason, I noticed.

Because Russia-controlled American operatives wanted to know my thoughts about palm readers, my mother was allowed to share helpful information as to how she met the operative the USSR used perfidy, a war crime, to destroy her quality of life and set her on the road to slavery, illegally owned by the Russian Mafia, forced by an enemy government to attack her own child. I'm sharing this experience so you'll learn how the Russian Mafia attacks the innocent, using their loved ones and friends against them.

Chapter 145.22 - The Russian Mafia attacks a person using someone already in the target's life[386]

The Russian government will entrap someone in your life who has at least one of the following characteristics: (1) you love them (romantically, as a friend, as a family member); (2) they love you (romantically, as a friend, as a family member); (3) you trust them (romantically, as a friend, as a family member,

---

386 I'm a person targeted by the Russian Mafia and have experienced many attacks over decades by people who I believe (1) <u>were already compromised by the Russian Mafia</u> before the attacker was used to harm and/or 'befriend' me because, among other reasons, of the way the 'co-worker pawns' insisted on being friends with me for no real reason, or (2) <u>may not have been previously compromised</u> by the Russian Mafia but were ambushed and forced by the Russian Mafia into a relationship with me because the Russian Mafia wanted to exploit that person's access to me. I've endured thousands of covert and/or overt attacks by people owned by the Russian Mafia, including attempts to influence me, as well as sexual, physical, and/or psychological attacks by work supervisors, co-workers, a friend of my aunt who accompanied her when my aunt visited me, neighbors I thought were my friends, strangers on the street, strangers on the bus, strangers in grocery stores, strangers at a hamburger joint, strangers at the pharmacy, hundreds of uniformed police, hundreds of plain clothed police, uniformed and plain clothed security, my parents, my half-siblings, my aunt, etc.

I believe the huge employee turnover our nation witnessed at the end of covid was the Russian Mafia stealth harassing innocent workers out of jobs. To get me to leave a company I liked and/or transfer to another department where the Russian Mafia would have another chance to entrap me, the Russian Mafia used my supervisors and/or coworkers over the majority of my work life, so for decades, the Russian Mafia pushed me from one job to the next. These were jobs I liked at companies I liked and where I wouldn't have left except the Russian Mafia pushed supervisors and/or coworkers to harass me until I found another job. I'm positive I'm not the only worker in America being attacked in this way. In my last two jobs I was mistreated and overworked, and before that job the one before that the temporary supervisor and a coworker harassed me and the job I'd been working part time at offered me a full time position. The Russian Mafia controls the ground here and our people don't know what a problem the Russian Mafia is here.

as a coworker); (4) they trust you; (5) you like them; (6) they like you; (7) you have something in common with them (ethnicity, the difficulty of trying to achieve in a racist, sexist, &/or homophobic world, etc.); (7) you see their vulnerability; (8) they are loyal to you; (9) they've been abused/mistreated; (10) they are not predatory; (11) they are not spiteful;(12) they are not cruel; (13) they are gentle; (14) they tell you you are one of their favorite people; (15) they never forget your birthday; (16) they always get you something for Christmas; (17) you've known them at least a year and likely for many years.

The Russian government is cruel. They love the idea of betrayal because they're all about destroying you because you live in America.

The Russian government has got issues. The person they've entrapped will be speaking the truth and the person won't know the Russian government is their handler. The slave will have been entrapped and won't know why they've been entrapped, they won't know who the puppet master is. The Russian government isn't big on transparency.

The person won't see themselves as betraying you or their country because they don't know they're part of a coup. To the person, they'll have been entrapped. In the case of the person I'm thinking of, the person will have been put through maybe seven surgeries in four years, painful, kneed surgeries, until all they know is pain. Their caregiver will have been

917

compromised and this is the person they depend upon emotionally.

The Russian government hates us. So it ha no problem using our most beloved, trusted people to betray us. They hate us, you must understand and the Russian mafia does not see anything it's doing as evil.

Chapter 146 - Conclusion

The Russian Mafia believes there can be only one lead country and they've chosen themselves. Everyone else is expendable. Exhibit A: Palestinian civilians, their precious babies & their extended family used as blood pawns by the Russian Mafia to unseat President Biden. We can not allow the Russian Mafia to seat any of their candidates in the White House in 2024 or beyond.

We must, as the Biden Administration is doing, as best we can continue working as hard as possible to ensure Palestinian people have their own nation, with quality security, to prevent terrorists entering and setting up and seizing control of their government. None of this is their fault. No one asked their permission to dig tunnels and bomb Israel.

We must prioritize our own survival, balance it with our need to help vulnerable people in the world used as blood pawns and bait by the Russian Mafia. The Russian Mafia has started their countdown to destroy us and seize control of the world, so anyone with accurate information, even information that doesn't appear on it's surface relevant, please come forward. Despite the Russian Mafia's endless propaganda and

The KGB and the Russian Mafia have worked Active Measures against us for roughly 60+ years – they're salivating in anticipation of destroying us. The Russian Mafia began salivating in 2001 after (I believe) they made 2001 happen, after which they went loud on me (attacking me in the grocery store, on the street, on the bus, etc.) They simply won't leave us alone, nor will they leave anyone else alone who has something they covet. Their unprovoked attack against innocent Ukrainians is another example of their willingness to steal anything they want from anybody.

For most of my life the KGB in varying reiterations has attacked me for no reason. I don't like it when anyone attacks children, destroys their parents, destroys their siblings, attacks our nation and our government, destroys other innocent children and their parents, and their homes as the Russian Mafia is doing against Palestinian civilians using proxies, all out of blind ambition for world dominance. I've never heard of anything so stupid and so evil. I'm positive the Russian Mafia has committed Crimes Against Humanity against Palestinians, against Israelis, against Ukrainians, and against Americans.

If we continue to refuse the admittedly horrific cost of ridding our species of the Russian Mafia, our life form as we know it, our families as we know them, our cultures, nations, everything we know as normal, will be destroyed. The Russian Mafia knows how to win a war - any mafia/government can ambush and attack children and secretly destroy their families. Sane governments don't do that but the Russian Mafia isn't sane. It is an anti-life government. Not just anti-American, not just anti-West alliance, but anti-life.

The Russian Mafia is destroying our people, and our allies' people (and murdering Ukrainian and Palestinian civilians) because they're willing to murder babies, (and in America - hook children on fentanyl). To seize world dominance the Russian Mafia is murdering Palestinian babies to use as blood pawns to install one of their puppets into our Presidency.

Any entity willing to slaughter babies to seize world dominance is unworthy of world dominance.

The Russian Mafia are experts on hating life not their own and will slaughter pretty much everyone to get what they want. What the Russian Mafia doesn't know or care about is how to keep a species thriving, how to keep a planet habitable. Say their coup is successful and they flip the world order. They won't bother to reassure stunned Americans and our allies. It's far easier for them to slaughter us, and they will. The Russian Mafia controls much of the world's media and has been pushing anti-American sentiment globally since the 1960s. There's a reason they've been doing that: it's so that if they successfully Close Active Measures in November 2024, when they start murdering us, no one will come to our aid.

If the Russian Mafia is successful this November, over the next decades they will stealth slaughter most Americans (already the Russian Mafia has a toilet paper weapon they've tried to give me for at least three years – I'm guessing it causes colon cancer).

They'll say that removing us will remove one of the major drivers of the climate catastrophe, so climate warming gases will be significantly reduced. The only reason those gases exist in the quantities they do is because the Russian Mafia has for decades prevented our government from effectively fighting the climate catastrophe (via blackmail and enslaving power players here and worldwide). The Russian Mafia will omit reporting that they prevented us from effectively mitigating the climate catastrophe.

Much of our reality over the last 60 years has been a setup by the Russian Mafia's as it works to Close Active Measures in November 2024.

We must be willing to out dominants who're trying to lure us into illegal behavior, we must tell everyone that this is part of the coup and genocide against us. Our people are being forced to entrap us, but they've been told by their handlers that our legitimate government is running this scam. That's a lie. The Russian Mafia depends on our not outing the important people in our lives who they've ens1laved, making it easier for the Russian Mafia to destabilize us.

We must warn our people and the world's innocent people as well.

www.ingramcontent.com/pod-product-compliance
Lightning Source LLC
Chambersburg PA
CBHW071129130626
46553CB00004B/1315